The biology of vines

THE BIOLOGY OF *Vines*

EDITED BY

FRANCIS E. PUTZ

Department of Botany,
University of Florida, Gainesville, Florida, USA

and

HAROLD A. MOONEY

Department of Biological Sciences,
Stanford University, Stanford, California, USA

The right of the
University of Cambridge
to print and sell
all manner of books
was granted by
Henry VIII in 1534.
The University has printed
and published continuously
since 1584.

CAMBRIDGE UNIVERSITY PRESS
Cambridge
New York Port Chester Melbourne Sydney

CAMBRIDGE UNIVERSITY PRESS
Cambridge, New York, Melbourne, Madrid, Cape Town, Singapore, São Paulo, Delhi

Cambridge University Press
The Edinburgh Building, Cambridge CB2 8RU, UK

Published in the United States of America by Cambridge University Press, New York

www.cambridge.org
Information on this title: www.cambridge.org/9780521107136

© Cambridge University Press 1991

This publication is in copyright. Subject to statutory exception
and to the provisions of relevant collective licensing agreements,
no reproduction of any part may take place without the written
permission of Cambridge University Press.

First published 1991
This digitally printed version 2009

A catalogue record for this publication is available from the British Library

Library of Congress Cataloguing in Publication data
The biology of vines / edited by Francis E. Putz and Harold A. Mooney,
p. cm.
1. Climbing plants. I. Putz, Francis E. II. Mooney, Harold A.
QK773.B58 1991
582.1′4–dc20 90-23763 CIP

ISBN 978-0-521-39250-1 hardback
ISBN 978-0-521-10713-6 paperback

Contents

Contributors

HERBERT G. BAKER
Department of Integrative Biology, University of California, Berkeley, CA 94720, USA

GUY CABALLÉ
Institut de Botanique, 163 rue Auguste-Broussonet, F-34000 Montpellier, France

SHERWIN CARLQUIST
Rancho Santa Ana Botanic Garden and *Department of Biology, Pomona College, Claremont, CA 91711, USA*

ALEJANDRO E. CASTELLANOS
CICTUS, Universidad de Sonora, Apdo Postal #54, Hermosillo, Sonora 83000, Mexico

FRANK W. EWERS
Department of Botany and Plant Pathology, Michigan State University, East Lansing, MI 48824, USA

KLAUS FICHTNER
Lehrstuhl Pflanzenökologie Universität Bayreuth, Postfach 10 12 51, D-8580 Bayreuth, Federal Republic of Germany

JACK B. FISHER
Fairchild Tropical Garden, 11935 Old Cutler Road, Miami, FL 33156, USA

IRWIN N. FORSETH
Department of Botany and *Maryland Agricultural Experiment Station, University of Maryland, College Park, MD 20742, USA*

GORDON W. FRANKIE
Department of Entomological Sciences, University of California, Berkeley, CA 94720, USA

TAMARA FRANKLIN
Laboratory of Biomedical and Environmental Sciences and *Department of Biology, University of California, Los Angeles, CA 90024, USA*

BARBARA L. GARTNER
Department of Biological Sciences, Stanford University, Stanford, CA 94305, USA

ALWYN H. GENTRY
Missouri Botanical Garden, PO Box 299, St Louis, MO 63166, USA

WARREN G. GOLD
Department of Botany and *Maryland Agricultural Experiment Station, University of Maryland, College Park, MD 20742, USA*

ELWYN E. HEGARTY
Department of Botany, University of Queensland, St Lucia, Brisbane, Queensland 4067, Australia

MERVYN P. HEGARTY
CSIRO, Division of Tropical Crops and Pastures, 306 Carmody Road, St Lucia, Brisbane, Queensland 4067, Australia

N. MICHELE HOLBROOK
Department of Biological Sciences, Stanford University, Stanford, CA 94305, USA

DAVID W. LEE
Department of Biological Sciences, Florida International University, Miami, FL 33199, USA

HAROLD A. MOONEY
Department of Biological Sciences, Stanford University, Stanford, CA 94305, USA

PAUL A. OPLER
US Fish and Wildlife Service, 1025 Pennock Place, Fort Collins, CO 80524, USA

OLIVER PHILLIPS
Department of Biology, Box 1137, Washington University, St Louis, MO 63130, USA

Contributors

FRANCIS E. PUTZ
Department of Botany, University of Florida, Gainesville, FL 32611, USA

JENNIFER H. RICHARDS
Department of Biological Sciences, Florida International University, Miami, FL 33199, USA

PHILIP W. RUNDEL
Laboratory of Biomedical and Environmental Sciences and *Department of Biology, University of California, Los Angeles, CA 90024, USA*

STEPHEN F. SIEBERT
Department of National Resources, Cornell University, Ithaca, NY 14853, USA

ALAN H. TERAMURA
Department of Botany and *Maryland Agricultural Experiment Station, University of Maryland, College Park, MD 20742, USA*

Foreword

Climbing plants – vines – are one of the most interesting, but also a very neglected group of plants. In the rainforests of the tropics, where they reach their greatest abundance and diversity, they climb into the crowns of tall trees, hang down in gigantic loops and often bind one tree firmly to several others. Their stems reach prodigious lengths and are often thicker than a man's thigh. Botanists have long been familiar with their curious stem anatomy and their varied means of attaching themselves to other plants. They are also of considerable economic importance, both as the most troublesome weeds with which the tropical forester has to contend and as sources of valuable drugs such as curare and strychnine. Though more numerous in the tropics, they are also common in temperate regions: in Britain ivy and traveller's joy are conspicuous features of the landscape as virginia creepers are in North America.

In spite of their varied interest and importance to man, vines have attracted relatively little scientific attention. In the nineteenth century Charles Darwin was fascinated by their structure and behaviour, which he described in his *Movement and habits of climbing plants* (1875). Later Schenck in Germany wrote two classical memoirs dealing mainly with the stem anatomy of climbers (1892–3). Since then no comprehensive work on them has appeared. Research on vines, particularly their general biology, is a conspicuous gap in modern plant science.

Now at last two editors, one a forest ecologist of wide experience, the other an ecological physiologist, have brought together eighteen chapters on aspects of vine biology ranging from photosynthesis and heteroblastic development to breeding systems and effects on other plants, as well as economic and ethnobotany. One chapter deals with the utilization and sylviculture of rattans, a group of climbing palms of great economic importance in the eastern tropics.

Foreword

This book provides access to a large amount of interesting and useful research which has not hitherto been easily available. It will no doubt stimulate much further work and ensure that in the future climbing plants will not be as neglected as they have been in the past. It should be warmly welcomed.

Paul W. Richards
Emeritus Professor of Botany
University College of North Wales, Bangor

Preface

Vines are plants that cannot remain free-standing to any appreciable height. There are both herbaceous and woody vines, the latter generally referred to as lianas or lianes. Using 'vines' to denote all climbing plants may initially confuse some readers from lands where, with due respect for wine, 'the vine' is used solely in reference to grapes. Terminological confusion aside, there are still some problems determining what is a vine and what is not. These problems derive from the fact that there is no clear distinction between self-supporting and non-self-supporting plants either ontogenetically or evolutionarily. Most vines do not require external support until they are a decimeter up to a meter or more tall. Under some conditions, normally climbing species seem to thrive in the absence of mechanical support and take on the appearance of rank shrubs or treelets. Some vines simply lean on neighboring plants without displaying any obvious 'adaptations' for climbing other than a tendency towards etiolation.

Although the climbing habit has evolved many times in lineages ranging from ferns and gymnosperms to palms and legumes (see Chapter 1 by A. H. Gentry), most vines share a suite of morphological, anatomical, and physiological characteristics. These shared, and primarily derived, characteristics are the primary focus of this volume but variations among climbing plants are also given due consideration.

It almost goes without saying that vines are long and slender; this simple observation seems to be coupled with distinctive anatomical and biomechanical features of vine stems (considered in Chapter 2 by S. Carlquist and Chapter 3 by F. E. Putz and N. M. Holbrook, respectively) as well as in stem repair mechanisms and xylem hydraulics (see Chapter 4 by J. B. Fisher and F. W. Ewers and Chapter 5 by F. W. Ewers, J. B. Fisher and K. Fichtner). Being long and slender and perhaps the environmental heterogeneity they experience seem to have inclined vines towards displaying profound developmental changes (see Chapter 8 on heteroblasty by D. Lee and J. Richards), but has led to neither uniformity in photosynthetic characteristics (see Chapter 7 by A. Castellanos and Chapter 9 by A. H. Teramura, W. G. Gold and I. N. Forseth), nor in secondary chemistry (see Chapter 10 by M. P. Hegarty, E. E. Hegarty and A. H. Gentry). Although not considered in this book, vine stems do not fare well in fires, perhaps because they are thin and not covered with thick layers of insulating bark and thus heat up rapidly. In regard to temporal patterns in leaf, flower, and fruit production, vines are fairly uniform in some forests and varied in others (see Chapter 14 by P. A. Opler, H. G. Baker and G. W. Frankie). Slender vine stems often support masses of leaves equivalent to those supported by much larger diameter trees but lack the trees' storage capacity; many vines, particularly those from arid environments, have storage tissues below ground in the form of modified stems and roots (see Chapter 6 by H. A. Mooney and B. L. Gartner and Chapter 12 by P. W. Rundel and T. Franklin). Thin vines with large leaf masses might also be constrained by lack of volume in which to include phloem tissue.

In order to climb, vines need to locate and somehow grasp, lean, or hook onto suitable supports. Failure to encounter a trellis leads to the demise of many forest vines. Their chances of success are improved by production of long, leafless leader shoots that circumnutate and tendrils that contract after clasping onto something (see Chapter 13 by E. E. Hegarty and Chapter 11 by E. E. Hegarty and G. Caballé). Vines that can climb up the sides of trees or even buildings with the aid of adventitious roots or adhesive tendrils do not seem constrained by lack of potential supports but nonetheless are rare in many forests for reasons that are not yet apparent.

The study of vine biology is important on economic grounds. Vines are among the most important agricultural and silvicultural weeds (see Chapter 18 by F. E. Putz and Chapter 9 by A. H. Teramura, W. G. Gold and I. N. Forseth). Vines are also of tremendous economic value as sources of pharmaceutical chemicals, fruit, and dyes (see Chapter 16 by O. Phillips); climbing palms provide the rattan canes of commerce (see Chapter 17 by S. F. Siebert).

Much remains to be learned about vines; hopefully this volume will provide a solid foundation upon which future studies will be based. In particular, information on the ecosystem function of vines is lacking. Given their abundance, rapid growth rates, and voluminous leaf production, vines certainly must play important roles in nutrient cycling. Environmental concerns about silvicultural prescriptions calling for vine removal also need to be considered in the light of their potential importance as food and inter-crown pathways for animals. Vines can be a nuisance or a godsend but regardless of your perspective, they are clearly worthy of further study.

Acknowledgements

Many of the chapters in this volume were presented at a symposium held at the Estacion de Biologia Chamela, Jalisco, Mexico. We acknowledge our hosts for providing a stimulating atmosphere with gracious hospitality. Many people have contributed towards the completion of this book but we particularly want to thank Stephen H. Bullock for his efforts in organizing the Chamela meeting, his thoughtful critiques of many of the chapters, and his insights into the biology of vines. This book has also benefitted from input from David Dobbins, Mark Matthews, Miguel Franco, and Javier Peñalosa.

PART I
INTRODUCTION

1
The distribution and evolution of climbing plants

ALWYN H. GENTRY

Introduction

The classic work on climbers by Schenck (1892, 1893), is focused especially on anatomical features of lianas, but also includes a taxonomic and geographic survey of the occurrence of climbing plants. Such 19th century luminaries as Charles Darwin (1867) were fascinated by the peculiarities of climbing plants. However, despite their obvious importance in the world's flora, especially in tropical forests, climbers have subsequently been generally neglected. As summarized by Jacobs (1976), 'The ecology of lianas is virtually a blank'. Indeed the significant ecological role played by lianas in tropical forests has only very recently begun to be investigated (e.g. Putz, 1984, 1985; Putz & Chai, 1987). Lianas have been no less neglected by plant collectors: quite probably lianas are the most undercollected of any major habit group of plants.

This overview of climbing plants is based largely on 0.1 ha data sets for plants > 2.5 cm in diameter at breast height (dbh). The sampling protocol under which these data were gathered was originally set up specifically to facilitate ecological sampling of lianas, which are notorious for clumped distributions related to rampant vegetative reproduction (Peñalosa, 1984; Putz, 1984; Gentry, 1985). Each sample consists of ten 2 × 50 m narrowly rectangular plots set up end to end or separated by *c.* (10–)20 m, thus covering a relatively large area of forest that exceeds in scale the frequent 'patches' of single liana species that presumably derive from vegetative reproduction. Altogether 1103 species and 3341 individual climbers and hemiepiphytes are included in the 56 available neotropical samples; for lowland neotropical sites these values are 860 species and 2636 individuals. An additional 600 species are represented in the 32 paleotropical samples, 4 island samples, and 13 temperate forest samples which have been analyzed. Altogether 82 climbing families are represented in these samples. Ecological data for most of the study sites are given in Gentry (1988a: Tables 1, 2); sites from which liana data were included in this analysis are indicated in Appendix 1.1.

This data set is supplemented by data from complete florulas (as summarized by Gentry & Dodson, 1987), tree plots (Gentry, 1988a, b), and from a literature survey. There is a strong neotropical bias inasmuch as my field experience is mostly in South and Central America. The lists of scandent genera from *Flora Malesiana* compiled by Jacobs (1976) and for Australia prepared by Hegarty & Clifford (1990) have been very helpful in compiling a very preliminary listing of Australasian climbing taxa. I have used traditional broad family concepts throughout, generally following the familial placements used in the Missouri Botanical Garden herbarium.

Kinds of climbers

There are several distinctly different kinds of climbers. Gentry (1985) recognized four fundamental climbing strategies differentiated by ecology as well as morphology: (i) Lianas are woody, relatively thick-stemmed climbers that begin life as terrestrial seedlings and are capable of growth in mature forests. (ii) Vines are thin-stemmed climbers or clamberers that begin life as terrestrial seedlings and generally grow in disturbed habitats or at the forest edge. Following the convention of this volume these are here referred to as 'herbaceous vines' even though many of them are subwoody. (iii) Woody hemiepiphytes, including stranglers, typically begin life as epiphytic seedlings with roots later reaching the ground. Other woody hemiepiphytes start out as terrestrial climbers, later sending out a system of adventitious roots and/or losing contact with the ground. (iv) Herbaceous epiphytes and hemiepiphytes include all herbaceous species that climb appressed to tree trunks and limbs, usually via adventitious roots, whether or not they ever establish contact with the ground. A fifth category of 'climber' might also be recognized, especially in the temperate zone, where numerous species have a sprawling or prostrate habit; I have excluded such species from this analysis unless they are clearly creeping (but not as hemiaquatics) or twining, in which case they are tabulated as herbaceous vines.

The hemiepiphyte categories, especially, are not altogether satisfactory. For example, it is not clear whether some climbing species begin life as epiphytes and even different individuals of the same species may be epiphytic or not. Although stranglers are here included with hemiepiphytes, their vine-like stems are actually descending roots. Some non-strangling *Clusia* behave in a similar fashion. Yet other species of *Clusia* and strangling figs are clearly climbing. I have tabulated stranglers and hemiepiphytes separately in Appendix 1.1. Non-climbing hemiepiphytes may be excluded from the floristic results reported here by excluding all Bombacaceae, Araliaceae, nearly all Moraceae and (except in cloud forests and pluvial forests) most Guttiferae from the various figures.

The four habit groups are treated differently in this review. The part of this analysis based on floristic lists includes all four of the climbing categories, except as otherwise noted. The part of the analysis that is based on samples of plants >2.5 cm dbh excludes vines on ecological (outside the forest) as well as morphological (stems <2.5 cm diameter) grounds. Small herbaceous epiphytic climbers (e.g. *Peperomia*) are also excluded from these analyses as are those canopy hemiepiphytes which attain 2.5 cm diameter only more than 1.37 m above the ground. In these analyses liana diameters are measured at the thickest part of the accessible stem (typically a node) but hemiepiphytes are included only when they attain 2.5 cm diameter at or below breast height. A few stranglers and non-climbing hemiepiphytes like some *Clusia* species are included in these analyses if their hanging roots reach 2.5 cm diameter.

Importance of lianas

Climbing plants are an important, though often neglected, part of tropical forests. The presence of woody lianas has been called the single most important physiognomic feature differentiating tropical from temperate forests (Croat, 1978). 43–50% of the trees over 15(–20) cm dbh on Barro Colorado Island, Panama (BCI) have lianas in their crowns (Montgomery & Sunquist, 1978; Putz, 1982, 1984) and 42% of the trees over 10 cm dbh at San Carlos de Rio Negro, Venezuela have lianas (Putz, 1982). Half the trees >20 cm dbh in Lambir National Park, Sarawak are liana-infested (Putz & Chai, 1987). Moreover, about a fifth (18–22%) of the upright plants of neotropical forest understories are usually juvenile lianas (Rollet, 1969; Putz, 1984).

Lianas are an important structural component of tropical forests, often literally tying the forest together (Fox, 1968; Jacobs, 1976; Gentry, 1983b; Putz, 1984; Appanah & Putz, 1984). Lianas have been estimated to account for 32–36% of the leaf litter of tropical forests in Thailand (Ogawa *et al.*, 1965) and Gabon (Hladik, 1974, 1978). In an extensive series of 0.1 ha plots of neotropical forest, lianas represent an average of 24% of the stems >2.5 cm dbh in dry forest and 18% of the stems in lowland moist and wet forests (Gentry, 1982a, 1986). In general there are about as many lianas >2.5 cm diameter as there are trees >10 cm diameter in lowland neotropical forests.

Lianas compete with trees for light, water, and nutrients. Ogawa *et al.* (1965) suggest that competition with lianas is an important cause of tree death. Putz (1980, 1984; Putz, *et al.*, 1984) has emphasized that competition with lianas has exerted strong selective pressure on tropical trees, with liana-infested trees suffering higher mortality rates than liana-free trees.

Lianas are also very important to forest animals, both as food and as a structural component of the habitat. The single most important factor affecting tree selection by two-toed sloths on BCI is the presence and density of lianas in their crowns (Montgomery & Sunquist, 1978). Mammals ranging from the African prosimian *Euoticus elegantulus* (Charles-Dominique, 1977) and the Amazonian pygmy marmoset to the African elephant (Short, 1981) are heavily dependent on lianas for food. An average of 21% of the plant species utilized for food by a wide variety of tropical primates are lianas (Emmons & Gentry, 1983). It has even been suggested that different liana densities on different continents may have been the key selective factor in determining distinctive vertebrate locomotor adaptations in the tropical forests of Africa, Asia, and the Neotropics (Emmons & Gentry, 1983).

Climbers are also very important floristically, making a major contribution to the taxonomic diversity of tropical forests. Even excluding epiphytic climbers and stranglers, the seven tropical field station florulas analyzed by Gentry & Dodson (1987) averaged 176 climbing species per site or 19% of each florula. Similarly, the dry forest in Chamela, Mexico has 181 climbing species or 24% of the total florula (Lott, 1985); interestingly, the two driest sites for which data are available (Chamela and Capeira, Ecuador) have the highest percentages of climber species of any site: 24% each.

Geographic perspective

Climbers are very unevenly distributed geographically. The great majority of woody lianas are restricted to tropical forests. Herbaceous vines are somewhat less restricted in occurrence, but are still overwhelmingly better represented in the tropics. I have quantitative data only for lianas and hemiepiphytic climbers > 2.5 cm diameter and will restrict this discussion of geographic distribution mostly to these groups.

It is well known that temperate zone forests have very few lianas and virtually no hemiepiphytes (e.g. Gentry, 1982a, 1985). Eastern North American forests average only five lianas > 2.5 cm diameter per 0.1 ha and European forests have even fewer (Table 1.1; Figure 1.1). North temperate forests thus differ in liana density from most tropical forests by about an order of magnitude. Curiously, south temperate forests have more lianas than north temperate ones at equivalent latitudes (Table 1.1; Dawson, 1980). For example, the Valdivian forests of Chile average about 30 lianas per 0.1 ha, 6–7 times as many as do their northern equivalents. One sampled Chilean forest, in Puyehue National Park (40°43′ S), had 52 lianas in 0.1 ha, more than in many tropical forests. Indeed, one Puyehue species, *Hydrangea integrifolia*, a giant hemiepiphytic climber, had the highest population density of any climbing species at any sample site.

Table 1.1. *Density of climbers (and hemiepiphytes) ⩾2.5 cm diameter in 0.1 ha*

Region	Number of sites	Number of climbers (or regional average)
Neotropics		
Lowland Amazonian moist	20	69
Trans-Andean moist + wet	9	69
Southern subtropics	5	77
Choco pluvial forest	2	67
Upland Andes (w/o Pasochoa)	7	64
Dry forest (w/o Chamela)	10	70
Chamela dry forest	3	78
Neotropical average	56	70
Africa		
Continental Africa	8	106
Madagascar	3	122
Australasia		
Continental Asia	5	94
Borneo	2	40.5
New Guinea	2	72.5
Taiwan	2	65.5
Davies River S.P., Queensland	1	35
Riviere de Pirogues, New Caledonia	1	43
Australasian average	13	70
Islands		
Round Hill, Jamaica	1	8
Brise Fer, Mauritius	1	19(+2)
Puerto Rico[a]	2	28
Island average	4	18
North Temperate		
North America	8	5
Europe	2	2
North temperate average	10	4
Valdivian forest		
Puyehue	1	52
Bosque de San Martin	1	28
Alto de Mirador	1	11
Average	3	30

[a]Sampled in 1989, subsequent to preparation of the graphs.

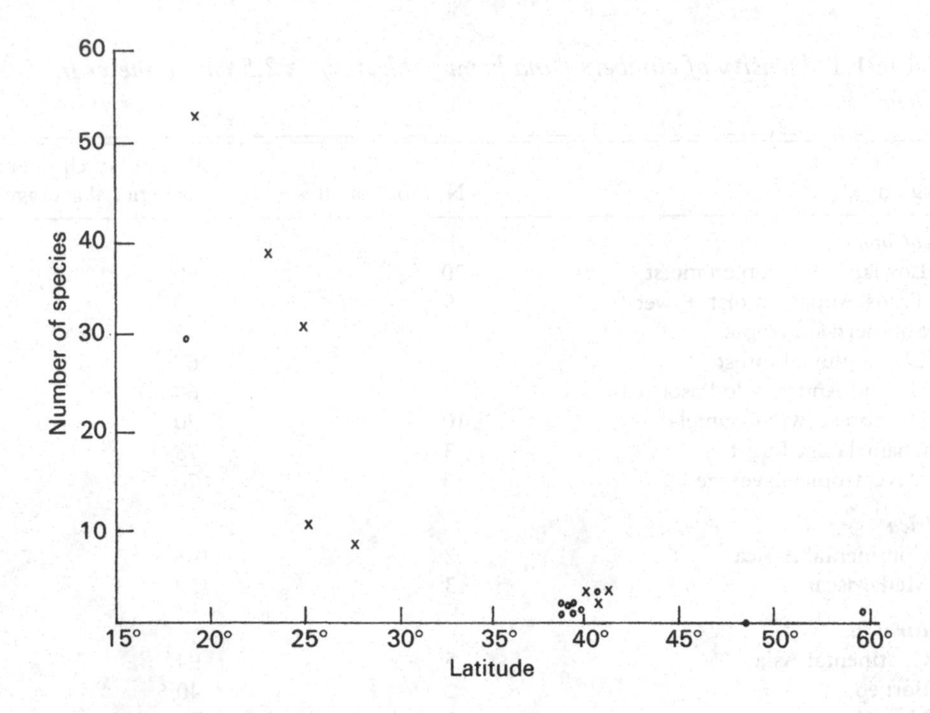

Figure 1.1. Decrease in liana diversity with increasing latitude for subtropical and temperate continental sites <1000 m elevation. ×, southern hemisphere; ○, northern hemisphere.

In a series ($N=18$) of tropical and temperate zone florulas (Table 1.2), tropical values averaging 176 climbing species per site (19% of each florula) contrast markedly with data from a set of eight comparable eastern North American florulas which averaged 37 climbing species per site (6% of each florula). If two larger-scale 'florulas' from the Smokey Mountains and Florida panhandle (White, 1982; Wilhelm, 1984) are excluded, they average 29 species, or 7% of each florula. Climber species richness on the tropical island, Jamaica, is intermediate. In three local florulas climbers average only 28 species, similar to temperate zone values; however, in percentage terms climbers constitute a near tropical average of 14% of these depauperate insular floras (Kelly *et al.*, 1988).

In the temperate forest florulas an average of 76% of the plant species are herbs (86% at a prairie site) (Table 1.2), and 19% woody (including lianas). In tropical forests only about half the species are herbaceous, even including herbaceous epiphytes. This tropical/temperate dichotomy between woody plants and herbs is also reflected in the climbers. In the temperate zone

Table 1.2. *Tropical vs. temperate habitats (% of species in local florulas). Data from Gentry & Dodson (1987) except as indicated*

Region	Average number of native species in florula	Average number of climbers in florula	Climbers (lianas, vines, & woody hemiepiphytes) (%)	Vines (%)	Lianas (%)	Trees (%)	Shrubs (%)	Woody (trees shrubs, lianas) (%)	Epiphytes (incl. Loranth.) (%)	Herbs (%)
Temperate forest[a] (7 sites)	634	41	7	5	2	12	6	19	0	76
Prairie[b]	368	17	5	4	1	6	3	10	0	86
Continental tropics[c] (7 sites)	963	176	19	10	10	21	9	40	13	41
Jamaica[d] (3 sites)	218	28	14	5	9	55	3	67	18	11

[a]Data from Mitchell, 1963; Stevens & Beach, 1980; White, 1982; Wilhelm, 1984; Fitch, 1966; Yatskievych & Yatskievych, 1987; R. Coles, personal communication.
[b]Konza Prairie, Kansas. Data from Freeman, 1980.
[c]Data from Gentry & Dodson, 1987 (includes hemiepiphytes with epiphytes rather than with climbers.).
[d]Data retabulated from Kelly *et al.*, 1988.

florulas an average of 2% of the species are lianas and 5% herbaceous or subwoody climbers, whereas in continental tropical florulas about 10% are lianas and 10% herbaceous vines.

This difference is also apparent on a larger scale. For example, Webb (1978) reports that the Flora Europaea includes 11 548 species (including 501 problematic taxa in the five large apomictic genera), of which the 24 species of woody and subwoody climbers (Baas & Schweingruber, 1987) constitute a mere 0.2%. In eastern North America there are more than twice as many species of woody climber as in Europe; Duncan (1975) lists 48 native species for the southeastern USA alone, despite excluding marginally lianescent species like *Rosa* which are included in the European data. Eight of Duncan's (1975) woody climbers do not reach north to the Carolinas, from where 40 species of woody vine constitute 1.3% of the native flora (based on data from Radford, Ahles & Bell (1968), in part compiled by D. Boufford). There are also at least 105 herbaceous climbers or scramblers in the Carolinas, constituting 3.5% of the flora. All climbing species together constitute almost 5% of the Carolina flora. The southwestern USA is even poorer in climbers, with only two lianas included among the 512 woody and subwoody species of southern California analyzed by Carlquist & Hoekman (1985); altogether there are four native liana species and 24 herbaceous vines (plus 12 species of *Cuscuta*) in the southern California flora (S. Carlquist, personal communication). In contrast, Jacobs (1976) suggests that, worldwide, 8% of all tropical plant species are lianas. My data (Tables 1.3, 1.4) indicate that scandent species constitute about 10% of the neotropical flora (assuming a neotropical flora of 90 000 species (Raven, 1976; Gentry, 1982b). Only about 200 of the 9140 scandent species listed in Table 1.4 are temperate North American.

Tropical islands, at least those that have received their floras via over-water dispersal, also tend to have very low liana densities (Table 1.1; Figures 1.1, 1.2). I have suggested that this may result from the prevalence of wind-dispersed pterochore seeds in lianas, whereas most long-distance island colonizers are bird-dispersed (Gentry, 1983a). Continental-fragment tropical islands that have never been submerged, like New Caledonia and Madagascar, have a full complement of lianas.

Different continental tropical forests also have different liana densities. On a regional scale, Australasian forests often have fewer lianas, and African and Madagascar forests more than do neotropical ones (Table 1.1; Figures 1.2, 1.3; Emmons & Gentry, 1983). The four Australasian sites reported by Gentry (1988a) average only 40 lianas >2.5 cm diameter per 0.1 ha plot, while eight continental African sites averaged 106 lianas and three on Madagascar averaged 122. Comparable neotropical samples average 69 lianas. Apparently continental Asian forests have as many lianas as neotropical forests, however, with 10 sites averaging 80 lianas >2.5 cm diameter, but

Table 1.3. *Largest New World scandent families*

Family	Number of climbing species in New World	Number of genera with climbing species in New World
Asclepiadaceae	1000	*c.* 40
Convolvulaceae	750	20
Leguminosae	720	37
Asteraceae	470	23
Araceae	400	7
Bignoniaceae	400	53
Sapindaceae	400	6
Malpighiaceae	400 ($\frac{1}{2}$ of 800)	30?
Passifloraceae	360	5
Apocynaceae	350	35
Cucurbitaceae	311	55
Ericaceae	300[a]	*c.* 25
Rubiaceae	220	13
Dioscoreaceae	200	2
Aristolochiaceae	180	3
Euphorbiaceae	170	12
Gesneriaceae	140	4
Solanaceae	130	7
Vitaceae	135	5
Marcgraviaceae	125	7
Piperaceae	125	3
Melastomataceae	120	8?
Amaryllidaceae	112	1
Smilacaceae	100	1
Hippocrateaceae	100	12
Loranthaceae	100	3
Tropeolaceae	90	1
Cyclanthaceae	84	4
Verbenaceae	80	4
Loganiaceae	70	2
Acanthaceae	70	3
Guttiferae	65	4
Palmae	65	2
Connaraceae	60	5
Sterculiaceae	60	1
Begoniaceae	60 ($\frac{1}{10}$ of 600)	1
Polygonaceae	60	6
Polygalaceae	50	5
Dillenlaceae	50	4

c. 39 families with 50 or more American climbers.
[a]No. of epiphytic species.

Table 1.4. *Categories of New World climbing families (genera with climbing species/estimated species of climbers). Old World taxa excluded*

Mostly climbing families		Large diverse families with major component of climbers	
Mostly vines			
Asclepiadaceae	40/1000	Apocynaceae	35/350
Basellaceae	2/6	Araceae	8/400
Convolvulaceae	20/650	Bignoniaceae	53/400
Cucurbitaceae	55/311	Combretaceae	2/35
Dioscoreaceae	2/200	Asteraceae	23/470
Smilacaceae	1/100	Cyclanthaceae	4/84
		Ericaceae	25/300
Mostly lianas			
Aristolochiaceae	3/180	Euphorbiaceae	12/170
Connaraceae	4/60	Gesneriaceae	9/150
Coriariaceae	1/1	Guttiferae	4/65
Dilleniaceae	4/50	Icacinaceae	3/9
Gnetaceae	1/6	Leguminosae	37/720
Hippocrateaceae	12/100	Malpighiaceae	30/400
Lardizabalaceae	2/3	Melastomataceae	8/120
Marcgraviaceae	7/125	Polygalaceae	5/50
Menispermaceae	18/120	Polygonaceae	6/60
Passifloraceae	5/360	Rubiaceae	13/220
Trigoniaceae	1/15	Sapindaceae	6/400
Schisandraceae	1/1	Solanaceae	7/130
Vitaceae	5/135		
Total	185/3513	Total	290/4533

these values may not be strictly comparable since many of the lianas in sites with high liana density are rattans (28 of 124 stems at Khao Yai, Thailand; an average of 23 of 102 stems in Peninsular Malaysia) and most of the sampled rattans are juveniles which have attained 2.5 cm basal diameter but have not yet begun to climb. It is curious that Madagascar forests, although floristically more like neotropical ones (Gentry, 1988a; A. Gentry & G. Schatz, unpublished data), are structurally like continental African ones in their high liana densities.

There is generally little difference in liana density among different neotropical forest types (Table 1.1). A series of 20 lowland Amazonian moist and wet forests averaged 69 climbers per 0.1 ha (range = 27–95), exactly the same as a series of nine trans-Andean lowland moist and wet forests (range = 45–123). The Choco pluvial forest averages 67 climbers in a similar sample, upland Andean forests 64, dry forests 70. Moreover, large lianas

Table 1.4 (*cont.*)

Families with 1(–2) predominantly climbing genera (plus miscellaneous species)		Families with miscellaneous climbing species but no predominantly climbing genus	
Acanthaceae	3/70	Annonaceae	2/6
Amaranthaceae	4/40	Araliaceae?	2/hemiep.
Amaryllidaceae	1/112	Begoniaceae	1/60
Anacardiaceae	1/10	Bombacaceae	1/strangler
Boraginaceae	2/30	Bromeliaceae	1/1
Cactaceae	7/40	Campanulaceae	2/10
Celastraceae	1/5	Caprifoliaceae	1/7
Dichapetalaceae	1/12	Caricaceae	1/1
Epacridaceae	1/1	Capparidaceae	1/2
Fumariaceae	1/1	Commelinaceae	1/1
Gentianaceae	1/4	Cornaceae	1/2
Gramineae	5/30	Cyperaceae	1/10
Hernandiaceae	1/13	Labiatae	1/2
Lauraceae	3/3	Lythraceae	1/3
Liliaceae	3/4	Malvaceae	1/1
Loasaceae	3/5	Marantaceae	1/2
Loganiaceae	2/70	Moraceae	3/3 (+ stranglers)
Loranthaceae	3/100	Nyctaginaceae	2/5
Orchidaceae	2/15	Olacaceae	1/1
Palmae	2/65	Onagraceae	1/26
Phytolaccaceae	3/35	Oxalidaceae	1/3
Piperaceae	3/125	Plumbaginaceae	1/2
Polemoniaceae	2/20	Rosaceae	2/10
Ranunculaceae	1/10	Turneraceae	1/1
Rhamnaceae	5/10	Ulmaceae	1/2
Saxifragaceae	2/9	Urticaceae	2/10
Scrophulariaceae	3/12	Valerianaceae	1/10
Sterculiaceae	1/59	Zygophyllaceae	2/prostrate
Thymelaeaceae	1/4		
Verbenaceae	4/80	Total	37/181[a]
Violaceae	3/15		
Total	73/989	Grand total	585/9216

[a]Excluding stranglers.

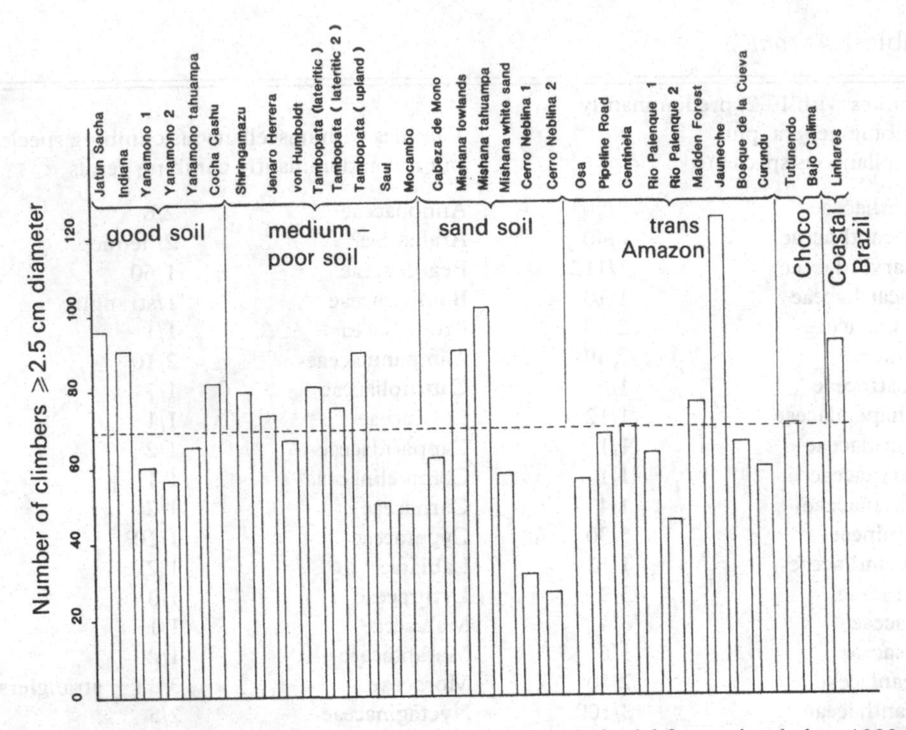

Figure 1.2. Liana density in neotropical moist, wet, and pluvial forest sites below 1000 m elevation. Broken line represents average density across all sites.

(> 10 cm diameter) are also about equally represented on different soil types, contrary to the situation for large trees, which are more prevalent on more fertile soils (Figure 1.4).

Despite the absence of overall trends in neotropical liana density, there is marked variation between individual sites (Figure 1.2), with more than a four-fold difference between the most liana-poor and liana-rich sites. The two most liana-poor samples are from the most extreme white sand site of all at Cerro Neblina, and other extremely nutrient-poor sandy soils tend to have fewer lianas than less extreme adjacent sites (e.g. Mishana lowlands vs. Mishana white sand and Cabeza de Mono vs. Shiringamazu). Excluding a single sample of seasonally inundated forest, the two Amazonian sites with the highest liana densities are perhaps the two sites with richest soils (Jatun Sacha, Ecuador and Indiana, Peru). Thus there is a very slight tendency for greater liana density on richer soils, as suggested by Putz & Chai (1987) and Proctor *et al.* (1983).

At least one correlation of liana density with climate is suggested by these data. It is probably not altogether coincidental that the neotropical site with highest liana density (Jauneche, Ecuador) occurs in an area with a marked dry season that is transitional between moist and dry forest. This is exactly the

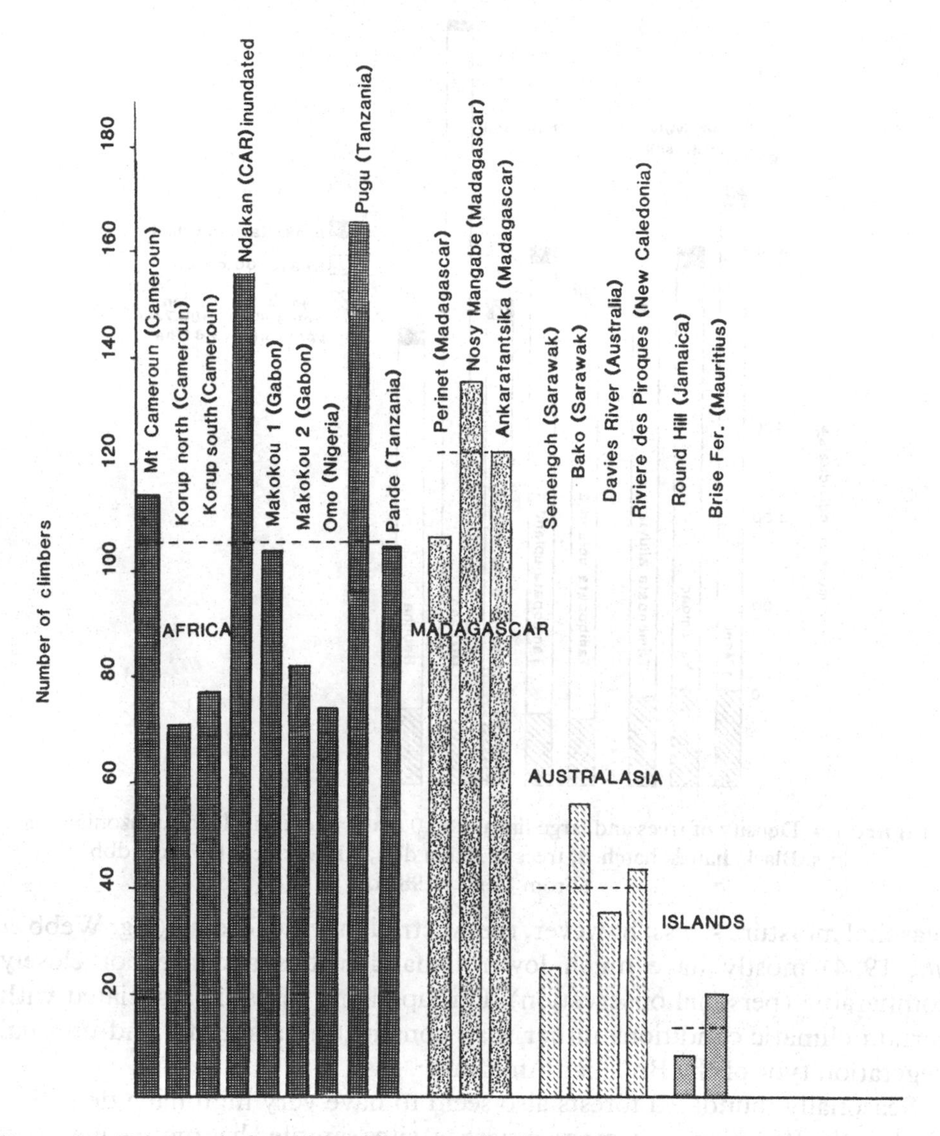

Figure 1.3. Paleotropical and island liana density. Broken lines represent average densities for each region.

same climatic regime in which the famed Brazilian liana forests occur (Hueck, 1981; Balee & Campbell, 1989). Moreover, an African forest with a similar climatic regime (Pugu, Tanzania) has the highest African liana density, and White (1983:200) reports that the South African Tongaland–Pandaland forests, which are similar to Pugu (Lovett, personal communication), probably had the highest liana diversity in Africa. In an analysis of Queensland forests Webb *et al.* (1967) found that liana abundance responded primarily to

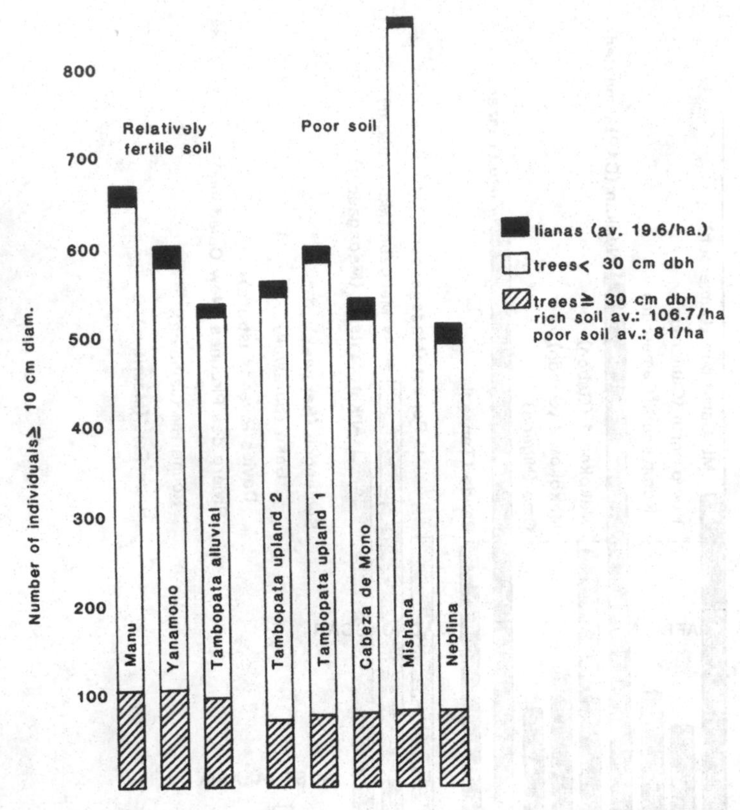

Figure 1.4. Density of trees and large lianas (> 10 cm diameter) in 1-ha Amazonian tree plots. Black, lianas; hatched, trees > 30 cm dbh; white, trees 10–30 cm dbh. (From Gentry 1988a.)

seasonal moisture stress; however, the Australian 'vine forests' (e.g. Webb *et al.*, 1984) mostly have much lower liana densities and are not closely comparable (personal observation). Perhaps liana forest is associated with certain climatic conditions rather than constituting a specific and unusual vegetation type of the Brazilian Amazon.

Seasonally inundated forests also seem to have very high liana densities. Each of the Peruvian tahuampa (= *varzea*) sites sampled has more lianas than do two nearby samples from non-inundated forests. The neotropical dry forest site with greatest liana density (Galerazamba, Colombia) is near the coast and apparently poorly drained. This trend is even more pronounced for large lianas (> 10 cm diameter); the Mishana tahuampa forest had more large lianas per 0.1 ha (13) than any other site and the Galerazamba sample was second. Moreover, the only available sample of seasonally inundated African forest, from Ndakan on the Central African Republic–Cameroun border, has many more lianas than an adjacent non-inundated forest and is the second most liana-rich African site (Figure 1.3).

Table 1.5. *Families with climbers only in Old World*

Actinidiaceae	Myrtaceae
Aizoaceae	Nepenthaceae
Ancistrocladaceae	Oleaceae
Austrobaileyaceae	Opiliaceae
Caryophyllaceae	Pandanaceae
Chenopodiaceae	Pittosporaceae
Chrysobalanaceae	Portulacaceae
Cunoniaceae	Rutaceae
Dioncophyllaceae[a]	Sabiaceae
Droseraceae	Santalaceae
Elaeagnaceae	Scytopetalaceae
Flacourtiaceae	Stemonaceae
Flagellariaceae	Stylidiaceae
Goodeniaceae	Theaceae
Lentibulariaceae	Tiliaceae
Linaceae	Tremandraceae
Monimiaceae	Trimeniaceae
Myrsinaceae	Umbelliferae

[a]Included in Flacourtiaceae in Missouri Herbarium.

Taxonomic composition

A total of at least 133 families includes at least a few climbers. This compares with 113 families (excluding ferns) reported to contain climbing species by Schenck (1892), but is still somewhat less than the half of all vascular plant families suggested by Putz (1984). At least 97 seed plant families (plus several ferns and fern allies) have climbing species in the New World (Tables 1.3, 1.4), most of these also containing at least some climbers in the Old World. There are at least an additional 36 seed plant families with climbers exclusively in the Old World (Table 1.5).

Even in the temperate zone, many plant families have scandent species. The 24 European vine species are distributed among 13 genera belonging to 10 different families (Baas & Schweingruber, 1987). The 48 woody climbers native to the southeastern USA belong to 25 different genera in 19 different families. The 146 climbing species of the Carolina Flora (Radford *et al.*, 1968) belong to no less than 54 different genera and 28 different families (and at least 80 more species belonging to an additional 37 genera and 12 families are prostrate or decumbent).

Evolution of scandent habit in a taxon does not automatically equate with evolutionary diversification of climbers in that taxon. Although many families include climbers, the great majority of climbing species belong to relatively few families. Thus 64% of the 9216 climbing neotropical species belong to only 12 families and 26 families include 85% of all New World climbers. The 39 families with 50 or more climbing species in the New World account for 95% of the total climbing species. The other 56 families with New World climbers altogether account for a mere 5% of the climbing species. Most of the New World climbers belong to large diversified families that also include many non-climbers. Families like Bignoniaceae, Apocynaceae, Leguminosae, Malpighiaceae, and Sapindaceae, with both several predominantly climbing genera and several non-climbing ones account for about half of the genera with climbers and of the climbing species (Table 1.4); predominantly non-climbing families with one or two genera of vines account for another 989 species. A quarter of the families listed as New World climbers in Table 1.4 include no predominantly climbing genera but these account for less than 2% of the climbing species. On the other hand, the 19 families (23 world-wide) composed entirely or almost entirely of climbers in the New World account for 38% of all climbing species.

The same generalities hold for subsets of the Neotropics. For example, only 27 of the 71 families with climbing species in Panama are predominantly climbing, but 21 of these 27 are entirely scandent in Panama (Gentry, 1985). Most of the evolutionary radiation of scandent species has taken place in relatively few taxa.

The two largest families of climbers in the New World, both overwhelmingly composed of vines, are Asclepiadaceae and Convolvulaceae. Third largest is Leguminosae, which contains many vines as well as woody lianas and many genera of erect plants. It is noteworthy that although climbers occur in all three legume subfamilies, the vast majority are papilionate; conversely, the majority of papilionate species are scandent. Papilionate legumes are by far the most successful temperate zone climbers, constituting over a third of the herbaceous vines of the southeastern USA. Asteraceae, although with few climbers relative to its size, still ranks fourth among neotropical families with climbers, thanks largely to the 300 species of *Mikania*, the world's sixth largest vine genus. The other four largest New World vine families each have about 400 climbing species in the Neotropics. Of these, Bignoniaceae, Malpighiaceae, and Sapindaceae have mostly lianas while Araceae climbers are overwhelmingly hemiepiphytic. The final four of the top dozen New World climbing families are overwhelmingly scandent Passifloraceae (mostly herbaceous vines but also with many lianas), and Cucurbitaceae (with few lianas), plus Apocynaceae and Ericaceae, both also with many erect genera, the former with free-climbing lianas belonging to at least 35 different scandent genera, the latter with at least 25 genera of mostly scandent woody hemiepiphytes. Of the other families with numerous scandent neotropical species listed in Table 1.3, three (Dioscoreaceae, Amaryllidaceae, Smilacaceae) contain mostly slender vines, five (Gesneriaceae, Marcgraviaceae, Piperaceae, Melastomataceae, Solanaceae) have mostly hemiepiphytic climbers, five (Rubiaceae, Aristolochiaceae, Euphorbiaceae, Vitaceae, and Hippocrateaceae) have mostly free-climbing lianas, and one (Loranthaceae) is parasitic.

If a similar analysis is conducted at the generic level, the two families Bignoniaceae (53 scandent New World genera) and Cucurbitaceae (55 genera) stand out. These are also the two largest angiosperm families with the fewest species per genus overall (Gentry, 1973). Asclepiadaceae, where generic level taxonomy is in a state of flux, has about 40(–57) scandent New World genera (and perhaps 200(!) worldwide: W. D. Stevens, personal communication), while Leguminosae, Malpighiaceae, and Apocynaceae each have about 30, Convolvulaceae 20, and Asteraceae about 23. Other families have a dozen or fewer genera with scandent New World species. In some of these families the climbing habit is clearly fundamental but in many others it has arisen independently in many different unrelated lineages.

Another noteworthy taxonomic pattern is that, although some families have numerous genera with climbers, the majority of scandent species of most of the large vine families belong to a very few large genera, many of which are pantropical (e.g. *Ipomoea* in Convolvulaceae, *Mikania* in Compositae, *Philodendron* in Araceae, *Dioscorea* in Dioscoreaceae, *Smilax* in Smilacaceae, *Combretum* in Combretaceae, *Strychnos* in Loganiaceae, *Serjania* and *Paullinia* in Sapindaceae) (Table 1.6). Conversely, few families have more than one or two large scandent genera. The 50 largest genera of climbers listed in Table 1.6 represent no fewer than 29 different families and of these only Leguminosae and Asclepiadaceae, each with eight large scandent genera, have more than two genera (plus Convolvulaceae with three, if *Cuscuta* is included) which have speciated extensively as climbers. Twenty-one of the 29 families represented in Table 1.6 have only one speciose genus of climbers.

Nor are the important scandent families related. They include both monocotyledons and dicotyledons, gymnosperms and angiosperms, primitive and advanced orders of flowering plants. Members of all of the major dicotyledon superorders except Hamamelidae are represented among the important families of climbers, and families with important complements of climbers are included in no fewer than 34 different Cronquistian angiosperm orders (29 of dicotyledons, 5 of monocotyledons). At least 133 families, or almost half of all seed plant families, have at least a few scandent species. Even of the *c.* 23 intrinsically scandent families of plants, only a very few seem

Table 1.6. *Some of the largest climbing genera. Species numbers mostly from Airy-Shaw (1973), except as otherwise noted. Sources are given in square brackets*

Genus	Number of climbing species	Family
Dioscorea	600	Dioscoreaceae
Ipomoea	500 (+ few trees)	Convolvulaceae
Calamus (*s.l.*)	375	Palmae
Passiflora	355 (+ few trees)	Passifloraceae
Cissus	350	Vitaceae
Mikania	300	Asteraceae
Rhynchosia	300	Leguminosae
Philodendron [1]	275	Araceae
Smilax	270	Smilacaceae
Combretum	230 (of 250)	Combretaceae
Jasminum	225 (¾ of 300)	Oleaceae
Serjania	215	Sapindaceae
Phaseolus	200	Leguminosae
Gonolobus	200	Asclepiadaceae
Convolvulus	200	Convolvulaceae
Mussaenda	200 (if all climbing)	Rubiaceae
Dalbergia	*c.* 200 (⅔ of 300)	Leguminosae
Matelea [2]	198	Asclepiadaceae
Aristolochia	180 (if all climbing)	Aristolochiaceae
Paullinia	180 (+ few treelets)	Sapindaceae
Cuscuta	170	Convolvulaceae
Strychnos [3]	150	Loganiaceae
Uvaria	150	Annonaceae
Vicia	150 (if all climbing)	Leguminosae
Salacia	150 (+ few trees)	Hippocrateaceae
Galactia	140 (if all climbing)	Leguminosae
Sabicea	130	Rubiaceae
Dichapetalum	130 (of 136)	Dichapetalaceae
Lathyrus	130 (if all climbing)	Leguminosae
Machaerium	120	Leguminosae
Bauhinia	*c.* 120 (of 350)	Leguminosae
Ceropegia [2]	*c.* 120 (¾ of 160)	Asclepiadaceae
Gonolobus	115	Asclepiadaceae
Bomarea [4]	112 (¾ of 150)	Amaryllidaceae
Cynanchum [2]	111	Asclepiadaceae
Byttneria [5]	111	Sterculiaceae
Marsdenia [2]	109	Asclepiadaceae
Metastelma [2]	106	Asclepiadaceae
Oxypetalum [2]	105	Asclepiadaceae
Mandevilla [6]	102 (+ 12 erect)	Apocynaceae
Medinilla	*c.* 100 (¼ of 400)	Melastomataceae
Parsonsia	100 (if all climbing)	Apocynaceae
Artabotrys	100+	Annonaceae

Genus	Number of climbing species	Family
Dalechampia	100 (of 110)	Euphorbiaceae
Tragia	100	Euphorbiaceae
Heteropteris	100 (if all climbing)	Malpighiaceae
Columnea	*c.* 100 (of 200)	Gesneriaceae
Peperomia	*c.* 100 (of 1000)	Piperaceae
Clerodendron [7]	100 (of 500)	Verbenaceae
Rourea	*c.* 100	Connaraceae
Vigna	*c.* 100	Leguminosae
Secamone [2]	100	Asclepiadaceae

Sources: 1. T. Croat (personal communication); 2. W. D. Stevens (personal communication); 3. Krukoff (1972); Leeuwenberg (1969); 4. R. Gereau (personal communication); 5. Cristobal (1976); 6. Woodson (1933); 7. R. Rueda (personal communication).

closely related. Perhaps only in Cucurbitaceae and Passifloraceae is the scandent habit a shared synapomorphy, although Asclepiadaceae undoubtedly evolved from scandent Apocynaceae, and Vitaceae and Trigoniaceae are related, respectively, to scandent Sapindaceae and Malpighiaceae. Even within a family, genera with scandent species typically belong to otherwise unrelated subfamilies or tribes.

Evolutionary diversification of climbers

From the above discussion, we may safely conclude that evolution of a scandent habit has taken place independently many times during the course of plant evolution. In an earlier paper (Gentry & Dodson, 1987), we made a similar claim for epiphytes. Therefore, it is instructive to compare evolution of the scandent habit with the epiphytic one. Worldwide, there are 83 families with epiphytes, representing some 876 genera and 29 000 species. Since my data for Old World climbers are incomplete, the neotropical subset of the epiphyte data – 42 families with 15 000 epiphytic species – can be extracted for comparison. It is apparently much easier to become a vine than an epiphyte: over twice as many neotropical families have scandent species as epiphytic ones. Moreover, the number of times it has arisen suggests that there must be strong selective pressures favoring evolution of a scandent habit.

Although the scandent habit has arisen more times than the epiphytic one, subsequent evolutionary radiation of scandent taxa has been less pronounced. There are half again as many epiphytic as scandent species in the

Neotropics. Over a quarter of climbing families have only 1–3 scandent species. Eighty-five per cent of neotropical climbers belong to only 26 families and 64% to 12 families. What features have led to success of the scandent habit in these few taxa?

Among lianas, the families that have most successfully speciated are generally those with the most specialized climbing mechanisms. Nearly all of the diversified families of climbers have developed specific climbing mechanisms in addition to twining, whereas most of the less speciose climbing families have not. In this respect, tendrils are the most important climbing specialization. For example, the 10 Panamanian families with tendrils account for 44% of that country's climbing species (Gentry, 1985). Eight climbing neotropical families have tendrils formed from modified branches or inflorescences, and of these Cucurbitaceae, Passifloraceae, Sapindaceae, Loganiaceae (*Strychnos*), and Vitaceae are among the most species-rich scandent taxa. Many scandent paleotropical Apocynaceae (along with neotropical *Pacouria*) also have inflorescences modified into tendrils as does *Iodes* of the Icacinaceae. On the other hand, similar tendrils in *Gouania* and *Lophopyxis* of the Rhamnaceae and *Antigonum* of the Polygonaceae have led to relatively low levels of speciation.

Five neotropical, and several paleotropical, families have taxa that climb by leaf tendrils, including Bignoniaceae, the pre-eminent neotropical liana family. In Bignoniaceae the terminal leaflet (or several leaflets in a few genera) of a compound leaf has been converted into a tendril. Paleotropical Bignoniaceae climbers, distantly related to most of the neotropical ones, lack tendrils (and perhaps as a result are not very numerous). *Cobaea* (Polemoniaceae) has similar tendrils to Bignoniaceae as does north temperate *Adlumia* (Fumariaceae). Other neotropical genera with leaf tendrils include *Mutisia* (Asteraceae) and several Leguminosae (*Vicia, Lathyrus, Entada*), where the apex of the leaf rachis becomes a tendril, and *Smilax* (Smilacaceae) where the stipules are modified as tendrils. The exclusively paleotropical families Nepenthaceae, Flagellariaceae (*Flagellaria* only), and Dioncophyllaceae also have tendrillate rachis apices, in Nepenthaceae with the tendril tip further modified into an insectivorous pitcher. At least four other families have twining petioles (Tropaeolaceae, plus several miscellaneous genera, e.g. *Clematis, Hidalgoa, Perianthomega*). Why taxa like *Cobaea* and *Entada*, that have tendrils similar to those of Bignoniaceae, have radiated relatively little is unclear.

Several liana families – Malpighiaceae, Hippocrateaceae, Connaraceae, Polygalaceae (*Securidaca*), Thymelaeaceae (*Craterosiphon*) – climb largely by means of sensitive young branchlets that bend back on themselves and function in a manner analagous to tendrils. Several paleotropical Annonaceae climb by short hook-forming branches, and in one of these, *Artabotrys*, this is modified into a very characteristic hook-like 'crochet'. Quite unrelated

Ancistrocladaceae have a very similar hook-like climbing mechanism, as does *Hugonia* of the Linaceae. Climbing palms are mostly spiny and have very characteristic grappling-hook leaf apices, both in the Old World rattans and in quite unrelated neotropical *Desmoncus*. Other adaptations associated with scandent habit – including adventitious roots, dimorphic juvenile forms, and striking cambial anomalies – are characteristic of various genera or families of climbers (see Carlquist, chapter 2; Lee & Richards, chapter 8).

Unlike the above taxa, a few large families which have successfully diversified as climbers are not characterized by a single effective climbing strategy. The most conspicuous of these are Leguminosae, which have genera that climb by leaf-tendrils (*Vicia*, *Entada*), branch tendrils (*Bauhinia*), hooked branches, twining branches, and spines. Asteraceae can have leaf-tendrils (*Mutisia*), or sensitive petioles (*Hidalgoa*), or merely be twiners. With the exception of *Mutisia* and the very large genus *Mikania*, which lacks obvious climbing adaptations other than twining, relatively few Asteraceae are climbers.

Only a few successful climbing families lack obviously specialized climbing organs. Convolvulaceae and Menispermaceae lack tendrils or tendril equivalents, but liana species mostly have pronounced cambial variance (cf. Carlquist, Chapter 2), in the former family associated with included phloem. The conspicuously successful diversification of Asclepiadaceae, which are twiners lacking specialized climbing organs, is likely to be related instead to their very specialized pollination strategy which features pollinia; in many ways, including the possession of pollinia and their extremely high species richness, Asclepiadaceae might appropriately be considered the equivalent among scandent plants to orchids among epiphytes. Other more or less important climbing families without obvious climbing organs are Dioscoreaceae, Amaryllidaceae, Aristolochiaceae, and perhaps Solanaceae (spines, often with adventitious roots), and Ericaceae (some with adventitious roots). Significantly, most of these are herbaceous vines or woody hemiepiphytes rather than lianas. Presumably the need to reach the mature forest canopy imposes more severe constraints on evolution of the liana habit than on other kinds of climbers.

Liana community ecology

Neotropics

At least on a given continent, liana communities are remarkably regular in floristic composition, a feature shared in common with other components of tropical lowland plant communities (Gentry, 1982a, 1988a). In neotropical lowland forests, Bignoniaceae and Leguminosae are nearly always the dominant climbing families. One of these is always the most speciose family of lianas in lowland Amazonia (21 of 21 0.1 ha samples: Figure 1.5),

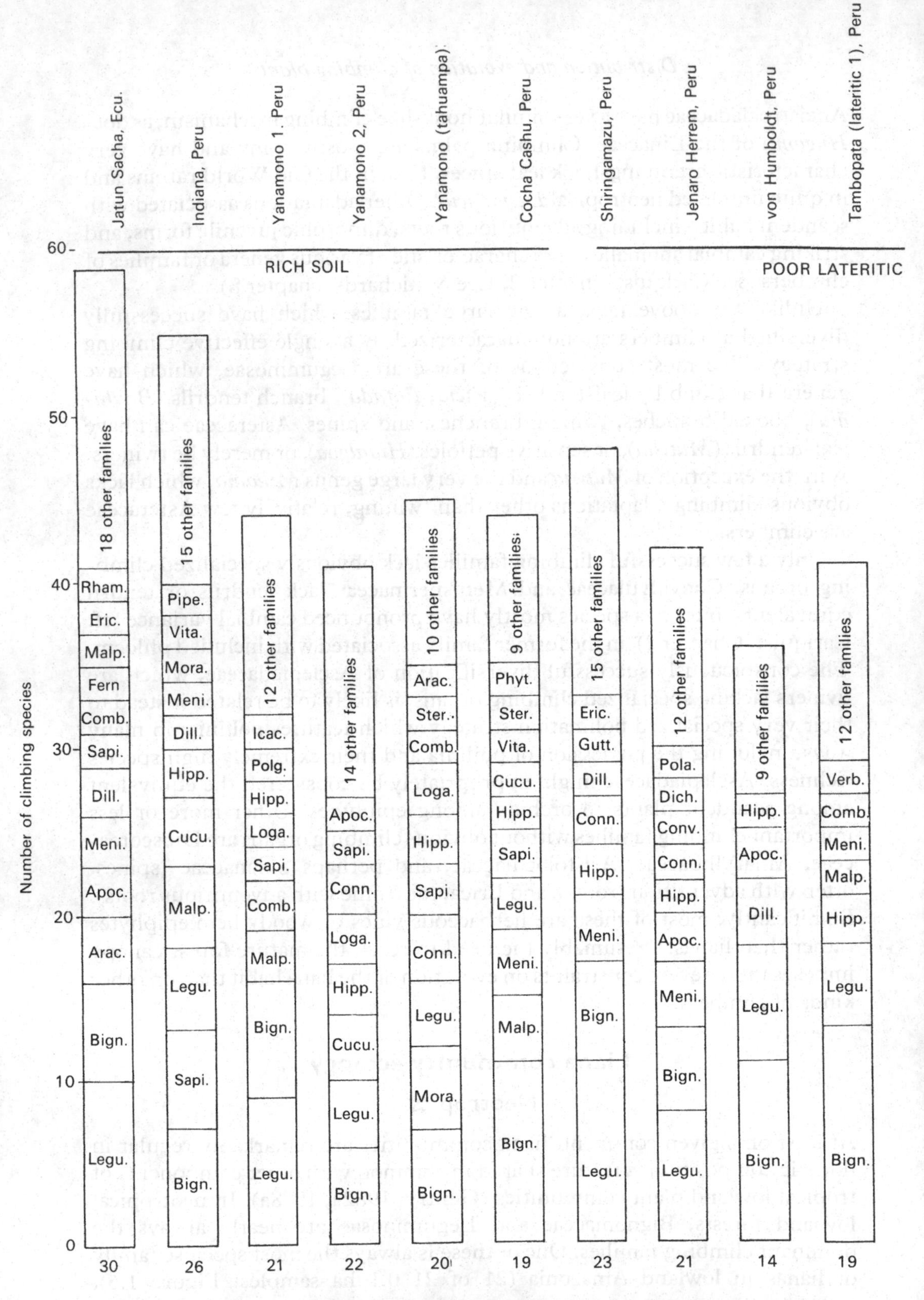

Figure 1.5. Species richness and familial composition of Amazonian liana communities (climbers > 2.5 cm diameter in 0.1 ha). Smallest bar segments represent 2 spp./family (1.5 spp./family in summary bar). Number below each bar is number of families represented by climbers in that sample.

Site	Soil	Families (top to bottom of bar)	Total
Tambopata (lateritic 2), Peru	SOIL	11 other families, Verb., Conn., Apoc., Malp., Comb., Sapi., Hipp., Legu., Cucu., Meni., Bign.	22
Tambopata (upland), Peru	SOIL	6 other families, Pipe., Conn., Comb., Malp., Hipp., Legu., Bign.	13
Saul, Fr. Gui.	SOIL	9 other families, Cycl., Pola., Gutt., Conv., Hipp., Legu., Apoc., Sapi., Bign.	18
Mocambo, Brazil	SOIL	5 other families, Acan., Pola., Conv., Loga., Conn., Legu., Bign.	12
Cabeza de Mono, Peru	SANDY SOIL	9 other families, Rubi., Dill., Conn., Hipp., Bign., Legu.	15
Mishana (lowland), Peru	SANDY SOIL	10 other families, Marc., Euph., Pola., Malp., Sapi., Conv., Dill., Conn., Legu., Meni., Hipp., Bign.	22
Mishana (tahuampa), Peru	SANDY SOIL	Pola., Apoc., Dill., Mora., Loga., Sapi., Bign., Hipp., Legu.	15
Mishana (white sand), Peru	SANDY SOIL	6 other families, Conv., Meni, Hipp., Bign., Malp., Legu.	12
Cerro Neblina 1, Ven.	SANDY SOIL	4 other families, Icac., Meni., Bign., Legu.	8
Cerro Neblina 2, Ven.	SANDY SOIL	5 other families, Malp., Sapi., Legu., Bign.	9
Average		Av. for 41 other families <1.5 spp./transect, Dill., Conn., Malp., Sapi., Meni., Hipp., Legu., Bign.	18 (49)

irrespective of soil type. Of six Amazonian samples from relatively rich soils, Bignoniaceae is the most speciose family in four and legumes (with bignons second) in the other two. Of eight Amazonian samples from forests on relatively infertile lateritic soil, bignons were most speciose in six and legumes (with bignons second) in the other two. Of seven sites on very infertile sandy soils, bignons were most diverse in three and legumes (with bignons second or third) in four. Altogether these 21 lowland Amazonian samples (0.1 ha each) averaged 39 species of climbers, including seven Bignoniaceae, six Leguminosae, two each of Hippocrateaceae, Menispermaceae, Sapindaceae, and Malpighiaceae, and 1.5 Connaraceae and Dilleniaceae. These eight families may be considered the predominant Amazonian liana families.

The 41 other families represented by climbers in the overall Amazonian sample series averaged less than 1.5 species per 0.1 ha. Other families which occasionally included more than three liana species in a sample are Araceae (4 spp. of hemiepiphytes in the wettest Amazonian site sampled), Cucurbitaceae (3 sites), Combretaceae (1 site), Moraceae (both seasonally inundated forests sampled: one with 5 strangler figs, the other with 3 stranglers), Loganiaceae (*Strychnos*: 3 sites), and Convolvulaceae (1 site).

A series of 10 lowland trans-Andean moist and wet forests distributed from Costa Rica to western Ecuador (Figure 1.6) showed essentially the same pattern, with Bignoniaceae the most speciose liana family in seven of the samples and runner-up in two others, legumes the second most speciose climbing family in five samples and third in another. In the three wettest trans-Andean lowland samples (> 3000 mm of annual precipitation) three other families (Sapindaceae, Malpighiaceae, and Moraceae) were the most speciose in climbers (including hemiepiphytes). This trend toward decreasing prevalence of Bignoniaceae and Leguminosae climbers in wetter forests is even more pronounced in the Choco area pluvial forests (see below). An average wet or moist trans-Andean lowland forest has slightly lower liana diversity than in Amazonia, including 32 scandent species per 0.1 ha, of which five are Bignoniaceae, three Leguminosae, two Sapindaceae, and 1.5 Dilleniaceae and Malpighiaceae. There are 40 additional families represented by < 1.5 climbing species per sample. Thus Connaraceae, Menispermaceae, and Hippocrateaceae are underrepresented when compared to equivalent Amazonia sites. Nevertheless, the floristic composition of the liana community is remarkably similar to that of Amazonia except at the three sub-cloud-forest sites.

The floristic composition of the liana community in neotropical dry forests is also similar in different forests (Figure 1.7). Although much poorer in species (an average of only 17 species per 0.1 ha vs. 39 for the Amazonian moist and wet forest samples), neotropical dry forest liana communities are

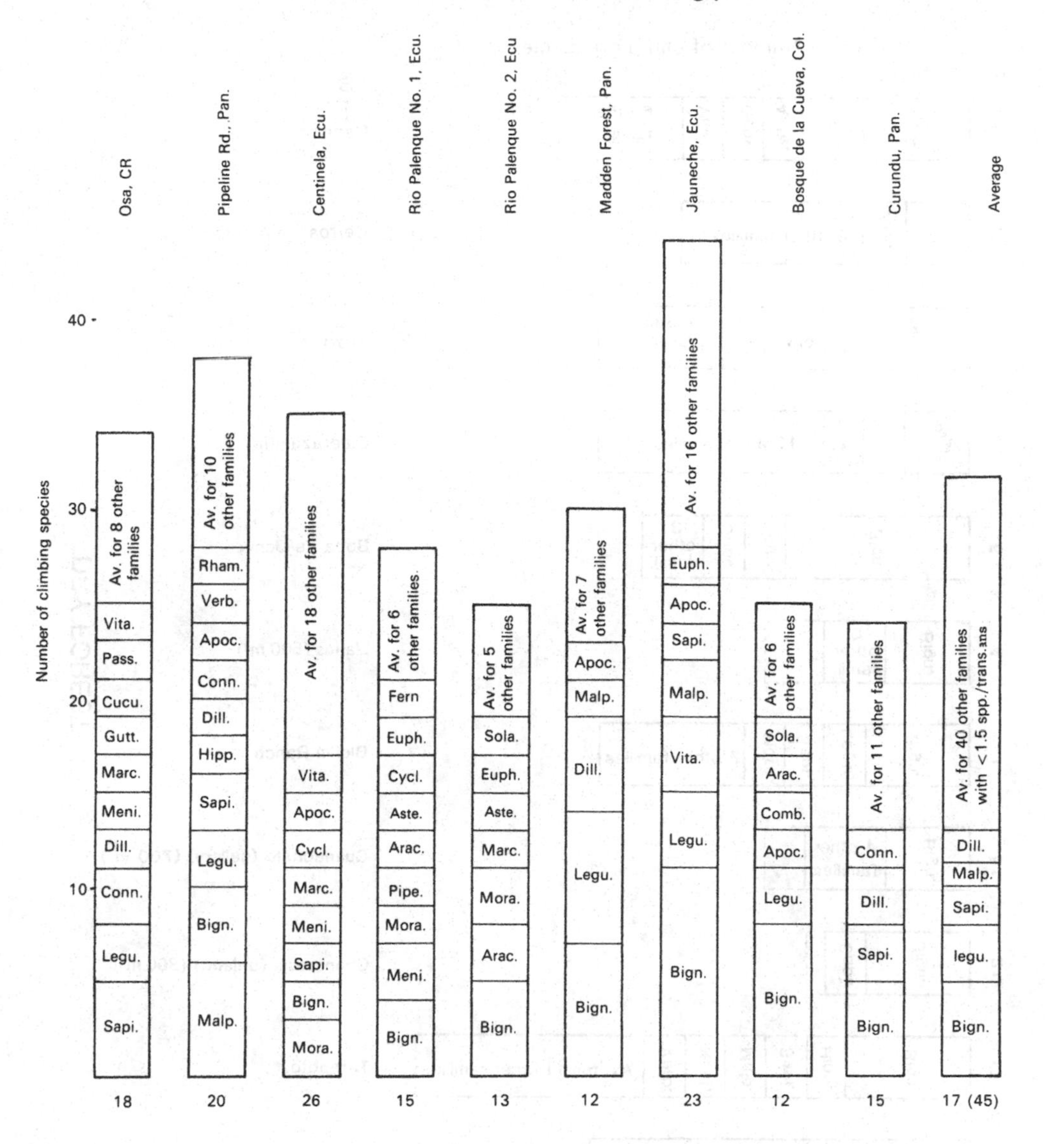

Figure 1.6. Species richness and familial composition of trans-Andean liana communities (climbers >2.5 cm diameter in 0.1 ha). Smallest bar segments = 2 spp./family. Number below each bar is number of families represented by climbers in that sample.

dominated by the same families. Bignoniaceae is the most speciose liana family in every sampled dry forest with an average of 5 species per 0.1 ha. Legumes and Sapindaceae average 1 species per 0.1 ha sample and are often represented by 2–3 species in a single sample. Thirty other families are represented at least once in the available sample series of 10 different lowland neotropical dry forests. Three subtropical Mexican dry forest samples (Lott

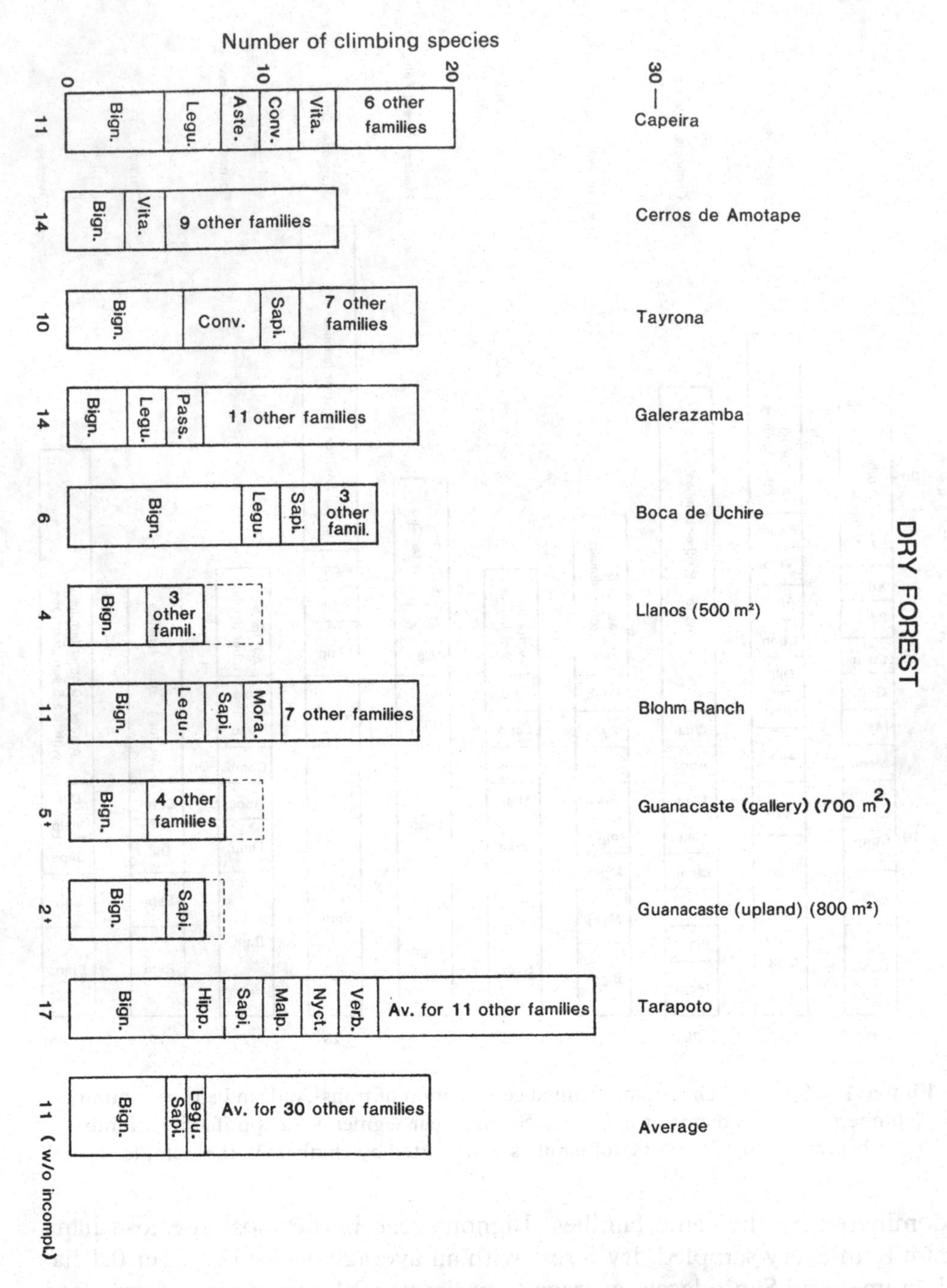

Figure 1.7. Species richness and familial composition of lowland neotropical dry forest (< 1000 m elevation) liana communities (climbers > 2.5 cm diameter in 0.1 ha). Smallest bar segments = 2 spp./family except in summary bar. Number below each bar is number of families represented by climbers in that sample.

et al., 1987) are poorer in liana species (averaging only 15 spp./0.1 ha) but still have Bignoniaceae and Sapindaceae, along with Malpighiaceae, as the most species-rich liana families.

Pluvial forests (>7400 mm annual precipitation), on the other hand, have a dramatically different familial composition (Figure 1.8). They are also richer in climbing species with an average of 51 species per 0.1 ha sample. The big difference in the climbing community is that in these forests hemiepiphytic climbers seem to have largely replaced free-climbing ones. Thus the most speciose family of pluvial forest climbers in my samples is hemiepiphytic Melastomataceae, followed by Guttiferae and Araceae, both also hemiepiphytic. Also prominent among pluvial-forest climbing families are other hemiepiphytic taxa like Marcgraviaceae, Ericaceae, Cyclanthaceae, and Moraceae. Even families like Bignoniaceae that are normally free-climbing are largely represented by hemiepiphytic genera (e.g. *Schlegelia*) in pluvial forests. In the prevalence of hemiepiphytic climbers, lowland pluvial forests are more similar to mid-elevation cloud forests than to other lowland forests (Gentry, 1988a).

Andean cloud forests (Figure 1.9) share with pluvial lowland forests the predominance of hemiepiphytic climbers and the prevalence of species of Araceae, Ericaceae, and Guttiferae (and below 2000 m Cyclanthaceae, Marcgraviaceae, and strangling Moraceae) among their climbers. The most unusual floristic feature of upland Andean forests is the prevalence of Asteraceae as the most speciose liana family, averaging 5 species per 0.1 ha sample. Another noteworthy feature of these forests is that their overall climber diversity, like that of trees (Gentry, 1988a), is generally low (only 22 sp./0.1 ha), and decreases more or less regularly with altitude.

Old World

I have fewer data available to analyze the liana communities of Old World forests, but some general trends seem evident. For a series of eight continental African lowland forests (Figure 1.10), liana species richness is almost exactly the same as in Amazonia (41 vs. 39 spp./0.1 ha). This contrasts with liana density which is greater in Africa than in the Neotropics (cf. above). On the other hand, the floristic composition of African liana communities is noticeably different from that in the Neotropics. The biggest difference is the lack of Bignoniaceae climbers in African forests. The other striking floristic feature of African climbing communities as compared to the Neotropics is the prevalence of Apocynaceae. Apocynaceae, averaging 7 species per 0.1 ha sample, are the most speciose liana family in five of the eight African samples and second in the other three. In a sense, Apocynaceae replace Bignoniaceae in Africa. Most of the other climbers in African forests are the same families,

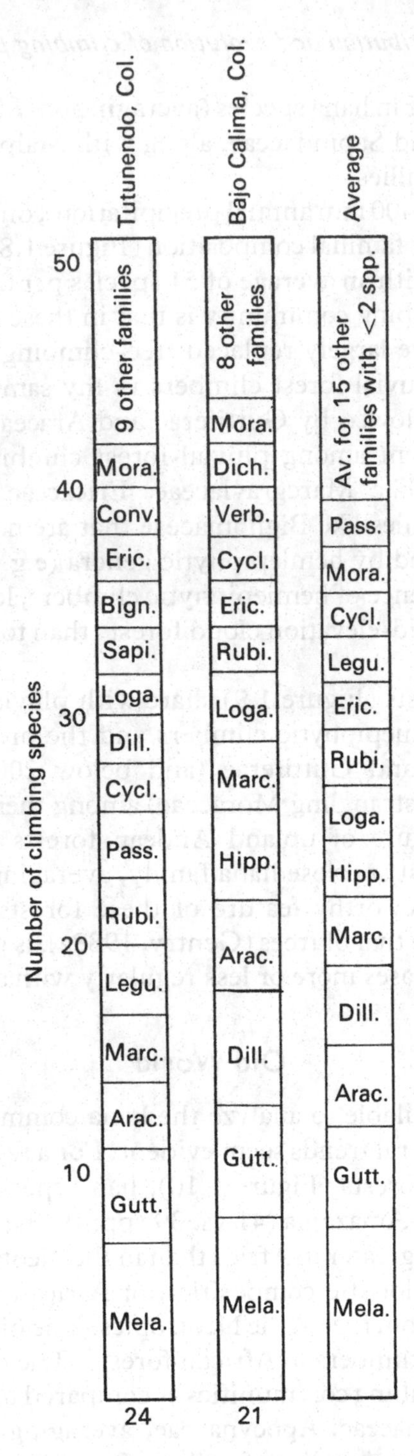

Figure 1.8. Species richness and familial composition in Choco area pluvial forest liana communities (climbers > 2.5 cm diameter in 0.1 ha). Number below each bar is number of families represented by climbers in that sample.

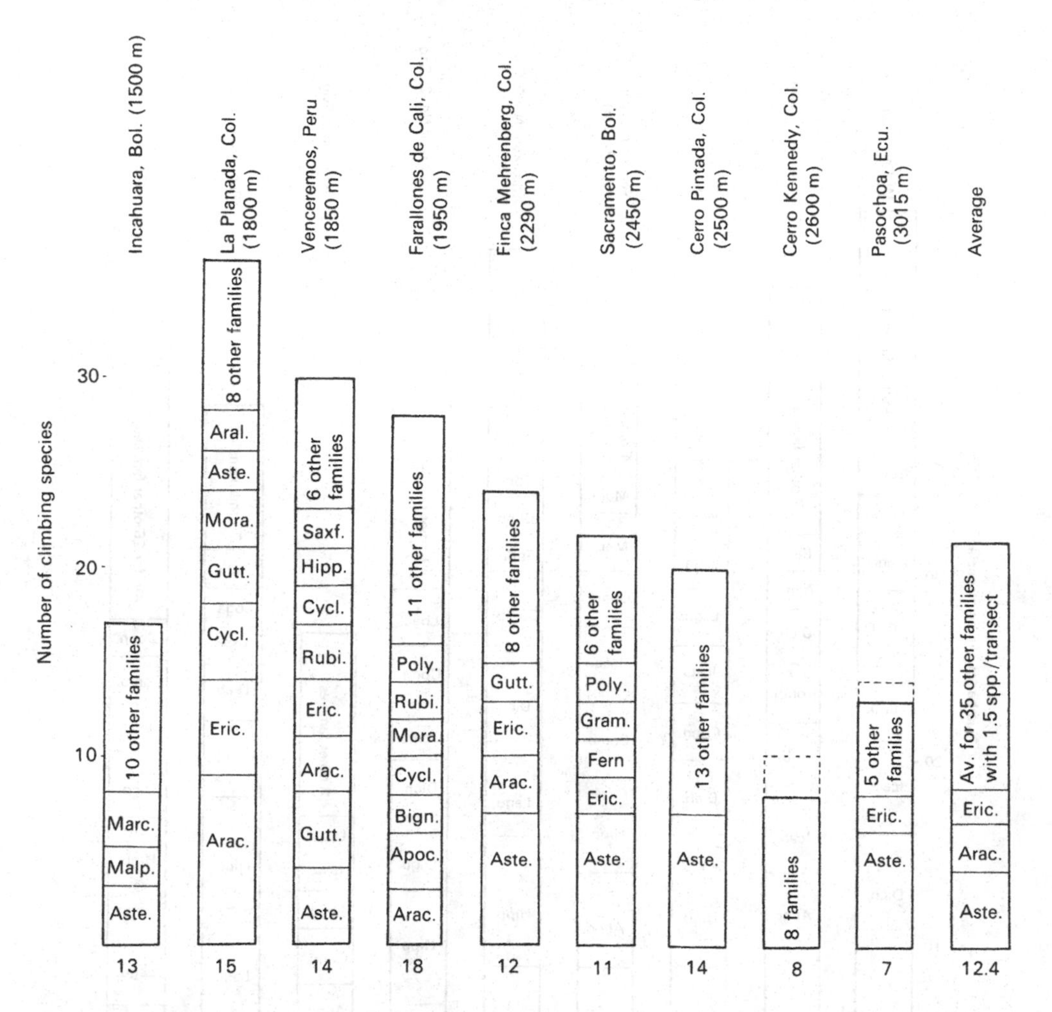

Figure 1.9. Species richness and familial composition of upland Andean liana communities (climbers >2.5 cm diameter in 0.1 ha). Smallest bar segments = 2 spp./family. Number below each bar is number of families represented by climbers in that sample. Dotted extensions of Cerro Kennedy and Pasochoa bars indicate anticipated species richness in complete 0.1 ha sample based on species-added curves for sampled subplots.

and even many of the same genera, as in the Neotropics. Leguminosae is runner-up in dominance with an average of 5 species per 0.1 ha sample, exactly as in the Neotropics. Hippocrateaceae (5 spp./0.1 ha), Dichapetalaceae (3 spp./0.1 ha), Icacinaceae (2 spp./0.1 ha), Combretaceae (2 spp./0.1 ha), Annonaceae (1.5 spp./0.1 ha), and perhaps Connaraceae (2 spp./0.1 ha) are more speciose in African forests. Of these, the greater role of Hippocrateaceae in African forests is most noteworthy; Hippocrateaceae is the most speciose climbing family at one site, is runner-up at three others, and is in the

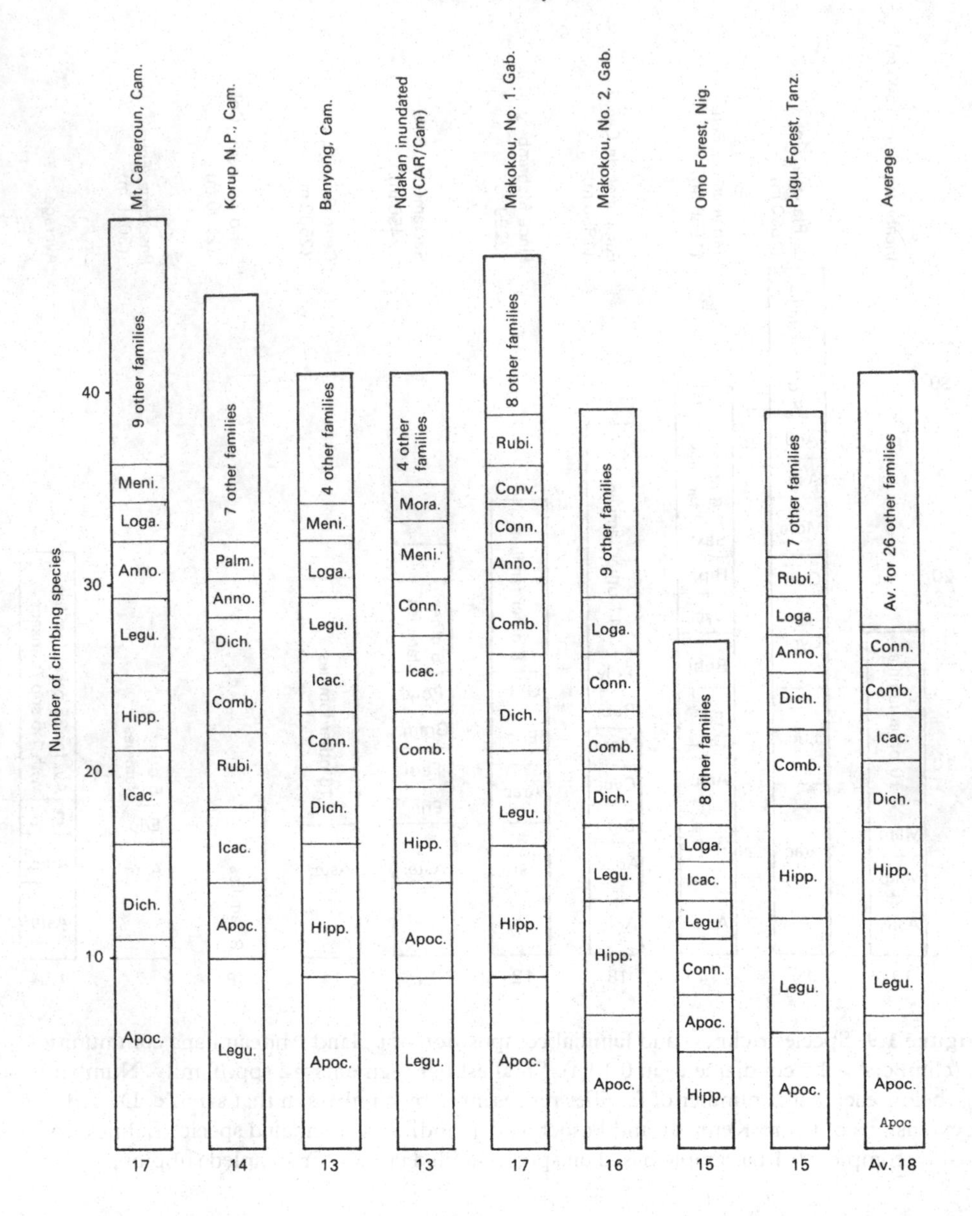

Figure 1.10. Species richness and family composition of continental African liana communities (climbers > 2.5 cm diameter in 0.1 ha). Smallest bar segments = 2 spp./family. Number below each bar is number of families represented by climbers in that sample.

top 3(–4) families at all sites except the one with the poorest soils (Korup, Cameroun).

In Madagascar (Figure 1.11) liana diversity is about the same as in Amazonia or continental Africa. Madagascar is also similar to Africa in overall high liana densities (Figure 1.3), dominance of Apocynaceae

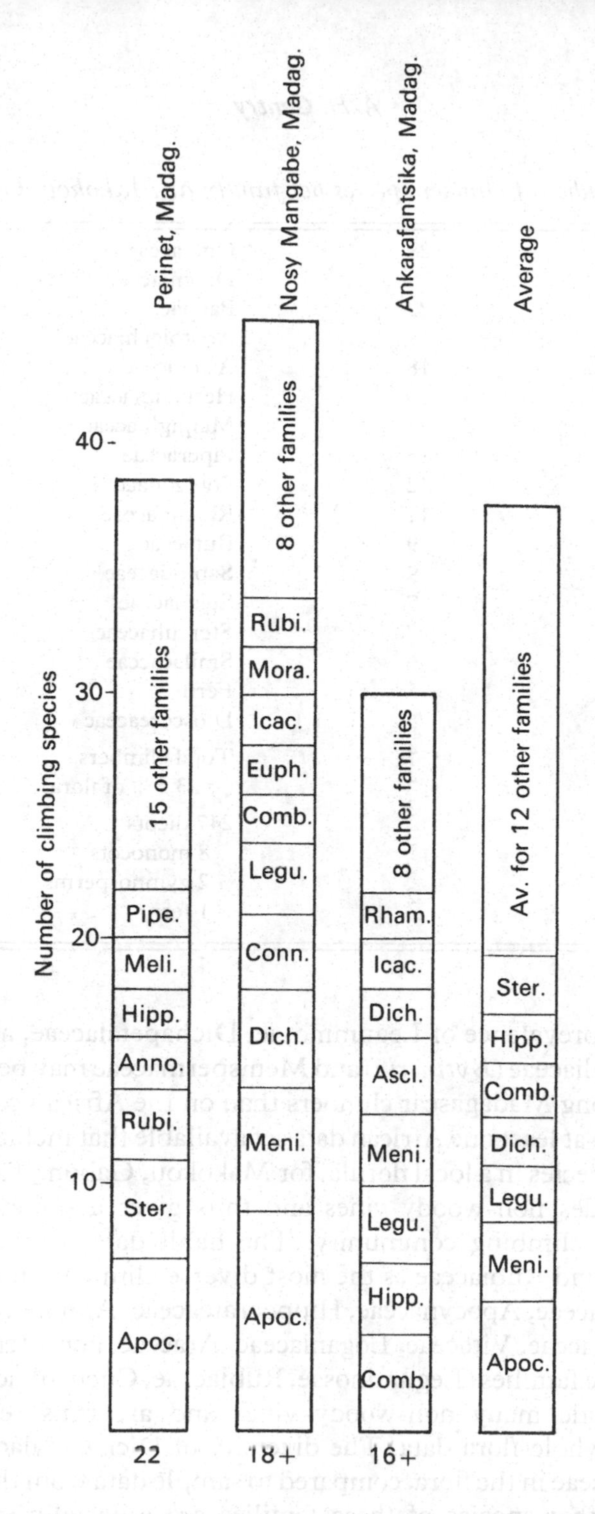

Figure 1.11. Species richness and familial composition of Madagascar liana communities (climbers > 2.5 cm diameter in 0.1 ha). Smallest bar segments = 2 spp./family. Number below each bar is number of families represented by climbers in that sample. (+ indicates a familial indet.)

Table 1.7. *Number of climber species per family at Makokou, Gabon*

Leguminosae	28	Gnetaceae	2
Rubiaceae	28	Dilleniaceae	2
Dichapetalaceae	22	Palmae	2
Apocynaceae	19	Aristolochiaceae	1
Hippocrateaceae	18	Asteraceae	1
Annonaceae	16	Hernandiaceae	1
Connaraceae	15	Malpighiaceae	1
Cucurbitaceae	13	Piperaceae	1
Vitaceae	12	Polygonaceae	1
Loganiaceae	11	Rhamnaceae	1
Araceae	9	Rutaceae	1
Menispermaceae	9	Sapindaceae	1
Combretaceae	7	Solanaceae	1
Verbenaceae	7	Sterculiaceae	1
Icacinaceae	6	Smilacaceae	1
Convolvulaceae	5	Fern	1
Linaceae	5	Dioscoreaceae	1
Asclepiadaceae	4	Total climbers	268
Marantaceae	4	(=23.5% of flora)	
Passifloraceae	3	247 dicots	
Euphorbiaceae	3	8 monocots	
Acanthaceae	2	2 gymnosperms	
Urticaceae	2	1 fern	

climbers, and prevalence of Leguminosae, Dichapetalaceae, and Hippocrateaceae. Sterculiaceae (*Byttneria*) and Menispermaceae may be slightly more important among Madagascar climbers than on the African continent.

There is also at least one African data set available that includes a complete listing of the species in a local florula, for Makokou, Gabon (Table 1.7). This data set includes non-woody vines and thus gives a somewhat different picture of the climbing community. The habit data for Makokou show Leguminosae and Rubiaceae as the most diverse climber families, followed by Dichapetalaceae, Apocynaceae, Hippocrateaceae, Annonaceae, Connaraceae, Cucurbitaceae, Vitaceae, Loganiaceae, Araceae, and Menispermaceae. Several of these families (Leguminosae, Rubiaceae, Cucurbitaceae, Vitaceae, Araceae) include many non-woody vines and are thus relatively more important in whole-flora data. The diversity of Dichapetalaceae, Annonaceae, and Vitaceae in the flora, compared to sample data from the same forest, may indicate that species of these families are especially sparse in their distributions. It is interesting to compare these data with a similar listing for Barro Colorado Island, Panama (Croat, 1978). The large climbing families on

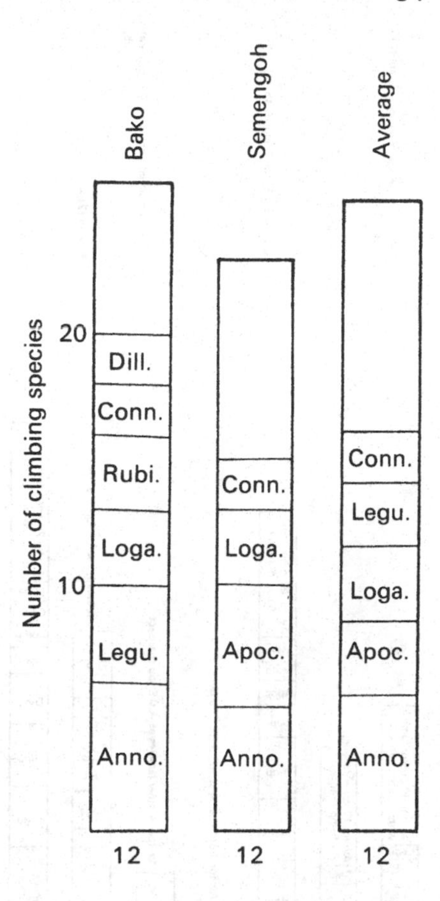

Figure 1.12. Species richness and familial composition of Borneo liana communities (climbers >2.5 cm diameter in 0.1 ha). Smallest bar segments = 2 spp./family. Number below each bar is number of families represented by climbers in that sample.

BCI are essentially the same ones as at Makokou except that Bignoniaceae, Sapindaceae, and Malpighiaceae (respectively 2nd, 3rd, and 4th in number of species) are much more important, and Rubiaceae, Dichapetalaceae, and Annonaceae much less important.

My data for Asia (Figures 1.12, 1.13) are even more limited than for Africa, but some general trends are evident. The most striking feature of Asian liana communities is the prevalence of Annonaceae and rattan palms. Annonaceae, averaging 6 species per 0.1 ha sample, are the most speciose climbing family in six of the eight lowland tropical Asian sites (including New Guinea) and second in the other two (but relatively unimportant in an intermediate elevation forest at Genting Highlands, Malaysia). Climbing palms are also far more speciose (and denser) in Asia, averaging almost 3 species per 0.1 ha sample. Otherwise the lianas of Asian forests seem similar to those of African

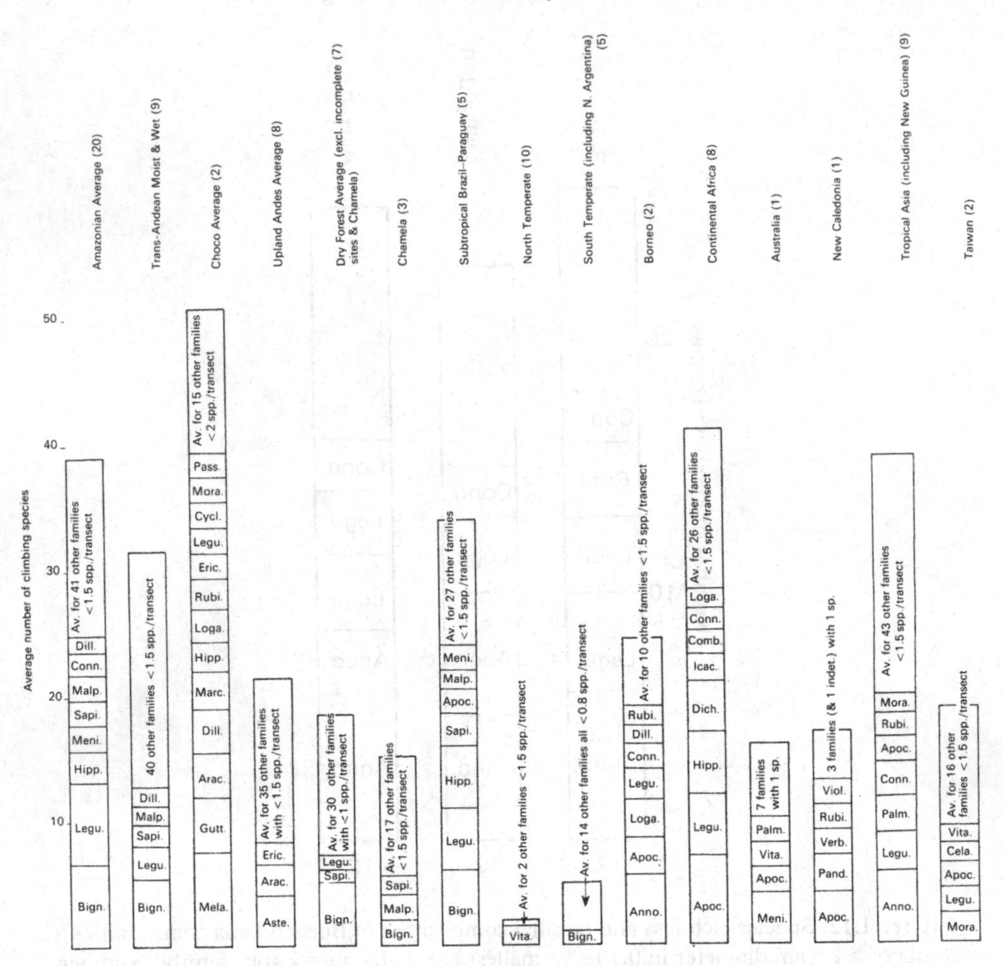

Figure 1.13. Comparative summary of average species richness and familial composition of different types of tropical liana communities (from Figures 1.5–1.12) compared with similar data sets for some temperate zone and subtropical sites. Numbers in parentheses indicate number of sample sites included in average for that forest type.

ones with legumes second only to Annonaceae and with Apocynaceae and Connaraceae the next most speciose families. Loganiaceae (*Strychnos*), Rubiaceae, and Dilleniaceae are well represented. Hippocrateaceae, Dichapetalaceae, and Icacinaceae may be somewhat less prevalent in Asia.

My single Australian and New Caledonian samples (Figure 1.13) are like most other paleotropical ones regarding the prevalence of Apocynaceae climbers. The Queensland sample is characterized by more menisperms than other samples and the New Caledonia one has more climbing Pandanaceae (*Freycinetia*) and Violaceae (*Agatea*). The prevalence of Menispermaceae lianas in Queensland forests, at least, is probably more than an artifact of inadequate sample size, to judge by the long list of Queensland Menispermaceae given by Hegarty & Clifford (1990). Both Australia and New Caledonia have low liana diversities (16 spp., including 1 strangler, and 17 spp./0.1 ha, respectively), similar to those of neotropical dry forests.

Temperate zone forests

Finally, a note on temperate zone liana communities seems appropriate (Figure 1.13). Their most obvious feature is extremely low diversity, paralleling their extremely low density. The ten north temperate samples average 1.6 liana species per 0.1 ha, only two samples having as many as three species. Only three families are represented in these samples. Vitaceae, with an average of four individuals representing 1.5 species per 0.1 ha in North America, is by far the most prevalent. The other north temperate liana families are Araliaceae (Europe only: *Hedera*) and Anacardiaceae (N. America only: *Toxicodendron radicans* at 3 sites). Note that several other north temperate woody climbers tend to be found in disturbed areas and forest edges, ecologically behaving more like herbaceous vines as here defined than like true lianas; *Campsis* is a good example, although in parts of northern Florida it does occur inside the forest (F. Putz, personal communication).

South temperate sites have a few more lianas, an average of 5 species per 0.1 ha in the five Chilean and northern Argentinian sites, representing 14 different families. No family averages even a single species per sample. If the somewhat subtropical northern Argentinian samples are excluded, the three Valdivian area samples average 2.7 liana species per 0.1 ha sample, with a total of six species belonging to six different families (Bignoniaceae, Gesneriaceae, Gramineae, Lardizabalaceae, Saxifragaceae, Vitaceae).

Conclusion

Clearly, there has been strong selection for the scandent habit, with perhaps half of all vascular plant families including climbing members. Although many unrelated taxa of plants have independently evolved a scandent habit, relatively few families have radiated extensively as climbers. Thus 26 families account for 85% of all neotropical climbers and the majority of scandent species of most families belong to only one or two large genera. The most speciose liana genera are generally those that have evolved the most specialized climbing mechanisms.

Liana density is relatively constant and little affected by environmental factors on a given continent, but changes greatly from region to region. Africa (and Madagascar) have more lianas than Asia or the Neotropics. Continental Asia may generally have more lianas than Borneo and New Guinea, at least when juvenile rattans are counted. Tropical islands have few lianas, temperate forests even fewer.

Liana diversity and floristic composition change greatly (though fairly predictably) between different forests in a given region or continent, but, with a few important exceptions, fairly little between continents. In the Neotropics, lowland pluvial forests have the greatest diversity of climbers, with increasingly drier forests having progressively decreasing species numbers. Lowland pluvial forests have around 50 climbing species, lowland moist and wet forests 30–40 species, lowland dry forests 15–20 species. Similarly, liana diversity decreases with altitude and to a lesser extent with decreasing soil fertility.

In most lowland neotropical forests Bignoniaceae is the most important liana family and Leguminosae the second most important. The other most important lowland neotropical families include (in order of decreasing importance) Hippocrateaceae, Menispermaceae, Sapindaceae, Malpighiaceae, Connaraceae, and Dilleniaceae. In pluvial forests hemiepiphytic climbers of Melastomataceae, Guttiferae and Araceae take over from Bignoniaceae and Leguminosae. In montane forests, Asteraceae become the predominant climbers. In the Paleotropics Bignoniaceae climbers are sparsely represented, being replaced in Africa by Apocynaceae and in Asia by Annonaceae (and palms) as the predominant lianas.

Lianas are a very important component of tropical forests. Moreover, that different liana communities seem to be put together in such predictable ways seems concordant with the idea of tropical forests as tightly integrated ecological systems, perhaps at ecological equilibrium.

Acknowledgments

Supported by a series of grants from the National Geographic Society and National Science Foundation. Many of the neotropical climber data sets were collected incidental to systematic and ecological studies of Bignoniaceae climbers funded by NSF BSR-8607113. Many of the Peruvian sites were sampled as part of inventory projects funded by the Mellon and MacArthur Foundations. I thank S. Carlquist for data on southern California climbers, F. Putz and W. D. Stevens for the use of unpublished data and for reviewing this manuscript, J. Miller and C. Blaney for additional review comments, R. Clinebell and J. Miller for assistance with data entry and analysis, and various colleagues for assistance with the field work at the sites listed in Appendix 1.1 (G. Schatz in Madagascar, L. Mwasumbe and students in Tanzania, D. Thomas and F. Namata in Cameroun, D. Harris and M. Fay in CAR, J. Ackerman and C. Taylor in Puerto Rico, H. Cuadros in Colombia, A. Custódio Filho and G. Franco in Brazil, C. Díaz, R. Vasquez, D. Gorchov, and O. Phillips in Peru, S. Mori and S. Beck in French Guiana, D. Neill, S. Manning, and P. Keating in Ecuador). I also thank various taxonomic specialists, and especially R. Gereau, for hazarding specific identifications of sterile vouchers, and B. Ursem for providing a critical reference.

References

Airy-Shaw, H. K. (1973). *A Dictionary of the Flowering Plants and Ferns*, 8th edition. Cambridge University Press, Cambridge.

Appanah, S., Gentry, A. H. & La Frankie, J. (1991). Asian forest structure: the liana connection. *Journal of Tropical Forest Science* 3 (in press).

Appanah, S. & Putz, F. E. (1984). Climber abundance in virgin dipterocarp forest and the effect of pre-felling climber cutting on logging damage. Malaysian Forester 47: 335–42.

Baas, P. & Schweingruber, F. (1987). Ecological trends in the wood anatomy of trees, shrubs, and climbers from Europe. *International Association of Wood Anatomists Bulletin, n.s.* 8: 245–74.

Balee, W. & Campbell, D. (1989). Evidence for the successional status of liana forest (Xingu River Basin, Amazonian Brazil). *Biotropica* 22: 36–47.

Carlquist, S. & Hoekman, D. A. (1985). Ecological wood anatomy of the woody southern California flora. *Int. Assoc. Wood Anat. Bulletin, n.s.* 6: 319–47.

Charles-Dominique, P. (1977). *Ecology and Behavior of Nocturnal Primates*. Duckworth's, London.

Cristobal, C. L. (1976). Estudio taxonomico del genero *Byttneria* Loefling (Sterculiaceae). *Bonplandia* 4: 1–428.

Croat, T. B. (1978). *Flora of Barro Colorado Island*. Stanford University Press, Stanford, Ca.

Darwin, C. (1867). On the movements and habits of climbing plants. *Journal of the Linnean Society (Botany)* 9: 1–118.

Dawson, J. W. (1980). Middle-latitude rainforests in the southern hemisphere. *Biotropica* 12: 159–60.

Duncan, W. H. (1975). *Woody Vines of the Southeastern United States*. University of Georgia Press, Athens, Ga.

Emmons, L. H. & Gentry, A. H. (1983). Tropical forest structure and the distribution of gliding and prehensile-tailed vertebrates. *American Naturalist* 121: 513–24.

Fitch, H. S. (1966). The University of Kansas Natural History Reservation in 1965. *University of Kansas Museum of Natural History, Miscellaneous Publications*.

Fox, J. E. D. (1968). Logging damage and the influence of climber cutting prior to logging in the lowland dipterocarp forest of Sabah. *Malayan Forester* **31**: 326–47.

Freeman, C. C. (1980). Annotated list of the vascular flora of Konza Prairie Research Natural Area. Kansas State University (unpublished).

Gentry, A. H. (1973). Generic delimitations of Central American Bignoniaceae. *Brittonia* **25**: 226–42.

Gentry, A. H. (1982a). Patterns of neotropical plant species diversity. *Evolutionary Biology* **15**: 1–84.

Gentry, A. H. (1982b). Neotropical floristic diversity: phytogeographical connections between Central and South America, Pleistocene climatic fluctuations, or an accident of the Andean orogeny? *Annals of the Missouri Botanical Garden* **69**: 557–93.

Gentry, A. H. (1983a). Dispersal ecology and diversity in neotropical forest communities. *Sonderbd. Naturwissenschaftlichen Vereins in Hamburg* **7**: 303–14.

Gentry, A. H. (1983b). Lianas and the 'paradox' of contrasting latitudinal gradients in wood and litter production. *Tropical Ecology* **24**: 63–7.

Gentry, A. H. (1985). An ecotaxonomic survey of Panamanian lianas. In *Historia Natural de Panama*, ed. W. G. D'Arcy and M. Correa, pp. 29–42. Missouri Botanical Garden, St Louis.

Gentry, A. H. (1986). Species richness and floristic composition of Choco region plant communities. *Caldasia* **15**: 71–91.

Gentry, A. H. (1988a). Changes in plant community diversity and floristic composition on environmental and geographical gradients. *Annals of the Missouri Botannical Garden* **75**: 1–34.

Gentry, A. H. (1988b). Tree species richness of upper Amazonian forests. *Proceedings of the National Academy of Sciences, USA* **85**: 156–9.

Gentry, A. H. & Dodson, C. (1987). Diversity and biogeography of neotropical vascular epiphytes. *Annals of the Missouri Botanical Garden* **74**: 205–33.

Hegarty, E. E. & Clifford, H. T. (1990). Climbing angiosperms in the Australian flora. In *The Rainforest Legacy*, ed. A. P. Kershaw and G. L. Warren. Australian Government Publishing Service, Canberra.

Hladik, A. (1974). Importance des lianes dans la production foliaire de la forêt equatoriale du Nord-Est du Gabon. *Comptes Rendus, Académie des Sciences (Paris), sér. D* **278**: 2527–30.

Hladik, A. (1978). Phenology of leaf production in rain forest of Gabon: distribution and composition of food for folivores. In *The Ecology of Arboreal Folivores*, ed. G. Montgomery, pp. 51–71. Smithsonian Institution Press, Washington, DC.

Hueck, K. (1981). *Vegetationskarte von Sudamerika: mapa de la vegetación*

de America del Sur. G. Fischer, Stuttgart.
Jacobs, M. (1976). The study of lianas. *Flora Malesiana Bulletin* **29**: 2610–18.
Kelly, D. L., Tanner, E. V. J., Kapos, V., Dickinson, T., Goodfriend, G. & Fairburn, P. (1988). Jamaican limestone forests: floristics, structure and environment of three examples along a rainfall gradient. *Journal of Tropical Ecology* **4**: 121–56.
Krukoff, B. A. (1972). American species of *Strychnos*. *Lloydia* **35**: 193–271.
Leeuwenberg, A. J. (1969). The Loganiaceae of Africa VIII. *Strychnos* III. *Mededeelingen vande Landbouw hoogeschool de Wageningen* **69**(1): 1–316.
Lott, E. J. (1985). *Listados floristicos de Mexico III. la Estación de Biologia Chamela, Jalisco*. Universidad Nacional Autónoma de Mexico, Mexico City.
Lott, E. J., Bullock, S. H. & Solis-Magallanes, J. A. (1987). Floristic diversity and structure of a tropical deciduous forest of coastal Jalisco. *Biotropica* **19**: 228–35.
Mitchell, R. S. (1963). Phytogeography and floristic survey of a relic area in the Marianna lowlands, Florida. *American Midland Naturalist* **69**: 328–66.
Montgomery, G. & Sunquist, M. (1978). Habitat selection and use by two-toed and three-toed sloths. In *The Ecology of Arboreal Folivores*, ed. G. Montgomery, pp. 329–59. Smithsonian Institution Press, Washington, DC.
Ogawa, H., Yoda, K., Ogino, K. & Kira, T. (1965). Comparative ecological studies on three main types of forest vegetation in Thailand. II. Plant biomass. *Nature and Life in South-East Asia* **4**: 49–80.
Peñalosa, J. (1984). Basal branching and vegetative spread in two tropical rain forest lianas. *Biotropica* **16**: 1–9.
Proctor, J., Anderson, J., Chai, P. & Vallack, H. (1983). Ecological studies in four contrasting lowland rain forests in Gunung Mulu National Park, Sarawak. I. Forest environment, structure, and floristics. *Journal of Ecology* **71**: 237–60.
Putz, F. E. (1980). Lianas vs. trees. *Biotropica* **12**: 224–5.
Putz, F. E. (1982). Natural history of lianas and their influences on tropical forest dynamics. PhD thesis, Cornell University, Ithaca, NY.
Putz, F. E. (1984). The natural history of lianas on Barro Colorado Island, Panama. *Ecology* **65**: 1713–24.
Putz, F. E. (1985). Woody lianas and forest management in Malaysia. *Commonwealth Forestry Review* **64**: 359–65.
Putz, F. E. & Chai, P. (1987). Ecological studies of lianas in Lambir National Park, Sarawak, Malaysia. *Journal of Ecology* **75**: 523–31.

Putz, F. E., Lee, H. S. & Goh, R. (1984). Effects of post-felling silvicultural treatments on woody vines in Sarawak. *Malaysian Forester* 47: 214–26.

Radford, A. E., Ahles, H. & Bell, C. R. (1968). *Manual of the Vascular Flora of the Carolinas*. University of North Carolina Press, Chapel Hill, NC.

Raven, P. H. (1976). Ethics and attitudes. In *Conservation of Threatened Plants*, pp. 155–79. Plenum, New York and London.

Rollet, B. (1969). Le régénération naturelle en forêt dense humide sempervirente de plaine de la Guyane Venezuelienne. *Bois et Forêts des Tropiques* **14**: 19–38.

Schenck, H. (1892). Beiträge zur Biologie und Anatomie der Lianen, im Besonderen der in Brasilien einheimischen Arten. 1. Beiträge zur Biologie der Lianen. In *Botanische Mittheilungen aus den Tropen 4*, ed. A. F. W. Schimper, pp. 1–253. G. Fischer, Jena.

Schenck, H. (1893). Beiträge zur Biologie und Anatomie der Lianen, im Besonderen der in Brasilien einheimischen Arten. 2. Beiträge zur Biologie der Lianen. In *Botanische Mittheilungen aus den Tropen* 5, ed. A. F. W. Schimper, pp. 1–271. G. Fischer, Jena.

Short, J. (1981). Diet and feeding behaviour of the forest elephant. *Mammalia* **45**: 177–85.

Stevens, W. D. & Beach, J. H. (1980). A study of Baker Woodlot III. Checklist of vascular plants. *Mich. Bot.* **19**: 51–69.

Webb, D. A. (1978). Flora Europaea – a retrospect. *Taxon* 27: 3–14.

Webb, L. J., Tracy, J. G., Williams, W. T. & Lance, G. N. (1967). Studies in the numerical analysis of complex rain-forest communities II. The problem of species sampling. *Journal of Ecology* **55**: 525–38.

Webb, L. J., Tracy, J. G. & Williams, W.T. (1984). A floristic framework of Australian rainforests. *Australian Journal of Ecology* **9**: 169–98.

White, P. (1982). *The flora of Great Smoky Mountains National Park: an annotated checklist of the vascular plants and a review of previous floristic work*. Research/Resources Management Report SER-55. National Park Service, Southeast Region, Gatlinburg, Tenn.

Wilhelm, G. S. (1984). Vascular Flora of the Pensacola Region. PhD dissertation, Southern Illinois University.

Woodson, R. E. (1933). Studies in the Apocynaceae IV. The American genera of Echitoideae. *Annals of the Missouri Botanical Garden* **20**: 605–790.

Yatskievych, G. & Yatskievych, K. (1987). A floristic survey of the Yellow Birch Ravine Nature Preserve, Crawford County, Indiana. *Indiana Academy of Sciences* **96**: 435–45.

Appendix 1.1. *Sites included in liana distribution analysis. See Gentry (1986, 1988a) for additional site description*

Site	Coordinates	Altitude (m)	Precipitation (mm)	Number of liana families	Number of lianas	Number of liana species
Temperate North America						
Burling Tract, VA	38°55′ N 77°10′ W	30	1053	1	1	1
Northwest Branch, MD	39°02′ N 77°02′ W	20	1060	1	1	1
Tyson Reserve, MO (oak woods)	38°30′ N 90°31′ W	150	932	2	9	3
Tyson Reserve, MO (chert glade)	38°30′ N 90°31′ W	150	932	1	2	2
Babler State Park, MO	38°32′ N 90°40′ W	150	930	1	8	2
Cuivre River State Park, MO	39°01′N 91°00′ W	140	930	2	8	2
Valley View Glades, MO	38°15′ N 90°37′ W	225	930	1	2	1
Indian Cave State Park, NE	40°30′ N 95°43′ W	320	900	2	9	3
Europe						
Süderhackstedt, West Germany	54° N 11° E	20	695	1	3	1
Allacher Löhe, West Germany	48°04′ N 11°30′ E	530	866	0	0	0
Temperate and Subtropical South America						
Rio Jejui-mi, Paraguay	24°42′ S 55°30′ W	150	1800	15	102(incl. 1 hemi.)	30(incl. 1 hemi.)
Parque El Rey, Argentina	24°45′ S 64°40′ W	1000	1500	7	44	10
Arroyo Riachuelo, Corrientes, Argentina	27°30′ S 58°50′ W	60	1200	7	110(incl. 1 str.)	8(incl. 1 str.)
Alto de Mirador, Chile	40°14′ S 73°18′ W	800	4000	2	11	2

Appendix 1.1. (*cont.*)

Site	Coordinates	Altitude (m)	Precipitation (mm)	Number of liana families	Number of lianas	Number of liana species
Bosque de San Martín, Chile	39°30′ S 73°10′ W	30	2316	3	28	3
Puyehue National Park, Chile	40°43′ S 72°18′ W	500	3000	3	52(incl. 45 hemi.)	3(incl. 1 hemi.)
'Subtropical' Central America						
Chamela, Mexico	19°30′ N 105°03′ W	50	733	17	145(incl. 2 str.)	23(incl. 1 str.)
Chamela, Mexico	19°30′ N 105°03′ W	50	733	9	42	12
Chamela, Mexico	19°30′ N 105°03′ W	50	733	7	47(incl. 1 str.)	9(incl. 1 str.)
Los Tuxtlas, Mexico	18°35′ N 95°08′ W	200	4953	14	55(incl. 3 hemi.)	29(incl. 3 hemi.)
Cerro Olumo, Nicaragua	12°18′ N 85°24′ W	750	2000	16	49(incl. 2 str., 10 hemi.)	25(incl. 1 str., 5 hemi.)
Cerro El Picacho, Nicaragua	13°00′ N 85°55′ W	1400	2000	9	16(incl. 5 hemi.)	10(incl. 4 hemi.)
Lowland Neotropics (12° N to 23° S, ⩽1000 m alt.)						
Corcovado, Costa Rica	8°30′ N 83°35′ W	30	3800	18	56(incl. 1 str., 8 hemi.)	34(incl. 1 str., 6 hemi.)
Guanacaste (upland), Costa Rica (700 m²)	10°30′ N 85°10′ W	100	1600	2+	81*	6+*
Guanacaste (gallery), Costa Rica (800 m²)	10°30′ N 85°10′ W	50	1600	5+	24*	8+*
Curundu, Panama	8°59′ N 79°33′ W	20	1830	15	59(incl. 1 str., 1 hemi.)	24(incl. 1 str., 1 hemi.)
Madden Forest, Panama	9°66′ N 79°36′ W	50	2433	12	76(incl. 1 str.)	30(incl. 1 str.)
Pipeline Road, Panama	9°10′ N 79°45′ W	300	3000	20	68(incl. 1 str.)	38(incl. 1 str.)

Galerazamba, Colombia	10°48′ N 75°15′ W	10	500	14	104	18
Tayrona, Colombia	11°20′ N 74°02′ W	50	1500	10	98	18
Bosque de la Cueva, Colombia	11°05′ N 73°28′ W	360	2000	12	65(incl. 1 str., 5 hemi)	25(incl. 1 str., 2 hemi)
Tutunendo, Colombia	5°46′ N 76°35′ W	90	9000	24	70(incl. 2 str., 33 hemi)	51(incl. 2 str., 22 hemi.)
Bajo Calima, Colombia	3°55′ N 77°02′ W	100	7470	22	64(incl. 2 str., 31 hemi)	51(incl. 2 str., 26 hemi.)
Boca de Uchire, Venezuela	10°09′ N 65°25′ W	150	1200	6	75	16
Blohm Ranch, Venezuela	8°34′ N 67°35′ W	100	1400	11	77(incl. 3 str.)	18(incl. 2 str.)
Estación Biológico de los Llanos, Venezuela (500 m²)	8°56′ N 67°25′ W	100	1312	4+	48	10+
Cerro Neblina, Venezuela (No. 1)	0°50′ N 66°11′ W	140	3000	8	32(incl. 1 hemi.)	21(incl. 1 hemi.)
Cerro Neblina, Venezuela (No. 2)	0°50′ N 66°11′ W	140	3000	9	27	14
Río Palenque, Ecuador (No. 1)	0°34′ S 79°20′ W	200	2980	15	65(incl. 4 str., 20 hemi)	28(incl. 2 str., 10 hemi.)
Río Palenque, Ecuador (No. 2)	0°34′ S 79°20′ W	200	2980	13	48(incl. 5 str., 15 hemi)	25 (incl. 3 str., 9 hemi)
Centinela, Ecuador	0°34′ S 79°18′ W	550	3000	26	76(incl. 5 str., 16 hemi)	35(incl. 3 str., 10 hemi.)
Jauneche, Ecuador	1°16′ S 79°42′ W	60	1855	23	123(incl. 1 hemi.)	44(incl. 1 hemi.)
Capeira, Ecuador	2°00′ S 79°58′ W	50	804	11	61	20
Jatun Sacha, Ecuador	1°04′ S 77°36′ W	450	4100	30	94(incl. 15 hemi, 1 str.)	59(incl. 12 hemi, 1 str.)
Saül, French Guiana	3°38′ N 53°12′ W	220		18	50(incl. 6 hemi.)	32–33(incl. 5 hemi.)
Mocambo, Para, Brazil	1°30′ S 47°59′ W	30	2760	12	49(incl. 2 hemi.)	27(incl. 2 hemi.)

Appendix 1.1. (*cont.*)

Site	Coordinates	Altitude (m)	Precipitation (mm)	Number of liana families	Number of lianas	Number of liana species
Linhares, Espirito Santo, Brazil	19°18′ S 40°04′ W	50	1403	18	91	52
Carlos Botelho, SP, Brazil	24°15′ S 26°56′ W	560		13	77(incl. 8 hemi.)	*c.* 24?(incl. 2 hemi.)
Boraceia, SP, Brazil	23°38′ S 45°50′ W	800		15	47(incl. 1 str., 12 hemi)	22(incl. 1 str., 2 hemi.)
Jacarepagua, RJ, Brazil	23°05′ S 43°25′ W	200	1500	16	61	38
Tarapoto, Peru	6°40′ S 76°20′ W	500	1400	17	85(incl. 7 hemi.)	27(incl. 2 hemi.)
Cerros de Amotape, Peru	4°09′ S 80°37′ W	830		11	37(incl. 3 str., 2 hemi.)	14(incl. 1 str., 1 hemi.)
Sucursari, Peru	3°15′ S 72°55′ W	140	3500	19	68(incl. 2 hemi.)	40(incl. 1 hemi.)
Yanamono, Peru (upland) (No. 1)	3°28′ S 72°50′ W	140	3500	21	60(incl. 2 hemi.)	44(incl. 2 hemi.)
Yanamono, Peru (upland) (No. 2)	3°28′ S 72°50′ W	140	3500	20	65(incl. 4 hemi.)	45(incl. 3 hemi.)
Yanamono, Peru (tahuampa)	3°28′ S 72°50′ W	130	3500	20	65(incl. 5 str., 4 hemi.)	45(incl. 4 str., 3 hemi.)
Mishana, Peru (flood plain)	3°47′ S 73°30′ W	130	3500	22	89(incl. 8 hemi.)	55(incl. 5 hemi.)
Mishana, Peru (tahuampa)	3°47′ S 73°30′ W	130	3500	15	100(incl. 3 hemi.)	44(incl. 3 str.)
Mishana, Peru (upland white sand)	3°47′ S 73°30′ W	130	3500	12	58(incl. 1 hemi.)	26(incl. 1 hemi.)
Bosque von Humboldt, Peru	8°50′ S 75°00′ W	270	2500	14	66(incl. 1 str.)	36(incl. 1 str.)
Cabeza de Mono, Peru	10°20′ S 75°18′ W	320	4000	15	62(incl. 2 hemi.)	31(incl. 2 hemi.)
Shiringamazu, Peru	10°20′ S 75°10′ W	300	5000?	23	79(incl. 9 hemi.)	48(incl. 5 hemi.)

Indiana, Peru	3°30′ S 73°03′ W	130	3500	26	90(incl. 1 str., 9 hemi)	55(incl. 1 str., 5 hemi.)
Jenaro Herrera, Peru	4°55′ S 73°45′ W	130	2521	21	73(incl. 1 str., 4 hemi)	*ca.*43(incl. 1 str., 2 hemi.)
Cocha Cashu, Peru	11°51′ S 71°19′ W	400	2000	19	81	44
Tambopata, Peru (lateritic No. 1)	12°50′ S 69°17′ W	260	2000	19	80(incl. 1 hemi.)	41(incl. 1 hemi.)
Tambopata, Peru (lateritic No. 2)	12°50′ S 69°17′ W	260	2000	22	75(incl. 1 str., 3 hemi.)	42(incl. 1 str., 3 hemi.)
Tambopata, Peru (sandy terra firme)	12°50′ S 69°17′ W	260	2000	22	75(incl. 1 str., 3 hemi.)	42(incl. 1 str., 3 hemi.]
Africa						
Makokou, Gabon (No. 1)	0°34′ N 12°52′ E	500	1755	17 (+1?)	103	47
Makokou, Gabon (No. 2)	0°34′ N 12°52′ E	500	1755	16 (+1?)	82	39
Omo Forest, Nigeria	7°00′ N 5°00′ E	50	1800	15	73	27
Mt Cameroun, Cameroun	4°00′ N 9°00′ E	230	8000	17 (+1?)	117(incl. 1 str., 1 hemi)	49(incl. 1 str., 1 hemi.)
Korup National Park, Cameroun	5°00′ N 8°31′ E	50	5460	14	71(incl. 8 hemi.)	45(incl. 2 hemi.)
Banyong, Cameroun	5°00′ N 9°10′ E	420		13	77*	*c.* 41*
Ndakan (inundated) CAR/Cameroun	2°22′ N 16°09′ E	390		13 (+1?)	156(incl. 2 str.)	*ca.* 41(incl. 2 str.)
Pugu Forest, Tanzania	6°50′ S 39°05′ E	180		15 (+1?)	168	39
Pande Forest, Tanzania (500 m²)	6°40′ W 39°05′ E	80		8 (+?)	104*(incl. 2 str.)	*c.* 22* (incl. 1 str.)
Perinet, Madagascar	18°55′ S 48°25′ E	950	1200	22	107(incl. 1 str., 3 hemi)	37(incl. 1 str., 3 hemi.)
Nosy Mangabe, Madagascar	15°30′ S 49°46′ E	300		18 (+3?)	137(incl. 3 str.)	*ca.* 45(incl. 2 str.)
Ankarafantiska, Madagascar	16°19′ S 46°49′ E	80		16 (+1?)	122(incl. 1 hemi.)	33(incl. 1 hemi.)

Appendix 1.1. (*cont.*)

Site	Coordinates	Altitude (m)	Precipitation (mm)	Number of liana families	Number of lianas	Number of liana species
Australia						
Davies River State Park, Queensland	17°05′ S 145°34′ E	800	2300	11	38(incl. 2 str.)	16(incl. 1 str.)
Asia[a]						
Pasoh (30), Malaysia	3°00′ N 102°20′ E	100	2000	21 (+1)	115(incl. 1 str.)	56(incl. 1 str.)
Pasoh (40), Malaysia	3°00′ N 102°20′ E	100	2000	20 (+1)	114(incl. 1 str., 1 hemi.)	57(incl. 1 str., 1 hemi.)
Genting, Malaysia	3°58′ N 101°38′ E	890	2000?	18	78(incl. 1 str., 3 hemi.)	34(incl. 1 str., 2 hemi.)
Nadugani, India	11°27′ N 76°23′ E	650	2500?	14	41	19
Khao Yai, Thailand	14°20′ N 101°50′ E	700	2300?	25	124(incl. 7 hemi.)	45(incl. 1 hemi.)
Nanjen Shan, Taiwan	22°00′ N 120°50′ E	310		11	64(incl. 5 hemi.)	13(incl. 2 hemi.)
Kenting N.P., Taiwan	21°55′ N 120°50′ E	200		12(–13)	67(incl. 3 str., 1 hemi.)	25(incl. 3 str., 1 hemi.)
Baitete, Papua New Guinea	5°10′ S	200		26	80(incl. 4 str., 2 hemi.)	48(incl. 4 str., 2 hemi.)
	145°48′ E	740		24	65(incl. 4 hemi., 2 str.)	36(incl. 2 str., 4 hemi.)
Varirata, Papua New Guinea	9°30′ S 147°30′ E					
Semengoh Forest, Sarawak	1°50′ N 110°05′ E	20	4000	12	25	23
Bako National Park, Sarawak	1°52′ N 110°06′ E	30	4000	12	56(incl. 1 str.)	26(incl. 1 str.)
Tropical Islands						
Riviere des Pirogues, New Caledonia	22°10′ S 166°50′ E	360	2200	8 (+1?)	43(incl. 13 hemi.)	17(incl. 3 hemi.)
Round Hill, Jamaica	17°50′ N 77°15′ W	40	1200	4	9(incl. 1 hemi.)	4(incl. 1 hemi.)

Luquillo, Puerto Rico	18°18′ N 65°50′ W	400	3000	6	21(incl. 15 hemi.)	6(incl. 3 hemi.)
Mogotes de Nevarez, Puerto Rico	18°25′ N 66°15′ W	50	1500	10	36(incl. 2 str.)	12(incl. 1 str.)
Brise Fer, Mauritius	20°30′ S 57°30′ E	600	2400	4	21(incl. 2 intro.)	4(incl. 1 intro.)
Montane Neotropics (≥1000 m, 12° N to 12° S)						
Cerro Kennedy, Colombia	11°05′ N 74°01′ W	2600		8	39(incl. 1 hemi.)	8(incl. 1 hemi.)
Cerro Espejo, Colombia	10°28′ N 72°50′ W	2560		14	87(incl. 2 hemi.)	20(incl. 1 hemi.)
Farallones de Cali, Colombia	3°30′ N 76°35′ W	1950		20	54(incl. 4 str., 9 hemi.)	30(incl. 3 str., 9 hemi.)
Finca Mehrenberg, Colombia	2°16′ N 76°12′ W	2290		11 (+1?)	62(incl. 1 str., 35 hemi.)	30(incl. 1 str., 9 hemi.)
La Planada, Colombia	1°10′ N 77°58′ W	1800		15	101(incl. 3 str., 78 hemi.)	36(incl. 3 str., 26 hemi.)
Pasochoa, Ecuador (400 m²)	0°28′ S 78°25′ W	3010		7	*c.* 200*	13+*
Venceremos, Peru	5°45′ S 77°40′ W	1850		14	60(incl. 3 str., 25 hemi.)	30(incl. 1 str., 13 hemi.)
Incahuara, Bolivia	15°55′ S 67°35′ W	1540		13	39(incl. 1 str., 8 hemi.)	17(incl. 1 str., 8 hemi.)
Sacramento, Bolivia	16°18′ S 67°48′ W	2450		11	101(incl. 23 hemi.)	21(incl. 4 hemi.)

*extrapolated from less than 0.1 ha sample.

[a] Asian data mostly from field identifications only and thus approximate; Malaysia data summarized from Appanah, Gentry & LaFrankie (1990).

PART II
CLIMBING MECHANICS AND STEM FORM

2

Anatomy of vine and liana stems: a review and synthesis

SHERWIN CARLQUIST

Introduction

In the 19th century, the work of Westermeier & Ambronn (1881) and Schenck (1893) initiated appreciation of the many ways in which stem and wood anatomy of climbing plants differs from that of self-supporting plants. During the first half of the 20th century, data on stem and wood anatomy accumulated in the form of systematic comparisons, as particular taxonomic groups were investigated. The present chapter, as well as others in this volume, attempts to review this literature and present a new synthesis of the characteristics of vines and how they are related to habit, function, and ecology. After Haberlandt (1914), interest in a synthesis between form and function lagged, perhaps mostly because experimental work in plant physiology of scandent dicotyledons was relatively dormant; the recent renewed interest represented by the work in this volume permits new syntheses. A review of anatomical data is presented here by way of showing what comparative anatomical patterns suggest in terms of function, and thereby what correlations can be demonstrated or may in the future be demonstrated.

Cambial variants and cambial activity

Since the pioneering work of Schenck (1893), Pfeiffer (1926), and Obaton (1960), there has been general appreciation that lianas and vines, much more frequently than self-supporting woody plants, are characterized by the presence of cambial variants (the term from Carlquist, 1988; 'anomalous secondary thickening' of other authors). At the outset, one must note that cambial variants are by no means unique to scandent woody plants. For example, successive cambia (the 'concentric interxylary phloem' of some earlier authors) occur in various families of Chenopodiales (Centrospermae):

Aizoaceae, Amaranthaceae, Basellaceae, Caryophyllaceae, Chenopodiaceae, Nyctaginaceae, Phytolaccaceae, and Rhabdodendraceae. Only a few of these (e.g. *Bougainvillea* of the Nyctaginaceae) are woody vines. Some of the complex forms of cambial variants (e.g. expansion of the woody stem by proliferation of parenchyma as in *Bauhinia, Flabellaria,* and *Mendoncia*) are found virtually only in lianoid plants (Pfeiffer, 1926). With attention to these facts, we may survey scandent dicotyledons with respect to cambial variants. The listing below is based on Carlquist (1988); genera of non-climbing plants have been omitted. For illustrations and detailed discussions of terminology and appropriate literature, see Carlquist (1988). Data are from compilations such as Solereder (1908) and Metcalfe & Chalk (1950) with additional sources as noted. These listings must be regarded as provisional, because more examples are likely to be discovered.

Successive cambia (centrifugal):

Amaranthaceae: *Chamissoa.*
Connaraceae: *Rourea* (Boureau, 1957), *Spiropetalum* (Obaton, 1960).
Convolvulaceae: *Argyreia, Calonyction, Convolvulus, Dicranostyles, Ericybe, Exogonium, Hewittia, Ipomoea, Maripa, Merremia,* and *Porana* (Carlquist & Hanson, 1991).
Cucurbitaceae: *Adenopus, Luffa, Melothria, Mormodica, Physedra, Sphaerosicyos* (Zimmermann, 1922).
Dilleniaceae: *Doliocarpus* (Chalk & Chattaway, 1937).
Dioncophyllaceae: *Dioncophyllum* (Boureau, 1957).
Fabaceae: *Machaerium, Milletia, Mucuna, Rhynchosia, Strongylodon, Wisteria* (Boureau, 1957).
Hippocrateaceae: *Cheiloclinium, Salacia, Salacighia* (Obaton, 1960).
Icacinaceae: *Icacina, Rhaphiostylis* (Obaton, 1960).
Menispermaceae: all lianoid genera.
Nyctaginaceae: *Bougainvillea* (Esau & Cheadle, 1969; Pulawska, 1973).
Passifloraceae: *Adesmia* (Ayensu & Stern, 1964), *Adenia* (Obaton, 1960).
Sapindaceae: *Serjania pinnata* (Pfeiffer, 1926).
Vitaceae: *Tetrastigma* sp. (Carlquist 1988).
In gymnosperms, lianoid species of *Gnetum* have successive cambia (Haberlandt, 1914; Martens, 1971).

Successive cambia (centripetal): This term is applied to cambia formed in the pith; Piperaceae and certain other families are excluded because they have bundles formed in the pith in the primary stem, and these may have cambial activity, but no cambia outside of these bundles are present in the pith.

Acanthaceae: *Afromendoncia* (Obaton, 1960).
Apocynaceae: *Willughbeia* (Scott & Brebner, 1891).

Asclepiadaceae: *Periploca* (Scott & Brebner, 1891).
Convolvulaceae: *Bonamia, Prevostea* (Obaton, 1960); *Exogonium* (Carlquist & Hanson, 1991).
Icacinaceae: a ring of inverted bundles with secondary growth between the first-formed secondary xylem and outer rings formed by successive cambial is illustrated for *Icacina mannii* by Obaton (1960).

Interxylary phloem derived from a single cambium:
Acanthaceae: *Mendoncia, Thunbergia* (Carlquist & Zona, 1987).
Asclepiadaceae: *Ceropegia, Leptadenia* (Singh, 1943).
Combretaceae: *Calycopteris, Combretum, Guiera, Thiloa* (van Vliet, 1979).
Convolvulaceae: *Ipomoea versicolor* (Solereder 1908).
Cucurbitaceae: *Cucurbita, Lagenaria.*
Fabaceae: *Entada* (Haberlandt, 1914), *Mucuna pruriens* (Obaton, 1960).
Icacinaceae: lianoid genera of the tribe Sarcostigmateae (Bailey & Howard, 1941a).
Loganiaceae: *Strychnos, Usteria* (Scott & Brebner, 1891; Obaton, 1960; Mennega, 1980).

Cambia normal in products but abnormal in conformation or dispersion:
Stems oval in transection because of differential amounts of xylem deposition (but other variants not present):
Celastraceae: *Celastrus.*
Combretaceae: *Quisqualis* (van Vliet, 1979).
Icacinaceae: *Iodes ovalis* (Bailey & Howard, 1941a).
Malpighiaceae: *Heteropteris.*
Marcgraviaceae: *Marcgravia.*
Moraceae: *Ficus.*
Piperaceae: *Piper.*
Polygonaceae: *Atraphaxis.*
Verbenaceae: *Lantana, Petrea.*
Stems markedly flattened in transection:
Apocynaceae: *Landolphia owariensis* (Obaton, 1960).
Aristolochiaceae: *Aristolochia, Pararistolochia.*
Fabaceae: *Abrus, Bauhinia, Rhynchosia.*
Vitaceae: *Cissus.*
Stems lobed in transection:
Celastraceae. *Celastrus.*
Combretaceae: *Combretum dolichandrone* (Obaton, 1960).
Fabaceae: *Acacia pennata* (Obaton, 1960).
Rubiaceae: *Mussaenda* (Obaton, 1960).
Verbenaceae: *Lantana* (Bhambie, 1972).

Stems with xylem furrowed (cambia becoming unilateral to various degrees at places on the stem, so that less secondary xylem than secondary phloem is produced at those places):
Bignoniaceae: *Adenocalymma, Anemopaegma, Arrabidaea, Bignonia, Callichlamys, Cuspidaria, Distictis, Doxantha, Fridericia, Glaziova, Haplolophium, Lundia, Melloa, Paragonia, Petastoma, Phaedranthus, Phryganocydia, Pithecoctenium, Pleonotoma, Pyrostegia, Stizophyllum, Tanaecium* (Schenck, 1893; Pfeiffer 1926).
Convolvulaceae: *Bonamia, Neuropeltis, Prevostea.*
Fabaceae: *Centrosema plumieri.*
Hippocrateaceae: *Hippocratea, Salacia.*
Icacinaceae: *Neostachyanthus, Pyrenacantha.*
Malpighiaceae: *Triapsis odorata.*
Passifloraceae: *Crossostemma, Passiflora* (Ayensu & Stern, 1964).

Xylem in plates (wide rays lacking in cell wall lignification and not altering during secondary growth, so that 'fibrous' xylem is separated by thin-walled ray cells):
Aristolochiaceae: *Aristolochia.*
Asclepiadaceae: *Ceropegia* (some species).
Asteraceae: *Mikania* (Pfeiffer, 1926; van der Walt *et al.*, 1973).
Cucurbitaceae: *Coccinia* (Fisher & Ewers, 1990).
Menispermaceae: *Menispermum* (Pfeiffer, 1926).
Vitaceae: *Cissus* (Obaton, 1960).

Wood parts ('fibrous' woody parts) dispersed by parenchyma proliferation:
Acanthaceae: *Afromendoncia, Mendoncia* (Schenck, 1895, Obaton, 1960).
Bignoniaceae: *Macfadyena mollis* (Schenck, 1895).
Convolvulaceae: *Ipomoea umbellata* (Schenck, 1895), *Merremia, Neuropeltis, Prevostea.*
Fabaceae: *Bauhinia* (*B. championi, B. japonica, B. langsdorffiana*: Scheck, 1895), *Kunstleria ridleyi.*
Icacinaceae: *Icacina, Iodes, Phytocrene, Pyrenacantha.*
Malpighiaceae: *Banisteria, Flabellaria, Mascagnia, Tetrapteris* (Schenck, 1895).
Vitaceae: *Cissus.*

Divided xylem cylinder (a furrowed xylem cylinder ontogenetically subdivides into segments, each of which becomes surrounded by a cambium):
Sapindaceae: *Serjania corrugata* and allied species.

Compound secondary xylem (in addition to a normal cylinder of secondary xylem, cambia form around cortical bundles and add secondary xylem to them):
Sapindaceae: *Paullinia* (16 of 122 species), *Serjania* (91 of 172 species).

Combinations of cambial variants:

Acanthaceae: *Afromendoncia, Mendoncia* (dispersed xylem segments plus interxylary phloem formed from a single cambium: Obaton, 1960; Carlquist & Zona, 1987).

Fabaceae: *Bauhinia* (flattened stems with dispersed xylem segments); *Machaerium, Milletia* (flattened stems with successive cambia).

Icacinaceae: *Icacina mannii* (successive cambia, with tendency of cambium to become unilateral in action in places, so that there are grooves in the xylem rings: Metcalfe & Chalk, 1983).

Menispermaceae: *Anomospermum* (flattened stems with successive cambia: Schenck, 1893).

Polygalaceae: *Securidaca* (successive cambia, grooved stems: Schenck, 1893).

Sapindaceae: *Thinouia* (lobed stem outline, successive cambia).

Why is the 'normal' cambium so pervasive in self-supporting trees and shrubs, but characteristic of a smaller proportion of vining and lianoid dicotyledons? The answer very likely involves several factors. For self-supporting trees and shrubs, a dense wood is achieved with the greatest certainty by means of a 'normal' cambium. Successive cambia, xylem dispersed (by parenchyma proliferation), and xylem in plates result inevitably in a greater proportion of parenchyma interspersed among 'fibrous' wood than is the case in most 'normal' cambia. A higher proportion of parenchyma in a dicotyledonous stem is not disadvantageous if there is a selective advantage for interspersed parenchyma. One possible selective advantage is 'cable construction' (flexibility of the stem without damage to vessels: Schenck, 1893; Putz & Holbrook, Chapter 3). Other potential advantages include sites for wound healing (Fisher & Ewers, 1990) or storage of water or photosynthates.

Vessel restriction patterns, and unusual ontogenetic change in vessel presence and diameter

In many dicotyledons, vessels tend to increase in diameter as a stem grows, and although this can be observed readily (Carlquist, 1984a), it is not often reported explicitly. Neither this tendency, nor the annual change in vessel diameter involved in ring porosity, is under discussion here. Rather, in certain lianas there are drastic and sudden increases in diameter of vessel elements shortly after the onset of secondary growth. This takes place in some lianas – for example, *Acacia pennata, Cissus afzelii, C. barteri, Salacia bipindensis*, and *Strychnos densiflora* (Obaton, 1960).

In a more extreme phenomenon, secondary xylem begins with a 'fibrous' wood in which vessels are either lacking or are of the same diameter as imperforate tracheary elements or apparently so (i.e. narrow vessels are very similar to imperforate tracheary elements as seen in transection). Later-formed secondary xylem – often abruptly initiated – contains wider vessels. Attention was called to this phenomenon by Haberlandt (1914), who considered this a way of achieving mechanical strength at first, followed by development of a wood efficient at water conduction. This can be considered a form of vessel restriction (non-random distribution of vessels within the secondary xylem), a term originally devised for a different pattern of vessel distributions: absence of vessels adjacent to rays, as in Papaveraceae (Carlquist & Zona, 1988). Examples of a vessel restriction pattern in which wood appears vessel-free at first but in which subsequent wood contains wide vessels can be found in Convolvulaceae (*Prevostea*: Obaton, 1960; *Exogonium*: Carlquist & Hanson, 1991) and Icacinaceae (*Iodes liberica, I. ovalis*: Bailey & Howard, 1941a; *Pyrenacantha klaineana*: Obaton, 1960). In Convolvulaceae and Icacinaceae, tracheids are the imperforate tracheary element type present (Carlquist, 1988), so that in the first-formed secondary xylem these cells offer a conductive potential absent in a wood in which the background to the vessels is composed of libriform fibers or fiber-tracheids.

A second type of vessel restriction pattern in scandent dicotyledons is constituted by presence of vessels in localized patches (with vessels apparently absent elsewhere) with no reference to ontogeny. Examples of this are illustrated photographically for *Pyrenacantha repanda, Hosiea sinense, Iodes liberica*, and *I. philippinensis* by Bailey & Howard (1941a). The explanation of this phenomenon is not readily apparent.

A third type of vessel restriction pattern is represented by lianas in which vessels are large and are sheathed by imperforate tracheary elements – libriform fibers in the examples cited here. Parenchyma rarely if ever occurs near a vessel. Instances of this have been figured for *Thunbergia grandiflora* of the Acanthaceae by Obaton (1960), for *T. alata* (Carlquist & Zona, 1987), and for *Mucuna pruriens* of the Fabaceae (Obaton, 1960). A similar condition is illustrated for *Afromendoncia* (Obaton, 1960), *Mendoncia gigas*, and *Thunbergia laurifolia* (Carlquist & Zona, 1987). In these, a few smaller vessels are present in addition to larger vessels, but vessel groupings are sheathed by libriform fibers. These conditions can be called vessel restriction patterns because vessels tend never to be adjacent to thin-walled parenchyma; only the sheathing libriform fibers abut on the thin-walled parenchyma. Because this pattern has not been recognized as a form of a vessel restriction pattern, further examples are likely to be found; a number of these may occur in lianas with xylem dispersed by parenchyma proliferation (e.g. *Bauhinia*). The significance of this type of vessel restriction, as in the first type mentioned above (*Thunbergia*) is very likely the protection of large vessels from injury by sheathing them with fibers.

Vessel element dimensions and morphology

Vessel elements in vines and lianas are notably wide. This condition was observed early by such authors as Westermeier & Ambronn (1881), who note that vessels in some species of *Passiflora* average more than 500 μm in diameter. The occurrence of notably wide vessels in lianas is mentioned prominently by Haberlandt (1914), who says, 'in such plants, the construction of the conductive system is governed by two factors, namely, the great length of the conductive region and the relatively small cross-sectional area available for the disposition of the conductive elements'. Haberlandt's (1914) idea that longer stems (such as those vines and lianas may have) produce friction, and that this friction must be compensated by wider vessels, is probably an incorrect concept. However, Haberlandt's idea that few, wide vessels produce less friction than more numerous, narrower vessels is certainly valid and operative in scandent plants.

Various authors since Westermeier & Ambronn (1881) have given data that dramatize the wideness of vessels in vines and lianas. Notable among these are Bailey & Howard (1941b), who contrast vessel diameters in three categories of Icacinaceae: trees and shrubs; scrambling shrubs; and vines and lianas. The scrambling shrubs are intermediate in vessel diameter, as one might expect if they precede vines and lianas in an evolutionary series. The fact that Icacinaceae contain all three categories is important, in that one is dealing with a monophyletic group that has evolved into different habitats, much like an experimental material subdivided into three experimental conditions, and thus one has a reliable measure of by how much vessels have widened as a result of evolution into scandent habits of various kinds and degrees. The comparison by Fisher & Ewers (Chapter 4) of shrubby and lianoid *Bauhinia* species is pertinent in this regard.

Quantitative vessel element figures on a sampling of vines and lianas (Carlquist, 1975, p. 206) showed mean vessel diameter to be greater than in any other category based on habit or habitat. The vessel density in climbing plants is accordingly low, but one must remember that vessel diameter figures may inadvertently omit small vessels that are similar in diameter to imperforate tracheary elements. Even despite a low figure for number of vessels per mm^2 of wood transection, the conductive area cited for the sampling of vines and lianas (mean vessel area multiplied by mean number of vessels per mm^2) is higher than for any other habit or habitat category. In that survey, the mean conductive area (0.36 mm^2 per mm^2 of transection) exceeds a third of the secondary xylem area. This figure is confirmed in Lardizabalaceae (Carlquist, 1984c), although a relatively high number of vessels per mm^2 characterizes the lianas in that family.

The implication in all of the above studies is clear that because vines are not self-supporting, 'fibrous' cells (imperforate tracheary elements) comprise a smaller proportion of wood of scandent dicotyledons than they do in other growth form categories. Related to this consideration is the tendency for the area of foliage of a liana to be comparable in quantity to that of a tree, whereas the transectional area of the liana stem is relatively smaller than that of a tree, as hinted by the Haberlandt quote cited above. Data have been provided by Putz (1983).

Emphasis on the notably wide vessels of vines and lianas is justifiable, but one must remember that throughout the mature wood of vines and lianas, one may find narrower vessels – often much narrower – as well. To call attention to this pheonomenon, the term 'vessel dimorphism' was used (Carlquist, 1981). In using this term, I did not intend to imply that vessel diameters in a scandent species would form a bimodal curve when graphed, although that occasionally does happen (Ewers & Fisher, 1989). Rather, I wished to include under this term any deviation from a normal distribution curve of vessel diameters. In fact, such attenuation of normal distribution curves is common in the data reported by Fisher & Ewers (1989). Vessel diameters in roots of lianas and vines are rarely recorded, but may be wider than in stems (Ewers, Fisher & Fichtner, Chapter 5). This follows the pattern for non-vining dicotyledons (Patel, 1965).

Narrow vessels (in wood of lianas in which wide vessels also occur) may take the form of fusiform cells, little wider than imperforate tracheary elements. In fact, Woodworth (1935) invented the term 'fibriform vessel elements' for narrow vessel elements of this kind in *Passiflora* wood. Fibriform vessel elements are not only fusiform in shape, they tend to have perforation plates placed laterally (but often near the tips) on the vessel elements instead of terminally. Fibriform vessel elements of this sort occur in Nepenthaceae (Carlquist, 1981) and other woods of vines (notably Convolvulaceae: Carlquist & Hanson, 1991). Fibriform vessel elements, although they will likely prove to be characteristic of vines and lianas when these groups are studied more thoroughly, may also occur in certain shrubby plants, such as *Eriodictyon* (Carlquist, Eckhart & Michener, 1983).

Narrower vessels in vines may function in a variety of ways in scandent dicotyledons: mechanical support, water storage, and conductive safety (redundancy that permits conduction to continue if larger vessels are embolized), and as juncture cells in the conductive system are among the ideas entertained by Ewers *et al.* (Chapter 5). The idea that the relatively high number of vessels per mm² in climbing Lardizabalaceae may represent adaptation to conductive safety was advanced earlier (Carlquist, 1984c).

The length of vessel elements in vines may not differ greatly from that in woody self-supporting plants. In some families, such as Trimeniaceae

(Carlquist, 1984a) and Loasaceae (Carlquist, 1984b), the vining representatives have vessel elements somewhat longer than those of most non-vining relatives. However, the mean vessel element length in a sampling of vines and lianas, 334 μm (Carlquist, 1975, p. 206) is appreciably lower than that of woody dicotyledons at large, 649 μm (Metcalfe & Chalk, 1950, p. 1360). To be sure, the Metcalfe & Chalk figure may be biased in favor of trees of wet tropical (and to a lesser extent, temperate) trees, but even if the figure represented trees exclusively, one would have to conclude that lianas as known at present have relatively short vessel elements. This trend is confirmed by the data of Bailey & Howard (1941c); in Icacinaceae, the scandent genera (tribes Iodeae, Phytocreneae, and Sarcostigmateae) have by far the shortest vessel elements in the family.

The length of vessels – the vertical series of interconnected vessel elements – is relatively great in lianas (Ewers, *et al.*, Chapter 5). On account of the exceptionally good conductive characteristics of scandent dicotyledons, these figures are not surprising.

Vessel element morphology

Perforation plates of vines and lianas are predominantly simple; this accords with the idea that simple plates offer the minimal impedance to water flow, and thereby promote conductive efficiency. One must note in this connection that considerations from flow physics suggest that bars in a conductive stream would have a slight negative impact at slow flow rates, but are of greater significance as flow rates accelerate. Vines and lianas do occur in some taxonomic groups in which scalariform perforation plates occur, and the nature of perforation plates in these groups is particularly instructive. In Trimeniaceae, the arboreal genus *Trimenia* has more numerous bars per perforation plate than does the scandent shrub *Piptocalyx*, and borders on perforations are reduced in *Piptocalyx* (Carlquist, 1984a). In Dilleniaceae, woods of shrubs and trees have scalariform perforation plates, but the three genera that are scandent (*Davilla, Doliocarpus*, and *Tetracera*) have simple perforation plates. In Lardizabalaceae, the single shrubby genus, *Decaisnea*, has scalariform perforation plates, but all the other genera, which are woody vines, have simple perforation plates (Carlquist, 1984c). In Actinidiaceae, the woody vine *Actinidia* has simple perforation plates on wider vessels and scalariform plates on narrower vessels, whereas the tree genus *Saurauia* has exclusively scalariform perforation plates (Metcalfe & Chalk, 1950). In the three tribes of Icacinaceae that are comprised of scandent plants (Iodeae, Phytocreneae, Sarcostigmateae), perforation plates are simple, whereas non-scandent genera of the family have scalariform perforation plates (Bailey & Howard, 1941b). Schisandraceae, a family of vines, has perforation plates with relatively few, non-bordered bars, whereas in the closely related family of trees and shrubs, Illiciaceae, perforation plates have more numerous, thicker bars with borders (Bailey & Nast, 1948).

With respect to lateral wall pitting of vessels, scalariform pitting occurs in a number of families with climbing representatives: Actinidiaceae, Aristolochiaceae, Dilleniaceae, Hydrangeaceae, Lardizabalaceae, Piperaceae, Trimeniaceae, and Vitaceae. This list may have little significance, because the vessels in these groups may have retained the scalariform condition (considered a primitive feature in dicotyledonous woods) independently. Scalariform lateral wall pitting may not be of negative selective value in scandent dicotyledons. A lateral wall pitting in which pits are circular and alternate is thought to form a mechanically stronger pattern than does scalariform pitting (Carlquist, 1975). Alternate pits may have greater selective value in self-supporting growth forms than in vines or lianas. Although deformation of vessels due to tensions in water columns may not be a problem, vines and lianas have lower tensions that do trees and shrubs, in general – in fact, positive pressures may be present (Ewers *et al.*, Chapter 5). If tensions in water columns of climbing plants were lower than those of non-climbing plants, scandent plants might not experience selection for maximally strong vessels walls. However, stronger vessel walls may result from selection for resistance to breakage and for support more than for resistance to deformation due to water column tensions.

Imperforate tracheary elements

A listing of vining or lianoid genera with true tracheids (Carlquist, 1985) is quite impressive: 28 families of dicotyledons were listed as containing scandent genera with true tracheids. To this list, *Durandea* of the Linaceae should be added. One should also mention in this connection the two genera of Gnetales with vessels in addition to tracheids: *Ephedra* (a few species scandent, such as *E. pedunculata*) and *Gnetum* (most species lianoid).

Vasicentric tracheids were once thought to be limited to a small number of families (about 30, according to Metcalfe & Chalk, 1950), but have now been found in more than 70 (Carlquist, 1988). Vasicentric tracheids occur in a high proportion of dicotyledon families that contain climbing genera, 24 (Carlquist, 1988). *Mendoncia* and *Thunbergia* of the Acanthaceae were not included in the latter listing, but do have vasicentric tracheids (Carlquist & Zona, 1987).

If one examines both the true tracheid and vasicentric tracheid lists (see Carlquist, 1988, for detailed discussion of these concepts), one finds numerous instances in which the scandent representatives have these cells, but in which non-scandent relatives have fiber-tracheids or libriform fibers instead.

For example, in Icacinaceae the scandent genera (tribes Iodeae, Phytocreneae, and Sarcostigmateae) have true tracheids whereas the non-scandent genera of the family have fiber-tracheids (Bailey & Howard, 1941c). *Hedera* is the only genus of Araliaceae in which vasicentric tracheids have been reported, just as *Beaumontia, Mandevilla,* and *Trachelospermum* have vasicentric tracheids but non-scandent Apocynaceae lack them. In the genus *Solanum*, only a few species have vasicentric tracheids; most of these (notably section *Basarthrum*) are scandent (Carlquist, 1991).

Thus, scandent dicotyledons have true tracheids and vasicentric tracheids to an extraordinarily high degree. The possible explanation cited (Carlquist, 1985) is that tracheids can serve as a subsidiary conductive system when vessels embolize, and on the basis of the abundance of tracheids in climbing dicotyledons, these species may have vessels that occasionally suffer air embolisms under extreme conditions. To be sure, the tracheids in these species could only account for a small proportion of the conduction that occurs, but they could, in case of embolism formation in the vessels, serve for a slow rate of conduction and maintain water columns to leaves until large vessels could become embolism-free and resume their role in conduction. The small collective lumen diameter of tracheids in a given scandent dicotyledon species makes it unlikely that tracheids serve for water storage to any marked extent.

Axial parenchyma

A survey that compares axial parenchyma in scandent species to that in non-scandent species is not at hand, and would be difficult to assemble. Nevertheless, patterns are evident in a number of cases, and these may point the way toward discovering other parenchyma distributions that may represent functional correlations in scandent dicotyledons.

Attention was called to the presence of starch-rich parenchyma adjacent to vessels in a number of scandent dicotyledons (Carlquist, 1985). To be sure, liquid-preserved specimens of these species permitted observation of starch, and comparable material of non-scandent dicotyledon wood would reveal more starch presence than is commonly reported. Starch may be less frequently present in wood prepared from dried specimens, and is often not reported in the anatomical literature in any case. However, the central theme with respect to vines and lianas is that there tends to be an abundance of paratracheal parenchyma, and that this may contain starch. Bailey & Howard (1941d) report that in 'Icacinaceae with a scrambling or climbing habit of growth . . . there is a more or less conspicuous reduction in the amount of banded apotracheal parenchyma. The associated paratracheal parenchyma tends to persist as the apotracheal is reduced'. Moreover, abundant paratracheal parenchyma (cited as having somewhat 'unstable' distribution: Bailey

& Howard, 1941d) occurs in the lianoid tribes Iodeae and Phytocreneae of Icacinaceae.

Parenchymatization is certainly an abundant byproduct in those species in which parenchyma proliferation breaks apart a woody cylinder. The listing above of genera in which this happens (e.g. *Mendoncia* of the Acanthaceae, or lianoid species of *Bauhinia*) is indicative of this phenomenon. Separation of the wood into strands of this sort was the basis for the concept of the cable principle of construction in lianas and vines by Schenck (1893, 1895) and Haberlandt (1914). Certainly Fisher & Ewers (1989) have demonstrated that axial parenchyma can serve a role in regeneration of vascular tissue following wounding, and Putz & Holbrook (Chapter 3) show how it contributes to liana stem flexibility.

Axial parenchyma has been hypothesized to play other possible roles related to conduction or storage (Carlquist, 1985). Scholander, Love & Kanwisher (1955) showed that positive pressures can occur in the xylem in late winter, although they did not determine the mechanism for this. Plumb & Bridgman (1972) thought that carbohydrates can form a mechanism for sap ascent. If starch is hydrolyzed into sugar (a process observable through phosphatase staining reactions), and the sugar is released into vessels, osmotic pressure in the vessels rises, pulling water into them. This idea has been accepted by others (Sauter, 1972, 1980; Czaninski, 1977). This idea may be applicable not merely to temperate trees like the maple, but to tropical trees (Braun, 1983) or herbaceous perennials (Carlquist & Eckhart, 1984). Whether this process occurs in scandent plants is worthy of investigation. Another possibility is that starch stored in parenchyma is related to sudden flushes of growth or flowering.

The conjunctive tissue (parenchyma between the concentric vascular rings) in species with successive cambia is not axial parenchyma, because the conjunctive tissue is not part of the xylem (for a discussion, see Carlquist, 1988). Conjunctive tissue parenchyma may have the effect of protecting vascular tissue from torsion or permitting regeneration when stems are wounded (Putz & Holbrook, Chapter 3). However, other functions must also be attributed to conjunctive tissue in species with successive cambia. For example, in *Beta*, the conjunctive tissue between successive cambia in roots is a storage tissue. This may be true in lianoid species with successive cambia also. For example, large quantities of starch may be found in conjunctive tissue in species of *Ipomoea* of the Convolvulaceae (Carlquist & Hanson, 1991).

Rays

Westermeier & Ambronn (1881) signalled the prominence of rays in lianas, citing *Aristolochia*. *Aristolochia* is indeed a good example of this tendency for the rays are wide and during secondary growth do not subdivide very much. More significantly for this discussion, the rays are very wide and may be thin-walled (Fisher & Ewers, 1990). Notably wide rays composed of thin-walled cells are involved in the 'xylem in plates' cambial variant discussed above. Indeed, there may be little difference between that concept and the presence of wide, tall rays in dicotyledonous woods: if the ray cells are thin-walled and the rays alter little in ontogeny, the xylem is effectively subdivided into plates. Dicotyledonous families that contain scandent genera and that have wide, tall rays (thin-walled in some cases) include Actinidiaceae, Araliaceae, Cactaceae, Dilleniaceae, Hippocrateaceae, Icacinaceae, Lardizabalaceae, Marcgraviaceae, Passifloraceae, Piperaceae, Schisandraceae, and Vitaceae. The significance of such wide rays in vines may be as a means whereby vascular segments of the wood can yield to torsion without sustaining damage to vessels (Putz & Holbrook, Chapter 3).

An interesting question is formed by the wood of *Cobaea*, which proves to be rayless (Carlquist, Eckhart & Michener, 1984). Raylessness would seem to be a condition that runs contrary to the flexibility required by vine stems. However, axial parenchyma is rather abundant in stems of *Cobaea*. More significantly, one notes that stems of *Cobaea* never become very large; in fact, lack of rays may limit the duration or size of *Cobaea* stems.

Phloem

Phloem of vines is notable for presence of sieve-tube elements that are notably wide, and that have simple sieve plates with large pores (see Carlquist, 1975). Although this has been observed occasionally with respect to scandent dicotyledons, there is no quantitative documentation of dimensions of sieve-tube elements in climbing versus self-supporting woody plants, and one would like to see a study involving this feature.

Sieve-tube elements seem vulnerable to deformation caused by torsion because of their thin walls. The strands of interxylary phloem formed from a single cambium (e.g. *Strychnos*, Combretaceae) may be protected by being embedded within secondary xylem.

Stem sclerenchyma

Because the nature of xylem in dicotyledonous vines and lianas has been a topic of predominant interest, we do not have an image of presence of sclerenchyma in their stems. Certainly in some vines, sclerenchyma outside the phloem is scarce or virtually absent (e.g. *Cobaea* of the Polemoniaceae; various Convolvulaceae). In lianoid Lardizabalaceae, however, a prominent cylinder of sclerenchyma lies between the phloem and the periderm. In the shrub *Decaisnea* of the Lardizabalaceae, fibers are present as strands, not in a continuous cylinder (Carlquist, 1984c). Stems in Lardizabalaceae increase in diameter slowly, according to my observations of *Akebia* in the field, so breakup of a sclerenchyma cylinder would be slow (allowing time for interpolation of new sclereids into the cylinder). Whether the sclerenchyma cylinder in Lardizabalaceae represents a deterrent to herbivores, or a mechanism for protection of xylem and phloem from crushing, would be interesting to know. Such scattered observations as these indicate that we do not have the beginnings of an understanding of how sclerenchyma fibers or brachysclereids function in stems of vines and lianas. Hopefully studies will address this question in the future.

Scandent monocotyledons

When one considers that monocotyledons have stems of rather finite duration (except for palms) and that none of the scandent species have addition of bundles by means of a lateral meristem, climbing species are more abundant in monocotyledons than one might expect. Families that contain genera notable for their climbing habit include Araceae, Alstroemeriaceae, Arecaceae, Flagellariaceae, Pandanaceae (*Freycinetia*), Philesiaceae, and various families now being recognized as segregates from Liliaceae (e.g. Asparagaceae, Smilacaceae).

The monocotyledonous stem has many features that form intriguing structural parallels to the dicotyledonous picture described above. To date, little attention has been devoted to comparing vining with non-vining monocotyledons, although one has seen various citations, from the time of De Bary (1877) onwards, of the fact that vining monocotyledons tend to have wide vessels. The marked differentiation between protoxylem and metaxylem vessels in monocotyledons, with metaxylem vessels notably wider, offers a mechanism for achievement of vessel diameters comparable to that of scandent dicotyledons. The scattered bundles of the monocotyledon stem offer a mechanism for separation of vascular strands by means of thin-walled parenchyma, a mechanism achieved in dicotyledons by rays or else by successive cambia in which each cambium segment extends only for short tangential distances (e.g. the stem of *Pisonia* of the Nyctaginaceae looks as if it is composed of numerous scattered bundles in some species: Metcalfe & Chalk, 1950, 1983). The fibrous sheath around vascular bundles in monocotyledon stems offers an ideal protection to vascular tissue from damage due to torsion, if such a sheath serves that function, as we have reason to believe it may. Anchoring (climbing) roots of Araceae are richer in sclerenchyma than are 'feeding' roots (roots that penetrate a substrate and function primarily in water absorption). Monocotyledons offer an interesting potential comparison to dicotyledons in terms of functional anatomy. If parenchyma in dicotyledons serves to increase ability of liana or vine stems to resist torsion with minimal damage (Putz & Holbrook, Chapter 3), one would expect that this function is also achieved by the scattered bundles ('cable construction') of monocotyledons. Fisher & Ewers (1989) have stressed the potential role of parenchyma in dicotyledons in initiating regeneration following wounding or girdling. However, one must note that climbing monocotyledons are incapable of adding vascular tissue, so perhaps in dicotyledons also, the damage resistance inherent in cable construction is of overriding importance, and regeneration possibilities offered by parenchyma may be secondary. As our understanding of functional anatomy in stems of scandent dicotyledons increases, we may be in a better position to analyze stems and roots of climbing monocotyledons with respect to anatomical adaptation to the climbing habit.

Concluding remarks: an overview

The material covered above shows that we have developed some concepts of (i) how vines and lianas modally differ from non-climbing plants in terms of stem anatomy; and (ii) what the functional significance of these differences may be. These concerns are not new, and can be found in research of a century ago as well as in the most recent studies. The distinctive adaptations represented by vines and lianas may have compelled workers to compare anatomy with function to a greater extent than one finds in literature on anatomy of shrubs, trees, or herbs. Indeed, work on vines and lianas may be a model of the synthesis we ultimately hope to achieve, a synthesis between structural and functional ways of looking at plants.

With respect to acquiring knowledge on anatomical characteristics of vines and lianas, several comments are in order. There are two ways of developing information. One is to sample vines and/or lianas, and then to compare the data obtained with data from similar samplings of non-scandent plants. This method is valuable to the extent that one defines categories of growth forms carefully, and has extensive samplings. A second type of comparison, represented in studies of Lardizabalaceae (Carlquist, 1984c) and *Bauhinia* (Fisher & Ewers, Chapter 4), compares wood of scandent species with closely related non-scandent species. This type of comparison ought to be very informative about the nature of anatomical adaptation in climbing plants, and such studies are to be encouraged wherever there is a phyla in which both types of growth forms are represented. However, the potential drawback in such studies is that each group studied in this respect may represent distinctive modes. For example, *Piptocalyx* is a climbing shrub not strongly different in habit from shrubby to arboreal species of *Trimenia*, in the same family. However, lianoid species of *Bauhinia* may be markedly different in construction from shrubby species of *Bauhinia*. This underlines the likelihood that each taxonomic group will represent adaptations in a slightly or appreciably distinctive way. Each study will suggest generalizations, but these generalizations cannot be applied to a particular group, only to a generalized image of scandent vs. non-scandent plants.

The generalizations obtained from comparative anatomy of vines suggest how those equipped to study physiological phenomena can design studies. With scandent plants, the cooperation between those who study structure and those who study function has been close and productive, as several of the co-authored chapters in this book illustrate. Collaborative efforts of this kind seem the optimal way of achieving a synthesis between anatomy and physiology. In such collaborative efforts, comparison between scandent and non-scandent plants, with appreciable contrast between the two categories, is of crucial importance. The synthesis between anatomy and function may not be as easy to achieve in other problems (e.g. functions of helical thickenings or vestured pits in wood), but data on comparative anatomy, collected with reference to ecological and habital categories, offer excellent points of departure for physiologists interested in building the structure–function synthesis.

References

Ayensu, E. S. & Stern, W. L. (1964). Systematic anatomy and ontogeny of the stem in Passifloraceae. *Contributions from the US National Herbarium* 34: 45–73.

Bailey, I. W. & Howard, R. A. (1941a). The comparative morphology of the Icacinaceae. I. Anatomy of the node and internode. *Journal of the Arnold Arboretum* 22: 125–32.

Bailey, I. W. & Howard, R. A. (1941b). The comparative morphology of the Icacinaceae. II. Vessels. *Journal of the Arnold Arboretum* 22: 171–87.

Bailey, I. W. & Howard, R. A. (1941c). The comparative morphology of the Icacinaceae. III. Imperforate tracheary elements and xylem parenchyma. *Journal of the Arnold Arboretum* 22: 432–42.

Bailey, I. W. & Howard, R. A. (1941d). The comparative morphology of the Icacinaceae. IV. Rays of the secondary xylem. *Journal of the Arnold Arboretum* 22: 556–68.

Bailey, I. W. & Nast, C. G. (1948). Morphology and relationships of *Illicium, Schisandra,* and *Kadsura*. *Journal of the Arnold Arboretum* 29: 77–89.

Bhambie, S. (1972). Correlation between form, structure, and habit in some lianas. *Proceedings of the Indian Academy of Sciences, ser. B* 75(5): 246–56.

Boureau, E. (1957). *Anatomie végétale, Vol. 3*, pp. 525–752. Presse Universitaire de France.

Braun, H. J. (1983). Zur Dynamik der Wassertransportes in Baumen. *Berichte der Deutschen Botanischen Gesellschaft* **96**: 29–47.

Carlquist, S. (1975). *Ecological Strategies of Xylem Evolution.* University of California Press, Berkeley, Ca.

Carlquist, S. (1981). Wood anatomy of Nepenthaceae. *Bulletin of the Torrey Botanical Club* **108**: 324–30.

Carlquist, S. (1984a). Wood anatomy of Trimeniaceae. *Plant Systematics and Evolution* **144**: 103–18.

Carlquist, S. (1984b). Wood anatomy of Loasaceae with relation to systematics, habit, and ecology. *Aliso* **10**: 583–602.

Carlquist, S. (1984c). Wood and stem anatomy of Lardizabalaceae, with comments on the vining habit, ecology and systematics. *Botanical Journal of the Linnean Society* **88**: 257–77.

Carlquist, S. (1985). Observations on the functional histology of vines and lianas: vessel dimorphism, tracheids, vasicentric tracheids, narrow vessels, and parenchyma. *Aliso* **11**: 139–57.

Carlquist, S. (1988). *Comparative Wood Anatomy: Systematic, ecological and evolutionary aspects of dicotyledon wood.* Springer-Verlag, Berlin.

Carlquist, S. (1991). Wood anatomy of Solanaceae. A survey. *Allertonia* (in press).

Carlquist, S. & Eckhart, V. M. (1984). Wood anatomy of Hydrophyllaceae. II. Genera other than *Eriodictyon*, with comments on parenchyma bands containing vessels with large pits. *Aliso* **10**: 527–46.

Carlquist, S., Eckhart, V. M. & Michener, D. C. (1983). Wood anatomy of Hydrophyllaceae. I. *Eriodictyon. Aliso* **10**: 397–412.

Carlquist, S., Eckhart, V. M. & Michener, D. C. (1984). Wood anatomy of Palemoniaceae. *Aliso* **10**: 547–72.

Carlquist, S. & Hanson, M. A. (1991). Wood anatomy of Convolvulaceae. *Aliso* (in press).

Carlquist, S. & Zona, S. (1987). Wood anatomy of Acanthaceae. A survey. *Aliso* **12**: 201–27.

Carlquist, S. & Zona, S. (1988). Wood anatomy of Papaveraceae, with comments on vessel restriction patterns. *International Association of Wood Anatomists Bulletin, n.s.* **9**: 253–67.

Chalk, L. & Chattaway M. M. (1937). Identification of woods with included phloem. *Tropical Woods* **50**: 1–31.

Czaninski, Y. (1977). Vessel-associated cells. *International Association of Wood Anatomists Bulletin,* 1977: 51–5.

De Bary, A. (1877). *Vergleichende Anatomie der Vegetationsorgane der Phanerogamen und Farne.* Verlag Engelmann, Leipzig.

Esau, K. & Cheadle, V. I. (1969). Secondary growth in *Bougainvillea*. *Annals of Botany, n.s.* 33: 807–19.

Ewers, F. W. & Fisher, J. B. (1989). Variation in vessel length and diameter in stems of some tropical and subtropical lianas. *American Journal of Botany* 76: 1452–9.

Fisher, J. B. & Ewers, F. W. (1989). Wound healing in stems of lianas after twisting and girdling injury. *Botanical Gazette* **150**: 251–65.

Haberlandt, G. (1914). *Physiological Plant Anatomy* (transl. M. Drummond). Macmillan, London.

Martens, P. (1971). Les Gnetophytes. In *Handbuch der Pflanzenanatomie XII(2)*, pp. 1–295. Borntraeger, Berlin.

Mennega, A. M. W. (1980). Anatomy of the secondary xylem. In *Angiospermae: Ordnung Gentianales, fam. Loganiaceae. Die natürlichen Pflanzenfamilien*, ed. 2., 28bI, ed. A. J. M. Leeuwenberg, pp. 111–61. Duncker & Humblot, Berlin.

Metcalfe, C. R. & Chalk, L. (1950). *Anatomy of the Dicotyledons*. Clarendon Press, Oxford.

Metcalfe, C. R. & Chalk, L. (1983). *Anatomy of the Dicotyledons*, 2nd edition. *Vol. II. Wood structure and conclusion of the general introduction*. Clarendon Press, Oxford.

Obaton, M. (1960). Les lianes ligneuses à structure anormales des forêts denses d'Afrique occidentale. *Annales des Sciences Naturelles, (Botanique), Sér. 12* 1: 1–220.

Patel, R. N. (1965). A comparison of the anatomy of the secondary xylem in roots and stems. *Holzforschung* **19**: 72–9.

Pfeiffer, H. (1926). Das abnorme Dickenwachstum. In *Handbuch der Pflanzenanatomie*, ed. 1., ed. K. Linsbauer, pp. 1–272. Borntraeger, Berlin.

Plumb, R. C. & Bridgman, W. B. (1972). Ascent of sap in trees. *Science* **176**: 1129–31.

Pulawska, Z. (1973). The parenchyma-vascular cambium and its derivative tissues in stems and roots of *Bougainvillea glabra* Choisy (Nyctaginaceae). *Ann. Soc. Bot. Pol.* **42**: 41–61.

Putz, F. E. (1983). Liana biomass and leaf area of a 'tierra firme' forest in the Rio Negro Basin, Venezuela. *Biotropica* **15**: 185–9.

Sauter, J. J. (1972). Respiratory and phosphatase activities in contact cells of wood rays and their possible role in sugar secretion. *Zeitschrift für Pflanzenphysiologie* **67**: 135–45.

Sauter, J. J. (1980). Seasonal variation of sucrose content in the xylem sap of *Salix*. *Zeitschrift für Pflanzenphysiologie* **98**: 377–91.

Schenck, H. (1893). Beiträge zur Biologie und Anatomie der Lianen im Besonderen der in Brasilien einheimische Arten. 2. Belträge zur

Anatomie der Lianen. In *Botanische Mittheilungen aus der Tropens*, ed. A. F. W. Schimper, pp. 1–271. G. Fischer, Jena.

Schenck, H. (1895). Ueber die Zerkluftungsvorgänge in anomalen Lianenstammen. *Jahrbucher für Wissenschaftliche Botanik* **27**: 581–612.

Scholander, P. F., Love, W. E. & Kanwisher, J. W. (1955). The rise of sap in tall grapevines. *Plant Physiology* **30**: 93–104.

Scott, D. H. & G. Brebner, (1891). On the anatomy and histology of *Strychnos*. *Annals of Botany* **3**: 275–302.

Singh, B. (1943). The origin and distribution of intraxylary and interxylary phloem in *Leptadenia*. *Proceedings of the Indian Academy of Sciences, ser. B* **18**: 14–19.

Solereder, H. (1908). *Systematic Anatomy of the Dicotyledons* (transl. L. A. Boodle and F. E. Fritsch). Clarendon Press, Oxford.

van der Walt, J. J. A., Schijff, H. P. & Schweickererdt, H. P. (1973). Anomalous growth in the stems of the lianes *Mikania cordifolia* (Burm. f.) Robins (Compositae) and *Paullinia pinnata* Linn. (Sapindaceae). *Kirkia* **9**: 109–38.

van Vliet, G. J. C. M. (1979). Wood anatomy of the Combretaceae. *Blumea* **25**: 141–223.

Westermeier, M. & H. Ambronn, (1881). Beziehungen zwischen Lebensweise und Struktur der Schling- und Kletterpflanzen. *Flora* **69**: 417–36.

Woodworth, R. H. (1935). Fibriform vessel members in the Passifloraceae. *Tropical Woods* **41**: 8–16.

Zimmermann, A. (1922). *Die Cucurbitaceen, Vols 1 & 2*. G. Fischer, Jena.

3

Biomechanical studies of vines

FRANCIS E. PUTZ AND N. MICHELE HOLBROOK

Although vines and self-supporting plants differ physiologically, anatomically, and phenologically, the primary differences are biomechanical. Lacking the capacity to hold themselves upright, climbing plants are constrained by their capacity to (i) encounter suitable structure(s) on which to climb; (ii) ascend efficiently; and, (iii) survive the inevitable mechanical demise of their supports. In this chapter, we address some of the biomechanical features of climbing plants within the context of their life history. We emphasize that vines have broken free from the constraints of having to be self-supporting, only to suffer new restrictions associated with mechanical dependence. Most of our comments pertain better to vines that climb by twining or with tendrils than to plants that climb with the aid of adventitious roots or other adhesive structures. Our bias towards woody vines (=lianas) in tropical forests (Figure 3.1) also will be apparent.

Locating suitable supports

Availability of suitable supports is a major constraint on the height growth of forest vines. The vast majority of vine 'seedlings' (and vines that have reached the top of their support) produce searcher shoots that grow up, bend over, and are successively replaced as each, in turn, fails to encounter a support (Palm, 1827; Darwin, 1867; Putz, 1984). In lowland forest in Panama, vines experimentally supplied with artificial trellises grew to be much taller than did control individuals (Putz, 1984). Vines can increase their likelihood of encountering a suitable support by: (i) actively searching; (ii) waiting (searching in time); and (iii) being able to ascend a wide range of support sizes. While the ability to persist until a suitable support is encountered lies outside the topic of this chapter, biomechanical properties of vine stems influence both searching behavior and the ascent of large diameter supports.

Figure 3.1. Form of a large woody liana.

Seedlings and seedling-sized plants of many climbing species, especially lianas, are initially self-supporting. We include the category 'seedling-sized' in recognition of the fact that many such individuals in the forest understory are not true seedlings at all, but sprouts from the roots of larger vines or from the stems of lianas which have fallen from the canopy (see below). In lowland tropical forest in both Southeast Asia and Latin America, liana 'seedlings' typically achieve 30–40 cm of vertical growth before becoming unstable, although some species are able to reach heights of 1–2 m before requiring mechanical support. Indeed, it is often impossible, based on growth forms, to distinguish trees and lianas during this phase – a serious difficulty in that in several tropical forests about 25% of the 'seedlings' are lianas (Rollet, 1969; Putz, 1984; Putz & Chai, 1987).

The climbing phase of a vine's life history is marked by radical increases in internode lengths and often at least temporary suppression of leaf expansion. Sachs (1875) describes this morphology as 'normally etiolated'. The probability that a liana shoot will encounter a suitable support is enhanced by the production of long whip-like leader (= searcher) shoots. Delayed leaf expansion or reduction of leaves to scales on leader shoots, suppression of lateral branch growth, and structural features of the stem itself, increase the maximum height to which the shoots can be held erect, a distinct advantage where the availability of supports is limited (Schenck, 1892; Raciborski, 1900; Gradman, 1929; Troll, 1937; Baillaud, 1962a; French, 1977). Raciborski (1900) demonstrated that in several species of twining and tendril-climbing plants, leaf expansion is delayed until the leader shoot contacts a support. In lowland tropical forest in Mexico, Peñalosa (1982) showed that *Marsdenia laxifora* (Asclepiadaceae), a species with leafless twining shoots and leaf-bearing short-shoots, was more successful at attaching to supports than *Ipomoea phillomega* (Convolvulaceae), a twiner whose leaves gradually expand as the shoot circumnutates.

Examination of unattached leader shoots in cross-section reveals anatomical features that probably contribute to their mechanical stability. Unlike the cambial variants (i.e. anomalous secondary growth) characteristic of the wood of many mature lianas (see Carlquist, Chapter 2), leader shoots generally have a large pith surrounded by densely packed xylem with small vessels (Courtot, 1966; Troll, 1937). The transition from typical leader wood to wood with extremely large vessels, intruded phloem, expanded rays, and other peculiarities of mature liana wood is generally quite abrupt (personal observation).

Leader shoots of vines and tendrils that are not anchored to a support generally are driven by endogenous growth-related rhythms to sweep through the air in arcs, a process called 'circumnutation' (for reviews see Baillaud, 1962a, b; Johnsson, 1979). When Charles Darwin published his well known paper 'On the movements and habits of climbing plants' (1867) there was already a substantial literature on the topic (see Baillaud, 1962b). Under controlled (i.e. still air) conditions, diameters of circumnutational spirals of leader shoots range from a few centimeters to more than a meter, with an average of about 40–50 cm. One rotation usually requires 1–2 h but both arc diameter and circumnutation rate vary with species and environmental conditions. Basically, conditions that promote vine growth also enhance circumnutation.

The actual mechanism by which circumnutation occurs is still under investigation. Earlier work considered it to be due to a pattern of differential growth across the cross-section of the stem (causing curvature) which is gradually propagated around the shoot (causing movement: e.g. Darwin, 1867). More recent studies at the cellular level, however, suggest that periodic changes in cell turgor may be the actual mechanism by which the curvature is achieved (Millet & Melin, 1978; Millet *et al.*, 1986). Millet *et al.* (1986) report differences in the osmotic potential of epidermal cells on the convex and concave surfaces of the stem, which result in changes in volume and shape of the cells during each revolution. Further studies are now focusing on the cell signaling which coordinates the propagation of the dissymmetry in osmotic potential around the shoot (Millet *et al.*, 1986).

While it seems likely that circumnutation increases the probability that a leader shoot or tendril will contact a support, all of the studies that we are aware of were conducted under laboratory conditions. A particularly intriguing series of laboratory investigations of circumnutation were conducted at the Université de Besançon, France (for a review, see Tronchet, 1977). According to these studies circumnutating stems and tendrils respond to the presence of a nearby support by modifying the rotational movement into an ellipse with the long axis oriented towards the support. The mechanism responsible for this phenomenon has not to our knowledge been elucidated but perhaps involves variations in the concentration gradient of the gaseous hormone ethylene (Tronchet, 1977). Darwin (1880) described a similar phenomenon in tendrils as 'apheliotropism' and suggested that climbers were dependent on this process for coming into contact with tree trunks.

The ecological significance of neither this 'searching' behavior nor circumnutation in general have been demonstrated in the field. The potential advantages of an effective searching method are great because climbing plants must search for supports throughout their life. Vines generally reach the forest canopy by ascending a succession of suitably sized structures (Putz, 1984). When the vertical limit of a structure is reached (for example the top of a sapling), an adjacent support of suitable diameter must be located for vertical growth to continue. Furthermore, this ability to move between supports allows vines to blanket the forest canopy by growing both from branch to branch and from tree to tree. On Barro Colorado Island, Panama, tree crowns separated from their neighbors' branches by distances of greater than 1–2 m lie outside the range of most searcher shoots and such trees are generally inaccessible to most lianas.

A second method of 'searching' consists of horizontal growth across the forest floor. This occurs both in flexible, herbaceous root climbers and vines of the Convolvulaceae that produce both twining shoots and leafless stoloniferous forms (Baillaud, 1958; Peñalosa, 1983). As an example of the former, *Monstera gigantea* (Araceae) seedlings produce slender, leafless shoots which locate a tree upon which to climb by growing over the ground towards the 'darkest sector of the horizon', a phenomenon termed skototropism (Strong & Ray, 1975). Such behavior promotes the efficient localization of the largest nearby host tree, perhaps important in adhesive root climbers which lack the ability to span gaps and hence cannot move from tree to tree. Lateral growth across the forest floor by *Ipomoea phillomega*, on the other hand, does not appear to be related to locating a suitable support trellis (Peñalosa, 1983). These stoloniferous sprouts, which presumably are fueled by photosynthate from shoots with leaves in the canopy, do not change into their vertical, twining form upon encountering a support of appropriate diameter. The horizontal shoots often pass by what appear to be suitable trellises and change over to vertical growth according to some combination of internal and external (perhaps light levels) cues (Peñalosa, 1983). A suitable support is then located via circumnutation as in other twining plants.

How vines climb

The majority of vines climb by twining, with the aid of tendrils, by attaching to supports with adventitious roots or adhesive tendrils, by grasping with tendril-like leaves or leaf-bearing branches, with hooks, or simply by sprawling over other plants (e.g. Darwin, 1867; Schenck, 1892). How a vine climbs determines both the maximum diameter support a vine can utilize as well as the trellis structure (distribution of supports) it requires. Support requirements for lianas have been studied in both Panama (Putz, 1984) and East Malaysia (Putz & Chai, 1987). In both areas the upper diameter limit for supports was greatest for root climbers and smallest for tendril climbers (Figure 3.2). Only the former could ascend anything > 30 cm in diameter. A seeming exception to this is that some tendril-climbers are able to ascend very large, rough-barked trees – but do so by attaching to small diameter irregularities in the bark itself (Darwin, 1867). Scramblers and palms that climb with the aid of hook-bearing leaves or modified inflorescences (i.e. *Desmoncus* and the lepidocaryoid rattans) climbed most successfully in dense clusters of small diameter supports, such as occur on the edges of treefall

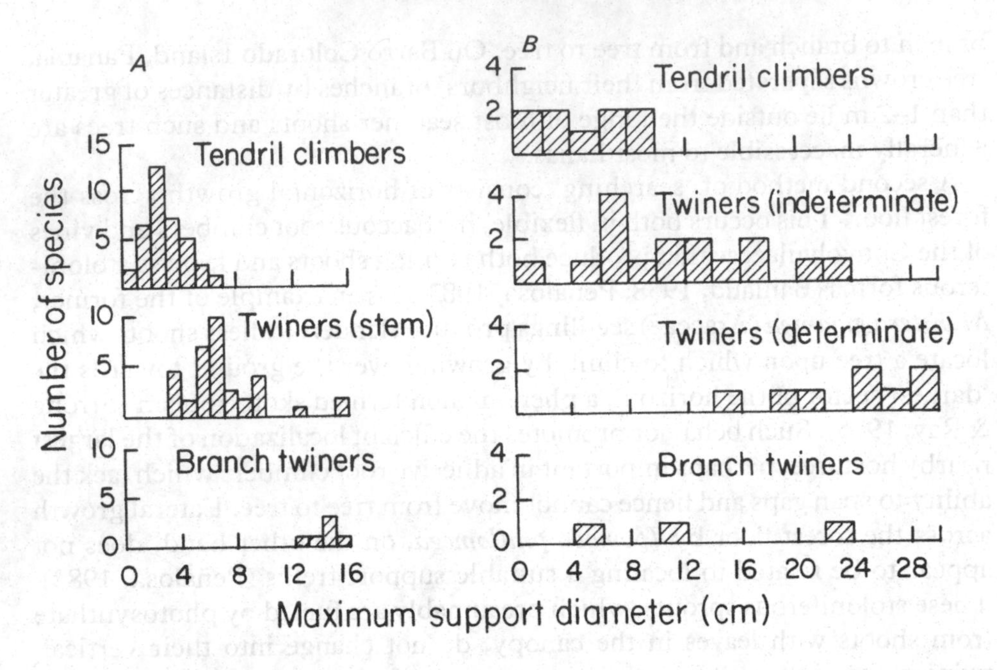

Figure 3.2. The maximum trellis (support) diameter grasped by lianas with different climbing mechanisms in *A*, Barro Colorado Island, Panama, and *B*, Lambir National Park, Sarawak, Malaysia. Only recently climbed supports were used. In Panama (*A*) data were collected in 1 cm intervals; in Malaysia (*B*) 2 cm intervals were used. In *B* lianas with stems that twine are divided into those with leader shoots that become tendril-like and are replaced (determinate) and those with leader shoots that continue to grow up the support (indeterminate). Rattans are not included because their ability to climb depends more on the distribution of supports than on the diameter of individual supports. All lianas can climb on supports smaller than those reported here. Root and adhesive-tendril climbers can climb on supports of any diameter. Figure redrawn from (*A*) Putz (1984) and (*B*) Putz & Chai (1987).

gaps. Malaysian lianas generally climbed trees larger in diameter than those climbed by lianas in Panama, perhaps a reflection of differences in forest dynamics (see Putz & Appanah, 1987). From comments in Darwin (1867), it seems that temperate zone climbers are restricted to growing on supports quite a bit smaller in diameter than has been reported for tropical climbers.

Stem-twiners

The mechanisms of climbing, particularly by twining, have attracted the attention of biologists for at least 150 years (e.g. Palm, 1827; von Mohl, 1827; Dutrochet, 1844). The subject was of sufficient interest during the latter part of the last century to be treated at length in botany textbooks (e.g. Sachs,

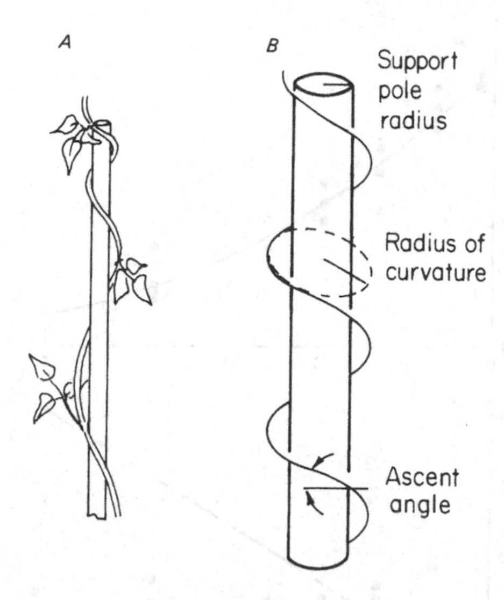

Figure 3.3. *A*, A twining vine. *B*, Diagram of the twining helix showing the angle of ascent, support radius, and radius of curvature.

1875; Vines, 1886). The tragic ballad of the woodbine that falls in love with a morning glory ('but she twines to the left and he to the right') is evidence that non-scientists were also fascinated by the mechanics of twining (note: according to the bard, the offspring of this unnatural union grew straight up and fell over).

Biomechanical aspects of twining that continue to intrigue biologists concern the generation of helical growth trajectories and how twining plants avoid slipping down their supports (Figure 3.3). Recent research by Silk (1989a, b) has employed the methods of tensor calculus for elucidating the phenomenology of twining. There is still much that can be learned about twining through observational studies. Bell (1958), for example, let *Humulus lupulus* (Moraceae) vines twine around poles of different diameters. She demonstrated that while the angle of vine ascent up the pole decreased with increasing support diameter, the radius of curvature (radius of the largest circle which can be placed inside a curve, Figure 3.3*B*) remained constant. Hop vines apparently cannot climb poles with a radius greater than 3.5 cm, which suggests that when the support radius approaches the radius of curvature of the vine helix, the coils become unstable. Our studies with *Dioscorea bulbifera* (Dioscoreaceae) confirm and extend these results. As in *H. lupulus*, the angle at which *D. bulbifera* twines around poles (of polyvinyl chloride) decreases with increasing pole diameter, but the radius of curvature of the vine helix remains relatively constant (Figure 3.4*A*, *B*). Individuals

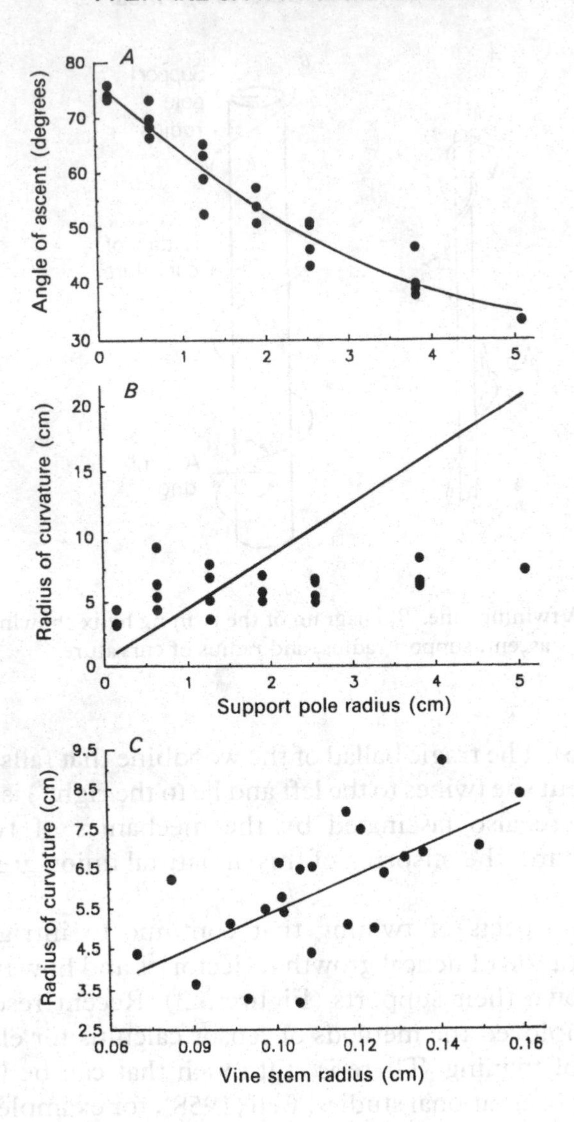

Figure 3.4. *A*, Relationship between ascent angle and support radius of *Dioscorea bulbifera* vines growing up smooth pvc poles. A second order polynomial ($Y = 76.89 - 14.32X + 1.191X^2$, $r^2 = 0.966$, d.f. = 21) was fitted to the data. *B*, Relationship between radius of curvature and support radius. The solid line is the relationship for a helix that maintained a constant ascent angle of 60°. *C*, Relationship between helix radius of curvature and vine stem radius. The regression equation is $Y = 0.757 + 4.446X$, $r^2 = 0.658$, d.f. = 21.

with larger diameter stems, however, ascend at steeper angles than thinner-stemmed plants (Figure 3.4*C*).

Although the physics of twining has been under investigation for more than a century, it is still not clear what keeps twining vines from slipping down their supports. What is in particular need of elucidation is the degree to which twiners avoid slipping by grasping on to their supports. The rough surface texture of many leader shoots undoubtedly increases friction (Sachs, 1882), but a force normal to the support needs to be generated for friction to act at all – at least until the coils are tightened by host tree diameter growth. Dependence on friction, however, is inherently unstable in that any slippage of the coils loosens the helix. Lower coils of *D. bulbifera* growing on large diameter poles frequently are quite loose with visible space between vine stem and pole. Such individuals appear to be, in part, hanging from the youngest 1–2 coils which are more appressed to the pole.

We investigated the importance of friction by modeling twining stems as loosely coiled, frictionless springs. In this case, the pole serves only to constrain how the vine can fail (i.e. it can only collapse upon itself rather than buckling to the side like a column). Springs are structures subject to both torsion and bending, with the relative stresses due to each being dependent on helix geometry. An ascent angle of 90° would denote a structure in pure bending, while an ascent angle of 0° would place the stem in pure torsion. We used formulas for loosely coiled springs (Sayre & deForest, 1936; Ancker & Goodier, 1958) to examine, as a function of ascent angle, the relationship between the strain energy stored in the stem due to loading by self-weight and the change in gravitational potential energy due to the resultant downward displacement. We used the support pole radius – ascent angle relationship determined for *D. bulbifera* (Figure 3.4), as well as material properties (modulus of elasticity (E) and torsion (G)) determined for this species. According to the assumption of no friction, vines climbing at angles less than 70° (= support pole radius of <0.5 cm in this species) should be unstable (defined to be the point at which the helix was shortened by 10%). Since *D. bulbifera* can grow at angles as small as 33° (= 5 cm support pole radius), friction obviously plays an important role. Note that the model is conservative in that it does not include the effects of changes in helical geometry on strain energy as the coils are vertically displaced. These results are supported by the observation that vines grown on small diameter poles retained their helical shape when the pole was removed, and were self-supporting if kept from buckling to the side; vines from large diameter poles collapsed vertically under their own weight.

Other evidence supports the existence of an inward force. For example, when twining vines are removed from their support pole and laid horizontally on a laboratory bench they tend to coil tighter (Vries, 1873; Silk & Hubbard,

1990). von Mohl (1827) found evidence for the inward force from the curvatures produced in a string upon which a vine was twining, while Pfeffer (1906) reports it sufficient to 'crush in a hollow paper cylinder'. More recently Silk & Hubbard (1990) have measured the normal forces generated by vine coils with a manometer-style pressure gauge fashioned from water-filled balloons. How the stem grows so as to generate these contact forces, however, is not yet clear. An anisotropically reinforced cross-section, however, could place the stem in tension, the 'inflation' (via cellular turgor pressure) of which could combine with the helix geometry to generate an inward force (Silk & Hubbard, 1990). Twisting of the shoot around its own axis, a phenomenon that has been recognized for many years (e.g. von Mohl, 1827; Hendricks 1919, 1923), may also play an important role.

Another as yet unexplained aspect of twining is how the plants alter their angle of ascent in such a precise manner. Although friction appears to play a part in keeping twining vines from slipping down their support, increasing the coefficient of friction (by coating the pvc pipes with sand) had no effect on helix geometry. Individuals of *D. bulbifera* growing on smooth and sand-covered poles had the same relationship between ascent angle and pole radius (F. E. Putz and N. M. Holbrook, unpublished data).

Tendril-climbers

Although tendrils are vegetative organs specialized for climbing, they have received little attention from biomechanics. Both early studies (e.g. Palm, 1827; von Mohl, 1827) and more recent work (Jaffe & Galston, 1968; Slatter, 1979) have focused primarily on contact-stimulated curvature, with the mechanical nature of tendrils being only secondarily considered. Tendrils are defined by their function (i.e. as climbing organs), and are variously derived from leaves, branches, inflorescences, leaflets, stipules, etc. (e.g. Darwin, 1867; Schenck, 1892). Tendrils begin as thin, relatively straight (or slightly bowed) structures (Figure 3.5). Although tendrils circumnutate in a manner similar to twining shoots, they are distinguished by their 'irritability'. Tactile stimulation near the apex of the tendril promotes a coiling response in which the tip grows in length around the object. This behavior can be stimulated by contact with something as light and flexible as spider's silk (MacDougal, 1896); a stream of water fails to elicit a response, although water containing a fine powder will (Pfeffer, 1906).

It is generally assumed that there is no lower size limit to what a tendril can clasp on to, while the upper limit is considered to be determined by the length of the tendril (e.g. MacDougal, 1893). A lower limit does, at least in theory, exist owing to the fact that the amount of extension (dorsal side) and compression (ventral surface) permitted by the flexibility (extensibility) of

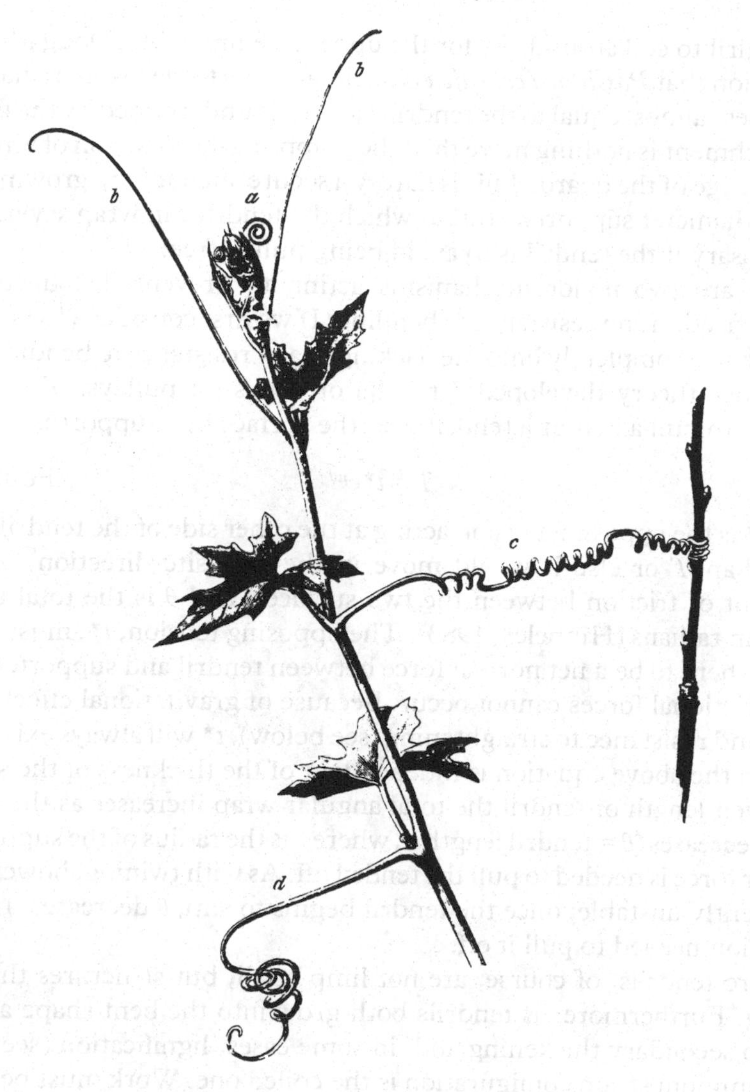

Figure 3.5. A tendril climber (*Bryonia dioica*) showing immature tendrils (*b*), an attached tendril with free coils (*c*), and a mature tendril which failed to contact a support (*d*). Figure reproduced from Pfeffer (1906).

the tissue may not allow the radius of curvature to approach the radius of the tendril itself. In other words, stout and/or inextensible tendrils would not be able to coil around extremely thin supports because they would not be able to bend tightly enough (Sachs, 1875; Vines, 1886). In nature, growing tendrils tend to be both thin and flexible and we know of no observation of a support for a tendril to coil around. As for the upper size limit, MacDougal's (1893) observation that *Passiflora caerulea* tendrils are able to clasp on to a flat board of thickness almost equal to the tendril's length is undermined by the fact that this attachment is nothing more than the grappling-hook action of tendril tip over the edge of the board. This is hardly a secure anchor for a growing stem. Smaller diameter supports, around which the tendril can wrap several coils, are necessary if the tendril is to avoid being pulled free.

There are two major mechanisms acting to prevent dislodgement of tendrils: friction and resistance to bending. If we first consider a hypothetical tendril that is completely limp (i.e. lacking in any resistance to bending), then we can use theory developed for belts or ropes on pulleys. The tension sufficient to pull a belt or a tendril over the surface of a support is

$$T = t^{\star} e^{\mu\theta} \qquad \text{(Equation 1)}$$

where $t^{\star}$ is the opposing tension acting at the other side of the tendril (which is less than T or else it would move in the opposite direction), μ is the coefficient of friction between the two surfaces, and θ is the total angle of contact in radians (Hibbeler, 1983). The opposing tension, $t^{\star}$, must be non-zero for there to be a net normal force between tendril and support, without which frictional forces cannot occur. Because of gravitational effects on the tendril and resistance to straightening (see below), $t^{\star}$ will always exist. Thus, although the above equation is independent of the thickness of the support, for a given length of tendril the total angular wrap increases as the support radius decreases (θ = tendril length/r, where r is the radius of the support) and a greater force is needed to pull the tendril off. As with twining, however, this is inherently unstable; once the tendril begins to slip, θ decreases, reducing the tension needed to pull it off.

Mature tendrils, of course, are not limp at all, but structures that resist bending. Furthermore, as tendrils both grow into the bent shape and then undergo secondary thickening and, in some cases, lignification (see below), their minimum strain configuration is the coiled one. Work must be done in order to straighten (uncoil) the tendril. Imagine a perfectly elastic, frictionless tendril wrapped around a smooth cylinder. The work that must be done in order to pull this tendril off is proportional to the change in curvature (where curvature at any point is defined as the inverse of the radius of the largest circle which can be placed tangent at that point, i.e. = 1/radius of curvature, see Figure 3.3*B*). A tendril that is wrapped around a smaller diameter support must undergo a greater degree of bending (in this case straightening after having coiled), and thus, for the same material properties, more work must be done.

Thickening of the tendril, especially in the coiled region, accompanies (and may aid) establishment of a firm connection (Worgitzky, 1887; Brush, 1912;

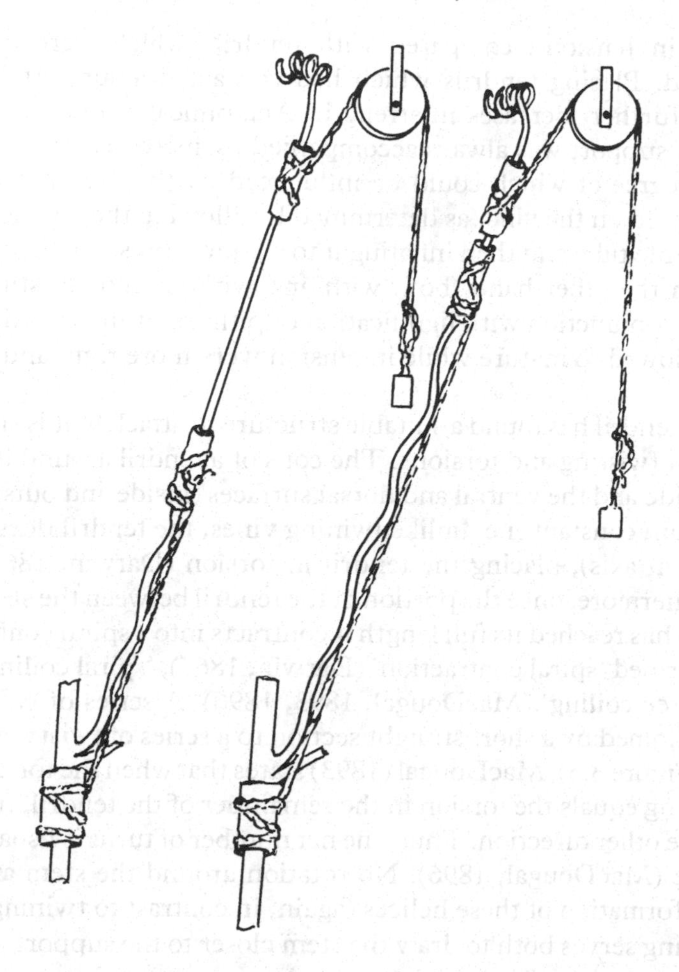

Figure 3.6. Diagram of the mechanism Brush (1912) used to study the effect of tension on tendril development. Figure redrawn from Brush (1912).

Lisk, 1924). Differences in tendril structure between the portion coiled around the support and the part which forms the free coils (see below) is not surprising, given that in the former stiffness would be advantageous, while in the latter elasticity would be more appropriate (Worgitzky, 1887). Anatomical studies have further shown that enlargement of the portion surrounding the support is attributable to cell expansion rather than cell division (Muller, 1887) and that development of a hard, brittle texture is accompanied by lignification of the pith (Muller, 1887; Worgitzky, 1887). Brush (1912) experimentally studied the effects of contact stimulation and tensile forces on this development (Figure 3.6). There was an increase in breaking strength of *P. caerulea* tendrils which were allowed to contact a support (without placing themselves in tension), compared with tendrils which were completely unstimulated. Placing tendrils which had contacted a support in tension resulted in further increases in strength. Anatomical studies showed that contacting a support was always accompanied by increased xylem development, the degree of which could be influenced by the pressure which the support exerted on the vine (as determined by allowing the vine to contact a hollow, flexible tube and then inflating it to a known pressure: Brush, 1912). Tension, on the other hand (both with and without contact stimulation), occurred in conjunction with lignification of pith cell walls. Tendrils which had been allowed to mature while in tension were more rigid and harder to cut.

Once the tendril has found a suitable structure to attach to it is still subject to additional twisting and torsions. The coils of a tendril around its support lie side by side and the ventral and dorsal surfaces (inside and outside) of the tendril remain constant (i.e. unlike twining vines, the tendril does not twist about its own axis), placing the tendril in torsion (Darwin, 1867; Raugh, 1979). Furthermore, once the portion of the tendril between the stem and the attached tip has reached its full length it contracts into a spiral configuration, variously termed 'spiral contraction' (Darwin, 1867), 'spiral coiling' (Sachs, 1882), or 'free coiling' (MacDougal, 1893, 1896). A series of twists in one direction is joined by a short straight section to a series of twists in the other direction (Figure 3.5). MacDougal (1893) states that when the torsion caused by free coiling equals the torsion in the remainder of the tendril, free coiling begins in the other direction. Thus, the net number of turns is usually zero or at most one (MacDougal, 1896). No rotation around the stem axis occurs during the formation of these helices (again, in contrast to twining stems).

Free coiling serves both to draw the stem closer to the support (indeed, to pull up the growing shoot) and to absorb energy (i.e. due to wind, self-weight) that might otherwise dislodge or rupture the tendril. MacDougal (1893) found that free coiling of *P. caerulea* tendrils brought the vine 1–6 cm (approximately $\frac{1}{3}$ the length of the tendril) closer to its support. He performed further experiments designed to measure the forces exerted by the tendril due to this coiling behavior. Attaching weights to the tips of young tendrils resulted in fewer and more open spirals until at 20 g the tendrils remained straight (Figure 3.7). *P. caerulea* tendrils attached to a 'dynamometer' (Figure 3.7) exhibited tensions of 3–10 g, while *Cucurbita* tendrils pulled with up to 30 g (MacDougal, 1893, 1896). *P. caerulea* tendrils must support static loads of only about 1 g in order to hold the shoot erect, yet it takes 350–750 g (i.e. the weight of several feet of vine) to rupture a tendril (MacDougal, 1893). These were presumably static loads; dynamic loading might indicate lower rupture thresholds (Wainwright *et al.*, 1976). On the other hand, a tendril must first be straightened (uncoiled) before it even experiences the full tensile

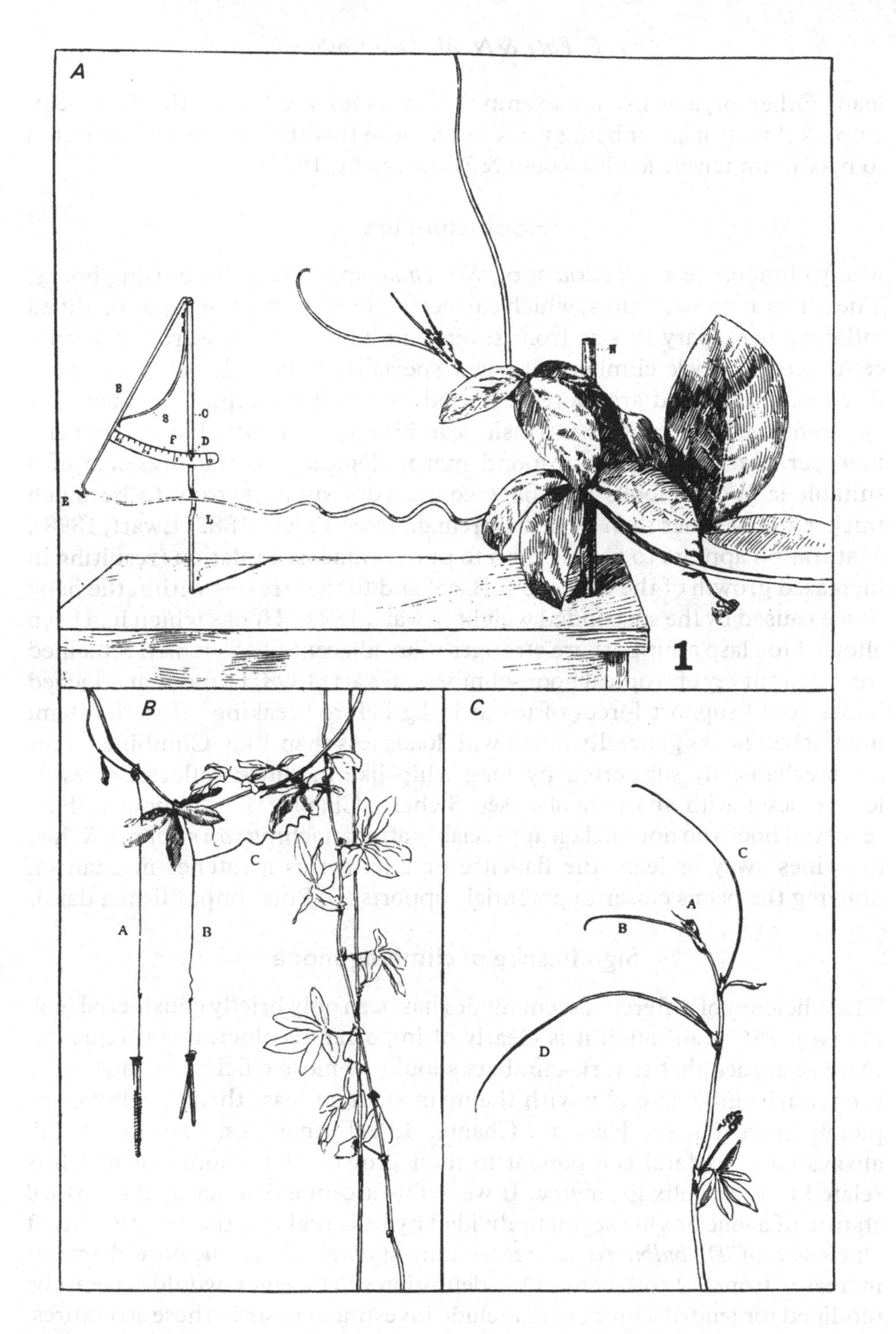

Figure 3.7. *A*, Line drawing of the 'dynamometer' used by MacDougal (1893) to measure tensile forces in attached tendrils. *B*, The effect of added weights on tendril development; tendril A supports nearly a 20 g weight and is almost straight, tendril B has a slightly lighter load and shows some sign of free coiling. *C*, An immature *Passiflora caerulea* apex. Figure reproduced from MacDougal (1893).

load. Other organisms, for example *Nereocystis leutkeana*, the bull kelp, employ this strategy of being weak but long so that they are rarely subjected to maximum tensile loads (Koehl & Wainwright, 1977).

Hook-climbers

Many climbers (e.g. *Uncaria* spp., *Strychnos* spp., etc.) climb using hooks. These recurved structures, which can occur along stems, leaves, or modified inflorescences, vary in size from several millimeters to upwards of several centimeters. Hook climbers are less specialized than the climber types discussed above and are more restricted in locating a support in that they apparently do not possess intrinsic searching movements. Hook-climbers, however, are still able to respond morphologically to the presence of a suitable support. Hooks that have contacted a support grow to be much thicker and woodier than those that remain free (Treub, 1882; Ewart, 1898). Maturation appears to be due both to pure contact stimulation (resulting in increased growth of the concave surface) and to the stresses within the hook tissue caused by the suspended weight (Ewart, 1898). Hooks which had been allowed to clasp a support are stronger than adjacent hooks which remained free. In a survey of tropical hook-climbers, Ewart (1898) found that attached hooks could support forces of up to 15 kg before breaking off at the stem; unattached hooks generally failed with loads less than 1 kg. Climbing palms are mechanically supported by long whip-like modified inflorescences or leaves beset with sharp hooks (see Siebert, Chapter 17), although these recurved hooks do not thicken appreciably after attaching to a support. When the vines sway or lean, the flagellae or cirri act as a ratchet mechanism drawing the palms closer to potential supports (F. Putz, unpublished data).

Significance of climbing mode

The efficiency of different ascent modes has been only briefly considered (e.g. Darwin, 1867), although it is clearly of important ecological consequence. Darwin argued that tendril-climbers should be more efficient because only the tendrils grow laterally with the main stem, at least theoretically, completely erect (but see Hegarty, Chapter 13). Twiners, on the other hand, always have a lateral component to their growth, the amount of which is related to their helix geometry. If we define ascent efficiency as the vertical stature of a vine or vine segment divided by its actual length, then the ascent efficiency of *D. bulbifera* decreases almost twofold as support diameter increases from 0.2 to 10 cm. This definition of efficiency would have to be modified for tendril-climbers to include investment costs in these structures. Comparative studies of ascent efficiency under natural conditions have not been made, although their results would be of interest in understanding the distribution of climbing modes within different habitats and successional stages.

Climbing mode is also significant in that the mechanism by which a vine climbs, in part, determines where it can climb in terms of community structure and the light environment in which its leaves occur. Tendril-climbers, which need a large number of small diameter supports, may be more suited to treefall gaps and forest edges where such trellis distributions more commonly occur, while stem-twiners may be more appropriate in forest understory as they can ascend smooth boles of saplings and small trees (Darwin, 1867; Putz, 1984; Hegarty, 1989; see also Hegarty, Chapter 13; Hegarty & Caballé, Chapter 11). Sprawling vines and hook climbers, on the other hand, may be common in early successional habitats owing to a need for dense vegetation.

How a vine climbs may also determine the photosynthetic environment in which it is able to display its leaf surfaces. Adhesive climbers, which cannot move between either branches or trees, are necessarily restricted to shadier microsites. Tendril-climbers, on the other hand, can ascend on the very outside of a forest edge and thus remain in a high-light environment. The ability to move laterally between supports affords twiners and tendril-climbers the freedom to, in Darwin's words (1867), 'securely ramble over a wide and sun-lit surface'. Examination of photosynthetic parameters in relation to climbing mode suggests that the two may be related (Carter & Teramura, 1988; see also Teramura, Gold & Forseth, Chapter 9).

Mechanical properties of vine stems

When, by whatever means, a liana finally reaches the canopy its problems with gravity are not over; a canopy vine suffers severely when its support falls. On Barro Colorado Island, Panama, for example, over a 10-year period 20 of the 193 canopy lianas included in a study of diameter growth rates fell from the canopy (Putz, 1990). Furthermore, most of the individuals which fell from the canopy survived and resprouted; only three actually died (Putz, 1990). The observation that lianas can survive falls is supported by other studies in both natural and selectively logged forest. In fact, many of the lianas that impede the passage of pedestrians and suppress tree regeneration are sprouts from fallen individuals (Putz, 1984; Appanah & Putz, 1984).

The capacity to survive falls from the canopy confers a considerable advantage on lianas in that they attain a degree of immortality – at least in a biomechanical sense. We conducted studies designed to document the flexibility and toughness of liana stems and to elucidate whether these stem properties derive from their being long and slender or from their material properties. In particular, we were interested in determining whether liana

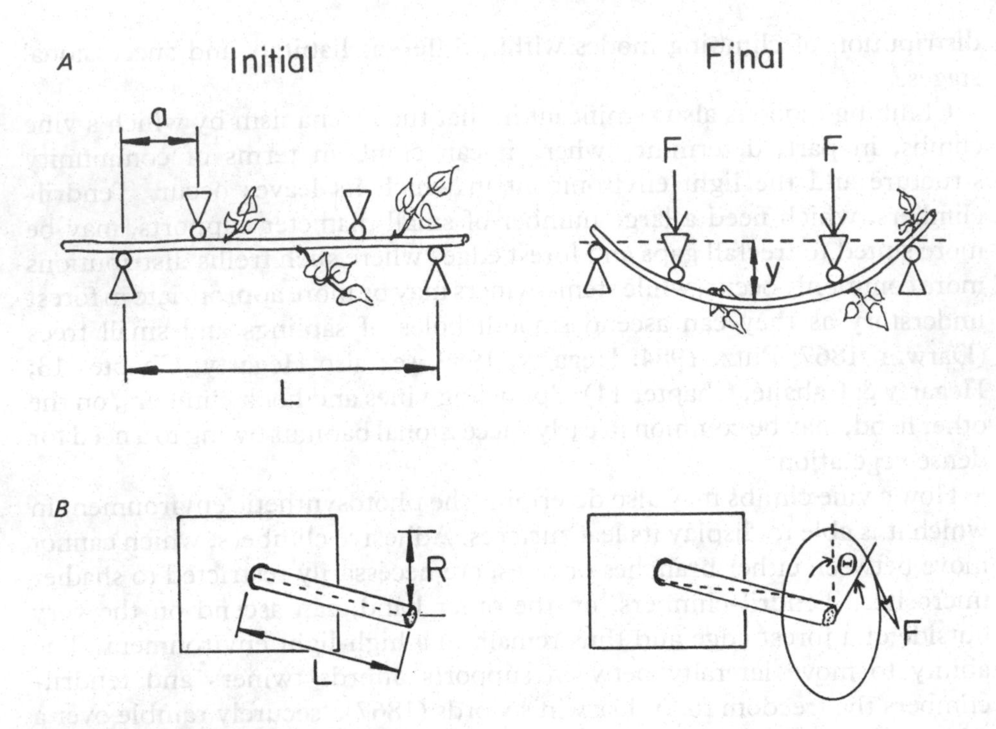

Figure 3.8. *A*, Determination of elastic modulus using four-point bending. The outer two supports are stationary. An equal force is applied to each of the inner supports. The stem section is not attached at the four contact points and hence assumes a smooth curvature throughout its length. The elastic modulus E (Pa) is calculated according to $E=(Fa(3L^2-4a^2)/(24yI)$, where F is the force applied to each of the inner contact points (N), a is the distance between the outer and inner supports (m), L is the distance between the outer two supports (m), y is the maximum vertical deflection (m), and I is the second moment of area assuming a central neutral plane (m^4, $\pi r^4/4$ for a circular cross-section). All measurements were made using small deformations ($y<0.1L$). *B*, Determination of the torsion modulus by applying a torque and measuring the angular deflection. The torsion modulus G (Pa) is calculated according to $G=FRL/\theta J$, where F is the force (N) applied at a distance R (m), L is the length of the stem segment from the fixed (non-rotational) end to the point where the torque is applied, θ is the angular rotation (radians) and J is (for a circular cross-section) the polar moment of inertia (m^4).

stem flexibility and toughness are related to the peculiar juxtapositions of hard and soft tissues characteristic of the stems of many lianas (see Carlquist Chapter 2; Fisher & Ewers, Chapter 4).

We conducted studies of stem properties with samples of lianas and tree saplings of similar diameter (2–5 cm) collected in the Luquillo Mountains, Puerto Rico. For each sample we determined stem density (dry weight/fresh volume) and estimated the apparent modulus of elasticity (E) in four-point bending (Young, 1989; Figure 3.8*A*). When lianas and their supporting trees fall, however, they are more likely to suffer severe torsional stress than pure bending (personal observation). We estimated the torsional modulus (G) of each sample with a torsionometer consisting of two clamps, one stationary and one free to rotate (Figure 3.8*B*). Torque was applied at a constant rate with a spring scale and rotational deflections measured with a protractor.

Stems from the 12 species of Puerto Rican lianas sampled differed substantially from tree saplings in both bending and torsion tests (Table 3.1). The mean apparent modulus of elasticity of lianas was almost an order of magnitude lower than estimated for tree saplings. In torsion, lianas were three times more flexible than the saplings. Stems of *Toxicodendron diversilobum*, a temperate species which can grow as either a vine or a shrub, showed a similar pattern (Gartner, 1988). Individuals or branches which were supported, and thus growing as vines, were significantly more flexible (E measured in bending) than self-supporting conspecifics. In addition to these differences in material properties, vine stems of *T. diversilobum* were more slender and significantly less tapered than shrub stems of the same age (B. Gartner, personal communication).

While great flexibility is undoubtedly an important attribute of liana stems, the probability of surviving tree falls may be more directly related to the magnitude of deformation that causes breakage, i.e. their toughness. Maintenance of functional xylem is probably a prerequisite for surviving tree falls and retaining the capacity to sprout. We defined stem toughness, therefore, as the angular rotation of stem segments at which flow in the xylem was completely disrupted. This point of failure was assessed by twisting stem segments of liana and tree saplings connected to a water-filled pressure chamber. Water was driven through the stem segments with 0.1 MPa pressure. Lianas proved to be uniformly tougher than tree saplings in this regard (Table 3.1). Whereas no tree stem sample maintained xylem flow for even a single rotation, most of the lianas revolved several times before flow stopped. In one case the last liana vessel ceased flow only when the two ends of the sample came apart; apparently the last thread holding the stem together was a functional xylem vessel.

The ability to withstand large deformations without losing function appears to be closely related to internal anatomy. Many lianas are characterized by the presence of expanded rays, large bands of wood parenchyma, or successive cambia (see Carlquist, Chapter 2). Compartmentalization of relatively inflexible xylem vessels within soft tissues may allow vines to function more as multistranded cables than as solid cylinders, and hence be strong as well as flexible (see also Obaton, 1960). Examination of sections of vines subject to torsion while dye was flowing through the vessels showed that large-scale structural damage was confined to the non-lignified regions between bands of xylem vessels (Figure 3.9). The presence of non-lignified

Table 3.1. *Wood density, bending modulus, torsion modulus, and toughness (rotation to break, see text for details) of tree and liana stems from the Luquillo Mountains, Puerto Rico. There was no significant difference in wood density between trees and lianas; all other comparisons were significant ($p < 0.001$)*

		wood density (g/cm³)	bending modulus (GPa)	torsion modulus (GPa)	rotation to break (degrees)
Trees					
Brunellia comocladifolia Humb. & Bompl.	Brunelliaceae	0.309	2.650	0.105	270
Dendropanax arboreus (L.) Decne. & Planch.	Araliaceae	0.441	12.000	0.179	300
Guarea trichilioides L.	Meliaceae	0.468	2.450	0.124	299
Inga vera Willd.	Mimosaceae	0.581	4.760	0.348	315
Ocotea floribunda (Sw.) Mez	Lauraceae	0.361	1.670	0.269	315
	Mean	0.432	4.710	0.169	300
	SD	0.093	3.800	0.124	16
Lianas					
Cissampelos pareira L.	Menispermaceae	0.403	0.462	0.031	1080
Cissus sicyoides L.	Vitaceae	0.228	0.146	0.008	1170
Forsteronia (?) *portoricesis* Woods	Apocynaceae	0.307	0.213	0.053	1980
Heteropteris laurifolia L.	Malpighiaceae	0.531	1.360	0.293	1350
Hippocratea volubilis L.	Hippocrateaceae	0.437	0.558	0.043	936
Ipomoea repanda Jacq.	Convolvulaceae	0.193	0.434	0.020	1170
Marcgravia rectiflora (Tr. & Pl.)	Marcgraviaceae	0.367	1.680	0.016	900
Mikania fragilis Urban	Asteraceae	0.272	0.189	0.013	1800
Paullinia pinnata L.	Sapindaceae	0.561	0.600	0.040	630
Rourea surinamensis Miq.	Connaraceae	0.484	0.825	0.078	1170
Schlegelia brachyantha Griseb.	Bignoniaceae	0.300	0.552	0.061	490
Securidaca virgata Sw.	Polygalaceae	0.352	0.202	0.023	1170
	Mean	0.374	0.608	0.055	1140
	SD	0.109	0.442	0.071	392

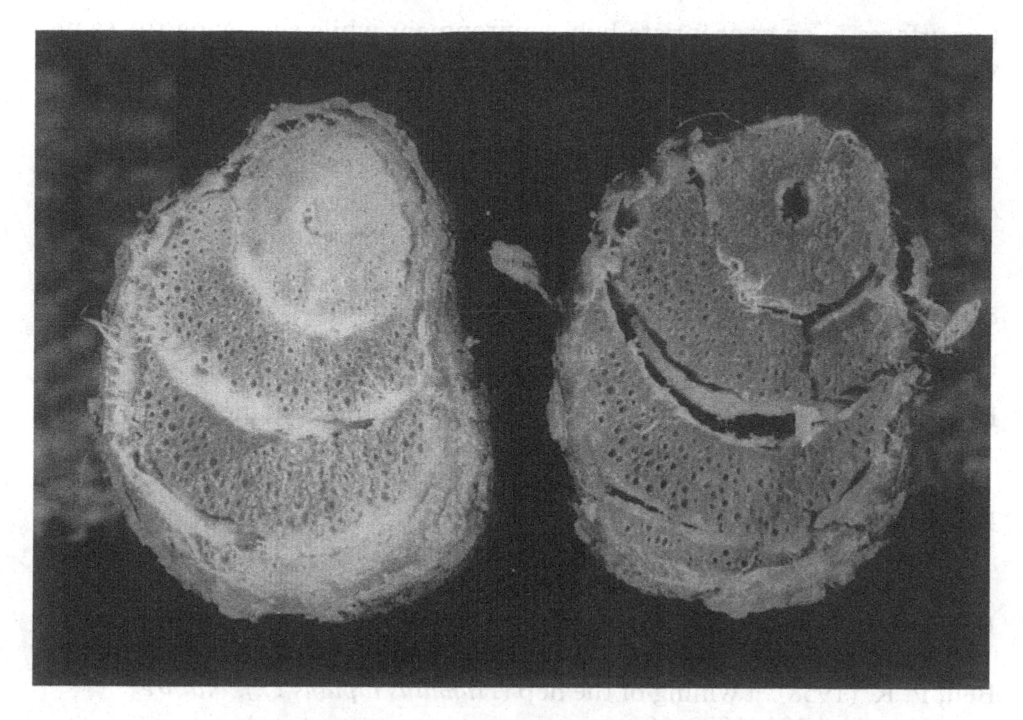

Figure 3.9. Photograph (2 ×) of *Securidaca virgata* stems before (left) and after (right) twisting.

tissue and multiple cambia may also aid in the repair of injuries sustained during tree falls (see Fisher & Ewers, Chapter 4).

Conclusions

Studies of the biomechanics of a group of plants which, by definition, are not self-supporting is primarily an examination of ways in which vines exploit other plants for support while attempting to remain as mechanically self-sufficient as possible. We have focused on specialized patterns of growth which allow vines to locate supports, adaptations which allow them to ascend, and features of the stems themselves that make them robust in the face of the mechanical failure of their support. In addition to the obvious presence of specialized climbing structures (e.g. hooks, tendrils), vines can be mechanically characterized by the occurrence of specialized 'searching' behaviors, attachment stimulated maturation patterns, and stem anatomies that allow them to withstand large physical traumas. Released from the constraint of having to support their own weight, vines enjoy greater freedom to exploit a wide array of mechanical solutions to the problem of displaying photosynthetic surfaces in an appropriate light environment while supplying the same with water and nutrients from the distant soil.

References

Ancker, C. J., Jr & Goodier, J. N. (1958). Pitch and curvature corrections for helical springs. *American Society of Mechanical Engineers Journal of Applied Mechanics* 37: 466–83.

Appanah, S. & Putz, F. E. (1984). Climber abundance in virgin dipterocarp forest and the effect of pre-felling climber cutting on logging damage. *Malaysian Forester* 47: 335–42.

Baillaud, L. (1958). Recherches sur les mouvements spontanées des plantes grimpantes. *Annales scientifiques de l'Université de Besançon (Botanique)* 11: 1–233.

Baillaud, L. (1962a). Mouvements autonomes des tiges, vrilles et autres organes a l'exception des organes volubiles et des feuilles. *Handbuch der Pflanzenphysiologie* 17: 562–634.

Baillaud, L. (1962b). Les mouvements d'exploration et d'enroulement des plants volubiles. *Handbuch der Pflanzenphysiologie* 17: 653–715.

Bell, P. R. (1958). Twining of the hop (*Humulus lupulus* L.). *Nature (London)* **181**: 1009–10.

Brush, W. D. (1912). The formation of mechanical tissue in the tendrils of *Passiflora caerulea* as influenced by tension and contact. *Botanical Gazette* **53**: 453–76.

Carter, G. A. & Teramura, A. H. (1988). Vine photosynthesis and relationships to climbing mechanisms in a forest understory. *American Journal of Botany* 75: 1011–18.

Courtot, Y. (1966). Structure anatomique des tiges volubiles de *Mandevilla suaveolens* (Apocynacées). *Annales scientifiques de l'Université de Besançon (Botanique)* 3: 24–7.

Darwin, C. (1867). On the movements and habits of climbing plants. *Journal of the Linnean Society (Botany)* 9: 1–118.

Darwin, C. (1880). *The Power of Movement in Plants*. D. Appleton and Company, New York.

Dutrochet, R. (1844). Recherches sur la volubilité des tiges de certains végétaux et sur la cause de ce phénomène. *Comptes Rendus, Académie des Sciences (Paris)* 17: 989–1008.

Ewart, A. J. (1898). On contact irritability. *Annales du Jardin Botanique de Buitenzorg* 15: 187–242.

French, J. C. (1977). Growth relationships of leaves and internodes in viny angiosperms with different modes of attachment. *American Journal of Botany* **64**: 292–304.

Gartner, B. L. (1988). Functional morphology and architecture in conspecific lianas and shrubs. *Bulletin of the Ecological Society of America* (Suppl.) **69**: 142.
Gradmann, H. (1929). Das Winden und Ranken der Pflanzen. *Ergebnisse der Biologie* **5**: 166–218.
Hegarty, E. E. (1989). The climbers – lianes and vines. In *Tropical Rain Forest Ecosystems: Biogeographical and Ecological Studies. Ecosystems of the World 14B*, ed. H. Lieth and M. J. A. Werger, pp. 339–53. Elsevier, Amsterdam.
Hendricks, H. V. (1919). Torsion studies in twining plants. *Botanical Gazette* **68**: 425–40.
Hendricks, H. V. (1923). Torsion studies in twining plants. II. *Botanical Gazette* **75**: 282–97.
Hibbeler, R. C. (1983). *Engineering Mechanics: Statics*. MacMillan Publishing Co., New York.
Jaffe, M. J. & Galston, A. W. (1968). The physiology of tendrils. *Annual Review of Plant Physiology* **19**: 417–34.
Johnsson, A. (1979). Circumnutation. In *Encyclopedia of Plant Physiology, n.s. Vol. 7*, ed. W. Haupt and M. E. Feinleib, pp. 627–46. Springer-Verlag, Berlin.
Koehl, M. A. R. & Wainwright, S. A. (1977). Mechanical adaptations of a giant kelp. *Limnology and Oceanography* **22**: 1067–71.
Lisk, H. (1924). Cellular structure of tendrils. *Botanical Gazette* **78**: 85–102.
MacDougal, D. T. (1893). The tendrils of *Passiflora caerulea*. II. External phenomena of irritability and coiling. *Botanical Gazette* **18**: 123–30.
MacDougal, D. T. (1896). The mechanism of curvature of tendrils. *Annals of Botany* **10**: 373–402.
Millet, B., Melin, D. & Badot, P. (1986). Circumnutation: a model for signal transduction from cell to cell. In *The Cell Surface in Signal Transduction, NATO ASI Series H, Vol. 12*, ed. E. Wagner, H. Greppin and B. Millct, pp. 169–80. Springer-Verlag, Berlin.
Millet, B. & Melin, D. (1978). Changes in cellular water content and circumnutation of twining plants. *Plant Physiology* (Suppl.) **61**(4): 30.
Muller, O. (1887). Untersuchungen über die Ranken der Cucurbitaceen. *Beiträge zur Biologie der Pflanzen* **4**: 97–143.
Obaton, M. (1960). Les lianes ligneuses a structure anormale des forêts denses d'Afrique Occidentale. *Annales des sciences naturelles (Botanique), Sér. 12*, **1**: 1–220.
Palm, L. H. (1827). *Über das Winden der Pflanzen*. Preisschrift, Stuttgart.
Peñalosa, J. (1982). Morphological specialization and attachment success in two twining lianas. *American Journal of Botany* **69**: 1043–5.

Peñalosa, J. (1983). Shoot dynamics and adaptive morphology of *Ipomoea phillomega* (Vell.) House (Convolvulaceae), a tropical rainforest liana. *Annals of Botany* 52: 737–54.

Pfeffer, W. (1906). *The Physiology of Plants, Vol. III*, 2nd edition, (transl. A. J. Ewart). Clarendon Press, Oxford.

Putz, F. E. (1984). The natural history of lianas on Barro Colorado Island, Panama. *Ecology* 65: 1713–24.

Putz, F. E. (1990). Liana stem diameter growth rates on Barro Colorado Island, Panama. *Biotropica* 22: 103–4.

Putz, F. E. & Appanah, S. (1987). Buried seeds, newly dispersed seeds, and the dynamics of a lowland forest in Malaysia. *Biotropica* 19: 326–33.

Putz, F. E. & Chai, P. (1987). Ecological studies of lianas in Lambir National Park, Sarawak, Malaysia. *Journal of Ecology* 75: 523–31.

Raciborski, M. (1900). Über die Vorläuferspitze. *Flora* 87: 1–25.

Raugh, M. R. (1979). Some Geometry Problems Suggested by the Shapes of Tendrils. PhD Dissertation, Stanford University, Stanford, Ca.

Rollet, B. (1969). La régénération naturelle en forêt dense humide sempervirente de plaine de la Guyane Vénézuélienne. *Bois et Forêts des Tropiques* 14: 19–38.

Sachs, J. (1875). *Textbook of Botany*. Clarendon Press, Oxford.

Sachs, J. (1882). *Vorlesungen über Pflanzen-Physiologie*. W. Englemann, Leipzig.

Sayre, M. F. & de Forest, A. V. (1936). New spring formulas and new materials for precision spring scales. *Transactions of the American Society of Mechanical Engineers* 58: 379–87.

Schenck, H. (1892). Beiträge zur Biologie und Anatomie der Lianen im Besonderen der in Brasilien einheimischen Arten, 1. In *Botanische Mittheilungen aus den Tropen, 4*, ed. A. F. W. Schimper, pp. 1–253. G. Fischer, Jena.

Slatter, R. L. (1979). Leaf movements and tendril curling. *Encyclopedia of Plant Physiology* 7: 442–84.

Silk, W. K. (1989a). On the curving and twining of stems. *Environmental and Experimental Botany* 29: 95–109.

Silk, W. K. (1989b). Growth rate patterns which maintain a helical tissue tube. *Journal of Theoretical Biology* 138: 311–27.

Silk, W. K. & Hubbard, M. (1990). Axial forces and normal distributed loads in twining stems of morning glory. *Journal of Biomechanics* (in press).

Strong, D. R. & Ray, T. S. (1975). Host tree location behavior of a tropical vine (*Monstera gigantea*) by skototropism. *Science* 190: 804–6.

Treub, M. (1883). Sur une nouvelle catégorie de plantes grimpantes. *Annales du Jardin Botanique de Buitenzorg* **3**: 44–75.
Troll, W. (1937). *Vergleichende Morphologie der Höheren Pflanzen. Band 1: Vegetationsorgane. Teil 3: Lieferung.* Gebrüder Bornträger, Berlin.
Tronchet, A. (1977). *La sensibilité des plantes.* Masson, Paris.
Vines, S. H. (1886). *Lectures on the Physiology of Plants.* Cambridge University Press, Cambridge.
von Mohl, H. (1827). *Über den Bau und das Winden der Ranken und Schlingpflanzen.* Tübingen.
Vries, H. de (1873). Längenwachsthum der Ober- und Unterseite krümmender Ranken. *Arbeiten des Botanischen Instituts zur Würzburg* **1**: 302–32.
Wainwright, S. A., Biggs, W. D., Currey, J. D. & Gosline, J. M. (1976). *Mechanical Design in Organisms.* Princeton University Press, Princeton, NJ.
Worgitzky, G. (1887). Vergleichende Anatomie der Ranken. *Flora* **70**: 2–86.
Young, W. C. (1989). *Roark's Formulas for Stress and Strain,* 6th edition. McGraw-Hill, New York.

4

Structural responses to stem injury in vines

JACK B. FISHER AND FRANK W. EWERS

Introduction

The stems of vines have a relatively small cross-sectional area in relation to the leaf area supplied by the stem compared with more typical herbaceous and woody plants. Non-climbing plants are self-supporting and, therefore, much of the stem tissue must serve a mechanical, non-conducting function. This relationship is discussed in detail by Ewers, Fisher & Fichtner (Chapter 5). Because a given area of sapwood and phloem supply a large leaf area in a vine, we assume that injury to a given area of a cylindrical stem surface or arc on a stem's circumference will have greater potential impact on the leaf area of the shoot than would an injury of similar size on a tree, shrub, or herb. We will consider several types of stem injury that a plant can experience in nature due to physical abrasion with itself or other objects, damage by animals, bending or twisting in wind or during limb and tree falls, and complete breakage during tree falls. Surface damage to the bark, including the functional inner secondary phloem and the vascular cambium, would interrupt the normal movement of organic compounds in the phloem and the production of new xylem and phloem. An encircling injury or bark girdle would completely disrupt these functions in a typical stem with a cylindrical vascular cambium. Stem twisting and bending can result in partial splitting of the stem or complete breakage. The xylem would then be partly or completely interrupted, thus breaking the water continuum from root to leaf.

Here we examine how vines respond to such real and potentially devastating injuries to their stems. New observations of stem healing in nature and previously published results of experimental injury will be used to test hypotheses about the functional significance of so-called 'anomalous' or 'variant' stem anatomy in vines, particularly lianas. The regenerative capacity of vines will also be related to their responses to injury.

Responses to whole plant injury

Vines have a great ability to resprout after damage to the shoot system. Most climbers produce numerous new shoots which are equivalent to the original leader. Such a repetition of the original main or leader shoot has been referred to as architectural reiteration (Hallé, Oldeman & Tomlinson, 1978) or traumatic reiteration (de Castro e Santos, 1980). These reiterations arise from lateral buds and are part of the normal architectural development of the intact plant (Cremers, 1973, 1974). After the leader is damaged or the plant is dislodged from its support, additional resting buds are released. In Panama, Putz (1984, p. 1722) reported that 90% of those lianas which fell together with their supporting trees sprouted vigorously 8–12 months after falling. Thus, few lianas died when their host trees fell. In recent treefall gaps (2–10 years old), 55% of independently climbing stems sprouted from fallen lianas. Appanah & Putz (1984) found resprouting in half the climbers which were on trees felled *c.* 6 years previously. They also reported that 33% of climbers cut from trees had resprouted within 3 months of cutting. In addition, 41% of stem pieces 1 m in length had sent out shoots and/or roots 3 months after cutting.

Root climbers among the Araceae, Cyclanthaceae, Piperaceae, Marcgraviaceae, Araliaceae, etc. regularly produce fine, holdfast roots along their stems. Thicker and longer aerial roots, so-called 'feeder roots', are produced additionally by some Araceae, Araliaceae, and Clusiaceae. Damage to the stem often allows the separated regions to grow independently as ramets.

The intact stems of many non-root-climbers also produce roots easily. Natural air layers or marcottages form when the stems, especially the nodes, touch the soil or moist forest litter. In old stems of numerous genera and families, this rooting can be detected in the field only after exposure of the buried base of what might otherwise appear to be the base of an old seedling axis (Caballé, 1977, 1980, 1986). Such widespread rooting permits extensive invasion and enlargement of an individual as noted by Hegarty & Caballé (Chapter 11). Vegetative propagation of a plant as separately rooted ramets can occur after the older stems die or are damaged. In a West African forest transect, Caballé (1986, p. 241) found that 90% of all lianas arose from adventitious roots rather than from seedling roots.

Damaged stems frequently produce roots at their proximal ends. Thin aerial roots literally 'rain down' from severed, succulent stems of *Cissus*. Some species are pests in orchards when rapid rooting defeats attempts at control by stem cutting. The lengths of aerial roots from the proximal end of a cut stem of *Cissus sicyoides* L. growing in Miami were measured over 6 h starting at 8 a.m. Average growth was an amazing 8.3 mm h^{-1} in one root. Like most adventitious roots, they are initiated at the vascular cambium.

Their initiation appears to be correlated with the higher levels of endogenous auxin at the basal ends and lower sides of stems.

Many climbers form large tubers which store water, carbohydrates, or both (see Mooney & Gartner, Chapter 6). As in tubers generally, the parenchymatous storage tissue is usually achieved by anomalous secondary growth. Species of Dioscoreaceae, Fabaceae, Cucurbitaceae, and Convolvulaceae form stem or root tubers which are of economic importance. Tubers serve as perennating organs after seasonal shoot die-back or after trauma caused by fire, drought, or herbivory. There is a considerable literature on wound healing and periderm formation in tubers of *Ipomoea* (see Walter & Schadel, 1983) and *Dioscorea* (Passam, Reed & Rickard, 1976; Knoblock *et al.*, 1989). New shoots may arise from pre-existing lateral buds at the hypocotyl node or shoot base. Adventitious shoot buds can arise from root tissue, especially from the distal ends of severed roots in *Ipomoea* and in several weedy *Convolvulus* spp.

Functional significance of anomalous stem structure

Early workers were struck by the frequency of anomalous stem anatomy in lianas and speculated on the adaptive significance of such atypical structure. Schenck (1893), Haberlandt (1914), and most later workers hypothesized that the different anomalous arrangements of secondary tissues greatly increase the mechanical flexibility of liana stems and prevent their breakage when stems are twisted, coiled, or bent. This has now been documented objectively by biomechanical experiments (see Putz & Holbrook, Chapter 3). Schenk and Haberlandt felt that unequal secondary growth results in flattened or fluted stems which might better lodge in supports. Haberlandt (1914) and Carlquist (1975) suggested that some anomalies offer mechanical protection to the soft phloem tissue when it is embedded within the xylem. Carlquist (1988) also felt that anomalies that result in phloem being distributed amongst xylem facilitate nutrient transport within the stem. Crüger (1850, 1851) suggested that living parenchyma tissues stay alive for a long time and aid in stem flexibility. Gentry (1985) noted that Acanthaceae (*Mendoncia*), Aristolochiaceae, Asteraceae, Cucurbitaceae, and Polygalaceae (*Securidaca*) all have complex anatomy and 'are among the softest-wooded and most flexible of all Panamanian lianas'. More recently, Dobbins & Fisher (1986) hypothesized that anomalous anatomy plays a role in promoting the healing of injured stems. In addition, the arrangement of cambia, and therefore young xylem and phloem, offers a degree of redundancy that could be significant when the vascular system is partly destroyed. They offered support for these views by observing the responses of liana stems after damage by experimental bark girdles. Carlquist (1988; p. 276) questioned whether the regeneration capacity shown by the experiments of Dobbins & Fisher (1986) would be effective after a stem falls, since 'such a liana could probably not be restored to a reproductive state (again growing to a canopy position) easily even if its conductive system were intact'. We hope to show in this chapter, and in more recent experimental studies (Fisher & Ewers, 1989), that lianas definitely can and do survive traumatic events and even flourish after major injuries, in part by means of flexible and quickly healing stems.

Anatomy and taxonomic distribution of anomalous stem structure

The diversity of stem structure among climbing plants is as great as the systematic position of these plants in a wide diversity of families and genera. Most vines have an arrangement of vascular tissues in the stem that is noticeably different from related plants.

In the stems of climbing monocotyledons, such as rattan palms, *Dioscorea*, *Smilax*, etc., separate vascular bundles are distributed throughout the stem. Because there is no secondary growth, no new vascular tissues form as the stem ages, except possibly at the nodes when lateral branches grow out. The late metaxylem vessels are wide, especially when compared to related, non-climbing species (Klotz, 1978). Wide vessels are far more efficient in carrying water than narrow vessels, on theoretical grounds (Zimmermann, 1983), and empirical observations of stem hydraulic conductivity support this view (see Ewers *et al.*, Chapter 5).

In dicotyledonous climbers, a cylindrical core of wood (secondary xylem) is usually surrounded by secondary phloem, similar to the stems of most herbs and woody trees. There is a continuum from thin-stemmed, herbaceous vines with limited lignification of secondary tissues to thick-stemmed, woody lianas. In transverse section the organization of tissues follows the typical sequence of central pith, xylem, vascular cambium, peripheral phloem, and periderm, e.g. Vitaceae (*Vitis*), Rubiaceae (*Uncaria*), Annonaceae (*Artabotrys*), Fabaceae (*Caesalpinia*), Hippocrateaceae (*Hippocratea*), etc. However, the vessels in the stems of most of these species are wider than in related non-climbers (see Ewers *et al.*, Chapter 5).

In a number of genera and species, the arrangement of tissues does not follow the typical pattern and has been refered to as 'anomalous'. Such anomalous (or 'cambial variants' in chapter 2) stem anatomy, especially lianas, has long been noted and variously classified (Schenck, 1893; Pfeiffer, 1926; Obaton, 1960; Carlquist, 1988). However, we know of only two quantitative reports on the frequency of anomalous stem structure in vines. Although Bamber (1984) states that approximately 10% of vines in Queensland had anomalous structure, he presents no supporting data. On the other hand, Caballé (1986, Tables 19, 20) surveyed 225 species from 35 families in the rainforest of Gabon. He found that 42% could be classified as having fragmented wood or wood with included phloem, based on his macroscopic observation of dried stem surfaces. If we transfer those species having highly lobed xylem, which he grouped with the entire (i.e. typical) wood class, into the anomalous class, then 56% of the liana species in this geographical region have some sort of anomalous stem structure. Other floristic surveys for frequency of anomalies are needed.

Some climbers show the same unusual anatomical features as the non-climbing species in the same family. For instance, secondary phloem is included within the secondary xylem of *Bougainvillea* and *Pisonia* of the Nyctaginaceae, and internal primary phloem from the bicollateral vascular bundles occurs in Apocynaceae, Asclepiadaceae, Convolvulaceae, Cucurbitaceae, and Solanaceae. Anastomosing phloem strands within the secondary xylem are also common in these same families. Other climbers have distinctive stem anatomy that can serve as a field characteristic for a family (multiple vascular cambia in Gnetaceae, Combretaceae, and Menispermaceae) or a genus ('cross vines' with phloem wedges in *Bignonia* and relatives, multiple steles of *Serjania* and *Paullinia*, flattened or lobed stems of *Bauhinia*).

We will examine the phenomenon of stem repair in vines that can be placed into one of the following broad categories of anomalous structure. These categories are used for convenience of data presentation and review; they do not necessarily correspond to the more detailed and comprehensive classification schemes given by Carlquist (1988 and Chapter 2) and earlier workers. Our categories are not mutually exclusive: *Coccinia* has conspicuously wide rays and also has internal and included phloem; stems with lobed xylem may eventually produce supernumerary cambia. The basic arrangement of tissues is described briefly for each category. Detailed examples are given in the next section of this chapter.

1. Typical arrangement of secondary vascular tissues. Cambium forms a complete circle in cross-section and produces secondary xylem to the inside and secondary phloem to the outside.
2. Secondary tissues typical and internal primary phloem. The primary vascular bundles are bicollateral with primary phloem on the interior and exterior sides of the primary xylem. Secondary growth is typical with the original primary phloem remaining between the pith and the xylem.
3. Included secondary phloem. The 'wood' is formed by a cylindrical 'cambial zone' which has been called a primary thickening meristem or successive cambia (see discussion in Carlquist, 1988). Small islands of phloem are produced by a short-lived cambium after an island of xylem is produced. This is repeated with an intervening conjunctive tissue of

fibers and parenchyma. Altogether the tissues form a woody core that is a mixture of xylem and phloem.

4. Relatively wide unlignified rays. The vascular cambium is typical in shape but wide rays are maintained by the intervasicular regions of cambium. The lignified regions of secondary xylem and the sieve tube regions of secondary phloem are discrete extensions of the original primary vascular bundles.
5. Lobed xylem with initially continuous cambium. The vascular cambium is continuous but is lobed or convoluted in cross-section. Secondary xylem and phloem are formed in the typical manner.
6. Lobed xylem with disjunct cambium. Arcs of an originally typical vascular cambium become physically disjunct when these regions produce more phloem than xylem. Wedges of secondary phloem form at these sites and divide the otherwise typical secondary xylem.
7. Separate vascular cylinders. The original primary stem has a central vascular bundle and several cortical bundles. Each vascular bundle develops its own vascular cambium, each of which behaves like a typical vascular cambium. As a result, old stems have separate cylinders of secondary xylem and phloem connected by a common matrix of cortex, phloem and periderm.
8. Supernumerary cambia. Older stems initiate new vascular cambia (also called successive cambia: discussed in Carlquist, 1988) in the outer regions of secondary phloem formed previously. Each additional cambium behaves like a typical cambium. Thus, the old stem has concentric rings of alternating xylem and phloem.

Responses to stem injury

Observations of stem healing after natural injuries were made on plants growing in Michigan (East Lansing), in Mexico (Jalisco), and in forests of Singapore and Malaysia (Johor, Pahang, Sabah, and Sarawak) as noted in the figure captions. If no location is given, the plants were growing in Miami, Florida. Stem samples were either fixed in FAA (formalin, alcohol, acetic acid), sectioned, and stained with safranin and fast green, or they were dried and later cut, sanded, and coated with polyurethane resin. We present these new observations below together with previously published information. All figures presented are original, except for Figure 4.7. The full nomenclature is noted for an original observation, but only the genus is given when we refer to examples from the literature.

Climbing monocotyledons all lack secondary growth, e.g. aroids, rattan palms like *Calamus, Luzuriaga* (Philesiaceae), *Smilax* (Smilacaeae), *Asparagus* (Liliaceae). Stem injuries heal without new vascular tissues being formed.

A type of wound periderm may form on the surface of injury by a suberization of parenchyma cells or their limited subdivisions that produce a storied cork (Esau, 1977). Experimental cuts into the stem of *Gloriosa*, a climbing Liliaceae, caused surface cells to enlarge radially and divide once. Parenchyma cells adjacent to the cut surface became sclerified (Swamy & Sivaramakrishna, 1975). Damaged stems of *Calamus* sp. observed in Malaysia were still alive and appeared healthy although about two thirds of the stem was destroyed. Thus, the remaining one-third of vascular bundles, each with xylem and phloem, was sufficient to bridge the wound site and supply the leaf crown. Scraped stem surfaces of *Smilax auriculata* Walt. in Florida had dark red-brown staining of all tissues adjacent to the damage, including cortex and outer vascular bundles. Both the xylem and phloem regions of the affected bundles were occluded with a dark, resin-like substance. Wide vessels in water mounted sections were filled with a mucilage-like substance. No periderm or subdivided parenchyma cells were found.

Climbing dicotyledons have secondary growth in which new vascular tissues are produced from one or more vascular cambia and periderm is produced by a cork cambium or phellogen. Herbaceous vines usually have a limited amount of secondary growth and little or no periderm. Lianas, which have more secondary growth, range as widely as do trees in bark characteristics. The bark or periderm of lianas can be green with a very thin layer of cork (phellem) as in many Fabaceae and Convolvulaceae or it may form thick and highly sculptured cork as in some Aristolochiaceae.

An immediate response to stem breakage or cutting in many species of monocotyledons and dicotyledons was a flow of latex, mucilage, or resin that can seal the surface. Secretion may be particularly important in fending off herbivores and in repairing insect damage. With time, wide vessels were sometimes occluded with unidentified gum or resin-like deposits, or by tyloses, but this varied widely with the species. Other structural responses to injury are reviewed in the following eight categories of vascular tissue arrangement.

Typical arrangement of secondary tissues

In *Actinidia arguta* (Siebold & Zucc.) Planch ex Miq. (Actinidiaceae), *Embelia* sp. (Myristicaceae) (Figure 4.1*A*), and *Campsis radicans* (L.) Seem. ex Bur. (Bignoniaceae) (Figure 4.2*A*), old sites of natural bark damage were partly healed by the outgrowth of new xylem, phloem and bark derived from the vascular cambium at the edges of the wound in a manner similar to stem healing in most woody plants. The exposed wood surface was discolored and had tyloses and deposits as in heart wood.

In old stems of *Vitis munsonia* Simpson ex Planc. (Vitaceae), natural

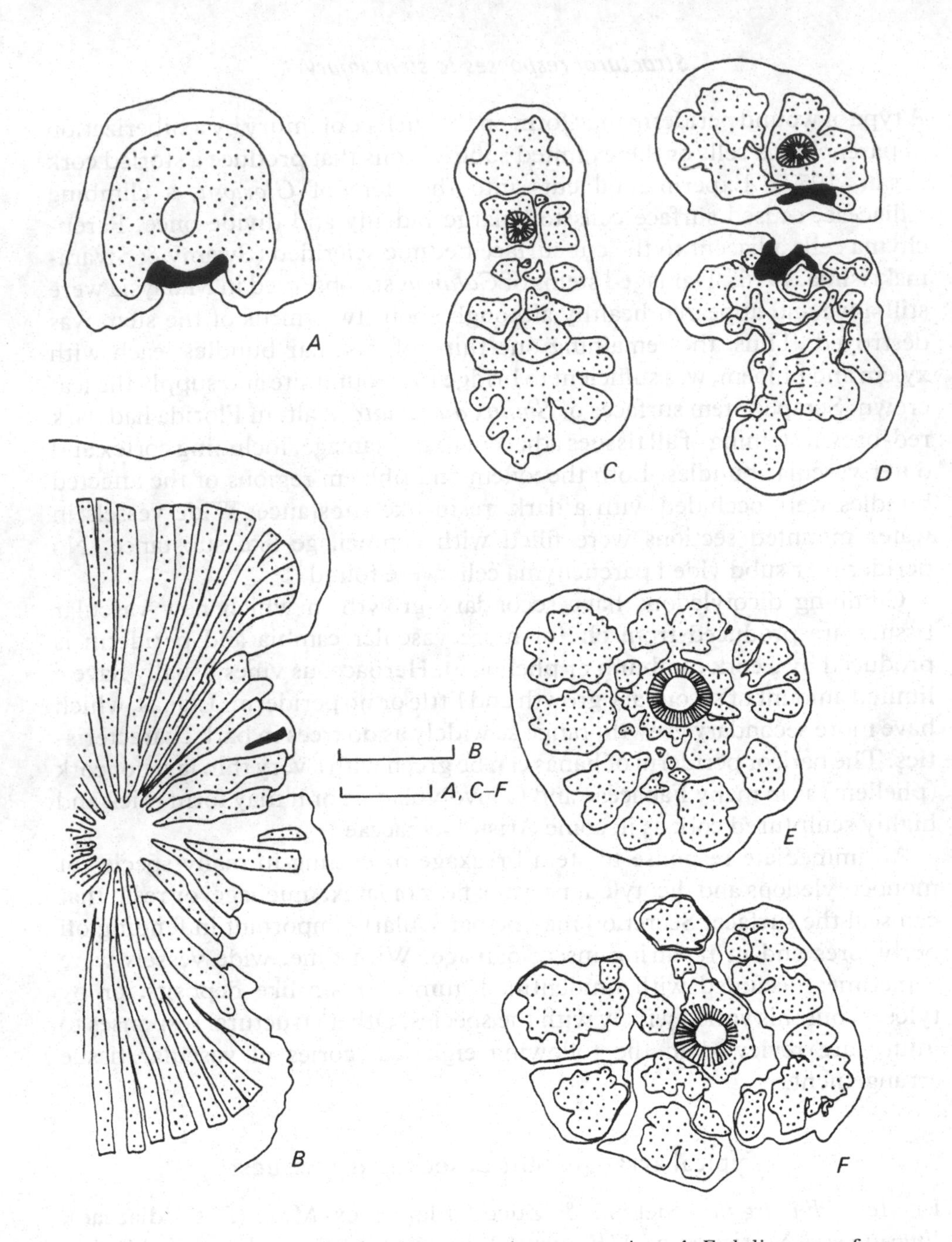

Figure 4.1. Liana stems injured in nature; in cross-section. *A*, *Embelia* sp., surface damage; wild collected in Johor. *B*, *Tinomiscium petiolare*, surface damage; wild collected in Singapore. *C*, *D*, *Bauhinia kochiana*, intact region and split stem with two segments taken 11 cm away, respectively; wild collected in Johor. *E*, *F*, Unidentified Malpighiaceae, intact undamaged region and crushed part of stem split into six segments taken 30 cm away, respectively; wild collected in Sabah. Entire stem shown in Figure 4.5*A* before sectioning. Solid area = dead xylem; stippled area = xylem; hatched area = dense inner xylem surrounding pith. Scale lines = 1 mm.

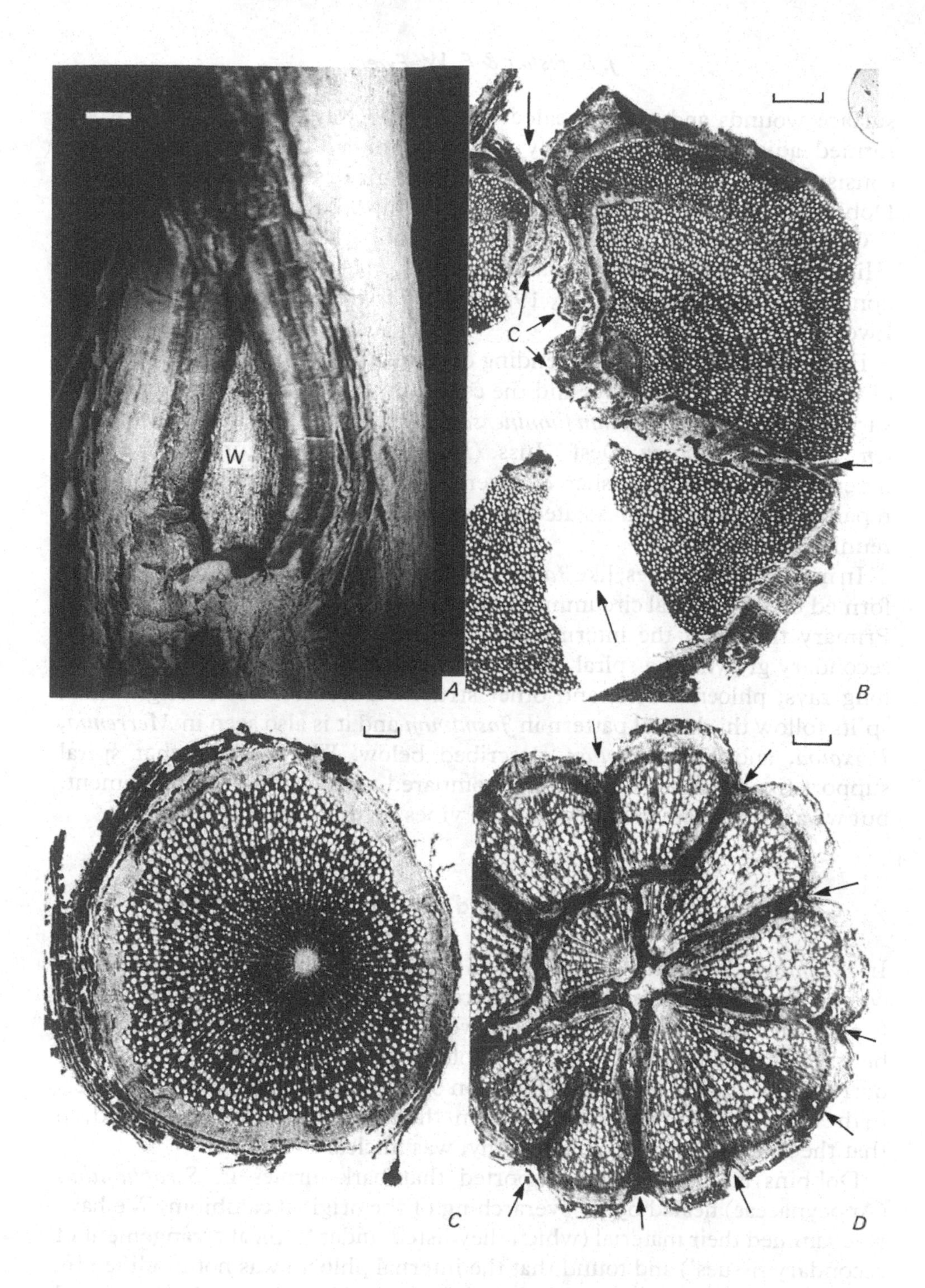

Figure 4.2. Liana stems with natural and experimental injuries. *A*, *Campsis radicans*, surface view of partly healed bark damage; wild collected in Michigan. *B*, *Jasminum fluminense*, cross-section of split stem 14 weeks after twisting. *C*, *D*, *Stigmaphyllon*, undamaged region and region of stem splitting taken 20 cm away, respectively. w = exposed surface of old wood; c = callus; arrows = sites of original splits or separation of bark from wood. Scale lines = 1 mm.

surface wounds and cracks healed in a similar way and discolored wood formed adjacent and proximally to the wound. These observations are consistent with the healing of artificial bark girdles in *Vitis* described by Dobbins & Fisher (1986) and Sidlowski, Phillips & Kuykendall (1971).

Other genera with similar typical healing of bark girdles are *Hippocratea* (Hippocrateaceae), *Passiflora* (Passifloraceae), *Podranea* and *Schlegelia* (Bignoniaceae) (Dobbins & Fisher, 1986), and *Jasminum* (Oleaceae) (Fisher & Ewers, 1989).

Experimental twisting and bending of stems caused longitudinal splitting of wood and bark along rays and the cambium. Parenchyma proliferated to varying degrees, in *Jasminum fluminense* Vell. (Figure 4.2*B*) and *Stigmaphyllon* cf. *periplocaefolium* (Desf.) Juss. (Malpighiaceae) (Figure 4.2*C*, *D*), as described in detail by Fisher & Ewers (1989). In the latter example, the repaired stem region had isolated sectors of new xylem and phloem within a reunited stem.

In many twining vines, like *Jasminum*, the stem has a gradual twist that was formed by the original circumnutation or searching movement of the leader. Primary tissues of the internode form a helix that is maintained during secondary growth as a spiral grain in the wood and helical arrangement of long rays, phloem fibers, and other structural features. The longitudinal splits follow this helical pattern in *Jasminum* and it is also seen in *Merremia*, *Peixotoa*, and *Stigmaphyllon*, described below. We presume that spiral support tissues increase flexibility, compared with a straight arrangement, but we are not aware of any work with vines on this topic.

Secondary tissues typical and internal primary phloem

In *Manihot chlorostica* Stanley & Goldman (Euphorbiaceae), a natural wound was healed by cambial growth from the edges of the wound, similar to healing in typical woody stems. However, wound periderm was present between the dead wood and the living old xylem. The internal periderm was derived presumably from a proliferation of xylem parenchyma. Most vessels in the old wood had tyloses. In this stem, the entire pith region had rotted, so that the role of internal phloem, if any, was unclear.

Dobbins & Fisher (1986) reported that bark girdles in *Strophanthus* (Apocynaceae) healed by an overarching of the original cambium. We have re-examined their material (which they listed under 'typical arrangement of secondary tissues') and found that the internal phloem was not modified by bark girdles. Girdles and injury from twisting and bending in *Aganosma* and *Allamanda* (both Apocynaceae) were repaired by new tissues growing over the wounds from the edges of the original cambium. Internal phloem showed no structural modification (Fisher & Ewers, 1989).

Included secondary phloem

In *Bougainvillea* (Nyctaginaceae) both girdles and injuries from twisting and bending healed as in species with typical arrangement but at a slower rate. New 'wood' with separate vascular bundles grew over the wound from the 'cambial zone' at the edges. Bark that was separated from the 'wood' reunited with the stem after some proliferation of the 'cambial zone'. Old wood and included phloem were unaffected. The stems were brittle and easily broke transversely when bent (Fisher & Ewers, 1989).

In *Pisonia* (Nyctaginaceae), the stems were also brittle. However, girdled and split stems developed callus from the parenchyma lenses in the 'wood' near an exposed surface or crack (Fisher & Ewers, 1989).

Relatively wide unlignified rays

In *Tinomiscium petiolare* Miers. (Menispermaceae), a deep wound to the side of a thick stem was partly healed by the proliferation of xylem rays (Figure 4.1*B*). The peripheral ends of the rays expanded, and callus overarched the adjacent exposed xylem which was darkened and dead. Many vessels had tyloses. The severed cambium at the edges of the wound slightly overgrew the damaged xylem.

In *Aristolochia* (Aristolochiaceae), a species with wide rays, bark girdles caused some ray cells on the surface to proliferate but had few other effects. Stem twisting and bending caused bark separation and ruptures along the rays (Figure 4.3*A*, *B*). These injuries healed by parenchyma proliferation and union of the separated tissues (Fisher & Ewers, 1989). However, in *Mikania cordifolia* (Asteraceae), girdles caused a major proliferation of all rays, new arcs of cambium, and new xylem and phloem throughout the stem (Dobbins & Fisher, 1986).

In *Cayratia* (Vitaceae), a succulent-stemmed species with wide rays, bark girdles caused rapid proliferation of ray parenchyma (Figure 4.3*C*). Twisting and bending caused stems to split, often completely through the pith, and the bark to separate at the cambium. Ray and pith parenchyma proliferated (Figure 4.3*D*). New vascular tissue formed within the partly reunited strips of bark. New radially aligned cambia in the rays produced files of xylem and phloem that were perpendicular to the old xylem (Fisher & Ewers, 1989), similar to what is described below in *Tetrastigma*.

In *Coccinia* (Cucurbitaceae), a succulent-stemmed species with wide rays, the stems have wide rays and also internal phloem and included or interxylary phloem (Figure 4.4*A*). Bark girdles and longitudinal stem splits from twisting and bending caused great proliferation of all ray parenchyma (Figure 4.4*B*). New arcs of cambium formed in rays and between the internal phloem and pith. Some new vessels were produced toward the pith center.

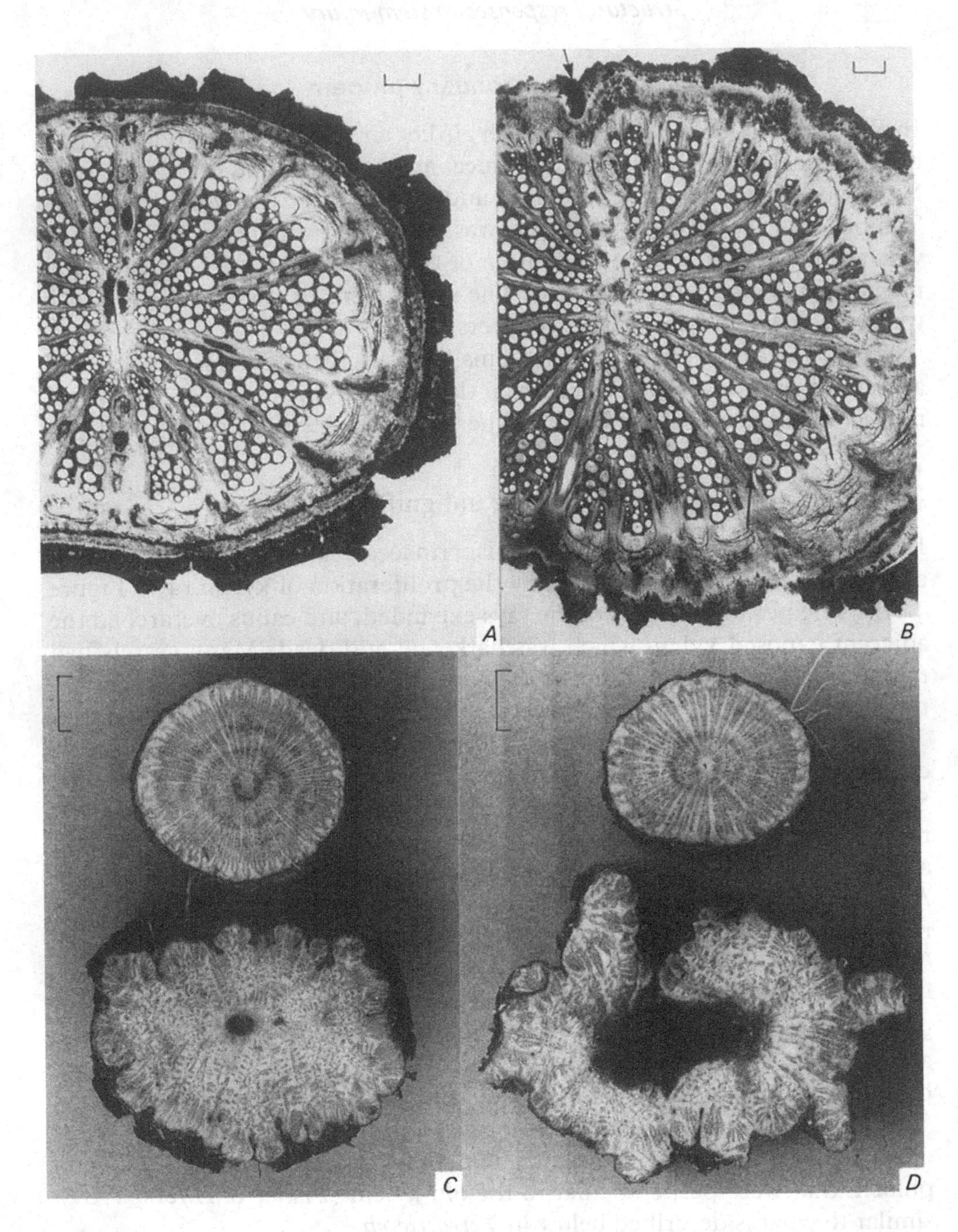

Figure 4.3. Liana stems; cross sections of stem 14 weeks after experimental bark girdle and twisting injury. *A*, *B*, *Aristolochia maxima*, cross-section of undamaged region and region in which the xylem split and the bark separated when twisted taken 20 cm away, respectively. *C*, *D*, *Cayratia trifolia*, two different stems showing undamaged region (top) and injured region (bottom) taken 20 cm away for bark girdle and twisted, split stem, respectively. Arrows = sites of original splits or separation of bark from wood. Scale lines = 1 mm in *A* and *B*; 1 cm in *C* and *D*.

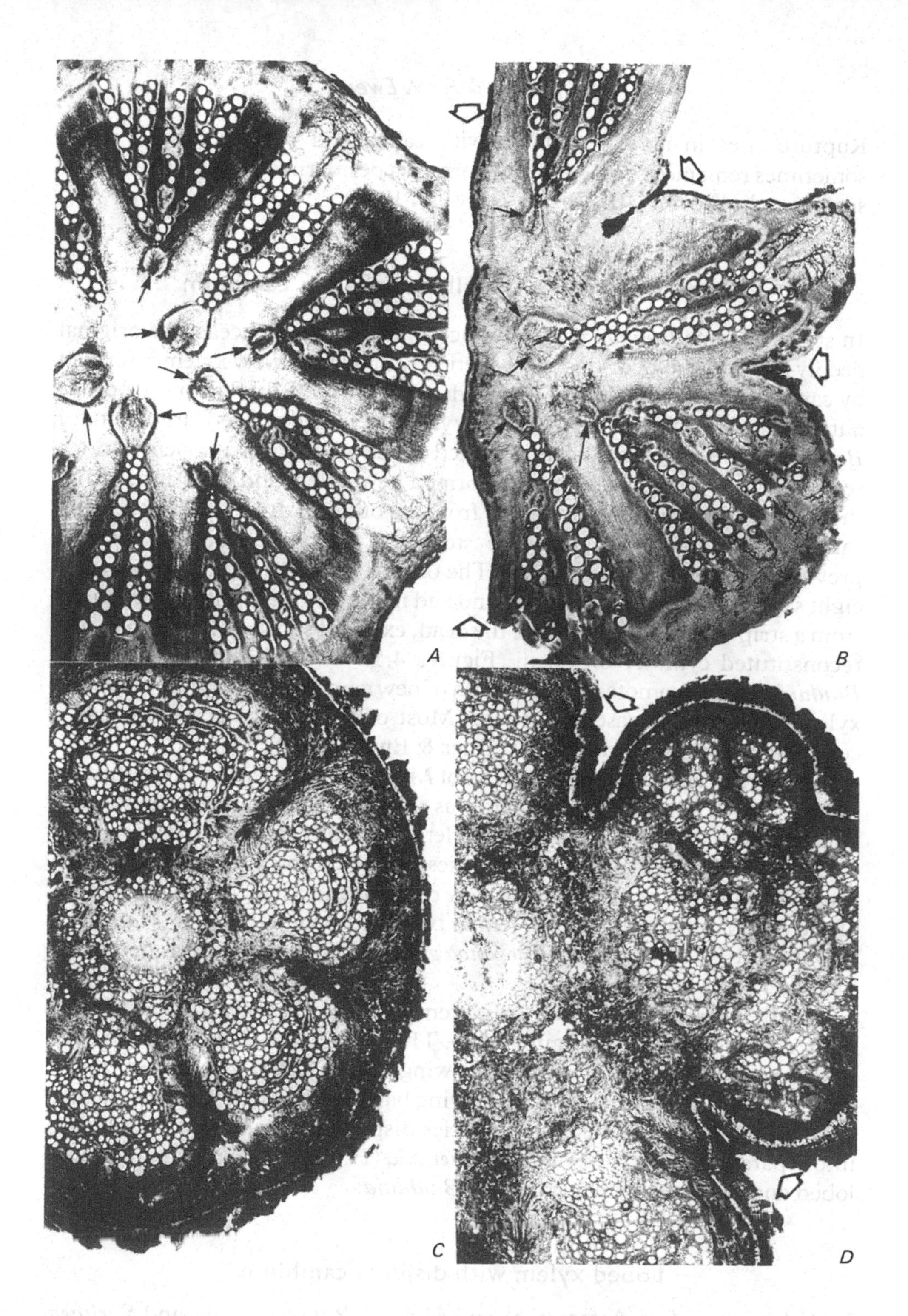

Figure 4.4. Liana stems; cross-sections of stem 14 weeks after experimental twisting injury. *A*, *B*, *Coccinia grandis*, cross-section of undamaged region and section showing one of two segments from the twisted stem region taken 20 cm away, respectively. *C*, *D*, *Peixtoa glabra*, intact undamaged region and region of stem splitting taken 20 cm away, respectively. Small arrows indicate internal phloem; large arrows indicate sites of stem splitting. Scale lines = 1 mm.

Rupture sites in rays were filled with callus and reunited. Split stems sometimes remained segmented and formed a periderm on the exposed inner surface as in Figure 4.4*B* (Fisher & Ewers, 1989).

Lobed xylem with initially continuous cambium

In stems of some *Bauhinia* spp. (Fabaceae) and Malpighiaceae, the original deeply lobed xylem becomes ontogenetically dissected. New xylem is formed by cambial activity between the inner dense xylem ring and the more porous outer xylem and within the rays of old xylem. Several damaged large stems of *Bauhinia kochiana* Korth. were collected in the wild. Parts of xylem became separated near a lateral wound and formed new xylem and bark either from the exposed xylem parenchyma or from an outgrowth of the pre-existing cambium (Figure 4.1*C*, *D*). One stem had been crushed and twisted previously, probably by a tree fall. The original stem was split into seven or eight separate strands, and each strand had new xylem and phloem produced from a strip of cambium opposite the dead, exposed surface of old xylem or a reconstituted cylinder of xylem (Figures 4.5*B*, 4.6*A*, *B*). Bark girdles in *Bauhinia vahlii* promoted the initiation of new cambia that split the old lobed xylem into separate vascular groups. Most of these vascular groups had a completely encircling cambium (Fisher & Ewers, 1989).

The stem of an unidentified species of Malpighiaceae had been crushed by a large branch fall. Its original stem was split into six separate strands, each with its own complete or nearly complete cambium and new bark (Figures 4.1*E*, *F*, 4.5*A*). Bark girdles and injuries from twisting and bending caused great proliferation of parenchyma, new cambia, and a general fragmentation of the original stem vascular system in two other Malpighiaceae: *Mascagnia* (Dobbins & Fisher, 1986) and *Peixotoa glabra* (Figure 4.4*C*, *D*) as described in Fisher & Ewers (1989).

Caballé (1986) illustrated a split stem of *Condylocarpon* (Apocynaceae) with highly lobed xylem (Figure 4.7*A*). The exposed surfaces appear to have new bark in these macroscopic drawings. In *Dalhousiea* (Fabaceae), a damaged stem is shown with the surviving half producing new xylem (Figure 4.7*B*). Caballé also showed many species displaying a natural stem splitting into separate strands. Of note is *Loeseneriella* (Hippocrateaceae), which has a lobed and dissected xylem similar to *Bauhinia*.

Lobed xylem with disjunct cambium

In *Mikania scandens* (Asteraceae), and *Mansoa*, *Pithecoctenium*, and *Saritaea* (all Bignoniaceae), bark girdles caused the disjunct cambia protected at the

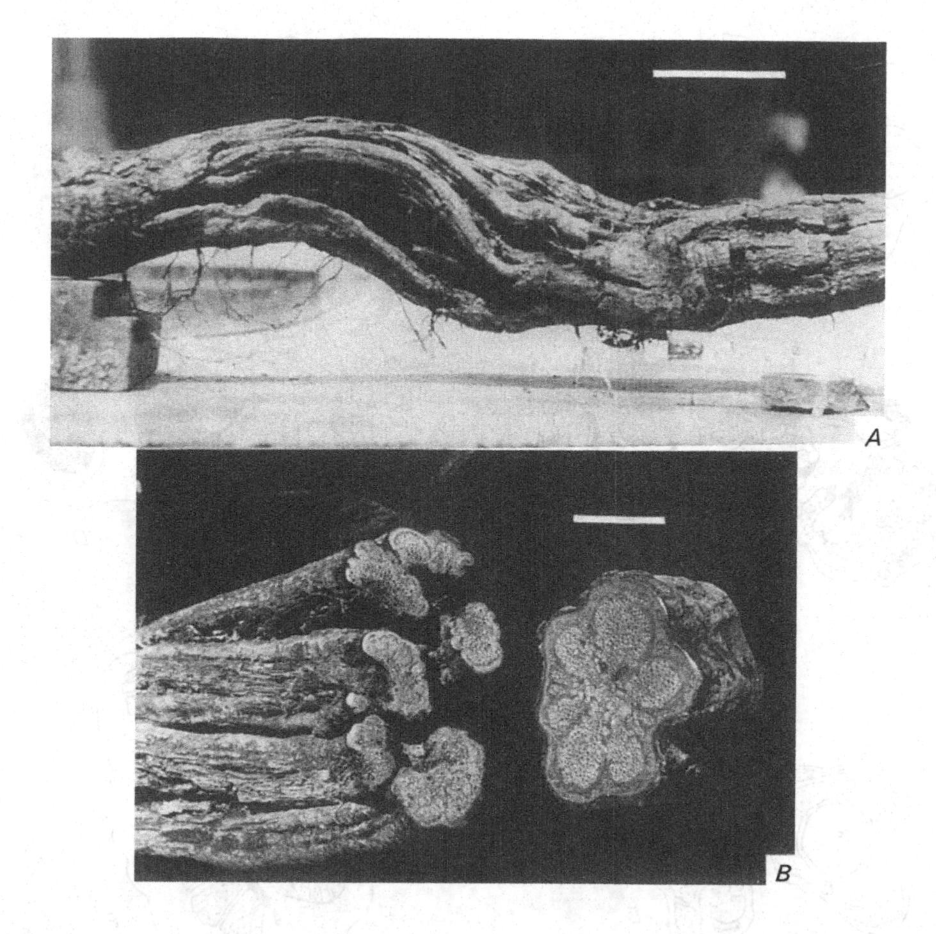

Figure 4.5. Liana stems injured in nature. *A*, Crushed region of an unidentified Malpighiaceae, wild collected in Sabah shown in cross-section in Figure 4.1 *E*, *F*. *B*, *Bauhinia* sp., split and intact regions of same stem; wild collected in Johor; also shown in Figure 4.6*A*, *B*. Scale lines = 5 cm in *A*; 10 cm in *B*.

interior side of the four phloem wedges to produce much callus and heal over the exposed wood (Dobbins & Fisher, 1986). Twisting and bending of *Pithecoctenium* split the xylem and bark longitudinally at sites of the phloem wedges in the xylem furrows. Both the disjunct cambium and the outer cambium segments contributed to healing (Fisher & Ewers, 1989).

In *Macfadyena unguis-cati* (L.) A. Gentry, eight phloem wedges developed in old stems (Figure 4.8*C*). A natural injury showed that the stem split into three strands (Figure 4.8*A*, *B*). Each strand developed a complete vascular cambium which produced new secondary xylem and phloem (Figure 4.8*C*).

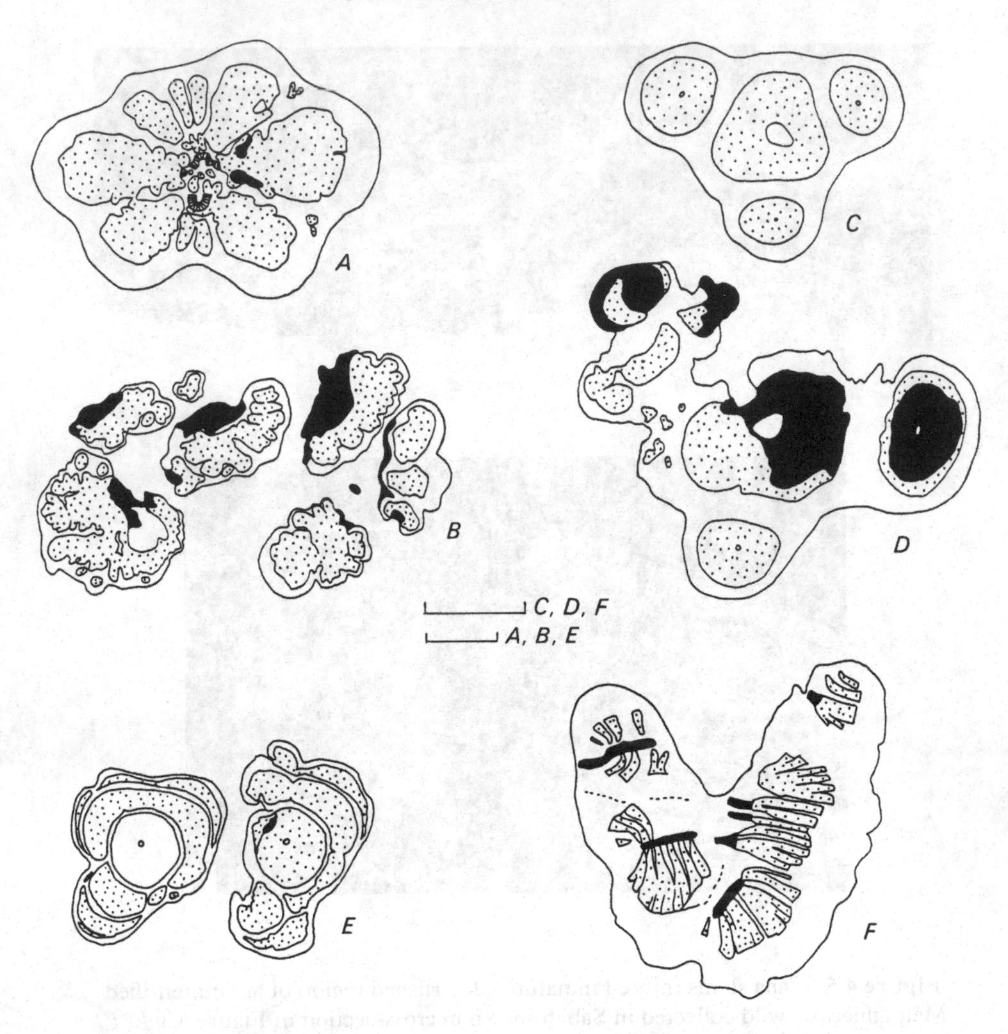

Figure 4.6. Liana stems injured in nature; in cross-section. *A*, *B*, *Bauhinia* sp., undamaged region and crushed bent stem region taken 52 cm away, respectively; wild collected in Johor; stem shown in Fig. 4.5*B*. *C*, *D*, *Serjania brachycarpa*, undamaged region and damaged region at base of split taken 21 cm away, respectively. Entire stem shown in Figure 4.8 *D* (right) before sectioning; wild collected in Jalisco. *E*, *Rourea mimosoides*, undamaged region (left) and surface injury (right) taken 8 cm away; wild collected in Singapore. *F*, *Tetrastigma* sp., one part of an old, hollow, and dissected stem shown in Figure 4.9*A*; wild collected in Pahang. Solid area = dead xylem; stippled area = xylem; hatched area (in *A* only) = dense inner xylem surrounding pith. Scale lines = 5 mm.

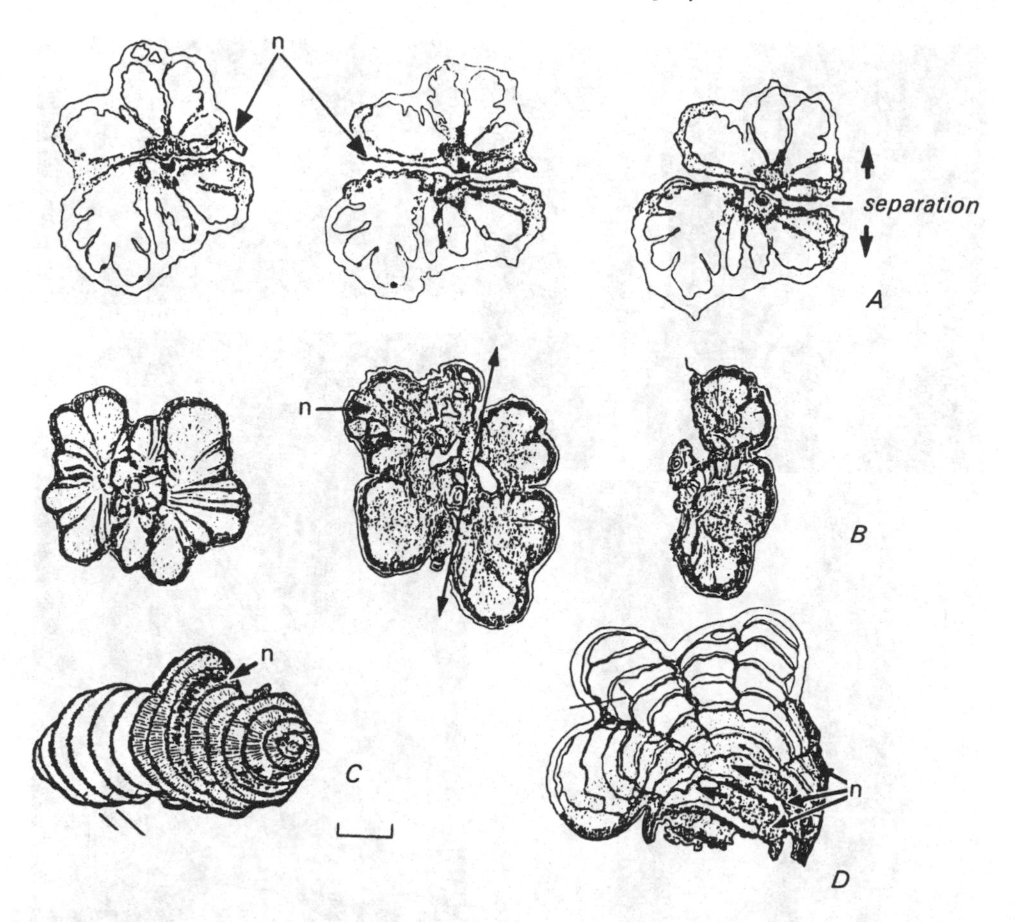

Figure 4.7. Liana stems injured in nature; cross-sections of dried stems; wild collected in Gabon, West Africa. *A*, *Condylocarpon guyanense* Desf. (Apocynaceae), three levels of a stem showing splitting. *B*, *Dalhousiea africana* De Wild (Fabaceae), three levels of a stem showing (left to right) healthy intact region, mainly necrotic stem, and only a functional half-stem. *C*, *Triclisia subcordata* Oliv. (Menispermaceae), older necrotic wood bounded by a supernumerary cambium. *D*, *Salacia* cf. *pyriformis* (Don) Steud. (Hippocrateaceae), older necrotic wood bounded by a supernumerary cambia. n = necrotic region. Scale line = 1 cm. Reproduced from Caballé (1986) with permission.

Separate vascular cylinders

In *Serjania brachycarpa* Cray (Sapindaceae), a natural crushing of the stem split the original axis into two strands with one and three vascular cylinders, respectively (Figure 4.8*D*). Much of the original old xylem was dead (Figure 4.6*C*, *D*), but the protected cambia in each of the three cortical cylinders and the central cylinder remained active.

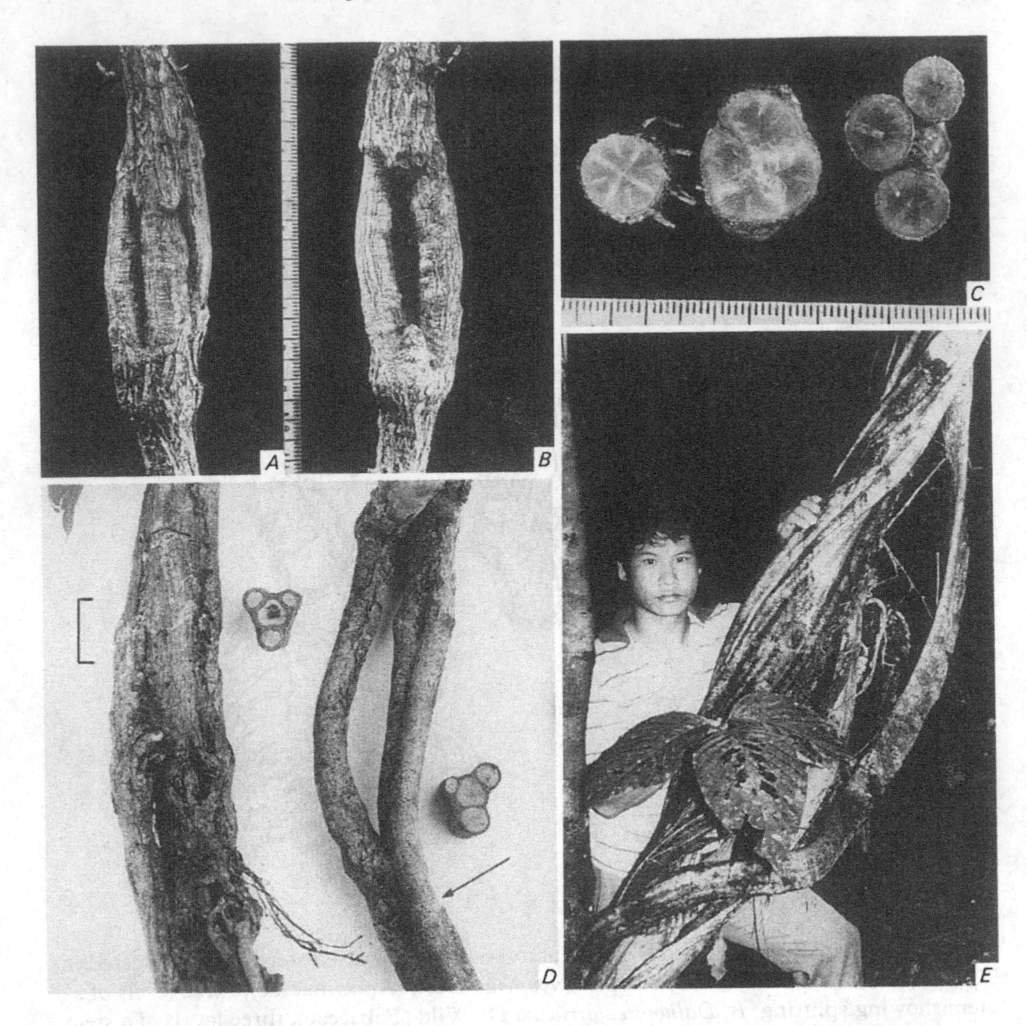

Figure 4.8. Liana stems injured in nature. *A–C*, *Macfadyena unguis-cati*. *A*, *B*, Two views of the same split stem. *C*, Cross-sections of stem in *A*; taken above injury, at upper edge, and within region of stem splitting, respectively from left. *D*, *Serjania brachycarpa*, two damaged and segmented stems with cross-sections of their adjacent intact regions. Arrow indicates location of section shown in Figure 4.6 *D*. *E*, *Spatholobus* sp., split into several strands. An assistant holds a shade leaf from this plant; wild collected in Sabah. Scale lines marked in mm in *A–C*; 1 cm in *D*.

In *S. polyphylla*, twisted and bent stems split longitudinally in a similar way. Sometimes all the four or five cylinders for this species were isolated between nodes as separate strands. Each cylinder continued to produce new xylem and phloem from its typical cambium (Fisher & Ewers, 1989). Bark girdles in this same species caused proliferation of the undamaged, inner arcs of cambium (Dobbins & Fisher, 1986).

Supernumerary cambia

In *Rourea mimosoides* Planch. (Connaraceae), a natural longitudinal stem injury left the wood exposed for over 9 cm. The two exposed cambia at the edges of the wound contributed to the healing over of each layer of exposed xylem (Figure 4.6*E*). In *Spatholobus* sp. (Fabaceae), large stems become ribbed by eccentric growth of new arcs and rings of cambium. Bent or twisted stems were found split into several segments which were longer than 1 m (Figure 4.8*E*). Here the rope-like appearance is due to the spiral course of the supernumerary cambia, presumably due to a twisting of the young axis of this twiner.

A massive, 20 cm diameter trunk of *Tetrastigma* sp. (Vitaceae) was growing on Pulau Tioman, Pahang and was infected by the parasitic dicotyledon, *Rafflesia* cf. *cantleyi* Solms-Laubach (Figure 4.9*A*). The center of the trunk had long ago rotted away, leaving only a cylinder of living tissue that was dissected into a mesh-like system of stem remains. The remaining stem had cambia on the inner and outer surfaces in some regions, apparently either the remains of two original cambial layers or an inner, additional cambium which was newly induced after the interior tissues rotted away. In cross-section each segment of the remaining stem system resembled either an entire stem with irregularly lobed cambium all around or a flattened stem segment with one side rotted away (Figure 4.6*F*). Often a new cambium formed in ray regions to produce files of new vascular tissue perpendicular to the old xylem file (Figure 4.9*B*). Bark girdles in *T. voinierianum* stimulated growth in the inner cambium which was protected by the xylem produced from the second, outer cambium in this stem (Dobbins & Fisher, 1986).

In *Merremia tuberosa* (L.) Radlk. (Convolvulaceae), bark girdles (Figure 4.9*C*) and injuries from twisting and bending (Figure 4.9*D*) stimulated ray parenchyma and arcs of cambium to proliferate. New cambia developed. Longitudinally split stems developed sectors of old xylem that were completely surrounded by new xylem and phloem (Fisher & Ewers, 1989).

Caballé (1986) illustrated stems of *Triclisia* (Menispermaceae), *Salacia* (Hippocrateaceae), and *Santaloides* (Connaraceae) with necrotic regions in the old xylem and bounded by younger vascular rings or supernumerary cambia (Figure 4.7*C*, *D*). New layers of xylem and phloem from supernumerary cambia produce external ridges or lobes.

Conclusions

Correlations with anomalous stem anatomy

Vines are particularly vulnerable to vascular interruptions, especially xylem dysfunction, because their narrow stems supply a large leaf area. Our review

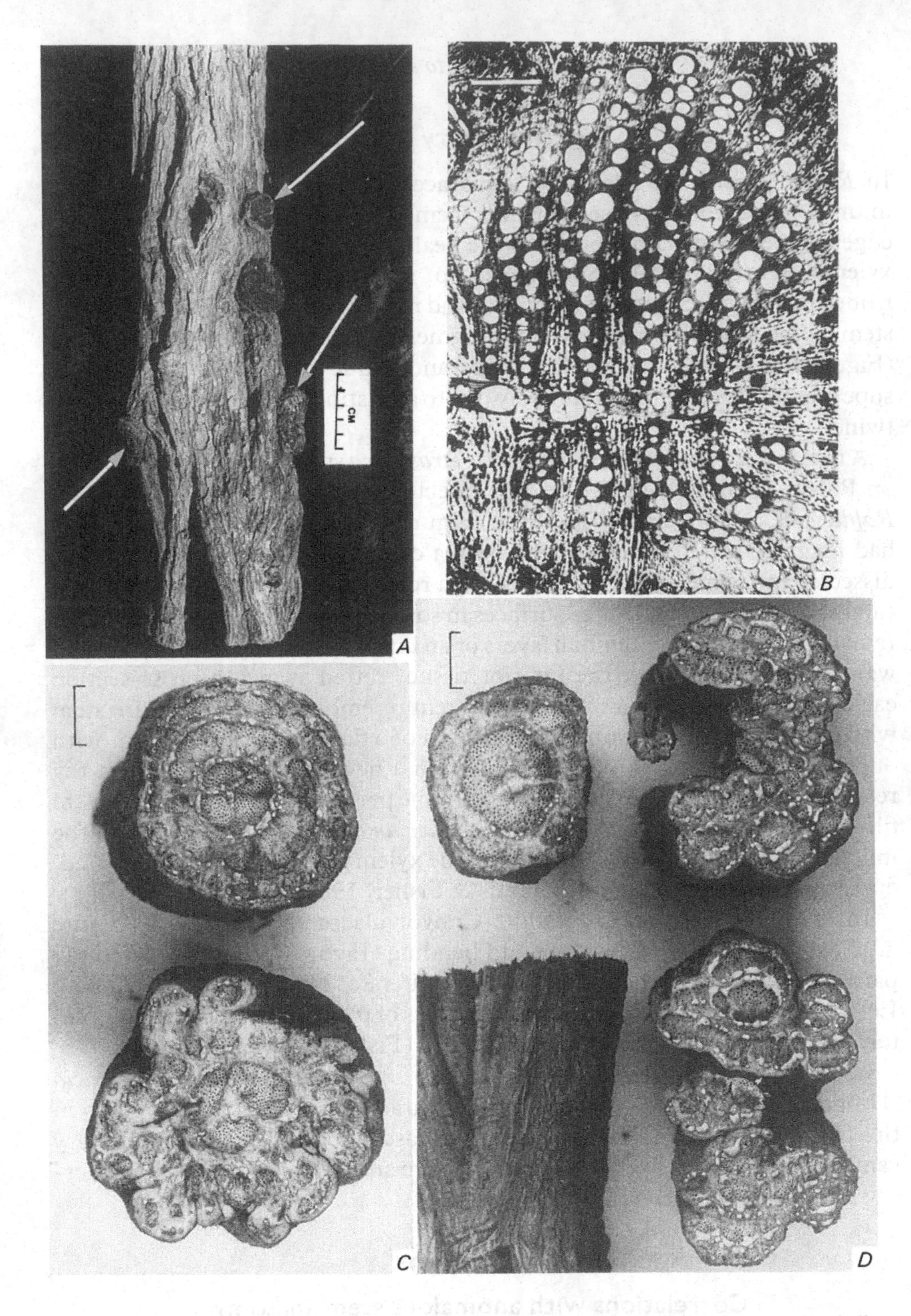

Figure 4.9. Liana stems with natural and experimental injuries 14 weeks after bark girdle or twisting injury. *A*, *B*, *Tetrastigma* sp. old stem with natural injuries. *A*, One of many strands that composed a hollow, dissected stem; wild collected in Pahang. Dark surface

of experimentally induced and natural injuries supports hypotheses that anomalous stem anatomy has two important functions: (i) various anomalies help to prevent stem breakage by increasing stem flexibility; (ii) anomalies allow orderly splitting of the stem and then rapid repair of vascular interruptions.

Caballé (1986) noted that many anomalies produce a dissected wood that is already physically compartmentalized and prepared to fight off infection in the same way that trees defend against rot by compartmentalization (Shigo, 1984). The zones of wood isolated by the living parenchyma of rays, cambium, or phloem limit the spread of infection and decay in the transverse plane. As in trees, a type of 'barrier zone' (Shigo, 1984) forms outside the injury and effectively isolates it. Adjacent old xylem is often pigmented, presumably by phenolics that tend to be antimicrobial. However, it is unclear how longitudinal spread is controlled. The long vessels are closed by tyloses or deposits in some vines. Yet other vines seem to have little structural blockage of vessels adjacent to wounds and infection appears to travel great distances, e.g. we have observed fungi-filled vessels in *Pithecoctenium*. Living, vasicentric parenchyma may well be functioning as barrier zones against infection. Generally, the mechanisms for protection against stem injury and the later healing process are both mechanical and physiological. However, any further generalizations about stem anatomy or responses to injury in vines as a life form would be misleading since the group has such structural and physiological diversity.

The stems of vines show a range in their internal tissue organization, from what is generally considered as typical for herbaceous or woody flowering plants to complex arrangements of tissues that have been termed 'anomalous'. The responses to stem injury show a similar wide range, from the typical callusing from the cambial zone at the edge of a wound and the covering or repair of injured wood by new secondary tissues, to extreme proliferation of living parenchyma cells throughout the stem. Anomalous stem anatomy is correlated with highly flexible stems and with rapid healing, or at least extensive reconnection of xylem and phloem after injury. Anomalous anatomy in many species appears to permit an orderly, longitudinal splitting of the stem rather than catastophic, transverse breakage. These anomalies involve either large areas of parenchyma (either xylem or phloem,

masses (arrows) are flower scars of the parasite, *Rafflesia*. *B*, Cross-section of one strand similar to one shown in Figure 4.6*F*; wild collected in Pahang. Files of new xylem radiate from single file of old original xylem. *C*, *D*, *Merremia tuberosa*, two different stems after experimental injury. *C* shows undamaged region (top) and full bark girdle (bottom) taken 20 cm away. *D* shows undamaged region (top left) and twisted, split region taken at two levels (right); at one level the split stem remained intact, and at another it consisted of three segments. Scale lines marked in cm in *A*; 1 mm in *B*; 1 cm in *C* and *D*.

or both in stems with multiple cambia) or sectored regions of xylem (irregular lobes, regular quadrats, or separate cylinders). In many anomalies the cambial zone (or zones) is physically protected from surface injury and remain(s) intact as longitudinal strips or arcs. *Stigmaphyllon* cf. *periplocaefolium* (Fisher & Ewers, 1989) was exceptional in having typical xylem anatomy with a slightly lobed cambium yet responding to injury by great proliferation of xylem parenchyma. Included phloem and internal phloem showed little or no histological response to stem injury, although the embedded living and presumably functional phloem remained uninterrupted.

The presence of much living xylem parenchyma and the absence of heart wood in some species of vines can be compared to sapwood trees (Ng, 1986) in which all the wood of the trunk is living sapwood. Such large forest trees are resistant to heart rot and insect damage although their wood is relatively soft and decays quickly after a tree is cut. All of the xylem parenchyma in these trees is capable of producing callus and a wound periderm when exposed (Fisher, 1981). In sapwood trees, the xylem parenchyma has important storage and regeneration functions in the same manner as in the stems of many vines.

Storage capacity

Many vine stems are succulent due to large amounts of unlignified parenchyma in the xylem rays or lenses, in the phloem and cortex, in the pith, or in concentric bands associated with multiple cambia. The living cells are storage regions for food, water, or both. The occurrence of dispersed conducting tissues, especially phloem strands, throughout the stem has implications for efficient translocation (Carlquist, 1988). Stored water may be very important to vines during midday water stress (see Ewers *et al.*, Chapter 5) or during seasonal droughts, particularly if new shoot growth occurs during such dry periods (Putz, 1984; Longino, 1986; Putz & Windsor, 1987). Carbohydrates stored in the aerial stems and underground tubers improve plant survival after stress and regeneration after fire or breakage. The storage capacity of stems and roots, as well as the morphogenetic ability to form roots and adventitious shoot buds, allow many vines to propagate vegetatively. The ability of cut stem segments to survive for long periods because of storage tissues was emphasized by Schenck (1893).

Is some anomalous structure equivalent to a wounding response?

Some of the structural oddities reported in the literature for old vine stems have identical or similar counterparts in injured stems. Crüger's (1850, 1851) descriptions of tangential proliferation of multiseriate rays which extend through the pith (his Taf. II, fig. 21 and Taf. IV, fig. 6) resemble our observations of healing after twisting and bending injuries. Fragmentation of wood in wounded narrow stems of Malpighiaceae, Convolvulaceae, and Bignoniaceae resembles the structure of thick stems of these very same species. Past workers have usually assumed that the anomalous structures of these thick stems were not the result of injury. However, Schenk (1893, pp. 36, 37) raised the question of intrinsic ('inner') or genotypic and extrinsic ('outer') or environmental factors that affect anomalous growth. He thought in terms of time of initiation and rate of anomalous growth in a stem, or the direction of eccentric growth. He noted that stems with strong torsions exhibit a stronger anomaly. We would expect that such stems also have a greater probability of injury. From our results, we suspect that some reported anomalous structures might have been modified or enhanced by earlier wounding.

The separation of the ring of dense, inner secondary xylem in *Bauhinia* and Malpighiaceae by later cambial development next to the pith in presumably undamaged stems also resembles the wound response in many species. In a detailed study of one large plant of *B. japonica* Maxim., Handa (1937) described the cleavage of the inner xylem at the stem base and the production of new vascular bundles internal and external to the inner xylem. He showed that 'dilation-parenchyma' developed from pre-existing pith and ray parenchyma with greater proliferation in older tissues toward the base of the stem. The cortex was described as unaffected, which would indicate no previous splitting injury. Although internal splitting and later healing cannot be ruled out as a cause for this anomaly, the frequent occurrence of such internal dilation in old xylem suggests a stronger correlation with stem age than with injury. Only long-term observations of stems protected from injury can definitively answer this question. We suggest that consistent anomalous organization of tissues in some species, as seen in cell proliferation and new cambial zones, can result from a 'physiological constriction' of intact stems. This postulated physiological constriction would disrupt the movement of growth regulators in the same way that a bark girdle or stem splitting does. Therefore, typical stem anatomy, anomalous anatomy, and wound response may represent a developmental continuum in some species. Further experimental studies of injuries to stems and roots and of the effects of growth regulators on vines will be needed to support our developmental interpretation.

Summary

Many vines have a well-developed capacity to heal after surface and deep-seated injuries to the stem. Dislodged and broken shoots regenerate vigorously by means of adventitious roots and shoot reiterations from either reserve buds or adventitious shoot buds. The histology of the wound responses was reviewed from the literature for monocotyledonous and dicotyledonous vines having either typical or anomalous arrangements of vascular tissues. Healing of damage to the stem caused by experimental bending, twisting, and bark girdles was described. Unpublished, original observations on the healing of liana stems with naturally occurring injuries were given for 11 genera in 9 families. Anomalies such as wide rays, much xylem parenchyma, lobed xylem, disjunct cambium (= phloem wedges), supernumerary cambia, and multiple steles participated in extensive callus formation and rapid healing of damaged vascular tissues. Included phloem participated in healing to a lesser degree. Although vines have a common life form, they vary widely in their stem structure. Because of this histological diversity, it is difficult to generalize about responses to injury in vines or to compare wounding in vines with other vascular plants. It appears that stems with anomalous anatomy diverge in their wound response from those with typical structure, both for vines (Dobbins and Fisher, 1986; Fisher and Ewers, 1989) and nonvines (Shigo, 1984; literature cited in Dobbins and Fisher, 1986). Unfortunately, we found no published information on the wound responses in non-vines with anomalous secondary growth. We concluded that the promotion of repair and prevention of total vascular disruption after injury are two of several adaptive advantages for anomalous stem anatomy in vines.

Acknowledgements

We thank Rolf Rutishauser for help in translating Crüger (1850, 1851), Sigfried Fink for help with Schenck (1893), Kai Larsen for identifying *Bauhinia*, and Rita Graham for technical help. This research was supported in part by National Science Foundation (NSF) grant BSR–8506370. Field collecting in Malaysia was supported by National Geographic Society grant 3158–85 and NSF grant INT–8416078 when J. B. F. was a visiting scientist in the Botany Department of the National University of Singapore. He thanks C. J. Goh and A. N. Rao for their hospitality and cooperation.

References

Appanah, S. & Putz, F. E. (1984). Climber abundance in virgin dipterocarp forest and the effect of pre-felling climber cutting on logging damage. *Malaysian Forester* 47: 335–42.

Bamber, R. K. (1984). Wood anatomy of some Australian rainforest vines. In *Proceedings of Pacific Regional Wood Anatomy Conference*, ed. Syoji

Sudo. pp. 58–60. Forest and Forest Products Research Institute, Tsukuba, Ibaraki, Japan.
Caballé, G. (1977). Multiplication végétative en forêt dense du Gabon de la liane *Entada sclerata* (Mimosoideae). *Adansonia, Sér. 2,* **17**: 215–20.
Caballé, G. (1980). Caractéristiques de croissance et multiplication végétative en forêt dense du Gabon de la 'liane à eau' *Tetracera alnifolia* Willd. (Dilleniaceae). *Adansonia, Sér. 2,* **19**: 467–75.
Caballé, G. (1986). Sur la biologie des lianes ligneuses en forêt gabonaise. Thèse, Doct. Etat, Université des Sciences et Technique du Languedoc, Montpellier, France.
Carlquist, S. (1975). *Ecological Strategies of Xylem Evolution.* University of California Press, Berkeley.
Carlquist, S. (1988). *Comparative Wood Anatomy: systematic, ecological and evolutionary aspects of dicotyledon wood.* Springer-Verlag, Berlin.
Cremers, G. (1973). Architecture de quelques lianes d'Afrique tropicale, 1. *Candollea* **28**, 249–80.
Cremers, G. (1974). Architecture de quelques lianes d'Afrique tropicale, 2. *Candollea* **29**: 57–110.
Crüger, H. (1850, 1851). Einige Beiträge zur Kentniss von sogenannten anomalen Holzbildungen des Dikotylenstammes. *Botanische Zeitung* **8**: 97–114, 121–8, 137–43, 161–7, 177–87, Taf. 2–4 and **9**: 465–73, 481–94, Taf. 7, 8.
de Castro e Santos, A. (1980). Essai de classification des arbres tropicaux selon leur capacité de réitération. *Biotropica* **12**: 187–94.
Dobbins, D. R. & Fisher, J. B. (1986). Wound responses in girdled stems of lianas. *Botanical Gazette* **147**: 278–89.
Esau, K. (1977). *Anatomy of Seed plants*, 2nd edition. Wiley, New York.
Fisher, J. B. (1981). Wound healing by exposed secondary xylem in *Adansonia* (Bombacaceae). *International Association of Wood Anatomists Bulletin, n.s.* **2**: 193–9.
Fisher, J. B. & Ewers, F. W. (1989). Wound healing in stems of lianas after twisting and girdling. *Botanical Gazette* **150**: 251–65.
Gentry, A. H. (1985). An ecotaxonomic survey of Panamanian lianas. In *The Botany and Natural History of Panama*, ed. W. G. D'Arcy and M. D. Correa A., pp. 29–42. Missouri Botanic Garden, St Louis, Mo.
Haberlandt, G. (1914). *Physiological Plant Anatomy*. Macmillan, London.
Hallé, F., Oldeman, R. A. A. & Tomlinson, P. B. (1978). *Tropical Trees and Forests: an architectural analysis.* Springer-Verlag, Berlin.
Handa, T. (1937). Anomalous secondary growth in *Bauhinia japonica* Maxim. *Japanese Journal of Botany* **9**: 37–53.
Klotz, L. H. (1978). Observations on diameters of vessels in stems of palms. *Principes* **22**: 99–106.

Knoblock, I., Kahl, G., Landré, P. & Nougarède, A. (1989). Cellular events during wound periderm formation in *Dioscorea bulbifera* bulbils. *Canadian Journal of Botany* **67**: 3090–102.

Longino, J. T. (1986). A negative correlation between growth and rainfall in a tropical liana. *Biotropica* **18**: 195–200.

Ng, F. S. P. (1986). Tropical sapwood trees, In *Naturalia monspeliensia – Colloque international sur l'Arbre*, ed. F. Hallé, pp. 61–7. Université de Montpellier, France.

Obaton, M. (1960). Les lianes ligneuses à structure anormale des forêts denses d'Afrique occidentale. *Annales des Sciences naturelles (Botanique), Sér. 12*, **1**: 1–220.

Passam, H. C., Reed, S. J. & Rickard, J. E. (1976). Wound repair in yam tubers: the dependence of storage procedures on the nature of the wound and its repair. *Tropical Science* **18**: 1–11.

Pfeiffer, H. (1926). Das abnorme Dickenwachstum. In *Handbuch der Pflanzenanatomie*, Bd. 9, Lief 15, ed. K. Linsbauer. Borntraeger, Berlin.

Putz, F. E. (1984). The natural history of lianas on Barro Colorado Island, Panama. *Ecology* **65**: 1713–24.

Putz, F. E. & Windsor, D. M. (1987). Liana phenology on Barro Colorado Island, Panama. *Biotropica* **19**: 334–41.

Schenck, H. (1893). Beiträge zur Biologie und Anatomie der Lianen, in Besonderen der in Brasilien einheimische Arten. 2. Beiträge zur Anatomie der Lianen. In *Botanische Mittheilungen aus der Tropen* 5, ed. A. F. W. Schimpers, pp. 1–271. G. Fischer, Jena.

Shigo, A. L. (1984). Compartmentalization: a conceptual framework for understanding how trees grow and defend themselves. *Annual Review of Phytopathology* **22**: 189–214.

Sidlowski, J. J., Phillips, W. S. & Kuykendall, J. R. (1971). Phloem regeneration across girdles of grape vines. *Journal of the American Society for Horticultural Science* **96**: 97–102.

Swamy, B. G. L. & Sivaramakrishna, D. (1975). Wound healing responses in monocotyledons. I. Responses in vivo. *Phytomorphology* **22**: 305–14.

Walter, W. M., Jr & Schadel, W. E. (1983). Structure and composition of normal skin (periderm) and wound tissue from cured sweet potatoes. *Journal of the American Society for Horticultural Science* **108**: 909–14.

Zimmermann, M. H. (1983). *Xylem Structure and the Ascent of Sap*. Springer-Verlag, Berlin.

PART III
VINE PHYSIOLOGY AND DEVELOPMENT

5

Water flux and xylem structure in vines

FRANK W. EWERS, JACK B. FISHER
AND KLAUS FICHTNER

Introduction

The xylem of plants has three major functions: transport of water and minerals, mechanical support of the plant body, and storage of water and nutrients. Since vines depend upon other plants (or trellises) for mechanical support of their plant body, the mechanical demands on the stem xylem of vines are much reduced compared with the situation in free-standing growth forms.

Vines in general have thin stems and a high ratio of supported leaf weight to transverse stem area (Schenck, 1893; Hallé, Oldeman & Tomlinson, 1978; Putz, 1983; Ewers, 1985). Putz (1983) showed that although lianas (woody vines) made up only 4.5% of the total above-ground biomass of a rainforest in Venezuela, they constituted 19% of the total leaf area.

In vines without secondary growth, the original primary xylem system must supply the leaves with water and minerals throughout the life of the plant. In contrast, most liana stems can add further xylem for water transport as the plant ages. Old liana stems can, in some instances, become quite wide. For instance, there is a report of a 433-year-old stem of *Hedera helix* that was over 60 cm in diameter (Schenk, 1893). Similarly, The Carpinteria Vine, a 51-year-old individual of *Vitis vinifera*, had a trunk that was over 80 cm in diameter (Winkler *et al.*, 1974).

Despite these extreme examples, the present evidence suggests that liana stems normally have much lower rates of secondary growth than do trees or shrubs. Schenck (1893) reported on the ages and stem diameters of eight species of temperate lianas, and concluded that liana stems were quite narrow for their age. However, his data set is difficult to interpret because growth conditions were not indicated, since only one or two specimens per species were examined, and because no direct comparisons were made to tree species.

Table 5.1. *Diameter growth in lianas ⩾100 mm dbh (1.3 m above ground) growing at La Selva, Costa Rica. Mean annual increment (MAI) is based upon repeated measurements of the same stems in 1969 and 1982. For purposes of calculating mean MAI values, negative increments were treated as zero.* N=*number of stems measured. These previously unpublished data were collected by D. Lieberman, M. Lieberman, G. Hartshorn, and R. Peralta with support from NSF grant BSR-8117507*

Taxon	*N*	dbh (mm) in 1969 (range)	MAI (mm) from 1969 to 1982	
			$x \pm$ SE	range
Bignoniaceae				
unidentified sp.	2	109–132	1.3	0.8–1.8
Dilleniaceae				
Doliocarpus coriacea and *D. multiflorus*	42	102–203	1.4±0.3	−0.7–7.5
Euphorbiaceae				
Anomospermum sp.	1	112	2.9	–
Omphalea diandra	1	114	0.0	–
Fabaceae				
Bauhinia sp.	4	107–124	4.3±1.1	2.3–6.7
Machaerium seemannii	1	107	5.2	–
Lauraceae				
Nectandra sp.	1	112	0.6	–
Marcgraviaceae				
Marcgravia spp.	2	122–124	2.5	1.6–3.3
Olacaceae				
Heisteria scandens	3	104–122	0.4±0.2	0.0–0.6
Sapindaceae				
Paullinia sp.	1	114	0.2	–
Unidentified liana sp.	2	107	1.0	0.9–1.1

We know of two data sets for growth rates of tropical lianas, one by Putz (1990), and the other by Lieberman *et al.* (Table 5.1). Both of these data sets indicate that the secondary growth rates of liana stems were much lower than for trees growing in the same location.

Putz (1990) reported that the mean annual increment (MAI) of stem diameter growth was 1.37 mm (SE=0.35) for 15 taxa of lianas growing in the canopy on Barro Colorado Island, Panama. This was much less than for canopy (30–50 cm diameter) trees in this forest, which, according to Lang & Knight (1983), had a MAI of 9.0 mm.

Lieberman *et al.* (1985) reported on the rates of secondary growth in forest trees and one liana genus (*Doliocarpus* spp.) growing at La Selva Biological Station, Costa Rica. On the basis of their growth simulation analyses, the median growth curve for the liana *Doliocarpus* was less than for 19 out of 22 canopy tree species, and the maximum curve was less than for 21 out of 22 canopy tree species. They have kindly provided us with previously unpublished data that they collected on 10 other taxa of lianas growing in this same forest (Table 5.1). Eight out of 60 stems they measured showed zero (or negative) growth over a 13-year period. Although the mean growth rates varied quite a bit among taxa, the raw data for all taxa fall within the range of results for *Doliocarpus*. The overall MAI for liana species in the La Selva forest was 1.80 mm. This may have been slightly inflated because some taxa had flattened or lobed stems; measurements with a diameter tape could thus overestimate diameter and diameter growth. It should also be pointed out that these data pertain to stems 100 mm in diameter or larger, which is a convenient cut-off point for studies of growth in large trees, but would include only liana stems with a tremendous leaf area.

In addition to having narrow stems, it has long been known that vines have wide vessels and unusual, often bizarre, types of xylem anatomy (e.g. Schenck, 1893; Haberlandt, 1914; Pfeiffer, 1926; Obaton, 1960; Carlquist, 1988). However, the effect of xylem anatomy on water flux through vines has been little studied.

For healthy plants in general, it has often been assumed that most of the resistance to water flow was located in the living tissues of the roots and leaves (Huber, 1956; Kramer & Kozlowski, 1979; Boyer, 1985). Liu *et al.* (1978) reported that for cultivated *Vitis labrusca* plants, the total stem resistance was much lower than in roots and leaves. However, at least for large trees which have great transport distances, there is growing evidence that the xylem transport system provides a significant portion of the total resistance to water flow (Hellkvist, Richards & Jarvis, 1974; Tyree, Caldwell & Dainty, 1975; Zimmermann, 1978; Tyree *et al.*, 1983; Ewers & Zimmermann, 1984a, b; Tyree, 1988). As will be summarized in this chapter, our recent studies suggest that a similar conclusion can be made for long liana stems.

Xylem anatomy of vines

Vessel dimensions

Most of the longitudinal water transport in vines occurs through xylem vessels. An exception is in viney pteridophytes, which lack vessels, with the transport instead occurring in tracheids.

Cell walls of tracheids as well as vessel members have pits which allow for lateral transport between cells. Each pit has a pit membrane which is a thin, porous area on the wall where secondary wall material has not been deposited. Water can move freely through the lumens of tracheids and vessel members, but for long-range transport in vessel-less xylem, water must pass through a series of many tracheids. Therefore, water must continually pass through pit pairs connecting adjacent tracheids.

A vessel is a series of vessel members (elements) stacked end to end. The members of a vessel are interconnected by perforations in the end walls, which offer less resistance to water flow than do pits. Tracheids lack perforations. However, since vessels, like tracheids, are of finite length, water must move from vessel to vessel or from vessel to tracheid through lateral pit pairs.

Vessel and tracheid diameter are important parameters in models of xylem transport (Carlquist, 1975, 1988; Zimmermann, 1983; Siau, 1984; Gibson, Calkin & Nobel, 1985). According to Poiseuille's law for ideal capillaries, K_h (hydraulic conductance per unit stem length in $m^3\,MPa^{-1}\,s^{-1}\,m = m^4\,MPa^{-1}\,s^{-1}$) is proportional to the summation of vessel or tracheid lumen diameters (d) each raised to the fourth power (Gibson *et al.*, 1985):

$$K_h\,predicted = \frac{\pi \Sigma d_i^4}{128\eta} \qquad \text{(Equation 1)}$$

where: η = dynamic viscosity of the fluid (MPa.s). As a result of the fourth power relationship, when vessel lumens are twice as wide, K_h *predicted* is 16 times as great.

Since vessels are not ideal capillaries of infinite length, the total length of vessels is also important in models of xylem transport. The vessel length represents the maximum distance that a water molecule can travel without passing through a pit membrane. Furthermore, vessel length information is relevant to studies of xylem dysfunction via embolization. When water within a vessel is under sufficient tension, any gas bubbles in the vessel will expand to the total size of the vessel lumen. Since a gas bubble cannot easily pass through a wet pit membrane (Zimmermann, 1983; Newbanks, Bosch & Zimmermann, 1983; Lewis, 1988; Sperry & Tyree, 1988), the longitudinal extent of xylem dysfunction due to an embolism is equal to the length of the vessel.

Vines are often said to have among the longest and widest vessels in the plant kingdom (Berger, 1931; Huber, 1956; Zimmermann & Brown, 1971; Carlquist, 1975; Kramer & Kozlowski, 1979; Zimmermann & Jeje, 1981). Vessel diameters of vines tend to be greater than in closely related species of trees (Ayensu & Stern, 1964; Carlquist, 1975; Klotz, 1978; Van Vliet, 1981; Bamber, 1984; Ter Welle, 1985). Similarly, Gartner *et al.* (1990) found that in Jalisco, Mexico, naturally growing lianes had greater maximum vessel diameters than unrelated trees growing in the same dry hillside forest.

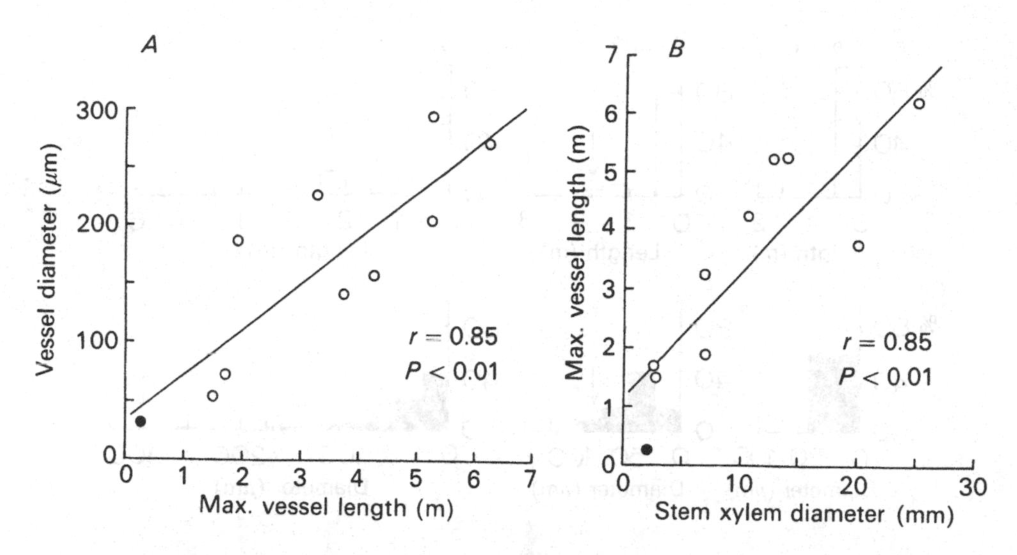

Figure 5.1. Vessel lengths in the liana *Pithecoctenium crucigerum* (Bignoniaceae). Correlations of maximum vessel length with vessel diameter (*A*) and stem diameter (*B*). Each point from a single stem, with the solid circle showing results for the inner secondary xylem only. From Ewers & Fisher (1989b).

The length of vessels can most easily be determined with the air or paint methods (Zimmermann & Jeje, 1981; Ewers & Fisher, 1989a). Latex paint, which can pass through vessel lumens but not through pit membranes, is perfused into the stem, and the stem is then examined at regular intervals along its length to determine the number and diameter of paint-filled vessels. The air method depends upon the fact that compressed air cannot pass through wet pit membranes and, hence, past vessel ends, except when very high pressures (> 2000 kPa) are used. A vessel that is cut open at both ends can pass gas even at low pressures (< 100 kPa).

Vessel length and vessel diameter appear to be correlated in lianas. This correlation seems to hold true within individual stems (Ewers & Fisher, 1989b), between different stems of the same species (Figure 5.1), and when comparing different liana species (Ewers & Fisher, 1987).

Scholander (1958) reported 'average vessel length' to be about 0.6 m in the temperate liana *Vitis labrusca* (Vitaceae) and over 1 m in the tropical liana *Tetracera* (Dilleniaceae). This was based upon measurements of water released by vertically held fresh stem segments trimmed back at regular intervals.

Maximum vessel lengths of 1.5 m (Sperry *et al.*, 1987), 3 m (Scholander, 1958), 3.8 m (F. W. Ewers and J. B. Fisher, unpublished data), and 7.75 m (Zimmermann & Jeje, 1981) have been found for species of *Vitis*. The short maximum vessel lengths reported by Sperry *et al.* (1987) were the result of

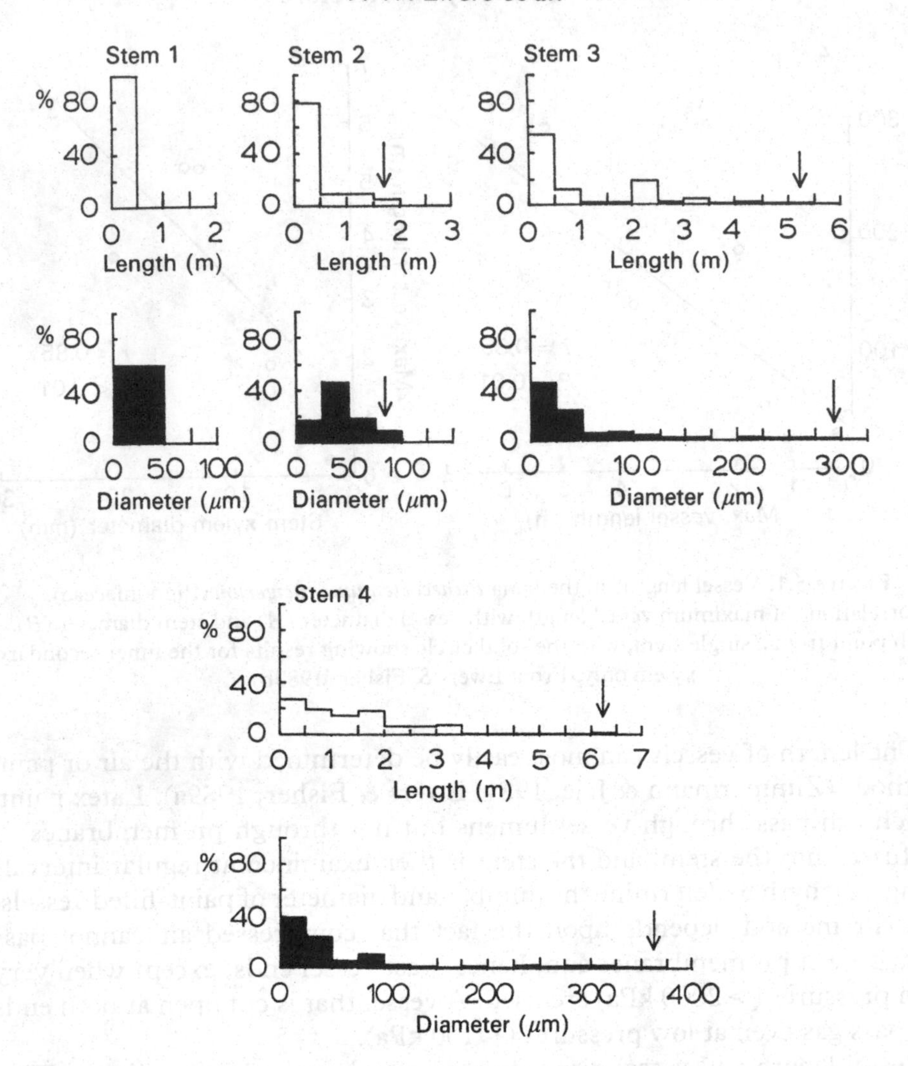

Figure 5.2. Frequency distributions of vessel length (open bars) and vessel diameter (solid bars) for four different stems of *Pithecoctenium crucigerum*. Stem xylem diameters were 2 mm (stem 1), 2.5 mm (stem 2), 14 mm (stem 3), and 25.5 mm (stem 4). Arrows indicate maximum values. From Ewers & Fisher (1989b).

using relatively narrow (5 mm diameter) stems. In a survey of the stems of 33 species of climbing plants from 26 genera in 16 families, we found the average maximum vessel dimensions to be 207 μm (SE = 13) for diameter, and 1.46 m (SE = 0.12) for length (Ewers & Fisher, 1987). The longest vessel we found, 7.73 m, was in a stem of *Pithecoctenium crucigerum* (Bignoniaceae). Maximum values should be approached with caution since, in most liana stems, as with plants in general, there are many more short and narrow vessels than long and wide ones (Figures 5.2 and 5.3; Skene & Balodis, 1968; Zimmermann & Jeje,

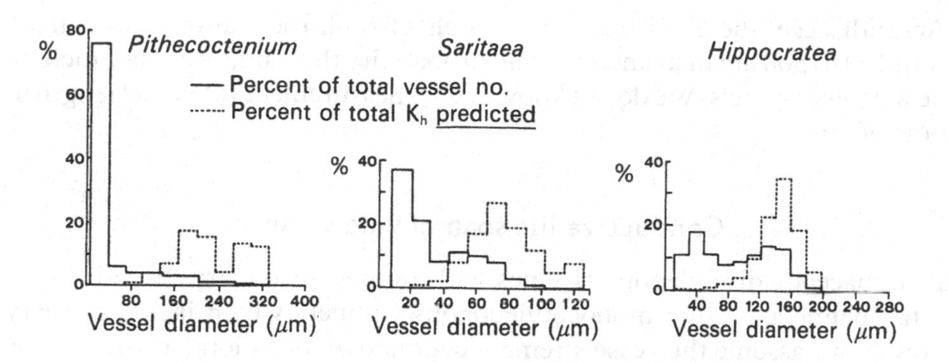

Figure 5.3. Vessel frequency and percentage of total K_h*predicted* for each vessel diameter class in a stem of *Pithecoctenium crucigerum*, *Saritaea magnifica* (Bignoniaceae), and *Hippocratea volubilis* (Hippocrateaceae). Modified from Ewers & Fisher (1989a).

1981; Sperry *et al.*, 1987; Ewers & Fisher, 1989a, b). In addition, it should be pointed out that the Ewers & Fisher (1987) survey was based upon relatively narrow stems, with a mean xylem diameter of 8.2 mm (SE = 0.06). Wider diameter stems tend to have longer and wider vessels (Figures 5.1, 5.2; Ewers & Fisher 1989a, b).

Contribution of narrow vessels

Although vines have many short and narrow vessels as well as long and wide ones, the role of the smaller vessels is unclear. The many narrow vessels contribute very little to the total K_h*predicted* in stems (Figure 5.3). The narrow vessels could be important for mechanical support, for water storage, for radial and tangential water movement, for conduction when larger vessels become embolized, and for making hydraulic linkups at junctions between plant parts (Zimmermann, 1983; Zimmermann & Sperry, 1983; Carlquist, 1985, 1988).

Xylem of vine roots

Root xylem anatomy is a relatively neglected area of research for plants in general. In plants where this has been studied, the vessels in roots are wider and longer than in the stem (Reidl, 1937; Fegel, 1941; Baas, 1982; Zimmermann & Potter, 1982; Zimmermann, 1983). However, there is much variation in vessel diameters among different sizes and types of roots on a particular plant. Reidl (1937) concluded that, at least for temperate trees, there is much more xylem parenchyma in woody roots than in woody stems.

McAneney & Judd (1983) reported that in the liana *Actinidia chinesis*

(Actinidiaceae: the Kiwi fruit) the vessels of both roots and stems ranged from 120 to 500 μm in diameter. In all probability this range does not include the narrower vessels. We do not know of any measurements of vessel length in roots of vines.

Conductive life span of vine vessels

In herbaceous dicotyledonous vines with limited secondary growth, or in rattan palms and other monocotyledonous climbers which lack secondary growth, we assume the vessels remain conductive for as long as the stem is alive. Vessels of the temperate liana *Vitis* remain conductive for up to 7 years (Smart & Coombe, 1983), after which time they become a part of the non-conductive heartwood. The conductive life span of tropical lianas is difficult to document since they lack reliable growth rings, but, based upon the broad xylem transverse conductive area detected with dye ascents (Putz, 1983; J. B. Fisher and F. W. Ewers, unpublished data), and upon the low secondary growth rates in liana stems (see Introduction, above), the vessels of tropical lianas presumably remain conductive for up to many decades. Similar wide vessels in temperate trees remain conductive for at most one growing season (Coster, 1927; Huber, 1935; Zimmermann, 1983; Ellmore & Ewers, 1986; Ewers & Cruiziat, 1990). A possible reason for the extended conductive lifespan of vine vessels is discussed in the section on root pressure, below.

Anomalous arrangements of secondary xylem

As discussed by Fisher & Ewers (Chapter 4), Putz & Holbrook (Chapter 3) and Carlquist (Chapter 2), many vines have anomalous forms of secondary growth. These anomalies often result in bundles, arcs, wedges, concentric cylinders, or other apparent discontinuities in the stele. However, the three-dimensional construction of these vascular systems, and the effect of the anomalous growth patterns on water transport, has been little studied.

Inner and outer secondary xylem

A striking ontogenetic change occurs in the stems of many species of lianas, which has great significance to conductivity. The first-formed, 'inner' secondary xylem is dense with thick-walled fibers and few, narrow vessels. The later-formed 'outer' xylem has many wide vessels and fewer fibers (e.g. Figure 5.4). This results in a much greater K_h *predicted*.

In *Bauhinia fassoglensis* (Fabaceae) and *Stigmaphyllon ellipticum* (Malpighiaceae) vessels of the inner xylem are much shorter as well as narrower than

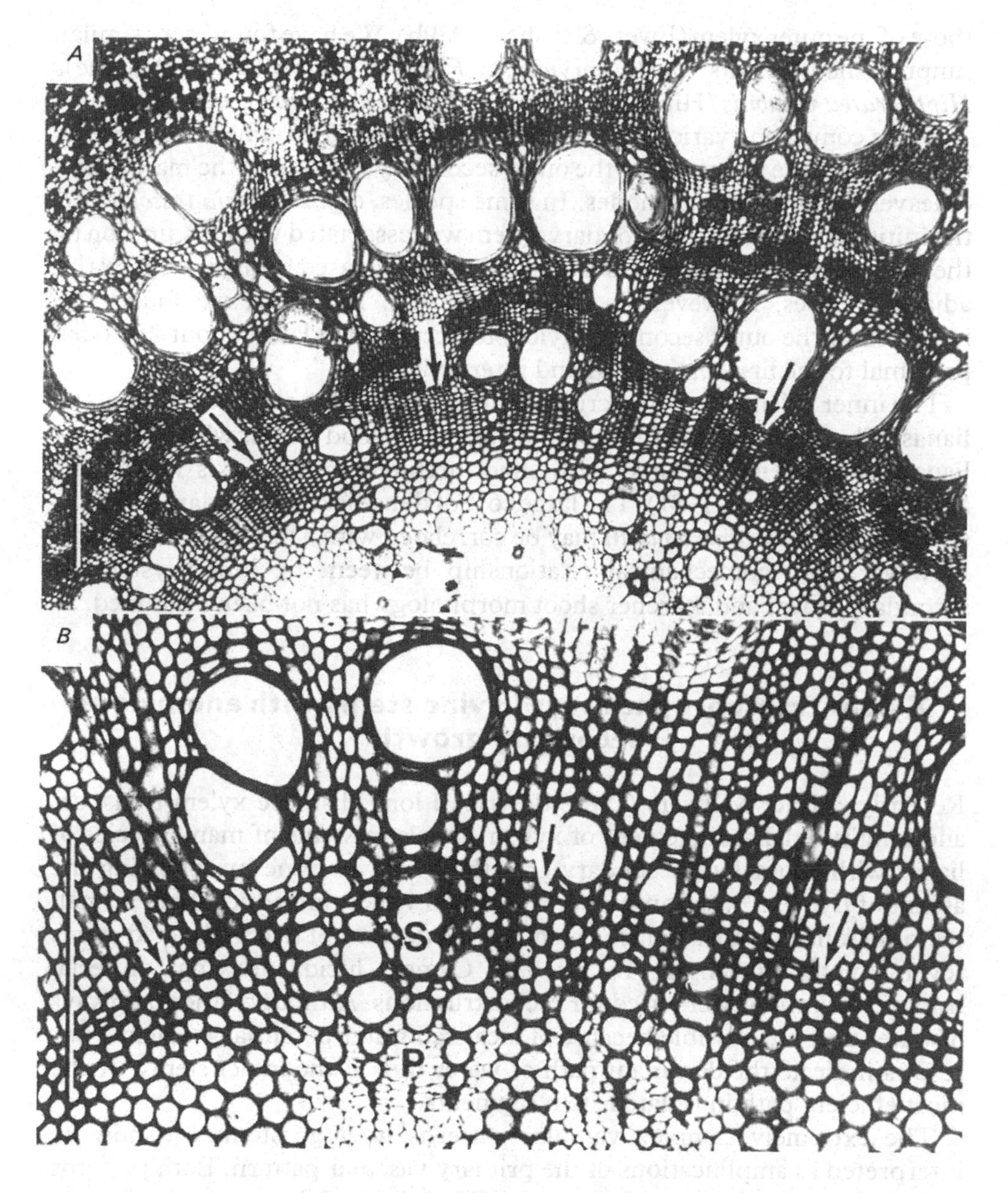

Figure 5.4. Stem transverse sections showing inner and outer xylem. *A*, *Stigmaphyllon ellipticum* (Malpighiaceae). *B*, *Bauhinia fassoglensis* (Fabaceae). Arrows show boundary between inner and outer secondary xylem. P, primary xylem; S, secondary xylem. Scale bars = 200 μm. From Ewers & Fisher (1989b).

those of the outer xylem (Ewers & Fisher, 1989b). We have found very similar (unpublished) results for *B. corymbosa, Peixotoa glabra* (Malpighiaceae), *Hippocratea volubilis* (Hippocrateacae), and *Pithecoctenium crucigerum*.

When comparing various species of lianas we found no consistent correlation between the transition to the outer secondary xylem and the maturation of leaves, tendrils, or internodes. In some species, e.g. *Bauhinia fassoglensis*, the initiation of the outer secondary xylem was associated with maturation of the leaf and internode, and was associated with increased transpiration by the adjacent leaves. However, in other cases, e.g. *B. vahlii*, we found the initiation of the outer secondary xylem to occur much later, about 23 nodes proximal to the first mature leaf and internode.

The inner xylem may play a critical role in the support of searcher shoots in lianas. The inner xylem often contains reaction wood fibers, which, in non-lianas, have been correlated with shoot architecture and axis movement (Fisher & Stevenson, 1981). The large core of dense inner secondary xylem in species such as *Bauhinia vahlii* may be correlated with a long searcher shoot. However, the biomechanical relationship between the inner system of secondary xylem and searcher shoot morphology has not been examined.

Xylem transport pathways in vine stems with anomalous secondary growth

Recently, Fisher & Ewers (1988) described long-distance xylem transport and longitudinal arrangement of xylem vessels in stems of many species of lianas with anomalous secondary growth. This was done by following the ascent of dye (0.5% safranin or 0.5% crystal violet) introduced into a small sector of xylem in a transpiring stem, or the descent of dye in isolated stems several meters in length (Figure 5.5). Camera lucida drawings of serial transverse sections were used for reconstructions. Although other pathways of water flow are possible because of localized water potential gradients, our results indicate the dominant path of water flux in the intact stems, or the most efficient path in isolated stem segments.

The extremely complex vascular patterns in large stems can now be interpreted as amplifications of the primary vascular pattern. Both patterns form an integrated conducting system. Regardless of the type of anomalous growth, the various portions of the stele are interconnected in three dimensions. Nodes and branch junctions are common regions for anastomoses, but the exact nature and frequency of the anastomoses vary among the different taxa.

Stems with wide rays (e.g. *Coccinia* of Cucurbitaceae) or narrow rays (e.g. *Cayratia* of Vitaceae) have xylem segments that may anastomose at nodes,

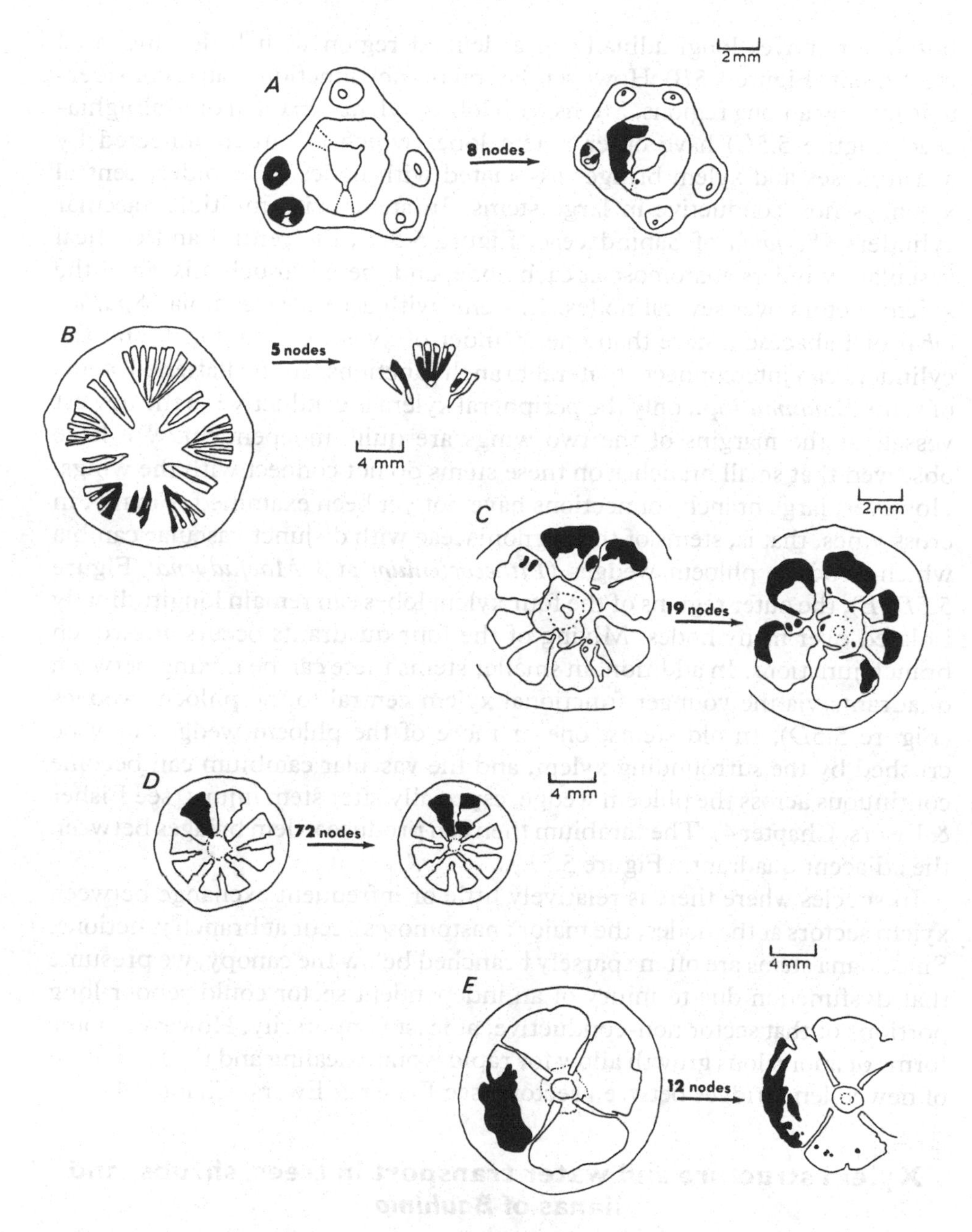

Figure 5.5. Dye experiments in stems with anomalous secondary growth. Dye (solid areas) was introduced into one or more sectors and observed at various distances from infusion port. Stem distance between diagramed transverse sections in brackets. *A*, *Serjania polyphylla* (Sapindaceae) [1.23 m]. *B*, *Coccinia grandis* (Cucurbitaceae) [0.53 m]. *C*, *Peixotoa glabra* (Malpighiaceae) [2.80 m]. *D*, *Macfadyena ungus-cati* (Bignoniaceae) [4.63 m]. *E*, *Pithecoctenium crucigerum* (Bignoniaceae) [3.79 m].

but water moves longitudinally in a defined region with little tangential movement (Figure 5.5*B*). However, lateral branch junctions cause considerable mixing among regions. Stems with lobed xylem (*Peixotoa* of Malpighiaceae, Figure 5.5*C*) have outer xylem lobes which are interconnected by anastomoses and xylem bridges associated with nodes. The older, central xylem is non-conductive in large stems. In stems with multiple vascular cylinders (*Serjania* of Sapindaceae, Figure 5.5*A*), the central and cortical vascular cylinders anatomose at each node, and there is much mixing of the xylem sectors over several nodes. In stems with accessory cambia (*Spatholobus* of Fabaceae), more than one cylinder of xylem is functional, and the cylinders can interconnect at lateral branch junctions. In the flattened stems of some *Bauhinia* spp., only the peripheral xylem is conductive so the newest vessels at the margins of the two wings are quite independent. We have observed that small branches on these stems do not connect with the wings. However, large branch connections have not yet been examined. Finally, in cross vines, that is, stems of the Bignoniaceae with disjunct vascular cambia which produce phloem wedges (*Pithecoctenium* and *Macfadyena*: Figure 5.5*D*, *E*), the outer regions of the four xylem lobes can remain longitudinally isolated over many nodes. Mixing of the four quadrants occurs at swollen branch junctions. In addition, in smaller stems there can be mixing between quadrants via the younger functional xylem central to the phloem wedges (Figure 5.5*D*). In old stems, one or more of the phloem wedges may be crushed by the surrounding xylem, and the vascular cambium can become continuous across the phloem wedge, especially after stem injury (see Fisher & Ewers, Chapter 4). The cambium then can produce xylem bridges between the adjacent quadrants (Figure 5.5*E*).

In species where there is relatively little or infrequent exchange between xylem sectors at the nodes, the major anastomoses occur at branch junctions. Since liana stems are often sparsely branched below the canopy, we presume that dysfunction due to injury of an independent sector could render long portions of that sector non-conductive, at least temporarily. However, some forms of anomalous growth allow for rapid wound healing and the formation of new xylem bridges between sectors (see Fisher & Ewers, Chapter 4).

Xylem structure and water transport in trees, shrubs, and lianas of *Bauhinia*

Recently, we made a comparative study of water transport in trees, shrubs, and lianas of the genus *Bauhinia* (Ewers, Fisher & Chiu, 1988). We used information on stomatal closure, transpiration rate (*E*), leaf area, stem xylem conductivity, stem pressure gradients, and xylem anatomy to determine

(i) whether stem xylem resistance was a limiting factor in the water relations of these taxa; (ii) whether the wide and narrow vessels of the liana stems compensated for their narrow stem diameters.

We used E and leaf-specific conductivity (LSC) to predict xylem pressure gradients in stems (from Zimmermann, 1978; Salleo, Rosso & Lo Gullo, 1982):

$$dp/dx = E/\mathrm{LSC} \qquad \text{(Equation 2)}$$

where dp/dx is the pressure gradient in MPa m^{-1}, E is in $m^3\ s^{-1}\ m^{-2}$ ($=m\ s^{-1}$), and LSC is in $m^4\ MPa^{-1}\ s^{-1}\ m^{-2}$ ($=m^2\ MPa^{-1}\ s^{-1}$).

LSC is equal to hydraulic conductance per unit stem length (K_h*measured*) divided by the leaf area (m^2) distal to the stem segment. K_h is equal to the volume flow rate of water ($m^3\ s^{-1}$) divided by dp/dx. Superimposed upon Equation 2 we added an additional -0.01 MPa for each meter in vertical distance to account for gravity. This equation is a simplification in that it does not account for water that goes in and out of storage in the plant tissue, and it assumes that the stem K_h is constant throughout the day (e.g. it assumes no additional embolisms occur during the day).

LSCs are shown for a shoot of *Bauhinia fassoglensis* in Figure 5.6. Notice the low LSCs at branch junctions (arrows), which, under periods of rapid transpiration, should result in particularly steep pressure gradients. Mean LSC values (excluding branch junctions) for six *Bauhinia* species growing at Fairchild Tropical Garden are shown in Table 5.2. The six species included two trees, two shrubs, and two lianas.

We made diurnal studies of E, stomatal conductance (g_s), and leaf water potential (ψ) on these six species (Figure 5.7). From diurnal studies we were able to determine the leaf ψ that would induce stomatal closure. These results correlated quite well with results where stomatal closure was experimentally induced by detaching the leaf from the plant.

Predicted maximum stem pressure gradients, based upon maximum transpiration rates and mean LSCs, are shown in Table 5.2. This does not include the steeper gradients at branch junctions. Nevertheless, the predicted maximum pressure gradients, from 0.05 to 0.12 MPa m^{-1}, are of a magnitude that should be significant in inducing stomatal closure in the longer stems of each of these species. Resistance to water flow offered by stem xylem in each of these species is, therefore, significant to the overall water relations of the plants. The predicted pressure gradients are less steep than what is often reported in herbaceous plants, but within the range of pressure gradients reported in trunks of coniferous and dicotyledonous trees (Hellkvist *et al.*, 1974; Boyer, 1985).

The predicted pressure gradients were roughly similar in different growth

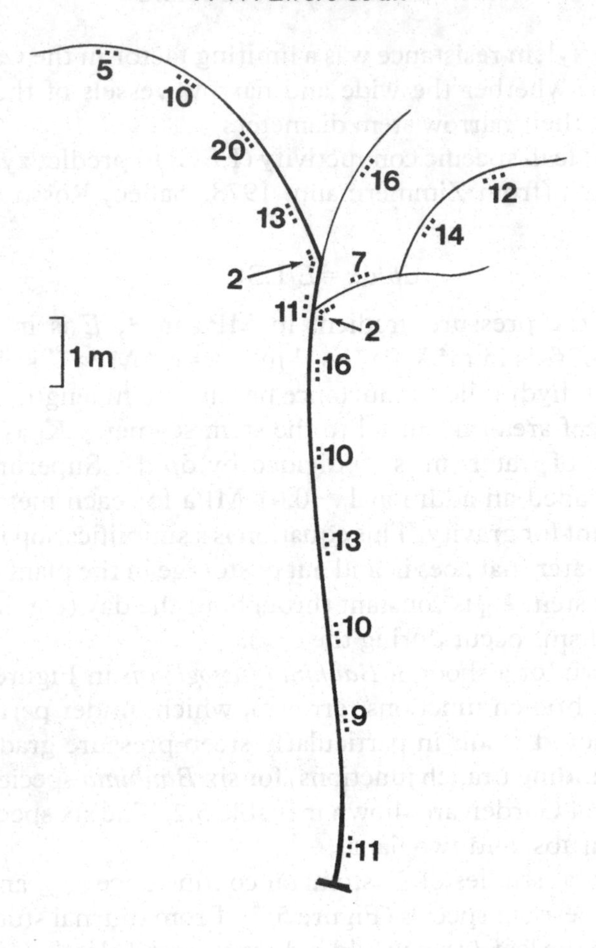

Figure 5.6. Leaf-specific conductivities (LSCs in 10^{-7} m² MPa^{-1} s^{-1}) for a shoot of *Bauhinia fassoglensis*. Note low LSCs at branch junctions (arrows). From Ewers *et al.* (1989).

forms of *Bauhinia* (Table 5.2). However, the stem xylem structure, the means by which the water flowed through the stems, differed considerably among the different taxa (Figure 5.8, Table 5.3).

Based upon the six *Bauhinia* species we have examined thus far at Fairchild Tropical Garden, shrubs have the greatest Huber values (transverse sapwood area divided by leaf area supplied), lianas the lowest values, and trees are intermediate (Table 5.3). Conversely, lianas have the greatest specific conductivity (K_h*measured* divided by sapwood transverse area) and shrubs the lowest values (Table 5.3). In the way we define these terms, LSC is equal to the Huber value multiplied by the specific conductivity. Since LSCs are as large or larger in the lianas as in the trees or shrubs we examined, it is fair to

Table 5.2. *Predicted maximum stem pressure gradients (*dp/dx*) in* Bauhinia *based upon maximum transpiration rates from diurnal studies (*E_{max}*) and mean stem leaf-specific conductivities (LSCs). Junctions excluded from these calculations. Predicted* dp/dx = *E/LSC*

Growth form and species	E_{max} (SE) (10^{-7}m s^{-1})	LSC (SE) (10^{-7}m^{2}MPa^{-1}s^{-1})	d*p*/dx (MPa m^{-1})
Lianas			
B. fassoglensis	1.11 (0.07)	11.0 (0.7)	0.10
B. vahlii	0.81 (0.07)	6.8 (0.7)	0.12
Trees			
B. blakeana	0.59 (0.09)	11.2 (1.4)	0.05
B. variegata	1.04 (0.06)	12.3 (1.2)	0.08
Shrubs			
B. aculeata	0.62 (0.10)	6.2 (0.5)	0.10
B. galpinii	0.48 (0.11)	5.8 (0.8)	0.08

say that the high specific conductivities of liana stems compensated for their low Huber values. So, in effect, the liana taxa produced 'little wood per leaf' in their stems, but the wood they produced was extremely efficient relative to the transverse stem area.

An analysis of vessel diameters is informative as to the way the different growth forms accomplished their stem conductive efficiency. Although mean vessel diameters were roughly similar for the three growth forms, maximum vessel diameters were greatest in lianas (Table 5.3). Based upon analyses of K_h *predicted* derived from Poiseuille's law, the high specific conductivities in the liana stems were due to the relatively few very wide vessels. The mean vessel diameter values are misleading since the liana stems we examined had a much more skewed distribution of vessel diameters than did the trees or shrubs. This is reflected in the fact that for liana stems the median vessel diameters were much less than the mean (Table 5.3). In the example shown in Figure 5.9, the segment from the tree *B. blakeana* had a similar K_h *measured* and K_h *predicted* as the stem segment from the liana *B. fassoglensis*, but the conductance was accomplished in very different ways. The tree stem had almost four times more conductive vessels than the liana, as seen in transverse view. In the liana, 52% of the vessels were from the smallest diameter class, but these contributed only 0.005% of the total K_h *predicted*. In contrast, the vessels that were wider than the widest vessel in the tree segment represented only 14% of the vessels in the liana, but these wider vessels contributed 95% of the total theoretical conductance.

Although the xylem anatomies of the various *Bauhinia* species differ

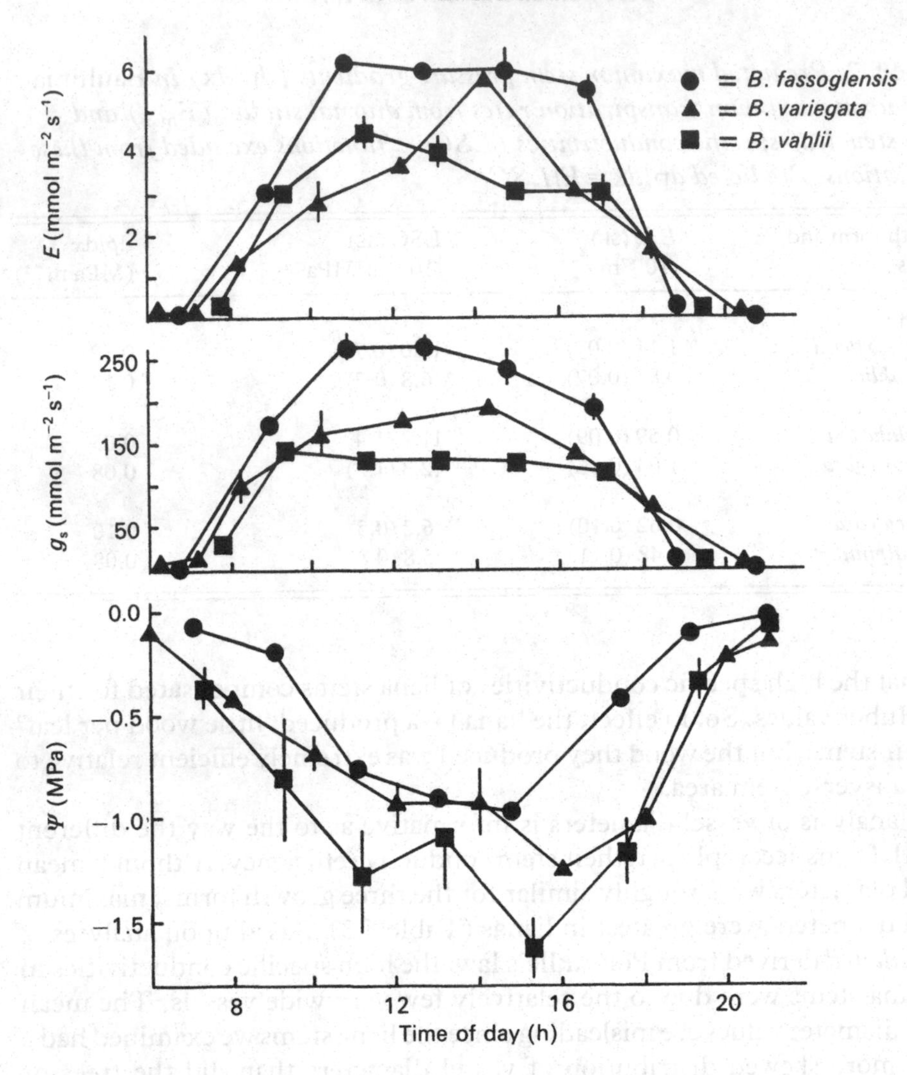

Figure 5.7. A diurnal study of transpiration (E), stomatal conductance (g_s), and leaf xylem water potential (ψ) for two lianas and for the tree *Bauhinia variegata*. Measurements were made on a hot sunny day during the rainy season. Bars = SE (not shown when less than size of symbol).

considerably, the predicted pressure gradients are rather similar (Table 5.2), and are of a magnitude that we would expect them to have a substantial effect on the water relations (Ewers, Fisher & Chiu, 1989). Therefore, we can say for the *Bauhinia* plants we have examined at Fairchild Tropical Garden, that the wide vessels in the liana stems appear to compensate hydraulically for the narrow stem diameters. With further research we hope to determine whether this applies to lianas in general.

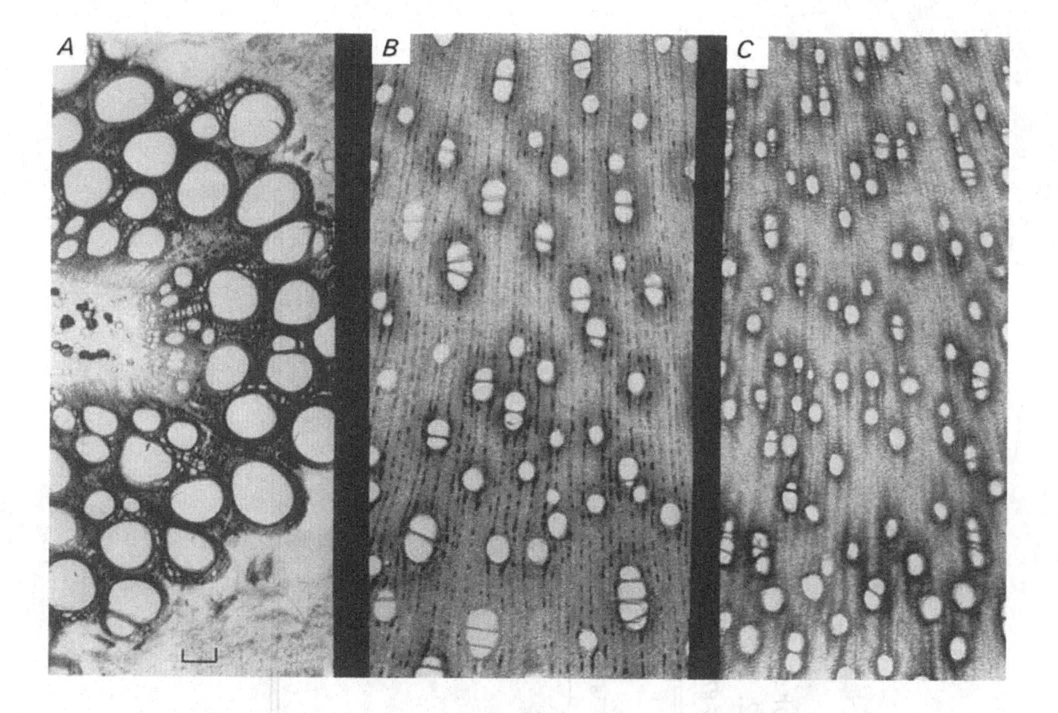

Figure 5.8. Comparison of stem transverse sections of *A*, the liana *Bauhinia fassoglensis*; *B*, the tree *B. blakeana*; and *C*, the shrub *B. galpinii*. All three at same magnification and all three supplying the same leaf area. Dyes surround the conductive elements. Scale bar = 200 μm.

Water storage

As noted earlier, Equation 2 ignores the fact that a certain amount of water may move in and out of storage. The importance of water storage is thought to be greatest in succulent plants. In non-succulent herbaceous plants, the amount of water in storage is usually considered to be quite small in relation to the total daily transpiration (Zimmermann & Milburn, 1982). Based upon measurements of water uptake and water content at wilting, in cultivated lianas of *Actinidia deliciosa*, a maximum of 18% of the daily transpiration could be from water coming out of storage (Judd, McAneney & Trought, 1986). In large trees, under certain conditions, from 14 to 50% of the total daily transpiration may be from water coming out of storage (Waring & Running, 1978; Waring, Whitehead & Jarvis, 1979; Braun, 1983; Schulze *et al.*, 1985).

Recently, we simultaneously measured water uptake and *E* in a shoot system of the liana *Vitis vinifera* with a leaf area of 2.67 m^2. Water uptake was measured with a potometer, and transpiration with a porometer. Over a 24 h period, the total measured transpiration was 1.23% less than total uptake.

Table 5.3. *Average results (standard errors in parentheses) for various anatomical and physiological parameters in some trees, shrubs, and lianas of* Bauhinia. *In each case* N > *12. The lianas have greater maximum vessel diameters and specific conductivities, but lower Huber values and specific gravities than the other growth forms*

Growth form and species	LSC ($10^{-7} m^2 MPa^{-1} s^{-1}$)	Huber value (10^{-5})	Specific conductivity ($10^{-2} m^2 MPa^{-1} s^{-1}$)	Specific gravity of xylem	Vessel frequency per mm^2	Vessel diameter (μm)		
						$\bar{x}$	Median	Max.
Lianas								
B. fassoglensis	11.0 (0.7)	1.4 (0.3)	17.1 (3.0)	0.18 (0.03)	51 (6)	54 (3)	24 (3)	290 (16)
B. vahlii	6.8 (0.7)	2.4 (0.6)	3.5 (0.4)	–	–	–	–	163 (8)
Trees								
B. blakeana	11.2 (1.4)	11.7 (3.1)	1.4 (0.2)	0.33 (0.08)	53 (5)	53 (5)	55 (7)	126 (15)
B. variegata	12.3 (1.2)	9.4 (1.0)	1.5 (0.2)	–	–	–	–	141 (6)
Shrubs								
B. aculeata	6.2 (0.5)	15.8 (6.9)	0.7 (0.1)	0.65 (0.09)	–	–	–	103 (6)
B. galpinii	5.8 (0.8)	6.0 (0.8)	1.1 (0.2)	–	24 (2)	49 (2)	42 (2)	100 (5)

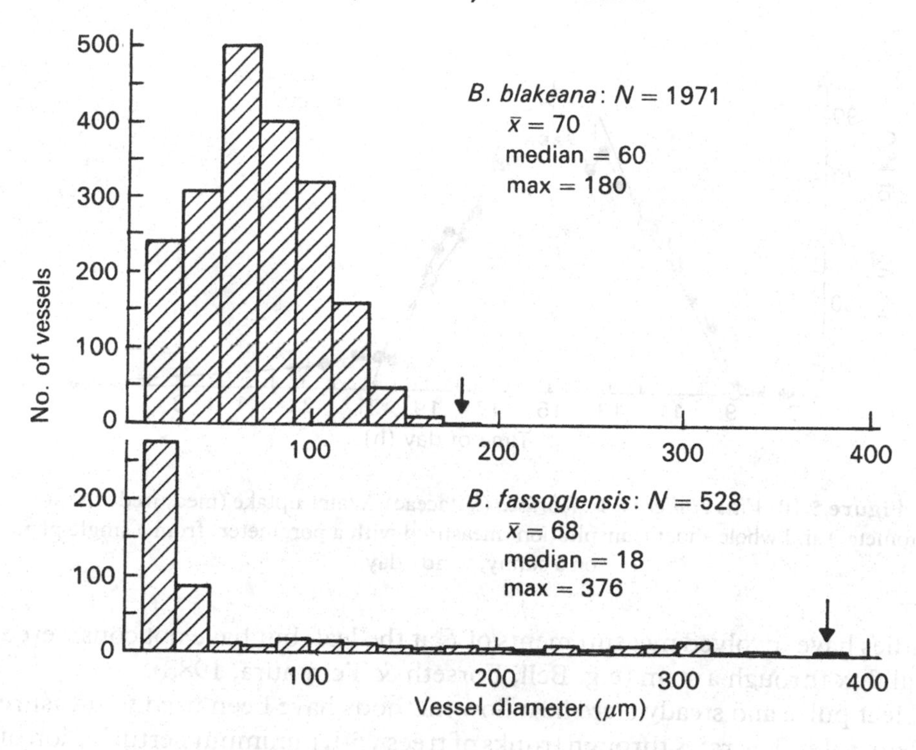

Figure 5.9. Vessel diameter frequency distribution for a stem segment of a tree (*Bauhinia blakeana*) and a liana (*B. fassoglensis*), each supplying the same leaf area.

During peak transpiration, uptake was about 85% of E (Figure 5.10). When E was greater than uptake, we assume that water was moving out of storage. For the entire 24 h period, 217 g of water moved out of storage. This would have been enough water to allow for about 40 min of transpiration at the maximum rate on that day. However, 245 g of water moved back into storage, most of this after sunset when E was zero.

If corrections were made for water coming out of storage, the predicted pressure gradients in Table 5.2 would be somewhat less. However, although trees, shrubs, and lianas differ considerably in their stem anatomy, and, presumably, in their water storage capacity, we know of no studies to compare water storage in these different growth forms.

Water flux through intact vines under natural conditions

Introduction

There is remarkably little information available on the actual rates of water movement through stems of naturally growing vines. Most water relations

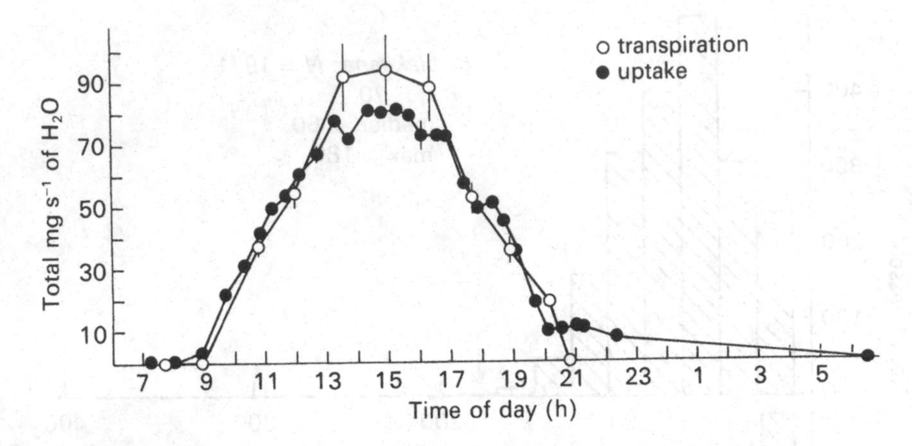

Figure 5.10. *Vitis vinifera* × *V. labrusca* (Vitaceae). Water uptake (measured with a potometer) and whole shoot transpiration (measured with a porometer) from a single plant on a sunny, windy day.

studies have involved measurements of E at the leaf, but have not considered total flux through a stem (e.g. Bell, Forseth & Teramura, 1988).

Heat pulse and steady state heat flow methods have been used to measure xylem water flow rates through trunks of trees with minimum perturbation of the plant. Heat pulse methods measure the water flow velocity, which has to be related accurately to water mass flow (Swanson & Whitfield, 1981; Edwards & Warwick, 1984; Green & Clothier, 1988), while steady state heat flow methods determine the mass flow of xylem water directly (Čermák, Deml & Penka, 1973; Sakuratani, 1984).

Recently, a principle introduced by Čermák *et al.* (1973) was applied to the study of water flow rate through stems of naturally growing lianas for the first time (Fichtner & Schulze, 1990). An insulated stem segment and its xylem water was heated up by 1 °C above the xylem water temperature. The energy supplied to maintain the constant elevated temperature was directly proportional to the mass flow multiplied by the specific heat of water. The constant energy loss by convection and conduction was detected at night under conditions of zero flow and subtracted from the energy input. In trees, plate electrodes can be inserted into the sapwood (Čermák *et al.*, 1973). This was not practical for narrow vine stems. Therefore, in our vines the energy was supplied by a heating coil at the stem surface, following the method of Sakuratani (1984). Simultaneous measurements of microclimate, E, and g_s were made with a LiCor 1600 porometer.

Results are presented below for two vines of the tropical deciduous forest at Chamela (Jalisco, Mexico) at the end of the rainy season of a rather dry year. *Entadopsis polystachya* (L.) Britt. (Fabaceae) is a liana with root tubers and

Cyclanthera multifoliolata Cogn. (Cucurbitaceae) is an annual, herbaceous vine with a shallow root system. In this forest type, water is the most limiting factor for plant growth (Fanjul & Barradas, 1987).

Diurnal study – *Entadopsis polystachya*

The leaves of this species showed a slight midday depression in g_s on a hot, sunny day with a water vapor pressure deficit (ΔW) up to 28 Pa kPa^{-1} (Figure 5.11). The trunk of the observed vine was bifurcated into two main stems. Xylem water flow rate was observed simultaneously on one of the main stems and on three of the seven branches attached to it. Xylem water flow started soon after sunrise (Figure 5.12). It reached a plateau on the early midday which started and ended according to the sun's orientation relative to the particular branch. Similarly, Daum (1967) and Čermák *et al.* (1984) observed a change in xylem water flow through branches of trees depending on the exposition of their leaf area to the sun. The main stem of the liana integrated all fluxes.

Xylem flow ended in the branches 1 h and in the main stem 2 h after sunset. After 16.00 (4 p.m.) the relative proportion of branch xylem flow to main stem xylem flow decreased (Figure 5.12), although it remained rather constant throughout the day. This means that from 16.00 h until the end of xylem flow there was an overproportional flow into the main stem. Between the measuring point on the main stem and the branches were 6 m of liana trunk, from which some storage water could be exploited during the day and refilled when water demand of leaves decreased.

The total xylem flow measured by the steady state heat flow method was about half of the flow determined with a porometer at the leaf surfaces ($E \times$ leaf area). We consider the heat flow method to be accurate and assume the porometer readings to be inflated. The porometer provides an artificial 'windy' microenvironment on the leaf surface. The instrument may be satisfactory on a windy day, but the reported measurements were madc on a day when wind velocity was close to zero.

The daily water use of the whole plant of *E. polystachya* (leaf area 23.6 m^2) was estimated at 47.3 kg. For comparison, lianas of *Actinidia deliciosa* with a projected leaf area of about 50 m^2 were reported to have a very similar water use of 80–100 kg per day. This was on dry and windy days under well-watered conditions (Judd *et al.*, 1986).

Diurnal study – *Cyclanthera multifoliolata*

This species grew on a steep slope, at a disturbed site, with many individuals forming a uniform, dense stand. A given branch could not be followed to its

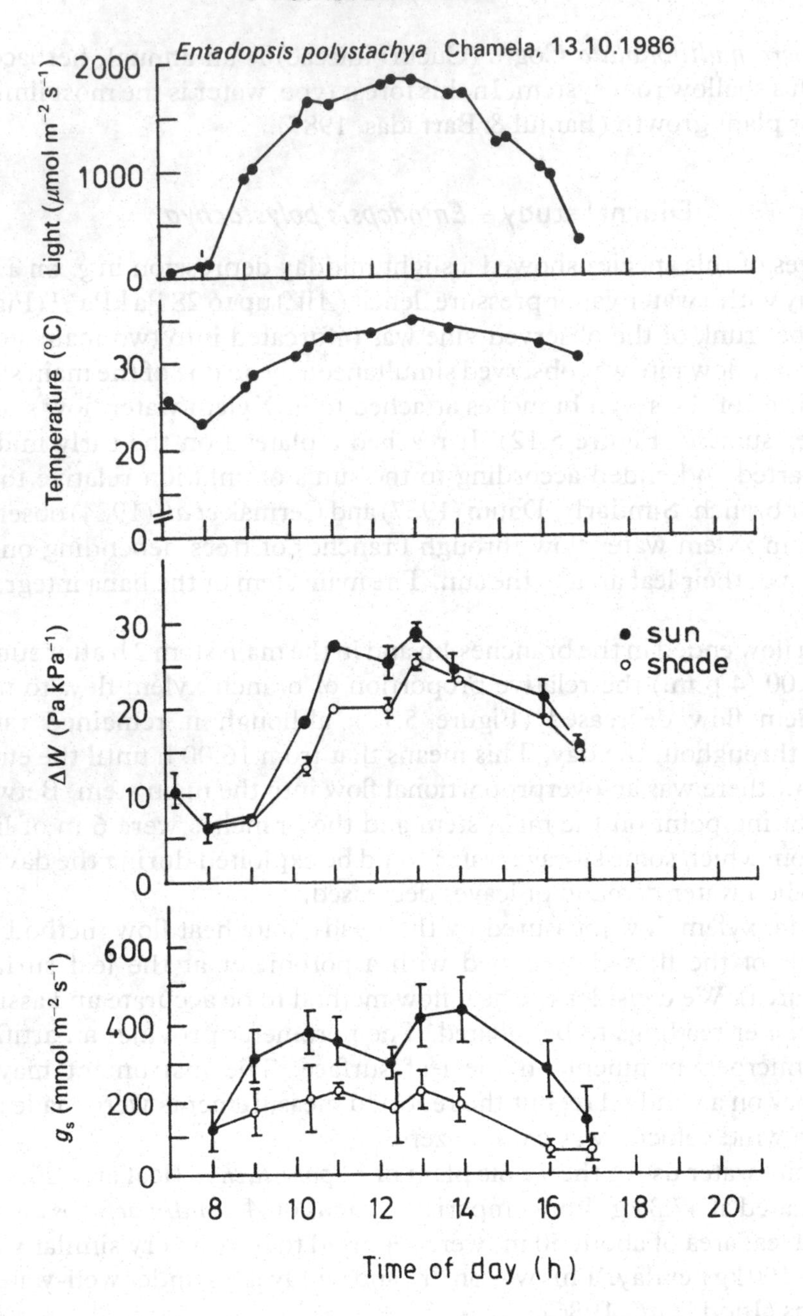

Figure 5.11. Diurnal time-course of light, air temperature, leaf to air water vapor pressure deficit (ΔW) and stomatal conductance (g_s) for sun and shade leaves of the liana *Entadopsis polystachya* (Fabaceae) in Chamela, Mexico. $N = 3$; bars indicate standard deviation (not shown when less than size of symbol). From Fichtner & Schulze (1990).

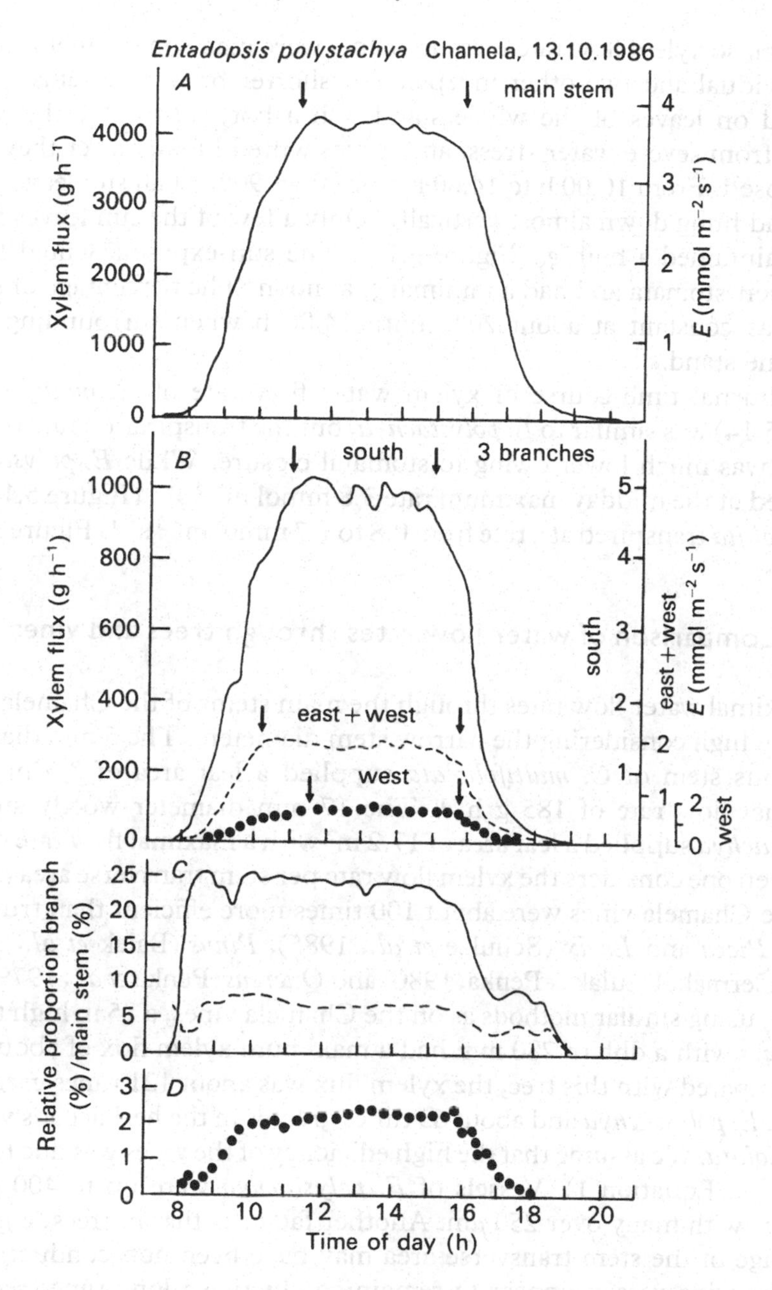

Figure 5.12. Measurements of water transport in the liana *Entadopsis polystachya*. Diurnal time-course of stem xylem water flow rate and transpiration (E) of the main stem (A) and branches (B), and the relative proportion of branch to main stem xylem water flow rate (C and D). The beginning and end of xylem water flux plateaux are marked with arrows. Measurements made by the steady state heat flow method, with E calculated as xylem flow rate/distal leaf area. From Fichtner & Schulze (1990).

main stem, so xylem flow rate measurements were made on the main stem of one individual and two other independent shorter branch systems. g_s was measured on leaves of the whole stand. On a hot, sunny day the plants suffered from severe water stress, and leaves wilted rapidly after they were sun-exposed. From 10.00 h to 14.00 h more than 90% of all sun leaves were wilted and hung down almost vertically. Only a few of the sun leaves in the stand maintained a high g_s (Figure 5.13). The sun-exposed, wilted leaves closed their stomata and had a minimal g_s at noon. The percentage of shade leaves was constant at about 20% until 14.00 h when surrounding trees shaded the stand.

The diurnal time-course of xylem water flow rate of *C. multifoliolata* (Figure 5.14) was similar to *E. polystachya*, but the transpiration rate per leaf area (E) was much lower owing to stomatal closure. While *E. polystachya* transpired at the midday maximum rate 3.8 mmol m^{-2} s^{-1} (Figure 5.12), *C. multifoliolata* transpired at a rate from 0.8 to 1.7 mmol m^{-2} s^{-1} (Figure 5.14).

Comparison of water flow rates through trees and vines

The maximal water flow rates through the main stems of the Chamela vines was quite high considering the narrow stem diameters. The 5 mm diameter herbaceous stem of *C. multifoliolata* supplied a leaf area of 3.6 m^2 with a maximal flow rate of 185 g h^{-1}. The 17 mm diameter woody stem of *E. polystachya* supplied a leaf area of 17.2 m^2 with a maximal flow rate of 4 kg h^{-1}. When one considers the xylem flow rate per stem transverse area (xylem flux), the Chamela vines were about 100 times more efficient than trunks of trees of *Picea* and *Larix* (Schulze *et al.*, 1985), *Pinus* (Balek *et al.*, 1983), *Betula* (Čermák, Hzulak & Penka, 1980) and *Quercus* (Penka *et al.*, 1979). For instance, using similar methods as on the Chamela vines, a 25 m high tree of *Picea abies* with a dbh of 280 mm had a maximum xylem flux of about 9 kg h^{-1}. Compared with this tree, the xylem flux was about 121 times greater in the liana *E. polystachya* and about 65 times greater in the herbaceous vine *C. multifoliolata*. We assume that the high efficiency of the vines was due to wide vessels (see Equation 1). Vessels of *E. polystachya* were up to 400 μm in diameter, with many over 250 μm. Another factor is that in trees, a greater percentage of the stem transverse area may have been non-conductive. In vines, the wide vessels appear to remain conductive a long time (see subsection on conductive life span, above).

The xylem flow rate data on naturally growing vines of *C. multifoliolata* and *E. polystachya* and the evidence from cultivated *Bauhinia* establish that narrow vine stems transport large volumes of water quickly. Further comparative studies on naturally growing trees, shrubs, herbs, herbaceous vines, and

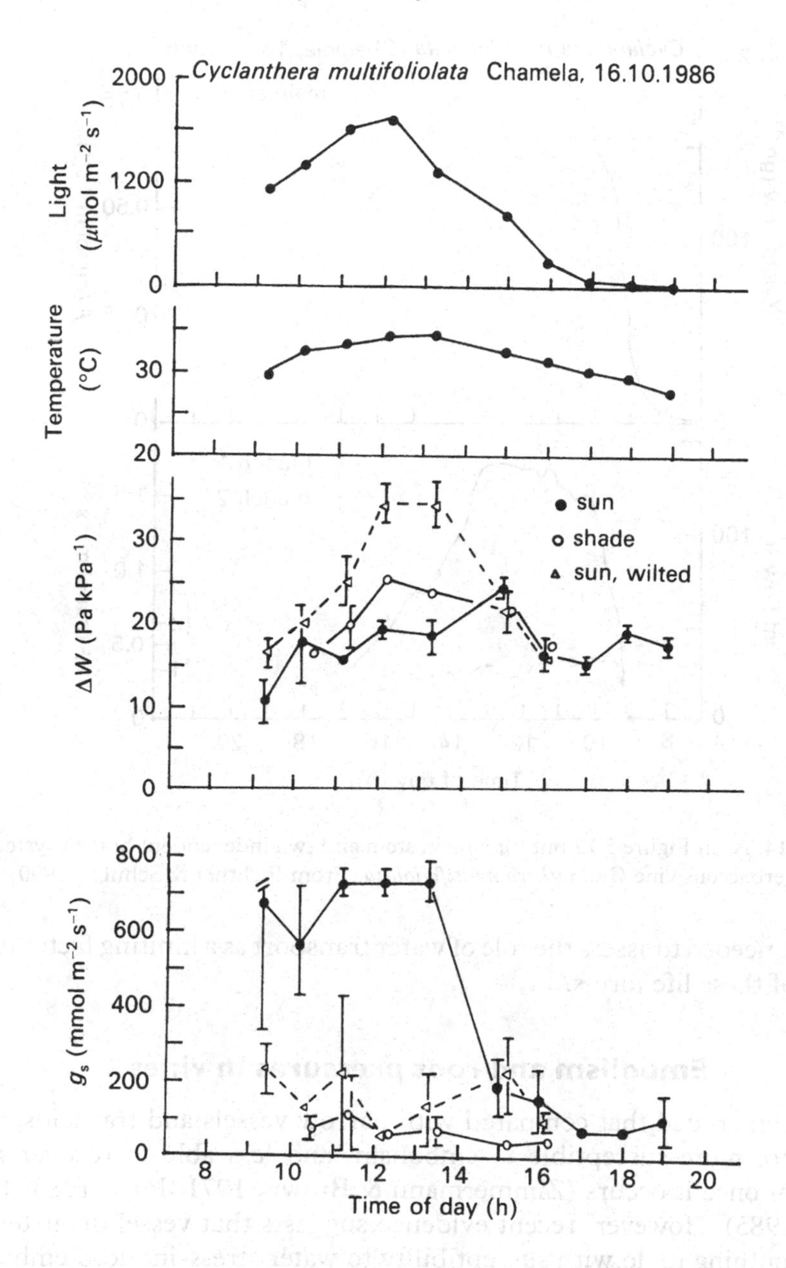

Figure 5.13. As in Figure 5.11 but for sun, wilted sun and shade leaves in the stand of the herbaceous vine *Cyclanthera multifoliolata* (Cucurbitaceae). From Fichtner & Schulze (1990).

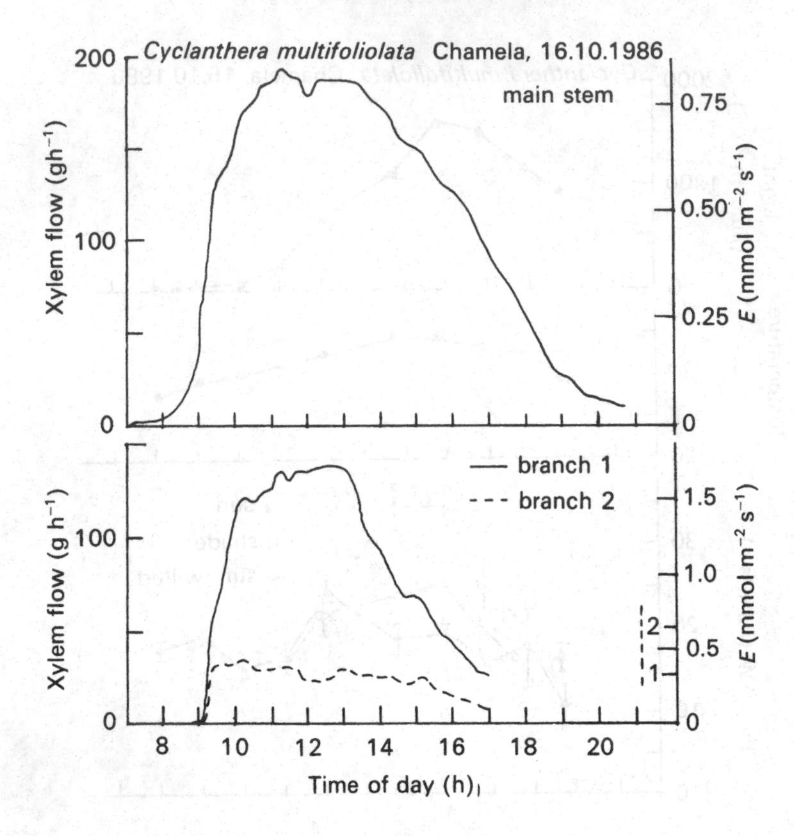

Figure 5.14. As in Figure 5.12 but for a main stem and two independent branch systems of the herbaceous vine *Cyclanthera multifoliolata*. From Fichtner & Schulze (1990).

lianas are needed to assess the role of water transport as a limiting factor in the growth of these life forms.

Embolism and root pressures in vines

It has been argued that compared with narrow vessels and tracheids, wide vessels are more susceptible to embolism, and less able to recover from embolism once it occurs (Zimmermann & Brown, 1971; Putz, 1983, 1984, Ewers, 1985). However, recent evidence suggests that vessel diameter has little or nothing to do with susceptibility to water stress-induced embolism (Crombie, Hipkins & Milburn, 1985; Sperry, 1986; Tyree & Dixon, 1986; Lewis, 1988; Sperry & Tyree, 1988). It is not known whether, in general, the wide vessels of vines avoid embolism altogether or whether they are able to refill their vessels following embolism.

Many vines produce positive root pressures at night or after rainstorms (Scholander, Love & Kanwisher, 1955; Scholander, 1958; Putz, 1983;

McAneney & Judd, 1983; Smart & Coombe, 1983; F. W. Ewers and J. B. Fisher, unpublished data). Such root pressures are extremely rare for trees and shrubs. As suggested by Putz (1983), root pressures might serve to refill cavitated vessels and may thus help extend their conductive life span. Diurnal vessel refilling, that is, dissolving of vapor bubbles under positive water potentials, is thought to occur in small herbs (Milburn & McLaughlin, 1974; Milburn, 1979) and in *Zea mays* (Tyree *et al.*, 1986). Seasonal vessel refilling seems to occur in certain tree species with rather narrow tracheary elements, for instance in the palm *Rhapis excelsa* (Sperry, 1986), in the dicotyledon *Acer saccharum* (Sperry, Donnelly & Tyree, 1988), and in conifers (Waring & Running, 1978). Among vines, vessel refilling has been clearly documented in the temperate lianas *Vitis labrusca* and *V. riparia*, where vessels are air-filled during the winter and become refilled via root pressures before the leaves emerge in the spring (Hales, 1727; Scholander *et al.*, 1955; O'Leary, 1965; Sperry *et al.*, 1987). There is also recent evidence that vessels of *Vitis vinifera* become embolized and refilled on a diurnal basis throughout the growing season (Schultz & Matthews, 1988).

We have observed positive root pressures in two large liana specimens of *Bauhinia* spp., *Mucuna biplicata* Teif & Binn. ex Kurz, and *Spatholobus* cf. *oblongifolius* Marr. (all Fabaceae) growing naturally in Borneo, and, at Fairchild Tropical Garden, in *Bauhinia fassoglensis*, *Mucuna sloanei* Fawcett & Rendle, and the climbing monocotyledon *Luzuriaga latifolia* (R.Br.) Poir. (Philesiaceae). However, except for the refilling of vessels in temperate species of *Vitis*, we do not know whether the root pressures in vines are sufficient to allow for repair of embolized vessels.

It is not clear whether morphological features of the root systems of vines make them particularly well adapted for water uptake or for the production of positive root pressures. Although *Vitis* is reported to have an extensive root system, in cultivated plants the root length per soil area and the root density per soil volume are actually lower than in pear trees, prune trees, conifers, and cereal grasses (Smart & Coombe, 1983). Although plants of *Vitis* exhibit 'bleeding' from positive root pressures in the winter and especially in the spring before bud break, root growth is much delayed (Smart & Coombe, 1983); it does not occur until 3–10 weeks after bud break. In contrast, most temperate deciduous trees commence root growth before renewed shoot growth in the spring.

The liana *Bauhinia fassoglensis*, for which we have measured positive root pressures (Ewers *et al.*, 1989), has enormous root tubers, up to 2.5 m long and 0.5 m wide (Brenan, 1967). The liana *Entadopsis polystachya* also has root tubers, which may explain, in part, why its leaves wilt much less readily than those of the herbaceous vine *Cyclanthera multifoliolata*, which has a shallow root system. We presume that root tubers are significant for water storage as well as for regeneration of the plant (see Fisher & Ewers, Chapter 4). It would be useful to know how common such tubers are in vines, and if there is a correlation with root pressures.

Summary and conclusions

Water conduction through the xylem may be one of the important factors limiting the growth and survival of vines in nature. Vines have particularly narrow stems in relation to the leaf surface area they supply. This is due in part to the reduced mechanical demands on the stem, since vines, by definition, are not self-supporting. In naturally growing vines of *Entadopsis polystachya* and *Cyclanthera multifoliolata*, the narrow stems exhibit a surprisingly high xylem flux. This may be because in vines in general, vessels are wider than in closely related trees. However, in addition to some wide and long vessels, vine stems each have many short and narrow vessels, the functional significance of which is uncertain. In the genus *Bauhinia*, which has trees, shrubs, and lianas, the relatively few wide vessels of the liana stems appear to compensate hydraulically for the narrow stem diameters. Anomalous patterns of secondary growth appear not to have much effect on water transport since, in three dimensions, the various portions of the xylem system are interconnected. In species with an inner/outer system of secondary xylem, the outer system is far more efficient in water transport than the inner system, because of longer and wider vessels. However, the functional and evolutionary significance of the inner/outer system may have more to do with searcher shoot biomechanics than with hydraulics. Although vines are often reported to have positive root pressures which could serve to refill cavitated vessels, it is not clear whether vessel refilling occurs in vines other than *Vitis*. In addition, it is not known if any morphological and/or anatomical peculiarities of vine root systems are responsible for the production of positive root pressures in the stem.

References

Ayensu, E. S. & Stern, W. L. (1964). Systematic anatomy and ontogeny of the stem in Passifloraceae. *Contributions from the US National Herbarium* **34**: 45–71.

Baas, P. (1982). Systematic, phylogenetic, and ecological wood anatomy. In *New Perspectives in Wood Anatomy*, ed. P. Baas, pp. 23–58. Nijhoff, Junk, The Hague.

Balek, J., Čermák, J., Kučera, J. & Prax, A. (1983). A direct method for forest transpiration measurement. *Journal of Hydrology* **66**: 123–31.

Bamber, R. K. (1984). Wood anatomy of some Australian rainforest vines.

In *Proceedings of Pacific Regional Wood Anatomy Conference*, ed. Syoji Sudo, pp. 58–60. Forest and Forest Products Research Institute, Tsukuba, Ibaraki, Japan.

Bell, D. J., Forseth, I. N., & Teramura, A. H. (1988). Field water relations of three temperate vines. *Oecologia (Berlin)* 74: 537–45.

Berger, W. (1931). Das Wasserleitungssystem von krautigen Pflanzen, Zwergsträuchern und Lianen in quantitativer Betrachtung. *Beih. Bot. Cbl.* 48 (I): 363–90.

Boyer, J. S. (1985). Water transport. *Annual Review of Plant Physiology* 36: 473–516.

Braun, H. J. (1983). Zur Dynamik des Wassertransportes in Bäumen. *Berichte der Deutschen Botanischen Gesellschaft* 96: 29–47.

Brenan, J. P. M. (1967). *Flora of Tropical East Africa.* p. 215.

Carlquist, S. (1975). *Ecological Strategies of Xylem Evolution.* University of California Press, Berkeley, Ca.

Carlquist, S. (1985). Observations on functional wood histology of vines and lianas: vessel dimorphism, tracheids, vasicentric tracheids, narrow vessels, and parenchyma. *Aliso* 11: 139–57.

Carlquist, S. (1988). *Comparative Wood Anatomy: systemic, ecological and evolutionary aspects of dicotyledon wood.* Springer-Verlag, Berlin.

Čermák, J., Deml, M. & Penka, M. (1973). A new method of sap flow rate determination in trees. *Biologia Plantarum (Praha)* 15: 171–8.

Čermák, J., Hzulak, J. & Penka, M. (1980). Water potential and sap flow rate in adult trees with moist and dry soil as used for the assessment of root system depth. *Biologia Plantarum (Praha)* 22: 34–41.

Čermák, J., Jeník, J., Kučera, J. & Židek, V. (1984). Xylem water flow in a crack willow tree (*Salix fragilis* L.) in relation to diurnal changes of environment. *Oecologia* 64: 145–51.

Coster, C. (1927). Zur Anatomie und Physiologie der Zuwachszonen- und Jahresringbildung in den Tropen. *Annales du Jardin Botanique de Buitenzorg* 38: 1–114.

Crombie, D. S., Hipkins, M. F. & Milburn, J. A. (1985). Gas penetration of pit membranes in the xylem of *Rhododendron* as the cause of acoustically detectable sap cavitation. *Australian Journal of Plant Physiology* 12: 445–53.

Daum, C. R. (1967). A method for determining water transport in trees. *Ecology* 48: 425–31.

Edwards, W. R. N. & Warwick, N. W. M. (1984). Transpiration from a kiwifruit vine as estimated by the heat pulse technique and the Penman-Monteith equation. *New Zealand Journal of Agricultural Research* 27: 537–43.

Ellmore, G. S. & Ewers, F. W. (1986). Fluid flow in the outermost xylem

increment of a ring-porous tree, *Ulmus americana*. *American Journal of Botany* 73: 1771–4.
Ewers, F. W. (1985). Xylem structure and water conduction in conifer trees, dicot trees, and lianas. *International Association of Wood Anatomists Bulletin, n.s.* **6**: 309–17.
Ewers, F. W. & Cruiziat, P. (1990). Measuring water transport and storage. In *Techniques and Approaches in Forest Tree Ecophysiology*, ed. J. P. Lassoie and T. M. Hinckley, Boca Raton, CRC Press, pp. 91–115.
Ewers, F. W. & Fisher, J. B. (1987). A survey of vessel lengths and diameters in stems of some tropical lianas. *American Journal of Botany* 75: 613 (abstract).
Ewers, F. W. & Fisher, J. B. (1989a). Techniques for measuring vessel lengths and diameters in stems of woody plants. *American Journal of Botany* 76: 645–56.
Ewers, F. W. & Fisher, J. B. (1989b). Variation in vessel length and diameter in stems of some tropical and subtropical lianas. *American Journal of Botany* 76: 1452–9.
Ewers, F. W., Fisher, J. B. & Chiu, S.-T. (1988). Xylem structure and water transport in *Bauhinia*. *American Journal of Botany* 75(6) part 2: 25–6 (abstract).
Ewers, F. W., Fisher, J. B. & Chiu, S.-T. (1989). Water transport in the liana *Bauhinia fassoglensis* (Fabaceae). *Plant Physiology* **91**: 1625–31.
Ewers, F. W. & Zimmermann, M. H. (1984a). The hydraulic architecture of balsam fir (*Abies balsamea*). *Physiologia Plantarum* **60**: 453–8.
Ewers, F. W. & Zimmermann, M. H. (1984b). The hydraulic architecture of eastern hemlock (*Tsuga canadensis*). *Canadian Journal of Botany* **62**: 940–6.
Fanjul, L. & Barradas, V. L. (1987). Diurnal and seasonal variation in the water relations of some deciduous and evergreen trees of a deciduous forest of the western coast of Mexico. *Journal of Applied Ecology* **24**: 289–303.
Fegel, A. C. (1941). Comparative anatomy and varying physiological properties of trunk, branch, and root wood in certain northeastern trees. *Bull. NY. State Coll. For. Syracuse Univ. Vol. 14(2b) Tech. Publ.* No. 55: 1–20.
Fichtner, K. & Schulze, E. D. (1990). Xylem water flow in tropical vines as measured by a steady state heating method. *Oecologia* **82**: 355–61.
Fisher, J. B. & Ewers, F. W. (1988). Water pathways in liana stems with anomalous secondary growth. *American Journal of Botany* 75(6) part 2: 26 (abstract).
Fisher, J. B. & Stevenson, J. W. (1981). Occurrence of reaction wood in

branches of dicotyledons and its role in tree architecture. *Botanical Gazette* **142**: 82–95.

Gartner, B. L., Bullock, S. H., Mooney, H. A., Brown, V. B. & Whitbeck, J. L. (1990). Water transport properties of vine and tree stems in a tropical deciduous forest. *American Journal of Botany* 77: 742–9.

Gibson, A. C., Calkin, A. C. & Nobel, P. S. (1985). Hydraulic conductance and xylem structure in some tracheid-bearing plants. *International Association of Wood Anatomists Bulletin, n.s.* **6**: 293–302.

Green, S. R. & Clothier, B. E. (1988). Water use of kiwifruit vines and apple trees by the heat-pulse technique. *Journal of Experimental Botany* **39**: 115–23.

Haberlandt, G. (1914). *Physiological Plant Anatomy*. Macmillan, London.

Hales, S. (1727). *Vegetable staticks, or an account of some statical experiments on the sap in vegetables.* J. Peele, London.

Hallé, F., Oldeman, R. A. A. & Tomlinson, P. B. (1978). *Tropical Trees and Forests: an architectural analysis.* Springer-Verlag, Berlin.

Hellkvist, J., Richards, G. P. & Jarvis, P. G. (1974). Vertical gradients of water potential and tissue water relations in Sitka spruce trees measured with the pressure chamber. *Journal of Applied Ecology* **11**: 637–68.

Huber, B. (1935). Die physiologische Bedeutung der Ring- und Zerstreutporigkeit. *Berichte der Deutschen Botanische Gesellschaft* **53**: 711–19.

Huber, B. (1956). Die Gefässleitung. In *Encyclopedia of Plant Physiology, Vol. 3*, ed. W. Ruhland. Springer-Verlag, Berlin.

Judd, M. J., McAneney, K. J. & Trought, M. C. T. (1986). Water use by sheltered kiwifruit under advective conditions. *New Zealand Journal of Agricultural Research* **29**: 83–92.

Klotz, L. H. (1978). Observations on diameters of vessels in stems of palms. *Principes* 22: 99–106.

Kramer, P. J. & Kozlowski, T. T. (1979). *Physiology of Woody Plants.* Academic Press, New York.

Lang, G. E. & Knight, D. H. (1983). Tree growth, mortality, recruitment and canopy gap formation during a 10-year period in a tropical moist forest. *Ecology* **64**: 1075–80.

Lewis, A. M. (1988). A test of the air-seeding hypothesis using *Sphagnum* hyalocysts. *Plant Physiology* **87**: 577–82.

Lieberman, D., Lieberman, M., Hartshorn, G. & Peralta, R. (1985). Growth rates and age-size relationships of tropical wet forest trees in Costa Rica. *Journal of Tropical Ecology* **1**: 97–109.

Liu, W. T., Wenkert, W., Allen, L. H. & Lemon, E. R. (1978). Soil-plant-water relations in a New York vineyard: resistances to water movement. *Journal of the American Society for Horticultural Science* **103**: 226–30.

McAneney, K. J. & Judd, M. J. (1983). Observations on kiwifruit (*Actinidia chinensis* Planch.) root exploration, root pressure, hydraulic conductivity, and water uptake. *New Zealand Journal of Agricultural Research* **26**: 507–10.

Milburn, J. A. (1979). *Water Flow in Plants*. Longman, London.

Milburn, J. A. & McLaughlin, M. E. (1974). Studies of cavitation in isolated vascular bundles and whole leaves of *Plantago major* L. *New Phytologist* **73**: 861–71.

Newbanks, D., Bosch, A. & Zimmermann, M. H. (1983). Evidence for xylem dysfunction by embolization in Dutch elm disease. *Phytopathology* **73**: 1060–3.

Obaton, M. (1960). Les lianes ligneuses a structure anormale des forêts denses d'Afrique occidentale. *Annales des Sciences naturelles (Botanique) Sér. 12*, **1**: 1–220.

O'Leary, J. W. (1965). Root-pressure exudation in woody plants. *Botanical Gazette* **126**: 108–15.

Penka, M., Čermák, J., Stepanek, V. & Palet, M. (1979). Diurnal courses of transpiration rate and transpiration flow rate as determined by the gravimetric and thermometric methods in a full-grown oak tree (*Quercus robur* L.). *Acta Universitatis Agriculturae (Brno), Series C* **48**: 3–30.

Pfeiffer, H. (1926). Das abnorme Dickenwachstum. In *Handbuch der Pflanzenanatomie*, Bd. 9, Lief 15, ed. K. Linsbauer. Borntraeger, Berlin.

Putz, F. E. (1983). Liana biomass and leaf area of a 'tierra firme' forest in the Rio Negro Basin, Venezuela. *Biotropica* **15**: 185–9.

Putz, F. E. (1984). The natural history of lianas on Barro Colorado Island, Panama. *Ecology* **65**: 1713–24.

Putz, F. E. (1990). Liana stem diameter growth and mortality rates on Barro Colorado Island, Panama. *Biotropica* **22**: 103–5.

Putz, F. E. & Windsor, D. M. (1987). Liana phenology on Barro Colorado Island, Panama. *Biotropica* **19**: 334–41.

Riedl, H. (1937). Bau und Leistungen des Wurzelholzes. (Structure and function of root wood). *Jahrbuchur für wissenschaftliche Botanik* **85**: 1–75 (Xerox copy of English translation available from National Translation Center, 35 West 33rd St., Chicago, IL 60616, USA).

Sakuratani, T. (1984). Improvement of the probe for measuring water flow rate in intact plants with the stem heat balance method. *Journal of Agricultural Meteorology* **40**(3): 273–7.

Salleo, S., Rosso, R. & Lo Gullo, M. A. (1982). Hydraulic architecture of

Vitis vinifera L. and *Populus deltoides* Bartr. 1-year-old twigs: I. Hydraulic conductivity (LSC) and water potential gradients. *Giornale Botanico Italiano* **116**: 15–27.

Schenck, H. (1893). Beiträge zur Biologie und Anatomie der Lianen, im Besonderen der in Brasilien einheimischen Arten, 2. Beiträge zur Anatomie der Lianen. In *Botanische Mittheilungen aus der Tropen, 5*, ed. A. F. W. Schimper, pp. 1–248. G. Fischer, Jena.

Scholander, P. F. (1958). The rise of sap in lianas. In *The Physiology of Forest Trees*, ed. K. V. Thimann, pp. 3–17. Ronald Press, New York.

Scholander, P. F., Love, W. E. & Kanwisher, J. W. (1955). The rise of sap in tall grapevines. *Plant Physiology* **30**: 93–104.

Schulze, E.-D., Čermák, J., Matyssek, R., Penka, M., Zimmermann, R. & Vasicek, F. (1985). Canopy transpiration and water fluxes in the xylem of the trunk of *Larix* and *Picea* trees – a comparison of xylem flow, porometer and cuvette measurements. *Oecologia* **66**: 475–83.

Schultz, H. R. & Matthews, M. (1988). Resistance to water transport in shoots of *Vitis vinifera* L. *Plant Physiology* **88**: 718–24.

Siau, J. F. (1984). *Transport Processes in Wood*. Springer-Verlag, Berlin.

Skene, D. S. & Balodis, V. (1968). A study of vessel length in *Eucalyptus obliqua* L'Hérit. *Journal of Experimental Botany* **19**: 825–30.

Smart, R. E. & Coombe, B. G. (1983). Water relations of grapevines. In *Water Deficits and Plant Growth, Vol. VII*, ed. T. T. Kozlowski, pp. 137–96. Academic Press, New York.

Sperry, J. S. (1986). Relationship of xylem embolism to xylem pressure potential, stomatal closure, and shoot morphology in the palm *Rhapis excelsa*. *Plant Physiology* **80**: 110–16.

Sperry, J. S. & Tyree, M. T. (1988). Mechanism of water stress-induced xylem embolism. *Plant Physiology* **88**: 581–7.

Sperry, J. S., Donnelly, J. R. & Tyree, M. T. (1988). Seasonal occurrence of xylem embolism in sugar maple (*Acer saccharum*). *American Journal of Botany* 75: 1212–18.

Sperry, J. S., Holbrook, N. M., Zimmermann, M. H. & Tyree, M. T. (1987). Spring filling of xylem vessels in wild grapevine. *Plant Physiology* **83**: 414–17.

Swanson, R. H. & Whitfield, D. W. A. (1981). A numerical analysis of the heat pulse velocity theory and practice. *Journal of Experimental Botany* **32**: 221–39.

ter Welle, B. J. H. (1985). Differences in wood anatomy of lianas and trees. *International Association of Wood Anatomists Bulletin, n.s.* **6**: 70 (abstract).

Tyree, M. T. (1988). A dynamic model for water flow in a single tree: evidence that models must account for hydraulic architecture. *Tree Physiology* **4**: 195–217.

Tyree, M. T., Caldwell, C. & Dainty, J. (1975). The water relations of hemlock (*Tsuga canadensis*). V. The localization of resistances to bulk water flow. *Canadian Journal of Botany* **53**: 1078–84.

Tyree, M. T. & Dixon, M. A. (1986). Water stress induced cavitation and embolism in some woody plants. *Physiologia Plantarum* **66**: 397–405.

Tyree, M. T., Fiscus, E. L., Wullschleger, S. D. & Dixon, M. A. (1986). Detection of xylem cavitation in corn under field conditions. *Plant Physiology* **82**: 597–9.

Tyree, M. T., Graham, M. E. D., Cooper, K. E. & Bazos, L. J. (1983). The hydraulic architecture of *Thuja occidentalis*. *Canadian Journal of Botany* **61**: 2105–11.

Van Vleit, G. J. C. M. (1981). Wood anatomy of the palaeotropical Melastomataceae. *Blumea* **27**: 395–462.

Waring, R. H. & Running, S. W. (1978). Sapwood water storage: its contribution to transpiration and effect upon water conductance through the stems of old-growth Douglas fir. *Plant, Cell and Environment* **1**: 131–40.

Waring, R. H., Whitehead, D. & Jarvis, P. G. (1979). The contribution of stored water to transpiration in Scots pine. *Plant, Cell and Environment* **2**: 309–17.

Winkler, A. J., Cook, J. A., Kliewer, W. M. & Lider, L. A. (1974). *General Viticulture*. University of California Press, Berkeley, Ca.

Zimmermann, M. H. (1978). Hydraulic architecture of some diffuse-porous trees. *Canadian Journal of Botany* **56**: 2286–95.

Zimmermann, M. H. (1983). *Xylem Structure and the Ascent of Sap*. Springer-Verlag, Berlin.

Zimmermann, M. H. & Brown, C. L. (1971). *Trees. Structure and Function*. Springer-Verlag, New York.

Zimmermann, M. H. & Jeje, A. A. (1981). Vessel-length distribution in stems of some American woody plants. *Canadian Journal of Botany* **59**: 1882–92.

Zimmermann, M. H. & Milburn, J. A. (1982). Transport and storage of water. In *Physiological Plant Ecology II, Encyclopedia of Plant Physiology, n.s., Vol. 12B*, ed. O. L. Lange, P. S. Nobel, C. B. Osmond and H. Ziegler, pp. 135–51. Springer-Verlag, Berlin.

Zimmermann, M. H. & Potter, D. (1982). Vessel-length distribution in branches, stem, and roots of *Acer rubrum*. *International Association of Wood Anatomists Bulletin, n.s.* **3**: 103–9.

Zimmermann, M. H. & Sperry, J. S. (1983). Anatomy of the palm *Rhapis excelsa*. IX. Xylem structure of the leaf insertion. *Journal of the Arnold Arboretum* **64**: 599–609.

6

Reserve economy of vines

H. A. MOONEY AND BARBARA L. GARTNER

Introduction

The principal advantage of the vine growth form is the potential for rapid growth rate. Carbon that would otherwise go into supporting stem tissue can go into leaves, thus yielding a compounding of the growth increment (Monsi & Murata, 1970). Although the difference in stem allocation between vines and other plants is clear, what has not been well documented is the proportional allocation of carbon and other resources in vines to non-productive compounds and structures, including roots, compared to other growth forms. Here we examine at least a part of the area by focusing on the pool sizes and utilization patterns of storage reserves in various vine species.

Why should we expect vining plants to have different storage cycles from self-supporting plants in the same habitat? Trees and shrubs often store much of their reserves of water, carbohydrates, and nutrients within the massive stem. As a consequence of the vine growth form, vines have much less stem-to-leaf mass than do self-supporting plants, and so may require specialized sites in which to store reserves, and cycles to deposit and withdraw these reserves. If the vining strategy results in greater risk of physical injury than other growth strategies, as discussed by Fisher & Ewers (Chapter 4), then vines may benefit from having greater energy stores for repair and regrowth than trees. In the case of herbaceous perennials that die back during the unfavorable period, vines may rely upon their stored reserves to regain quickly the canopy position before deterioration of the light climate due to leaf out by overstory trees. Moreover, as will be shown, reserve carbohydrates may also play a unique role in the water balance of certain vine species, rather than only in supporting growth.

Most of our information comes from vines of economic importance that live in seasonal habitats of some sort. Data on wild plants are not abundant.

These constraints limit our ability to compare vines with plants of other growth forms from the same habitat. However, vines are well represented in agronomic literature and because of their carbohydrate storage capacity, vines have been of importance to humans through time. As an example, Pate & Dixon (1982) reviewed the biology of over 200 species of Western Australian plants that have fleshy underground storage organs. They noted that the storage organs of 31 genera of these plants were utilized as food plants by aborigines. Nearly half of the plants studied were vines and included both woody and herbaceous species. More globally, vines have been domesticated for carbohydrate including *Ipomoea batatas* (sweet potato) and *Dioscorea* spp. (yams).

In this survey we will focus principally on carbohydrate compounds that may be considered storage reserves.

Generalized storage cycles of woody and herbaceous perennials

The general pattern of carbon storage in perennial plants is detailed elsewhere (i.e. Priestley, 1962; Larcher, 1980) and, thus, will be simply outlined here. A woody plant increases its above-ground biomass from year to year with most of the carbon serving structural functions and unavailable for growth after it has been deposited. However, some carbon and other elements may be stored in non-structural compounds and utilized as reserves. In trees, stems generally serve as the most important storage depot, although evergreens often use leaves for storage of minerals, carbohydrate and lipids. Storage materials deposited in the stem, either cortex or xylem, will remain accessible as long as the cells containing them are alive. In woody plants carbohydrate can also be stored below ground in roots or rarely in specialized storage organs.

The leaves of evergreen plants generally can fix carbon and thus provide some of the energy for growth before the new flush of leaves appears. As the growing season progresses, reserves that were stored in leaves and stem, as well as newly produced photosynthate, are directed to new stem (both primary and secondary), leaf, and root growth. When new leaves have expanded partway they become self-sufficient for carbon. When meristematic activity declines but photosynthesis continues, the leaves begin to replenish their reserves. If reproductive growth occurs, reproductive organs become carbon sinks. The energy sources can be either the current year's photosynthate or stored material, depending on the species. As the favorable season for growth ends, carbohydrate and lipid pools are formed and stored once again in various plant organs. Some carbon is depleted during the dormant season for maintenance respiration from all living organs.

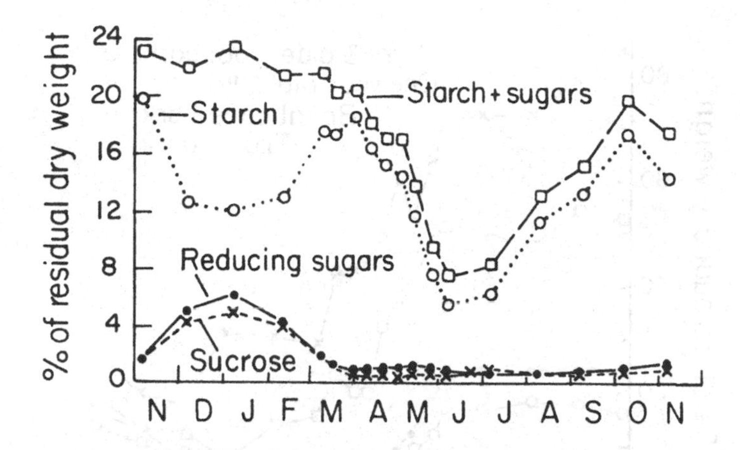

Figure 6.1. Seasonal change of carbohydrate fractions in basal cane wood of *Vitis vinifera* cultivar 'Carignane'. From Winkler & Williams (1945).

In deciduous trees there are obviously no functioning leaves to produce photosynthate until new ones are constructed and no leaves persisting in the dormant season in which to store or respire materials. Herbaceous perennials that die back each year differ from the above patterns in several ways. During the establishment phase they build up large storage organs, unlike trees which are investing in above-ground structures. Also, they have no above-ground storage pools during the dormant season, and so all material must be stored below ground. Maintenance respiration below ground can be substantial. As with deciduous woody plants, all very early growth comes from stores.

Storage cycles of the basic vine types

Woody vines

Carbohydrate storage cycle: *Vitis vinifera* (grape). Our knowledge of the carbohydrate economy of woody vines is limited. By far the greatest available information relates to *Vitis vinifera*. Winkler & Williams (1945) described in unusual detail the seasonal carbohydrate cycle of the cultivar 'Carignane' of this species. The sampling program was carried out at Davis, California, USA in plants with trunks 4 years of age.

The generalized pattern of the seasonal carbohydrate utilization cycle is shown for basal cane wood (Figure 6.1). Relatively high concentrations of starch are present in this tissue at the beginning of the dormant period. Concentrations go down slowly during the winter due to respiratory losses.

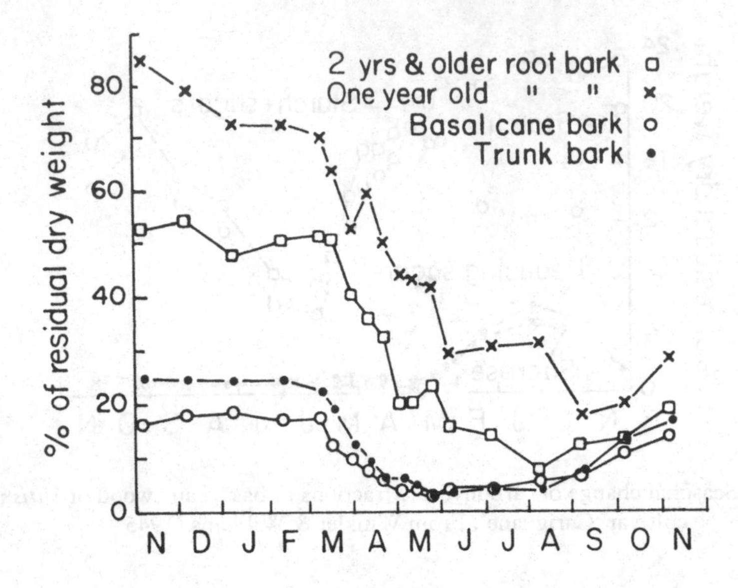

Figure 6.2. Seasonal change in bark carbohydrate contents (starch + sugar) in various positions on *Vitis vinifera* plants. From Winkler & Williams (1945).

Conversions from starch support this respiratory activity. Starting with March bud break, growth reserves begin to be depleted. This depletion continues until flowering at the end of May. When stem growth ceases in early July reserves begin to be restored.

Relatively high concentrations of total non-structural carbohydrate (TNC, the carbohydrate not bound into structural compounds such as cellulose) are present in all tissues but differences exist in the absolute amounts and in the degree of their seasonal depletion. The bark on older roots consists of more than 50% TNC (Figure 6.2). This tissue becomes nearly depleted by the end of the growth period. The younger bark also has a very high carbohydrate content. The above-ground bark tissues have lesser absolute amounts of TNC but the deep depletion cycles are comparable with those of the roots.

The TNC in the above-ground wood shows an increasingly dampened cycle with age (Figure 6.3), perhaps indicating decreasing availability of carbohydrate stored in older wood. The root wood shows a strong depletion cycle (Figure 6.4).

It would thus appear from the Winkler & Williams study that carbohydrate reserves, principally starch, play a role in supporting rapid spring stem growth in grape. Unfortunately, the lack of absolute values of carbohydrate stores in the various plant parts makes it impossible to develop a quantitative picture of dry matter conversions. Buttrose (1966), however, presents data on

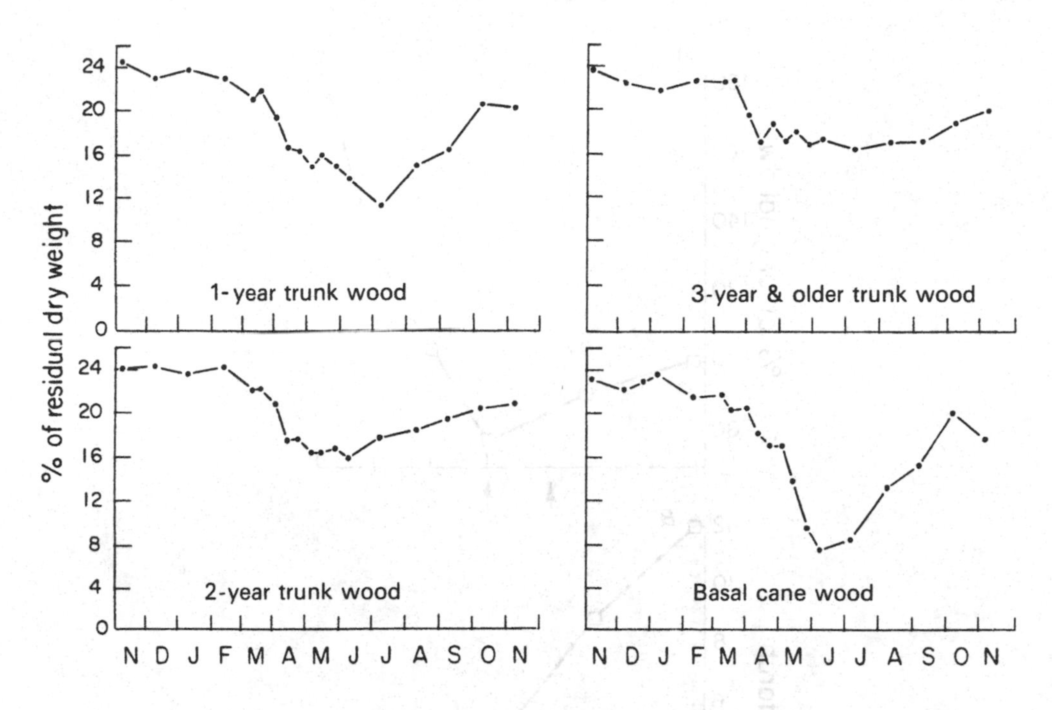

Figure 6.3. Seasonal change in carbohydrate contents (starch + sugar) of trunk and cane wood of different ages. From Winkler & Williams (1945).

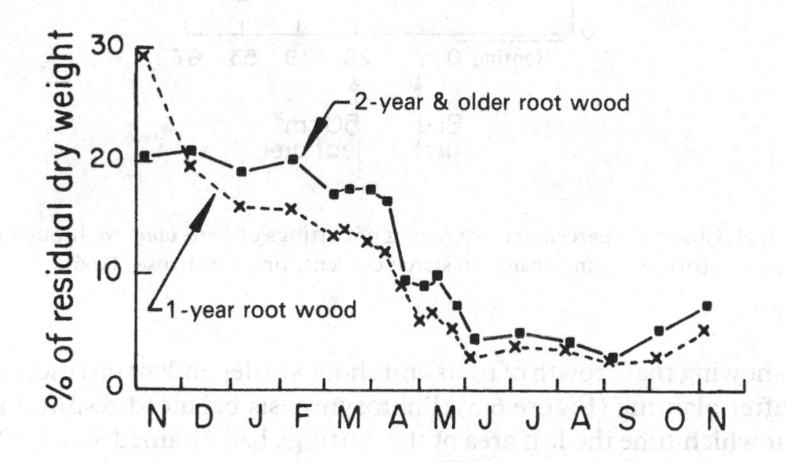

Figure 6.4. Seasonal change in carbohydrate contents (starch + sugar) of *Vitis vinifera* root wood of different ages. From Winkler & Williams (1945).

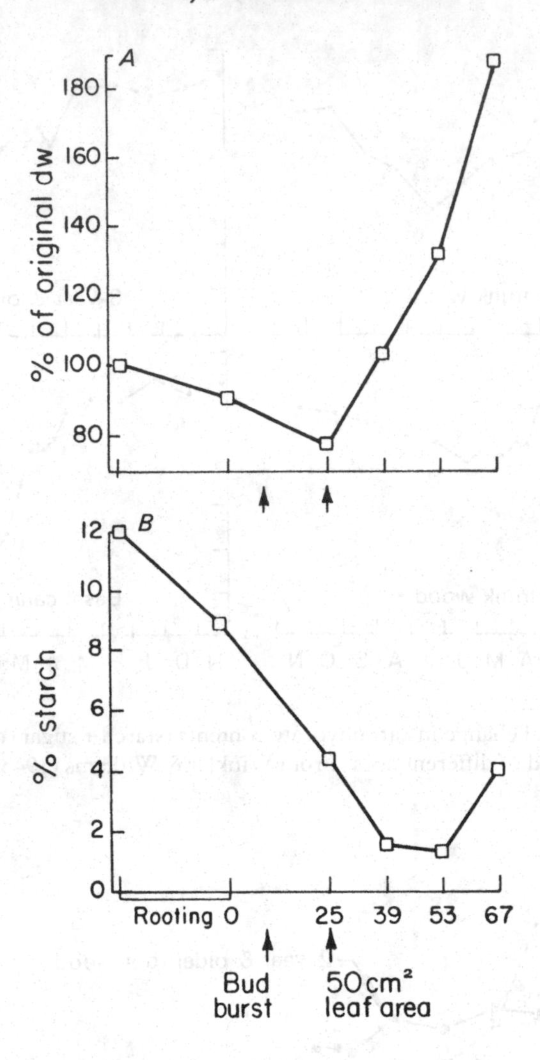

Figure 6.5. *A*, Change in percentage dry weight of cuttings of *Vitis vinifera* through time and *B*, corresponding change in starch content. From Buttrose (1966).

cuttings showing that growth of roots and shoot are dependent on reserves for 40 days after planting (Figure 6.5). Photosynthesis balanced respiration by 25 days at which time the leaf area of the cuttings had attained about 50 cm^2 (about four leaves formed with the oldest half-expanded). This is comparable to the stage when carbohydrates begin to be exported from the leaves as described below.

The cycles described above would differ in plants that were not subjected to the normal pruning cycle that is utilized in the cultivation of grapes.

Pruning obviously makes different demands on reserves for subsequent new growth than would non-pruning. The degree of pruning also affects the storage cycle. Leaving only a few buds on canes results in less reserve depletion than does leaving many (Gutierrez, Williams & Kido, 1985). This is not true when dealing with cuttings where more reserves are available from stem segments with two nodes than from those having only one node (Buttrose, 1966).

Input and movement of carbohydrate within shoots. The photosynthetic capacity of young expanded leaves of *Vitis* has been measured at nearly 19 μmol m^{-2} s^{-2} (Kriedeman, Kliewer & Harris, 1970). This value is typical of temperate-climate vine species (see Teramura, Gold & Forseth, Chapter 9), vines of tropical dry forests (see Castellanos, Chapter 7), and sun leaves of seasonally deciduous trees (Larcher, 1980). It is lower than many C_3 crop plants (average maximum values of 20–45 μmol m^{-2} s^{-2}: Larcher, 1980). With increasing leaf age photosynthetic capacity, and hence the supply of carbohydrate to growth and stores, declines.

Once fixed, carbon is translocated as sucrose from leaves through the phloem to growing and storage centers (Swanson & El-Shishiny, 1958). The pattern of photosynthate movement in the whole shoot is shown in Figure 6.6. During early growth all carbohydrate movement is toward the tip (Hale & Weaver, 1962; Currle *et al.*, 1983). This pattern continues for 2–3 weeks until the oldest leaf starts exporting photosynthate. As the shoot grows, the basal leaves export basipetally and the upper leaves acropetally. During fruit development reserves from the basal leaves are again utilized.

Isotope tracers, on which these conclusions are based, do not provide quantitative information on the amounts of carbon being moved during these periods. These results also do not relate well to the overall carbohydrate cycle given by Winkler & Williams (1945) that would indicate a longer period of reserve utilization. However, Freeman & Smart (1976) have shown that rapid root growth, which may call on reserves, does not begin until 10 weeks after bud burst when shoot growth is about half-completed (Figure 6.7).

Studies of xylem sap also indicate a relatively brief temperature-dependent period of supply of carbohydrate from the rootstock (Reuther & Reichardt, 1963; Marangoni, Vitagliano & Peterlunger, 1986). In the xylem sap soluble carbohydrates are present only prior to bud break (Figure 6.8). Subsequent to bud break nitrogen becomes available, perhaps related to the delayed root growth described above. The presence of leaves is required for this movement as indicated by the time of the appearance of N in the sap and the reduction in its quantity in defoliated stems (Marangoni *et al.*, 1986). The seasonal maximum concentration of organic and inorganic components in xylem sap for another perennial vine, Kiwi fruit, *Actinidia deliciosa* (Ferguson,

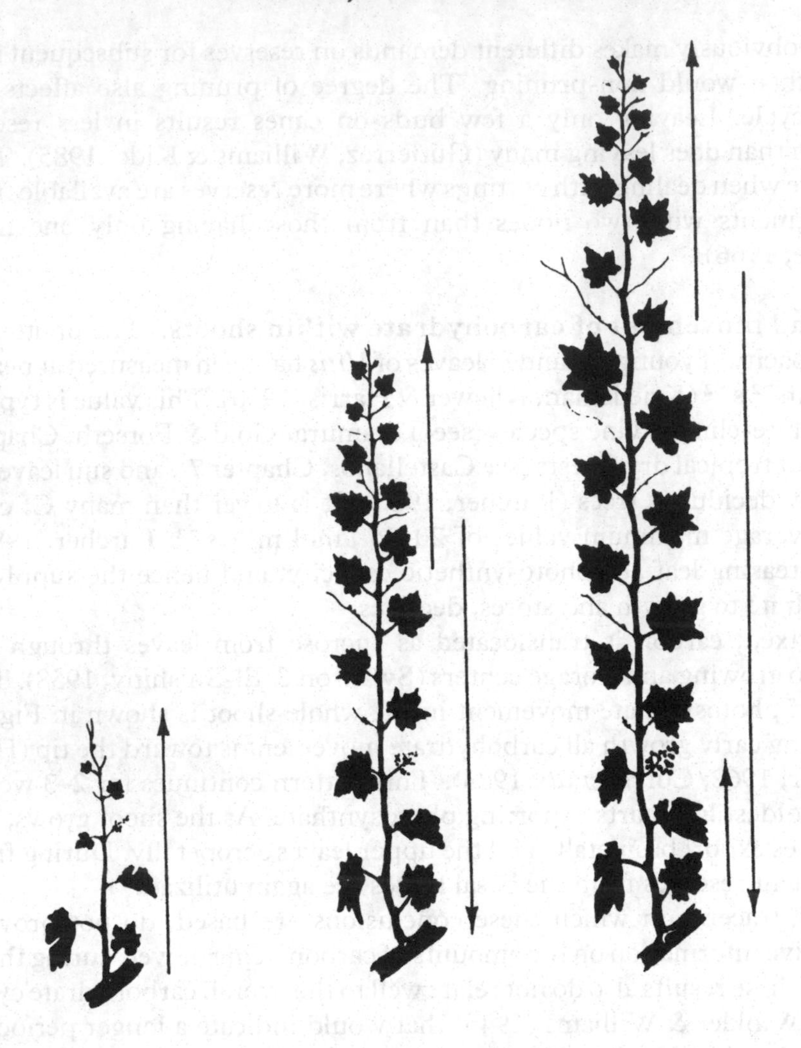

Figure 6.6. Direction of translocation of photosynthate of *Vitis vinifera* through different developmental stages. From Hale & Weaver (1962).

Eiseman & Leonard, 1983) is in the brief period between bud break and leaf burst.

Carbon storage in herbaceous perennials

Kudzu vine (*Pueraria lobata*). This semi-woody leguminous vine, a native of Japan and a weed in the southeastern USA, has large storage tissues in its tuberous roots (Sasek & Strain, 1988) and a photosynthetic capacity of about 18 μmol m^{-2} s^{-1}, comparable to *Vitis* and that of early successional species in

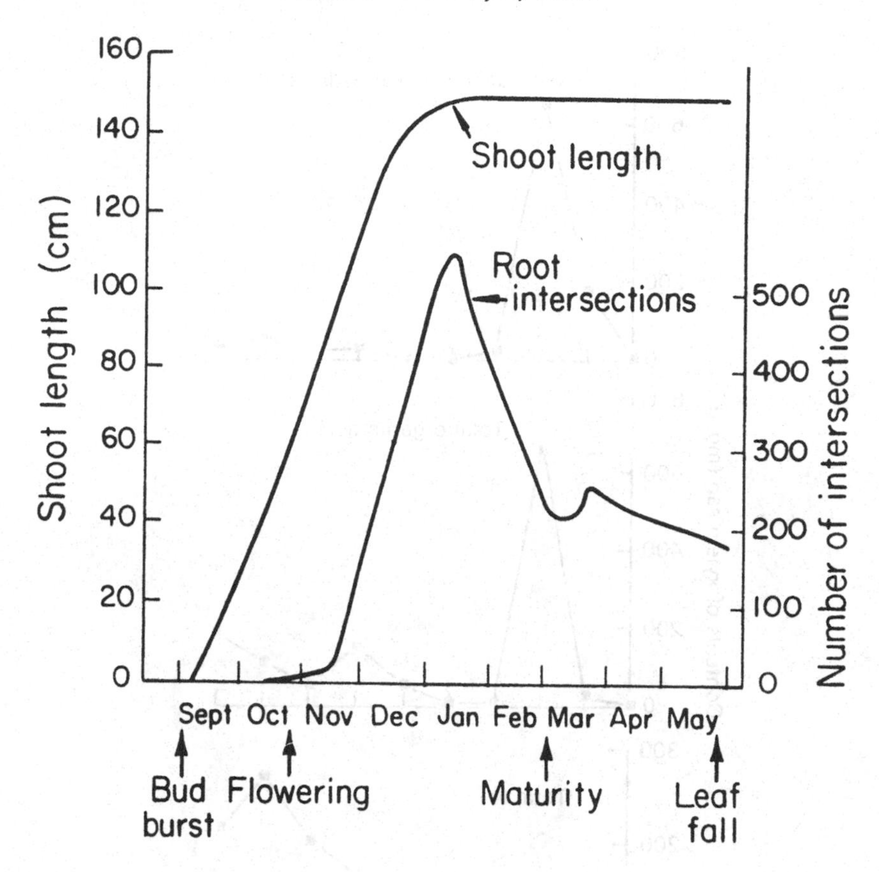

Figure 6.7. Relationship between periodicity of root and shoot growth in *Vitis vinifera*. From Freeman & Smart (1976).

general (Forseth & Teramura, 1987). The roots have been utilized commercially for starch production (Tanner *et al.*, 1979).

Buffalo gourd (*Cucurbita foetidissima*). Berry, Scheerens & Bemis (1978) give an analysis for the root reserve status of buffalo gourd (*Cucurbita foetidissima*) under cultivation in Tucson, Arizona, USA. Under these conditions this arid zone native maintains high levels of reserve starch for most of the year (*c.* 50% dry weight). During the brief springtime growth period starch reserves drop a few per cent and again much more substantially during fruit set to less than 20% (Figure 6.9). At the peak of the reserve status period in September, 54% starch, 3.2% sugar, 9.7% lipid, and 10.9% protein were found. Thus 78% of the dry weight of the root can be considered reserves.

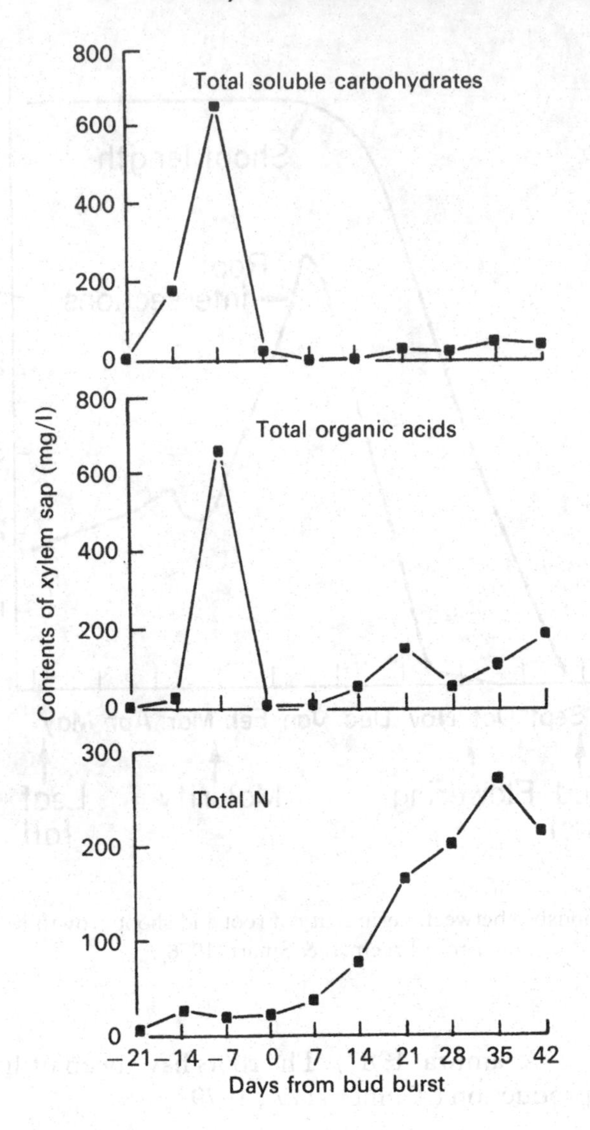

Figure 6.8. Changing carbohydrate, organic acid, and total nitrogen contents of xylem sap of *Vitis vinifera* during early growth. From Marangoni *et al.* (1986).

Bindweed (*Convolvulus arvensis*). Of the world's 18 most serious agronomic weeds there is a single vine, *Convolvulus arvensis* (Holm *et al.*, 1977). This herbaceous perennial vine, native to Eurasia, has prodigious underground productivity reaching 1300 kg ha^{-1}. Probably because of this substantial underground allocation and, hence, loss of sustained growth potential, this vine is not a particularly good competitor for light and vigorous closed-canopy crops can exclude it. The deep primary roots of this weed are rich in

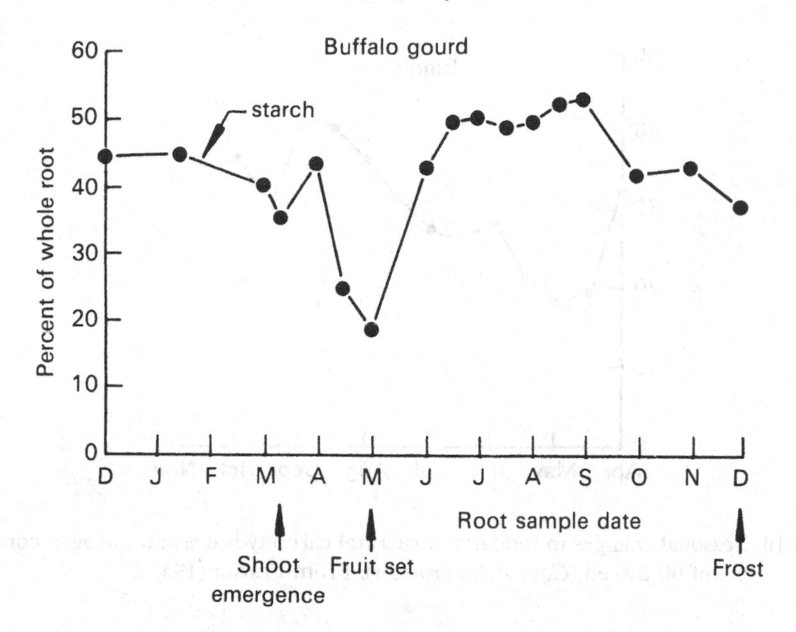

Figure 6.9. Seasonal change in starch contents of buffalo gourd (*Cucurbita foetidissima*). From Berry *et al.* (1978).

carbohydrate, containing over 40% by dry weight during the peak period in fall (Frazier, 1943). Three-quarters of the carbohydrate is starch. During the first weeks of April the reserves are depleted to just below half of the full fall complement (Figure 6.10).

Wild cucumber (*Marah fabaceus*). *Marah fabaceus* is an herbaceous perennial cucurbit found in the winter-wet, summer-dry mediterranean-climate regions of central California. It commences its annual growth in midwinter during the wet season and dries up completely above ground by the end of spring at the beginning of the drought (Figure 6.11). As is characteristic of arid zone cucurbits (Dittmer & Talley, 1964), *Marah* has an extensive storage root. Tubers as large as 90 kg fresh weight have been measured and a plant with an estimated age of 14 years attained a tuber fresh weight of 58 kg (Stocking, 1955).

Initial stem growth is rapid, with extension growth preceding leaf development (E. D. Schulze and H. A. Mooney, unpublished data). During this time the stem has a very low mass to length ratio (Figure 6.11). Growth at this time must be supported by reserves. Carbohydrate content drops from about 39 to 32% during the initial growth period. Because of the very large mass of the tubers this could represent a significant amount of carbohydrate. In this study the largest tuber found was 12.5 kg fresh weight (2.28 kg dw). The

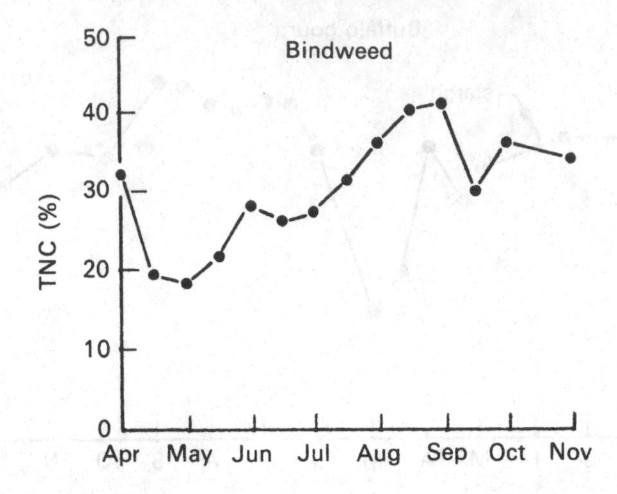

Figure 6.10. Seasonal changes in total non-structural carbohydrates and nitrogen contents of bindweed (*Convolulus arvensis*). From Frazier (1943).

carbohydrate loss would be 160 g during the growth period. The maximum amount of shoot growth produced during the whole season on an individual of a sample of 12 studied was 240 g. Since the biomass produced prior to leaf expansion is relatively small, the reserves utilized are quite adequate to support this growth. The *Marah* study shows that even though the depletion level of carbohydrates may be relatively small on a percentage basis it can be substantial on an absolute basis. Further, it shows that the reserve status maintained in the plant exceeds needs for rapid establishment of a canopy after a dry period. It may be that surplus reserves are maintained to accommodate regrowth after episodic catastrophes, such as unseasonable freezing.

Co-occurring tropical vines

The growth-form analysis given above indicates that both woody and herbaceous perennials, from diverse habitats, have considerable carbohydrate reserves that are utilized to varying degrees to support new stem growth and reproduction.

We now compare the carbohydrate storage patterns of both woody and herbaceous perennials growing together in the same natural habitat, a tropical dry forest in western Mexico. The carbohydrate contents of the roots of these forest plants are shown for the dry season in March and the early rainy season in July (Figure 6.12). As can be seen three of the vine species have TNC contents during the dry period exceeding 40%. These include two

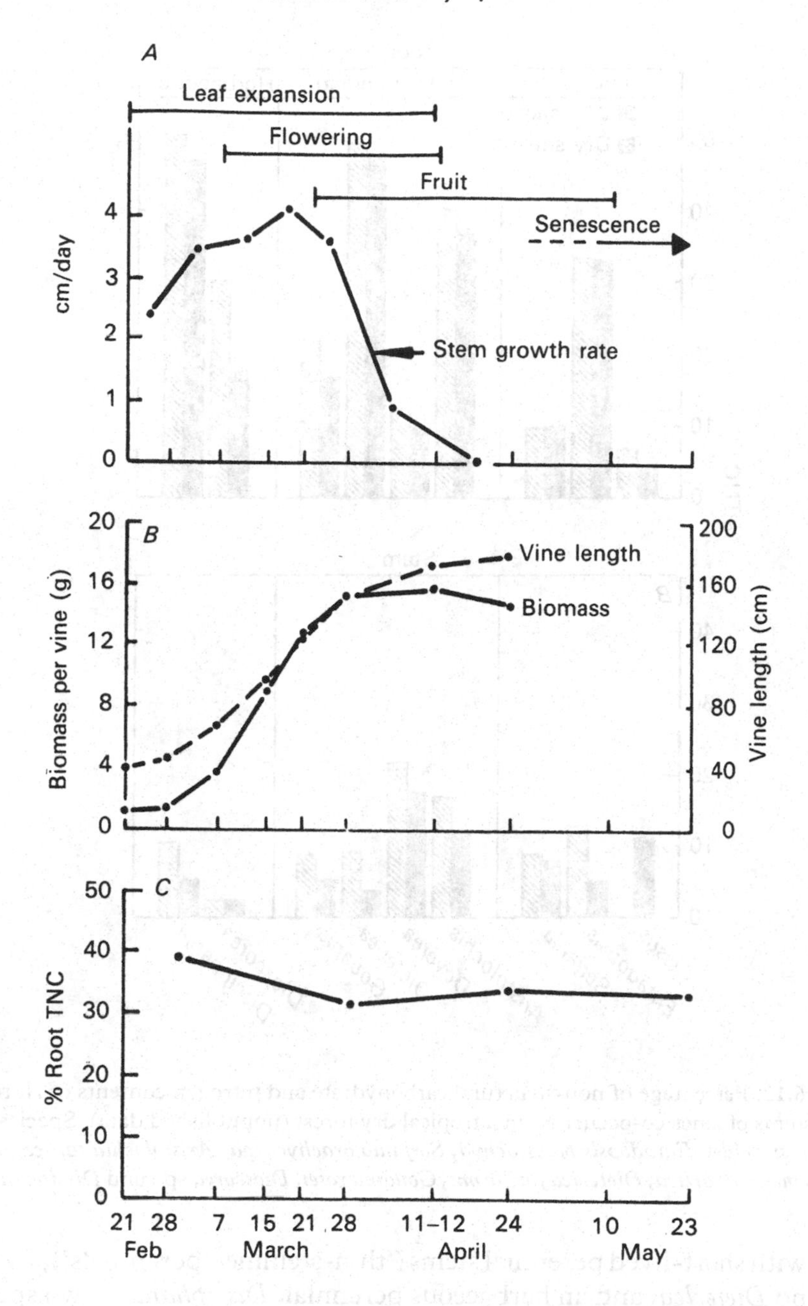

Figure 6.11. *A*, Phenology of *Marah fabaceus* in Central California; *B*, seasonal change in vine length and biomass; and *C*. seasonal change in tuber total non-structural carbohydrate (TNC) (unpublished data).

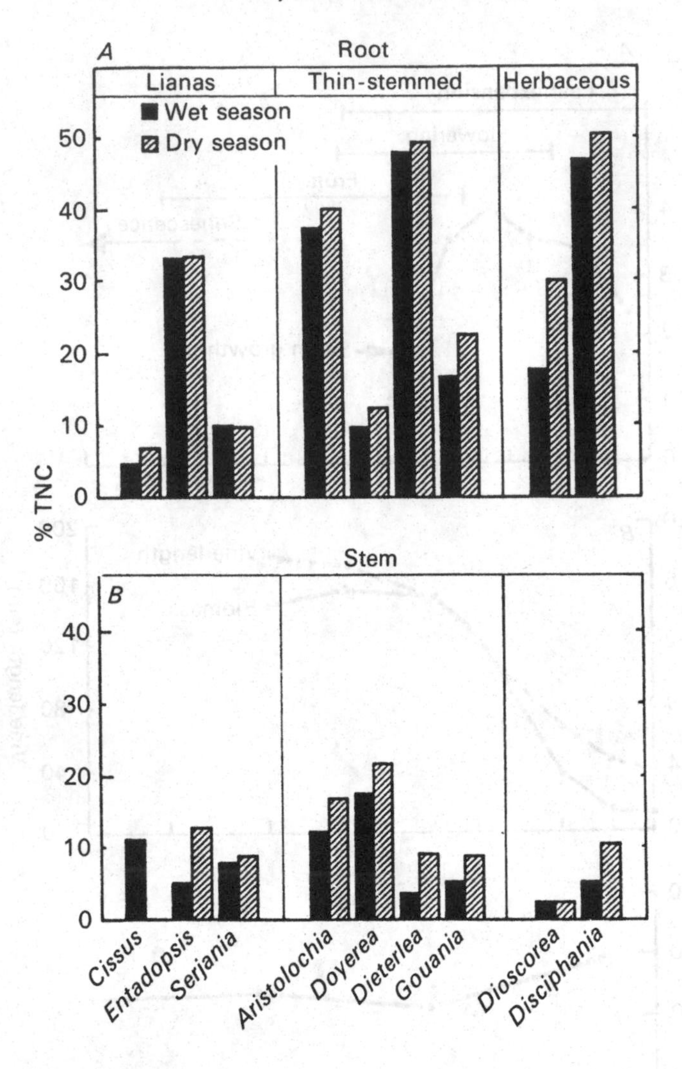

Figure 6.12. Percentage of non-structural carbohydrate and nitrogen contents of *A*, roots and *B*, stems of vines co-occurring in a tropical dry forest (unpublished data). Species are *Cissus sicyoides*, *Entadopsis polystachya*, *Serjania brachycarpa*, *Aristolochia taliscana*, *Doyerea emetocathartica*, *Dieterlea fusiformis*, *Gouania rosei*, *Dioscorea* sp., and *Disciphania* sp.

species with short-lived perennial stems ('thin-stemmed perennials'), *Aristolochia* and *Dieterlea*, and an herbaceous perennial, *Disciphania*. Two species have dry season contents between 30 and 40%, the woody perennial, *Entadopsis*, and the herbaceous perennial (stems entirely renewed yearly), *Dioscorea*. Thus, those species having quite high contents of carbohydrates during the dry season include herbaceous and thin-stemmed perennial vines as well as a liana. Species with about 10% or less carbohydrate in their roots during the dry season include two lianas, *Serjania* and *Cissus*, and the thin-stemmed perennial, *Doyerea*. *Cissus* has tuberous stem segments and frequently survives epiphytically with no connections to the ground, and *Doyerea* is semi-deciduous. All of the species except *Doyerea* are deciduous during the dry season.

Judging from concentrations of storage materials, without knowing total pool sizes, we can make the following observations. The differences in carbohydrate contents between the wet and dry seasons are very small, indicating that the root reserves are not utilized extensively in seasonal regrowth. Differences between dry and wet season carbohydrate contents of stem tissues were generally greater (Figure 6.12), indicating that these tissues may serve as the principal seasonal reserve depots.

Storage location

Parenchyma serves as the principal location for storage materials in plants and hence it is of interest to determine if vine species have a greater proportion of parenchyma than other plants. Beakbane & Thompson (1939) suggested that the content of storage material in a parenchyma cell may be relatively constant and thus an indication of the capacity for carbohydrate storage would be the number of living medullary ray cells. There is no information comparing bark parenchyma of vines with other plants, but there are some generalizations about xylem parenchyma. It has been noted that vines generally have wider and/or taller ray tissues than non-vines (Carlquist, 1984a, b). Further, they characteristically have more parenchyma tissue in contact with vessels (Carlquist, 1985, Chapter 2). Both the larger amount of parenchyma and its position in vines may be distinctive. Carlquist (1988) discusses four hypotheses regarding the function of the abundant parenchyma in stems of vines and lianas. One of the hypotheses is that parenchyma provides the only ample site for starch storage. Much of the unusual structure involves the presence of axial parenchyma adjacent to vessels which may serve in plant nutrient and water balance, as described below. The importance of other aspects of anomalous growth in relation to regeneration and injury resistance is discussed in this volume by Carlquist (Chapter 2), Fisher & Ewers (Chapter 4), and Putz & Holbrook (Chapter 3).

Herbaceous perennial vines store all of their compounds below ground. There it can be in living cells of the cortex or xylem of the root, or it can be in specialized organs such as tubers or corms. Such specialized organs appear quite common.

Storage function

It is generally assumed that lipids and non-structural carbohydrates, located in non-photosynthetic tissues, serve as storage compounds that can be mobilized to provide energy for growth following a dormant period. Support for this concept resides in, among other phenomena, seasonal depletion cycles of starch accompanying growth as shown above for grape. Another use of reserves is in the production of reproductive tissue as shown for buffalo gourd. Sugars may also serve as reserves, but to a lesser extent, since they are osmotically active.

There are plants which accumulate large reserves that do not appear to deplete them substantially during the annual growth or reproductive cycle. Vines of the dry tropical forest, as discussed above, may be examples of this type. How is this pattern explained? First, reserves may be utilized as insurance against loss of tissue due to catastrophes, including grazing, burning, and injury from tree fall. Second, the possibility exists that these compounds represent surplus carbohydrates that cannot be utilized in current growth due to seasonal limitations of either water or nutrients. They, then, are true surplus (Jenner, 1982).

Carbohydrates stored in stem and especially root parenchyma may serve the dual role of supplying carbohydrate, on new growth demand, but also in providing, on release into the xylem, an osmoticum to generate root pressure (von Fink, 1982), thus overcoming xylem embolisms (Sperry *et al.*, 1987). These dual functions could be important simultaneously during the initiation of new growth of vines after either a dry or freezing period.

This hypothesis, that reserves are used to generate root pressure, has been considered for more than 80 years (references in Maximov, 1929). Evidence supporting this hypothesis includes abundance of sugars in the vessels in early spring (e.g. Ziegler, 1956; Sauter, 1980), radial movement of sugars through living ray parenchyma in spring, and the existence and spring activity levels of specialized 'contact cells' in the rays adjacent to and touching the vessels (Braun, 1964; Sauter, 1966, 1972; Sauter, Iten & Zimmermann, 1973). These findings have been mainly in deciduous trees and shrubs of the seasonal temperate forest, but may equally be important in vines. Indeed, several of the classic examples of root pressure were demonstrated in vines (Hales, 1727; White, 1938). The location and abundance of parenchyma often surrounding vessels in vines, as mentioned above and in Carlquist (Chapter 2), could function for storage of starches and their subsequent transport as sugars into the hydrosystem.

Summary

The seasonal pattern of carbohydrate storage and mobilization in vines is similar to the general pattern of trees having the same phenology. There is really only one scheme a plant can use for the timing of storage and withdrawal of its carbohydrates for growth for a given phenology. An important research goal is to quantify the sizes of these pools and the magnitude of their fluxes for different plant growth forms.

From a very limited data base it appears that vines often have considerable storage of carbohydrate reserves in their stems and roots. These reserves may be in excess of those utilized in annual growth and reproductive cycles and could serve to support growth subsequent to tissue damage. Further, these reserves may play a role in maintaining plant water balance by osmotically causing embolized vessels to refill. All the plants discussed in this chapter live in highly seasonal habitats where the importance of storage may be greater than in less seasonal habitats. More data on vines from other habitats are needed to generalize further the role of storage in plants with the vining habit.

References

Beakbane, A. B., & Thompson, E. C. (1939). Anatomical studies of stems and roots of hardy fruit trees. II. The internal structure of the roots of some vigorous and some dwarfing apple root-stocks and the correlation of structure with vigour. *Journal of Pomology and Horticultural Science* 17: 141–9.

Berry, J. W., Scheerens, J. C. & Bemis, W. P. (1978). Buffalo gourd roots: chemical composition and seasonal changes in starch content. *Journal of Agricultural and Food Chemistry* 26: 354–6.

Braun, H. J. (1964). Zelldifferenzierung im Holzstrahl. *Berichte der Deutschen Botanische Gesellschaft* 77: 355–67.

Buttrose, M. S. (1966). Use of carbohydrate reserves during growth from cuttings of grape vine. *Australian Journal of Biological Sciences* 19: 247–56.

Carlquist, S. (1984a). Wood anatomy of Trimeniaceae. *Plant Systematics and Evolution* 144: 103–18.

Carlquist, S. (1984b). Wood and stem anatomy of Lardizabalaceae, with comments on the vining habit, ecology and systematics. *Botanical Journal of the Linnean Society* 88: 257–77.

Carlquist, S. (1985). Observations on functional wood histology of vines and lianas: vessel dimorphism, tracheids, vasicentric tracheids, narrow vessels, and parenchyma. *Aliso* 11: 139–57.

Carlquist, S. (1988). *Comparative Wood Anatomy: systematic, ecological and evolutionary aspects of dicotyledon wood.* Springer-Verlag, Berlin.

Currle, O., Bauer, O., Hofacker, W., Schumann, F. & Frisch, W. (1983). *Biologie der Rebe: Aufbau Entwicklung Wachstum.* Meininger, Neustadt.

Dittmer, H. J. & Talley, B. P. (1964). Gross morphology of tap roots of desert cucurbits. *Botanical Gazette* 125: 121–6.

Ferguson, A. R., Eiseman, J. A. & Leonard, J. A. (1983), Xylem sap from *Actinidia chinensis*: seasonal changes in composition. *Annals of Botany* **46**: 791–801.

Forseth, I. N. & Teramura, A. H. (1987). Field photosynthesis, microclimate and water relations of an exotic temperate liana, *Pueraria lobata*, kudzu. *Oecologia* **71**: 262–7.

Frazier, J. C. (1943). Amount, distribution, and seasonal trend of certain organic reserves in the root system of field bindweed, *Convolvulus arvensis* L. *Plant Physiology* **18**: 167–84.

Freeman, B. M. & Smart, R. E. (1976). Research note: a root observation laboratory for studies with grapevines. *Am. J. Enol. Viticult.* **27**: 36–9.

Gutierrez, A. P., Williams, D. W. & Kido, H. (1985). A model of grape growth and development: the mathematical structure and biological considerations. *Crop Science* **25**: 721–8.

Hale, C. R. & Weaver, R. J. (1962). The effect of developmental stage on direction of translocation of photosynthate in *Vitis vinifera*. *Hilgardia* **33**: 89–131.

Hales, S. (1727). *Vegetable staticks, or an account of some statical experiments on the sap in vegetables*. J. Peele, London.

Holm, L.G., Plucknett, D. L., Pancho, J. V. & Herberger, J. P. (1977). *The World's Worst Weeds*. University Press of Hawaii, Honolulu.

Jenner, C. F. (1982). Storage of starch. In *Plant Carbohydrates I, Encylopedia of Plant Physiology, Vol. 13A*, ed. F. A. Loewus and W. Tanner, pp. 700–47. Springer-Verlag, Berlin.

Kriedeman, P. E., Kliewer, W. M. & Harris, J. M. (1970). Leaf age and photosynthesis in *Vitis vinifera* L. *Vitis* **9**: 97–104.

Larcher, W. (1980). *Physiological Plant Ecology*, 2nd edition. Springer-Verlag, Berlin.

Marangoni, B., Vitagliano, C. & Peterlunger, E. (1986). The effect of defoliation on the composition of xylem sap from cabernet franc grapevines. *Am. J. Enol. Viticult.* **37**: 259–262.

Maximov, N. A. (1929). *The plant in relation to water: a study of the physiological basis of drought resistance*, ed. and transl. R. H. Yapp. George Allen and Unwin, London.

Monsi, M. & Murata, Y. (1970). Development of photosynthetic systems as influenced by distribution of matter. In *Prediction and Measurement of Photosynthetic Productivity*, ed. I. Setlik, pp. 115–29. PUDOC, Wageningen.

Pate, J. S. & Dixon, K. W. (1982). *Tuberous, Cormous and Bulbous Plants*. University of Western Australia Press, Nedlands, Western Australia.

Priestley, C. A. (1962). *Carbohydrate Resources Within the Perennial Plant: their utilization and conservation*. Commonwealth Agricultural

Bureaux, Farnham Royal, UK.

Reuther, G. & Reichardt, A. (1963). Temperature inflüsse auf Blutung und Stoffwechsel bei *Vitis vinifera*. *Planta* **59**: 391–410.

Sasek, T. W. & Strain, B. R. (1988). Effects of carbon dioxide enrichment on the growth and morphology of kudzu (*Pueraria lobata*). *Weed Science* **36**: 28–36.

Sauter, J. J. (1966). Untersuchungen zur Physiologie der Pappelholzstrahlen. II. Jahresperiodische Anderungen der Phosphataseaktivitat im Holzstrahlparenchym und ihr mögliche Bedeutung fur den Kohlenhydratstoffwechsel und den aktiven Assimilattransport. *Zeitschrift für Pflanzenphysiologie* **55**: 349–62.

Sauter, J. J. (1972). Respiratory and phosphatase activities in contact cells of wood rays and their possible role in sugar secretion. *Zeitschrift für Pflanzenphysiologie* **67**: 135–45.

Sauter, J. J. (1980). Seasonal variation of sucrose content in the xylem sap of *Salix*. *Zeitschrift für Pflanzenphysiologie* **98**: 377–91.

Sauter, J. J., Iten, W. & Zimmermann, M. H. (1973). Studies on the release of sugar into vessels of sugar maple (*Acer saccharum*). *Canadian Journal of Botany* **51**: 1–8.

Sperry, J. S., Holbrook, N. M., Zimmermann, M. H. & Tyree, M. T. (1987). Spring filling of xylem vessels in wild grapevine. *Plant Physiology* **83**: 414–17.

Stocking, K. M. (1955). Some taxonomic and ecological considerations of the genus *Marah* (Cucurbitaceae). *Madroño* **13**: 113–44.

Swanson, C. A. & El-Shishiny, E. D. H. (1958). Translocation of sugars in the Concord grape. *Plant Physiology* **33**: 33–7.

Tanner, R. D., Hussain, S. S., Hamilton, L. A. & Wolf, F. T. (1979). Kudzu (*Pueraria lobata*): potential agricultural and industrial resource. *Economic Botany* **33**: 400–12.

von Fink, W. (1982). Histochemische Untersuchungen über Starkeverteilung und Phosphatase Aktivität im Holz einiger tropischer Baumarten. *Holzforschung* **36**: 295–302.

White, P. R. (1938). 'Root pressure' – an unappreciated force in sap movement. *American Journal of Botany* **25**: 223–7.

Winkler, A. J. & Williams, W. O. (1945). Starch and sugars of *Vitis vinifera*. *Plant Physiology* **20**: 412–32.

Ziegler, H. (1956). Untersuchungen uber die Leitung und Sekretion der Assimilate. *Planta (Berlin)* **47**: 447–500.

7

Photosynthesis and gas exchange of vines

ALEJANDRO E. CASTELLANOS

Introduction

Vines allocate proportionally larger amounts of biomass to photosynthetic biomass than do self-supporting plants (Suzuki, 1987). Most forest vines have been described as 'light hungry' (Putz, 1984) and, in reaching for high light environments on top of the canopy, and at their edges, vines have developed a variety of climbing systems and associated characteristics (Schenck, 1892; Castellanos *et al.*, 1989). Climbing in vines is mainly constrained by support availability (Putz, 1983, 1984; Teramura, Gold & Forseth, Chapter 9), attachment success to the substrate (Peñalosa, 1982; Putz & Holbrook, Chapter 3) and, as a consequence of support attachment, by light availability.

Although ubiquitous in most communities worldwide, vines in general are most strongly represented within highly heterogeneous light habitats (Caballé, 1984). Vines have developed several strategies for 'foraging' for light and support, growing horizontally and vertically across varying levels of resource availability (Peñalosa, 1983; Putz, 1984). As such, vines may represent the end-point in a series of growth forms adapted to exploit extensive horizontal and vertical above-ground space volumes, without being dimensionally limited by root 'foraging' sites as in the case of clonal (stoloniferous or rhizomatous) species (Harper & Bell, 1979; Bell, 1984; Slade & Hutchings, 1987).

Exploitation of sites varying greatly in light and other available resources may be enhanced in vines by combining a number of physiological and morphological traits in their leaves. Our knowledge is still meager about different aspects of the leaf carbon economy in vines such as photosynthetic capacity, cost of construction and maintenance of leaf biomass, photosynthetic acclimation potential, leaf life span and environmental constraints on their carbon balance, although these data should provide the formal basis for understanding population dynamics studies of shoots and individuals (Peñalosa, 1983).

In this chapter we review the available information on the effects of different environmental factors and resource availabilities, particularly light, on the photosynthetic and gas exchange responses of vine species and related aspects affecting the carbon economy of their leaves. By summarizing the available information for vines, we point to some interesting aspects that are perhaps unique to this growth form, provide ideas for future research directions, and serve as a reference for comparisons with other growth forms in reviews in which they have not normally been included (Larcher, 1980; Jones, 1983). A companion chapter (Teramura *et al.*, Chapter 9) focuses on the physiological ecology of temperate vines.

Vine characteristics affecting the photosynthetic surface

Vines not only allocate a larger proportion of their biomass to leaves, in many cases resulting in higher growth rates, but also have faster leaf turnover rates and short leaf longevities (Hladik, 1974; Gentry, 1983; Putz & Windsor, 1987; Hegarty, Chapter 13). As a consequence, vines have greater amounts of leaf biomass per stem diameter compared with tree species (Putz, 1983), and a high leaf productivity. Above-ground three-dimensional mapping of individuals of *Liabum caducifolium* and *Heteropteris palmeri* shows how leaf area index in the two liana species is distributed along their stem and within favorable light environments (A. E. Castellanos *et al.*, unpublished data). Specific leaf weight in vines may be similar or lower, compared with tree species (Castellanos *et al.*, 1989), suggesting that a greater allocation of vines to leaf biomass may result in larger leaf areas compared to other growth forms present within the same site. The greater abundance of vine species in tropical forests may be responsible for the observed lower wood production efficiencies that appear to be present in tropical compared to temperate forests (Jordan & Murphy, 1978; Gentry, 1983).

Vines have developed a number of leaf size and shape characteristics that may serve to minimize the physiological constraints that arise from having a large photosynthetic surface in spatially and temporally heterogeneous environments. Biophysical models relating photosynthesis to leaf temperature and energy load (Taylor, 1975), water use efficiency (Parkhurst & Loucks, 1972), and cost–benefit analyses of photosynthesis and water loss (Givnish & Vermeij, 1976) have been used to predict optimal leaf size for individuals from different habitats. Developed in particular for vine leaves, a model by Givnish & Vermeij (1976) assumes the predominating effect of leaf temperature on carboxylation efficiency, and relates transpiration (water loss) to carbon costs by their effect on leaf temperature. Based on their model, Givnish & Vermeij (1976) found that as leaf size increases, transpirational costs will rise more slowly than photosynthetic carbon gains, thus predicting large leaves for vines growing at intermediate-light and mesic environments. Intermediate size of leaves was predicted in more mesic environments as photosynthetic light saturation diminished the advantages of larger leaves by decreasing the marginal profits of the photosynthesis–transpiration relationship. Based on similar principles, smaller leaves were predicted on vine species within sunny and xeric environments. Some discrepancies between the model and the leaf size and shape patterns found in nature were thought to arise from assumptions in the model that did not take into account the structural costs involved in leaf support needed to increase leaf area display. Further studies, as discussed below, indicate that photosynthesis in vines is a highly regulated process adapted to maximize the carbon gain capacity of leaves in heterogeneous light environments.

Leaf photosynthetic characteristics of vine species

Much of the literature on leaf photosynthesis and stomatal conductance, and some of the best known systems in the study of photosynthesis, relate to commercial vine species (beans, grapes, cucurbits, etc.); however, no attempt is made here to cover this literature in any great detail. It needs to be said that many of the studies covered are not recent and in many cases data are not fully comparable because of the different techniques and measurement units employed. Reported units of photosynthesis and conductance were converted using equivalences from Nobel (1974) and Jones (1983) for molar units and those given by Larcher (1980) for illuminance and radiation to photosynthetically active radiation (PAR). In all cases the highest conversion equivalence was used and only the highest value reported for the species is given (Table 7.1).

Photosynthetic pathways of vines

As is the case in the majority of higher plants, most vine species fix carbon during daylight hours utilizing the C_3 biochemical pathway of photosynthesis. There are no reports of C_4 vine species at present, although C_4 species and the vine growth form may be found separately within the Asteraceae.

A number of vine species have increased nocturnal acid accumulation and strongly negative $\delta^{13}C$ ratios that are characteristic of plants with Crassulacean Acid Metabolism (CAM) pathway for CO_2 fixation (Table 7.2). Vine

Table 7.1. *Net photosynthesis of vine species grouped by degree of woodiness. Tropical and temperate species refer to latitudinal and/or habitat characteristics of species in each category. When several values were found for any one species, only the highest reported photosynthetic rate was chosen. Sources are given in parentheses*

Tropical vine species	Maximum photosynthesis[a]	Temperate vine species	Maximum photosynthesis[a]
Herbaceous		Herbaceous perennials	
Cucumis sativus	11.8 (11)	*Pueraria lobata*	18.7 (4)
Momordica charantia	5.2 (2)	*Smilax rotundifolia*	8.0 (1)
Phaseolus vulgaris	17.6 (12)	*Solanum dulcamara*	21.7 (6)
Vicia faba	21.2 (16)		
Herbaceous perennials		Woody	
Cardiospermum halicacabum	13.0 (2)	*Actinidia arguta*	12.6 (19)
Cucurbita pepo	6.9 (20)	*Actinidia deliciosa*	17.5 (7)
Macroptilum atropurpureum	25.0 (18)	*Celastrus orbiculatus*	5.7 (15)
Phaseolus lunatus	12.5 (2)	*Clematis virginiana*	12.9 (1)
Phaseolus viridissimus	18.3 (17)	*Hedera helix*	14.1 (10)
Vigna unguiculata	20.0 (8)	*Lonicera japonica*	9.0 (1)
		Lonicera japonica var. *aureoreticulata*	5.0 (19)
Woody		*Lonicera sempervirens*	1.9 (19)
Caesalpina scortechinie	9.9 (9)	*Parthenocissus quinqueifolia*	7.8 (1)
Calamus muelleri	3.8 (9)	*Rhus radicans*	5.5 (1)
Cudronia cochinchinensis	8.8 (9)	*Rubus corylifolius*	15.0 (13)
Melodinus australis	2.5 (9)	*Vitis californica*	14.5 (5)
Parsonsia lilacina	2.9 (9)	*Vitis vinifera*	10.1 (14)
Ripogonum elseyanum	3.4 (9)	*Vitis vinifera* var. *amarela*	7.8 (3)
Sephania japonica	6.3 (9)	*Vitis vinifera* var. *periquita*	6.3 (3)
		Vitis vulpina	11.1 (1)

[a] μmol CO_2 m^{-2} s^{-1}.

Sources: 1. Carter & Teramura (1988); 2. A. Castellanos (unpublished); 3. Chaves *et al.* (1987); 4. Forseth & Teramura (1986); 5. J. Gamon and R. Pearcy (unpublished); 6. Gauhl (1976); 7. Greer & Laing (1988); 8. Hall & Schulze (1980); 9. E. Hegarty (unpublished); 10. Hoflacher & Bauer (1982); 11. Hopkinson (1964); 12. Keuneman *et al.* (1979); 13. Küppers (1984); 14. Loveys & Kriedemann (1973); 15. Patterson (1975); 16. Pearson (1974); 17. Saeki (1959); 18. Sheriff & Ludlow (1984); 19. Shim *et al.* (1985); 20. Turgeon & Webb (1975).

Table 7.2. *Vine species reported with CAM metabolism. Measurements of their photosynthetic activity by any of three different methods: $\delta^{13}C$ value, nocturnal acid accumulation and gas exchange (CO_2 or O_2). For the most part, species are from desert or rainforest habitats*

Species	$\delta^{13}C$	$\mu eq\ g^{-1}$ fresh wt	$\mu mol\ CO_2\ m^{-2}\ s^{-1}$	Source
Cissus antarctica			4.1	E. Hegarty (unpublished)
Cissus quadrangularis	−15.0			Ziegler *et al.* (1981)
Cissus quadrangularis		35.9		Virzo de Santo *et al.* (1980)
Cissus quadrangularis		62.6		Virzo de Santo *et al.* (1980)
Cissus sicyoides			2.5	A. Castellanos (unpublished)
Cissus trifoliata		27.7	4.5	Olivares *et al.* (1984)
Clusia flava	−14.8			Ting *et al.* (1987)
Clusia rosea	−17.7			Ting *et al.* (1987)
Clusia witana	−21.2			Ting *et al.* (1987)
Cyphostemma curorii	−12.2			Mooney *et al.* (1977)
Cyphostemma quinatum	−14.8			Mooney *et al.* (1977)
Hoya australis		12.3	12.2[a]	Adams *et al.* (1988)
Hoya australis		20.1	12.0[a]	Adams *et al.* (1988)
Hoya nicholsoniae			5.0[a]	Winter *et al.* (1986)
Hoya nicholsoniae			2.5[a]	Winter *et al.* (1986)
Rhipsalis sp.	−13.3			Ting *et al.* (1987)
Seyrigia humbertii		24.0		de Luca *et al.* (1977)
Tribulus rajasthanensis	−11.9			Ziegler *et al.* (1981)
Xerosicyos danguyi		67.0		de Luca *et al.* (1977)

[a] $\mu mol\ O_2\ m^{-2}\ s^{-1}$.

species with CAM photosynthetic pathway do not seem to be restricted to particular biomes or types of habitats. They have been reported from deserts (de Luca, Alfani & Virzo de Santo, 1977; Mooney, Troughton & Berry, 1977, Virzo de Santo *et al.*, 1980; Ziegler *et al.*, 1981) and tropical rainforests (Winter *et al.*, 1983, Olivares *et al.*, 1984; Ting *et al.*, 1987; Adams *et al.*, 1988.). Vines with CAM metabolism may be successful in a wide variety of habitats, some probably by means of CAM-idling. CAM-idling has been found in species of *Cissus*, a genus in the Vitaceae that occurs from desert to tropical rainforest habitats (Virzo de Santo *et al.*, 1980; Olivares *et al.*, 1984; Hegarty & Caballé, Chapter 11). *Cissus sicyoides* grows within the tropical deciduous forests near the coast in western Mexico to the highlands around Mexico City. This species shows some degree of morphological differentiation in its vegetative parts and leaves, perhaps as a result of its CAM-facultative metabolism, when growing in the 'pedregal' (rocky lava outcrops) under fully exposed sun conditions or under partial shade when climbing on trees.

Vines with CAM metabolism within rainforests normally occur as epiphytes or on rocky substrates, exposed to some degree of drought or to high light conditions, although there are some reports of CAM vine species growing under deep-shade conditions. CAM species normally grow under high light conditions without apparently experiencing any signs of photoinhibition, but CAM vines growing within rainforest habitats under deep-shade conditions, as is the case for *Hoya nicholsoniae* and *H. australis*, do experience photoinhibition under strong light (Winter, Osmond & Hubick, 1986; Adams *et al.*, 1988). Current hypotheses to explain the lack of protection against photoinhibition in these species are based on the presence of constraints for internal CO_2 recycling or some measure of CAM-idling (Osmond, Björkman & Anderson, 1980; Adams *et al.*, 1988). The ecophysiology of the photosynthetic responses of CAM species growing under deep-shade habitats is not well known at present.

Environmental limitations to photosynthesis and gas exchange of vine leaves

Heterogeneous light conditions may be the most prevalent characteristic of vine microhabitats. For a fast-growing vine, such spatial and temporal heterogeneity may rapidly change. Although the study of the photosynthetic response to dynamic light environments is just beginning (Gross & Chabot, 1979; Gross, 1982; Pearcy, Osteryoung & Calkin, 1985; Chazdon & Pearcy, 1986), some steady state photosynthetic measurements may be indicative of the dynamic characteristics present in vine species.

Maximum leaf photosynthetic rates per unit area in vines measured under normal air conditions cover a range of values from around 2 to 25 μmol CO_2 m^{-2} s^{-1}, or an order of magnitude difference (Table 7.3). Photosynthetic rates in vines range from those found for shade-evergreens to those for sun-deciduous species (Larcher, 1980; Mooney, 1981). In general, the highest photosynthetic rates of vines are greater than those found for deciduous shrub and tree species, similar to those present in old-field annuals and lower than those of desert annuals (Field & Mooney, 1986). Although photosynthesis is often inversely related to leaf specific weight, no particularly high photosynthetic rates appear to be found in vine species, as would be expected based on their low specific leaf weight, low dry matter allocation to stems or fast growth rates (Castellanos *et al.*, 1989). Different values found for the same species may depend also on habitat and growth conditions, type of measurement (field or laboratory) and methodology employed.

Table 7.3. *Mean and standard error of the mean for net photosynthetic capacity, stomatal conductance and dark respiration in tropical and temperate vines with different degrees of woodiness. Only the highest reported value for the species was used for calculations. Differences in units were transformed as explained in the text. Numbers in brackets refer to sources*

	Net photosynthesis[a]		Stomatal conductance[b]		Dark respiration[a]	
	Tropical ± SE	Temperate ± SE	Tropical ± SE	Temperate ± SE	Tropical ± SE	Temperate ± SE
Herbaceous	13.9 ± 4.0 [3,13,14,19]	–	0.63 ± 0.19 [3,14,17,20]	–	2.2 ± 1.3 [1,24]	–
Herbaceous perennials	16.0 ± 2.9 [3,9,10,22,24]	16.1 ± 5.1 [2,6,8]	0.56 ± 0.35 [3,10,22]	0.53 ± 0.12 [2,6]	–	1.50 ± 0.9 [2,6]
Woody/lianas	5.2 ± 1.1 [11]	10.3 ± 1.2 [2,4,7,9,12,15,16,18,23]	0.48 ± 0.24 [5,17]	0.36 ± 0.06 [2,4,12,15,18]	–	1.68 ± 0.3 [2,18,23]

[a] $\mu mol\ CO_2\ m^{-2}s^{-1}$.
[b] $cm\ s^{-1}$.

Sources: 1. Burnside & Böhning (1957); 2. Carter & Teramura (1988); 3. Castellanos (unpublished); 4. Chaves *et al.* (1987); 5. Fetcher (1981); 6. Forseth & Teramura (1986); 7. Gamon & Pearcy (unpublished); 8. Gauhl (1976); 9. Greer & Laing (1988); 10. Hall & Schulze (1980); 11. Hegarty (unpublished); 12. Hoflacher & Bauer (1982); 13. Hopkinson (1964); 14. Kueneman *et al.* (1979); 15. Küppers (1984); 16. Loveys & Kriedemann (1973); 17. Mooney & Castellanos (unpublished); 18. Patterson (1975); 19. Pearson (1974); 20. Peet *et al.* (1977); 21. Saeki (1959); 22. Sheriff & Ludlow (1984); 23. Shim *et al.* (1985); 24. Turgeon & Webb (1975).

Light saturation of photosynthesis in vines is consistently achieved above or close to peak photon flux density levels to which the leaf has been exposed. Photosynthetic light saturation for most of the vine species studied was attained at photon flux densities below 25% and a still larger group was saturated below 15% of peak daily maximum (< 300 μmol quanta m^{-2} s^{-1}) (Figure 7.1). An example is seen in the tropical rainforest vine *Hoya nicholsoniae*, living in deep-shade habitats with PAR photon flux densities below 50 μmol quanta m^{-2} s^{-1}, in which photosynthesis light saturation is obtained at about 100 μmol quanta m^{-2} s^{-1} (Winter *et al.*, 1986). Ferns compared in the same study had light saturation lower than peak growing light levels. Similarly, photosynthetic light saturation close to peak light levels is found in temperate vines as in *Pueraria lobata*, when grown under contrasting light conditions (Forseth & Teramura, 1987; Carter & Teramura, 1988), and in tropical vine species (Figure 7.1). Light saturation of photosynthesis close to higher levels for the habitat ensures higher photosynthetic light efficiency throughout most of the light regimes that the leaf will experience during its life span.

Extremely low light compensation points have been observed in many tropical and temperate vines and positive net photosynthetic rates are found at light intensities below 50 μmol quanta m^{-2} s^{-1}. Species characteristically growing under strong light conditions, as in the case of *Pueraria lobata*, a weedy vine introduced to the USA from Asia (Carter & Teramura 1988), *Phaseolus lunatus*, a native from open sites in tropical deciduous forests, and *Cardiospermum halicacabum*, a widespread weedy heliophyte, also have low light compensation points (Figure 7.1).

Both internal and external conditions affect the net photosynthetic capacity of leaves during the day. In the case of grape cultivars, carboxylation efficiency increases in the morning hours and declines during the day (Chaves *et al.*, 1987). These changes in carboxylation efficiency result in higher net photosynthetic capacity at favorable hours during the day and a metabolic instead of diffusional regulation, associated with carboxylase activation and deactivation during stressful hours. These changes have been reported also for a number of mediterranean-climate species (Tenhunen *et al.*, 1984; Weber, Tenhunen & Lange, 1985) and *Eucalyptus* (Küppers *et al.*, 1986) as a response to changing stressful environmental conditions.

Differences in leaf photosynthesis and photoinhibitory susceptibility have been found for individuals of different growth forms growing under contrasting light environments (Boardmann, 1977; Björkman, 1981), for leaves growing under different light exposure on the same individual (Benecke *et al.*, 1981; Björkman, 1981) and for leaves at different developmental stages (Jurik, Chabot & Chabot, 1979). Vines have developed at least two morphological mechanisms, reversible leaf movements and leaf dimorphism, to avoid

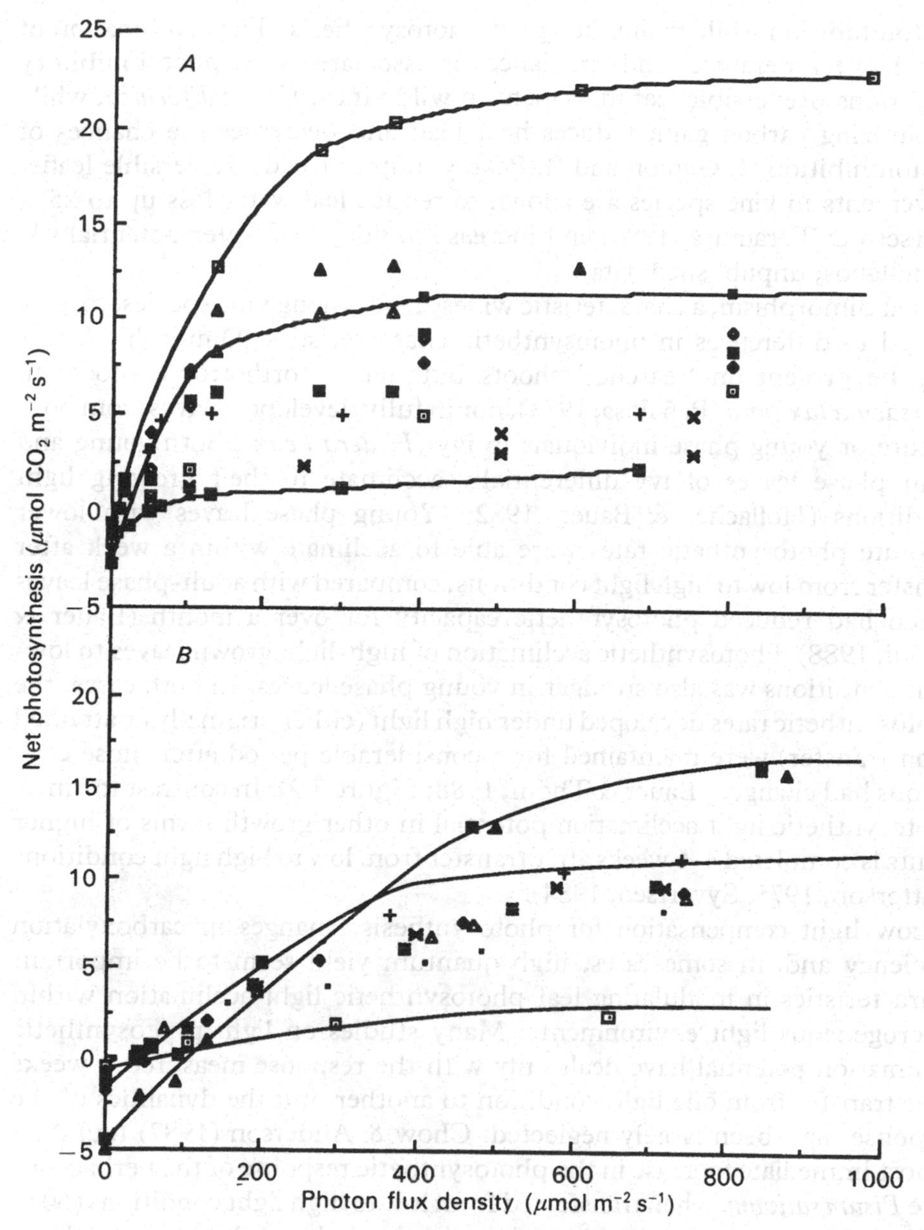

Figure 7.1. Net photosynthesis for *A*, temperate and *B*, tropical species of vines. Only three representative curves in each group are drawn for clarity. Curves for temperate species, from highest to lowest, are for *Hedera helix*, *Clematis virginiana* (Carter & Teramura, 1988) and *Hoya nicholsoniae* (Winter *et al.*, 1986). In tropical species *Vigna unguiculata* (Hall & Schulze, 1980), *Phaseolus lunatus* and *Cissus sicyoides* (A. E. Castellanos, unpublished) correspond to highest, medium and lowest values respectively.

photoinhibition while maintaining net photosynthesis. The combination of high leaf temperatures and irradiances is associated with photoinhibitory conditions. Reversible leaf movement in wild vines, *Vitis californica*, while maximizing carbon gain, reduces heat load and decreases the chances of photoinhibition (J. Gamon and R. Pearcy, unpublished). Reversible leaflet movements in vine species are known to reduce leaf water loss up to 25% (Forseth & Teramura, 1986) and increase midday leaf water potential (A. Castellanos, unpublished data).

Leaf dimorphism, a characteristic widespread among vine species, may be related to differences in photosynthetic characteristics. Dimorphic leaves may be present on 'searcher' shoots but not on orthotropic shoots in *Marsdenia laxiflora* (Peñalosa, 1983), nor in fully developed leaves from both mature or young phase individuals in ivy (*Hedera helix*). Both young and adult phase leaves of ivy differentially acclimate to their growing light conditions (Hoflacher & Bauer, 1982). Young phase leaves with lower absolute photosynthetic rates were able to acclimate within a week after transfer from low to high light conditions, compared with adult-phase leaves which had reduced photosynthetic capacity for over a month (Bauer & Thöni, 1988). Photosynthetic acclimation of high-light grown leaves to low-light conditions was also stronger in young phase leaves. In both cases, the photosynthetic rates developed under high light (either originally or attained upon transfer) were maintained for a considerable period after those conditions had changed (Bauer & Thöni, 1988; Figure 7.2). In contrast to vines, photosynthetic light acclimation potential in other growth forms of higher plants is completed 4–6 weeks after transfer from low to high light conditions (Patterson, 1975; Syvertsen, 1984).

Low light compensation for photosynthesis, changes in carboxylation efficiency and, in some cases, high quantum yield seem to be important characteristics in modulating leaf photosynthetic light acclimation within heterogeneous light environments. Many studies on light photosynthetic acclimation potential have dealt only with the response measured 4 weeks after transfer from one light condition to another, but the dynamics of the response have been largely neglected. Chow & Anderson (1987) found an almost immediate increase in the photosynthetic response of the herbaceous vine *Pisum sativum*, when transferred from low to high light conditions (60 to 390 μmol quanta $m^{-2} s^{-1}$). Changes in the dynamics of the photosynthetic response were the result of almost immediate increases in Rubisco activity and photosystem II electron transport while whole-chain electron transport increased more slowly. Total chlorophyll remained the same throughout all the acclimation period, contrasting with the rapid increase in the chlorophyll a/chlorophyll b ratio.

Although data are limited it appears that under good conditions, dark

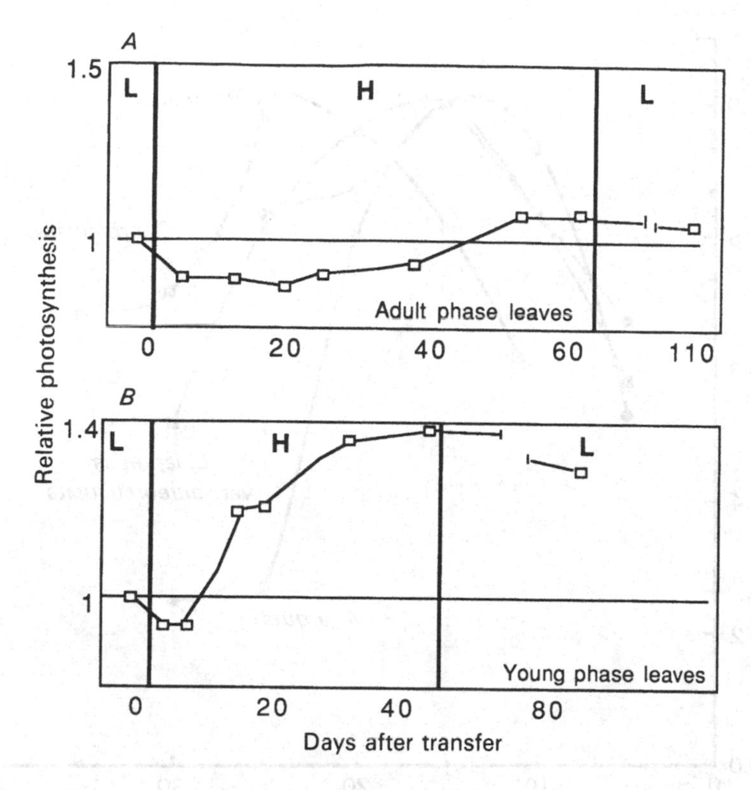

Figure 7.2. Proportional change in net photosynthesis of leaves from *A*, adult and *B*, young phase individuals when individuals are transferred from a low light environment (L) to high light (H) and later returned to original growing conditions. Time after initial transfer is indicated in days. (Modified from Bauer & Thöni, 1988).

respiration of some vine species may exceed 20% of the maximum net photosynthetic rate and only in a few cases does not exceed 10% (Table 7.3), a figure roughly comparable with that found in other growth forms (Larcher, 1980). High dark respiration rates in vine leaves may result from higher maintenance respiration costs. One possible cause may be the existence of excess photosynthetic capacity maintained in their leaves in order to maximize carbon gain within the highly heterogeneous environments where vines live. In a survey of vines from a tropical deciduous forest in Mexico, it was found that leaf specific weights were among the lowest reported in the literature while leaf nitrogen concentrations were among the highest (Castellanos *et al.*, 1989). These traits are generally associated with high photosynthetic capacity, but apparently not in vines.

Vines have temperature responses to photosynthesis similar to other C_3 growth forms. Differences from their photosynthetic maximum close to 30% were found for several temperate and tropical deciduous perennial vine

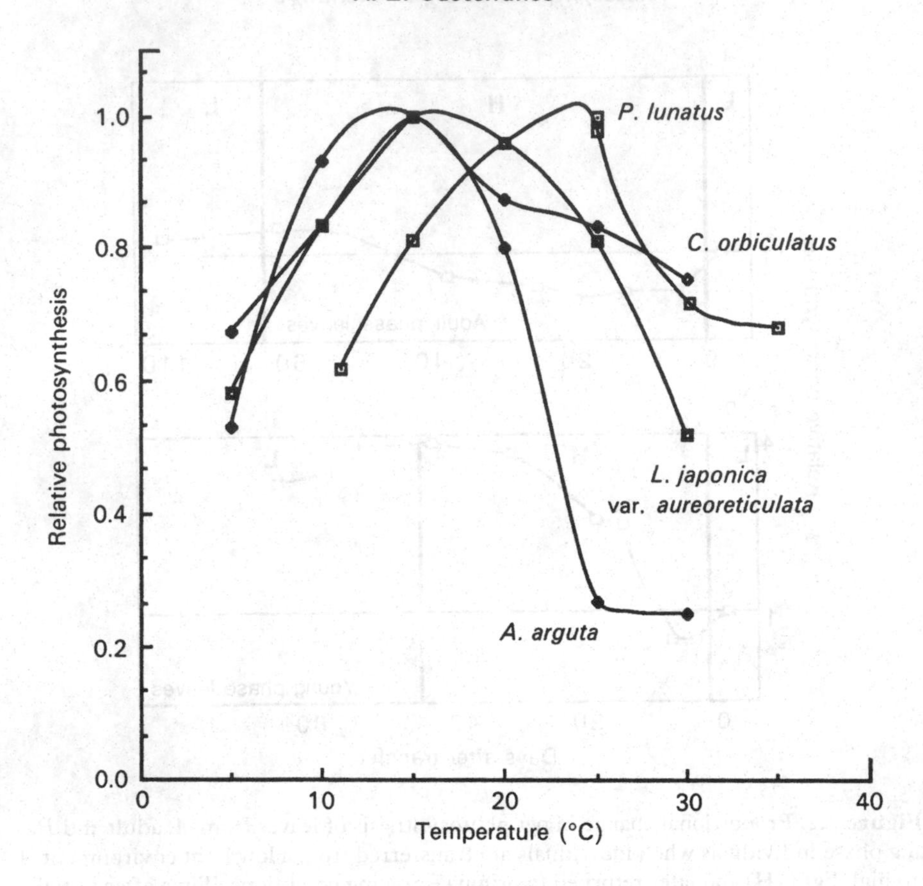

Figure 7.3. Net photosynthesis in relation to temperature in three temperate vines, *Celastrus orbiculatus*, *Lonicera japonica* var. *aureoreticulata*, *Actinidia arguta* (Shim *et al.*, 1985), and a tropical vine, *Phaseolus lunatus*, (Castellanos, unpublished). Changes are proportional to the maximum net photosynthetic rate found at optimum temperature.

species when measured 10 °C below and above their optimum temperature, and up to 75% decrease in leaf photosynthesis for *Actinidia arguta* over the same range of temperatures (Shim, Ahn & Yoo, 1985, Figure 7.3). Leaf temperature effects on photosynthesis are expected to be of more importance in vines of open habitats (Forseth & Teramura, 1986) but less within closed canopy communities. In a study of five species of vines in lowland tropical forests of Costa Rica, Fetcher (1981) found no evidence of a relationship between leaf size, leaf temperature, stomatal conductance or vertical gradient within the forest. Although no studies on photosynthesis were made, Fetcher (1981) suggests that photosynthetically active radiation flux density may be the principal constraint to leaf photosynthesis in vines instead of temperature-related carboxylation efficiency as suggested by the Givnish & Vermeij (1976) model.

Because of the generally large vessel sizes of vines and the wide range of environments which this growth form inhabits, water stress effects on photosynthesis may be the second most important constraint on gas exchange and photosynthetic performance of vines. Increased leaf to air water vapor pressure deficit has a negative effect on leaf photosynthesis; the largest effect is, however, on stomatal conductance (Hall & Schulze, 1980; Forseth & Teramura, 1987). Stomatal conductance to water vapor in vines covers a wide range of values and consistently higher values are found for more herbaceous species (Table 7.3). Consistent with data on photosynthesis, no exceptionally high conductances are normally found on vines. Data on daily patterns of conductance for several tropical deciduous vine heliophytes in a Mexican tropical dry forest during the wet season show a marked midday depression in stomatal conductance (Figure 7.4). While no seasonal patterns of photosynthesis or gas exchange are available for the same species, stomatal conductances in *Aristolochia* sp., measured during the dry season in March, are not drastically different from the other species measured during the rainy season except for a lower recovery in the afternoon (Figure 7.4).

A leaf carbon balance approach

The discussion so far has focused on characteristics of the photosynthetic responses and their dynamics in variable light environments. In this section the dynamic aspects of an integrated carbon balance are discussed.

Throughout any period of time, carbon fixed during photosynthesis (P_g) is partially invested by the leaf to contribute more carbon to the rest of the individual, or is lost and represents a cost (C). Simply stated,

$$G = \sum P_g - \sum C \qquad \text{(Equation 1)}$$

where G is the amount of carbon gained by a leaf. Costs result from diverting some of the carbon budget to defense, growth and maintenance activities.

Leaves are thought to maximize total net carbon profits over their life span, in terms of limiting resources or their gain ratios, by modifying their costs of carbon acquisition and loss (Mooney, 1972; Chabot & Hicks, 1982; Bloom, Chapin & Mooney, 1985). In heterogeneous light environments, maximization of carbon gain will be affected by habitat conditions during the leaf life span, with the total amount of carbon gained by a leaf over its life span, the net carbon gain (G') expressed by

$$G' = \sum P_f F + \sum P_d D - \sum (M'_v + R'_v) \qquad \text{(Equation 2)}$$

where $P_f F$ refers to the amount of carbon fixed during favorable periods and $P_d D$ is the amount of carbon fixed during less favorable periods. M'_v and R'_v

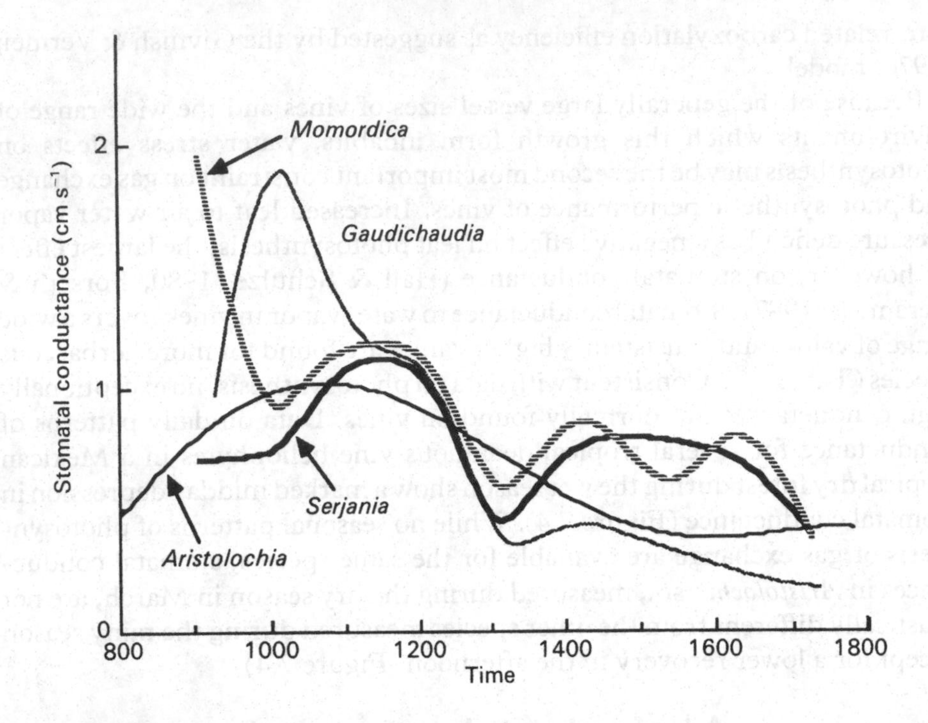

Figure 7.4. *In situ* diurnal changes of stomatal conductance to water vapor for herbaceous (*Momordica charantia*, *Aristolochia* sp.) and woody (*Gaudichaudia* sp., *Serjania brachycarpa*) vines in a tropical deciduous forest at Chamela, Jalisco, Mexico. Measurements were made in August 1985 during the wet season, except for *Aristolochia* which was measured in March 1987 at the end of the dry season.

are variable amounts of carbon loss by maintenance and growth, and by construction respiration over the leaf life span, that will depend on changing environmental and ontogenetic conditions. Compared with evergreens, deciduous species will have shorter leaf life spans which in turn relate to larger investments in photosynthetic machinery and lower investment on secondary defensive compounds and construction costs to increase carbon gain (Chabot & Hicks, 1982). From Equation 2 it is seen that G' measures the actual amount of carbon gained by a leaf, according to the maximization hypothesis, and is not necessarily a measure of the contribution of a leaf to the rest of the plant.

Short leaf life span in vine species (Peñalosa, 1985; Putz & Windsor, 1987; Hegarty, Chapter 13) seems to be an adaptation of individuals from heterogeneous environments to minimize the pay-back period while at the same time maximizing carbon contribution. Deciduousness in vine species may be interpreted as a low initial investment of carbon although only indirect inferences can be made since cost accounting for vine species is lacking. In variable light environments, such as those where vines are most abundant, leaf life span may be instead an ecophysiological strategy to minimize the time needed by a leaf to pay back its own investment and start contributing carbon to the rest of the plant. A model by Chabot & Hicks (1982) emphasizes the importance that different environmental factors have in affecting the potential carbon gain of a leaf in terms of future reductions of the carbon gain of the same leaf over its life span. A modification to their model is expressed by

$$G' = \sum P_f F - \sum \{P_u U + (M'_v + R'_v) + W + H + S\} \quad \text{(Equation 3)}$$

where $P_f F$ is the amount of actual and future carbon fixed during favorable periods and $P_u U$ is the amount of future carbon fixation lost because of unfavorable periods; W and H represent the actual loss of biomass and future photosynthetic contribution by stress and herbivores respectively, and S is the amount of carbon stored in the leaf but not transported and made available to the rest of the individual.

During leaf ontogeny, net photosynthesis increases to a distinct peak from unfolding to near full leaf expansion, and then declines until leaf death. Decrease of net photosynthesis during leaf ontogeny in some herbaceous vines may result from the interplay of changes in photorespiration, carboxylation efficiency and stomatal conductance (Sesták *et al.*, 1985). In several species of vines this occurs about a week after leaf unfolding but it is largely a characteristic related to leaf life span (Figure 7.5). Under field conditions, leaf ontogenetic pattern of photosynthesis is overridden by microclimatic conditions which makes the pay-back period for any given investment costs of the leaf an important ecophysiological characteristic.

Leaves function initially as sinks for carbon, with peak sink activity in common bean attained at 10% final leaf size. Bean leaves start exporting and importing photosynthate for growth and construction before and until about 45% final leaf size, when leaves act as a photosynthate source exclusively (Swanson & Hoddinott, 1978). Carbon profit returns are shorter on leaves of individuals growing under high light environments compared with leaves on individuals growing in the shade (Jurik & Chabot 1986). In *Rubus corylifolius*, a pioneer vine species, pay-back period was the shortest (22 days) compared with other successional species in a Central European hedgerow community (Küppers, 1984). In *Cucumis sativus* leaf pay-back occurred 9–13 days after initiation (Hopkinson, 1964).

Some advantages of shorter pay-back periods for given costs may relate to both animal–vine and plant–vine interactions. Larger standing levels of leaf damage appear to be present on vine seedlings from tropical rainforests than on those of trees (R. Dirzo, personal communication). A trade-off between seemingly low investment costs for construction of secondary defensive

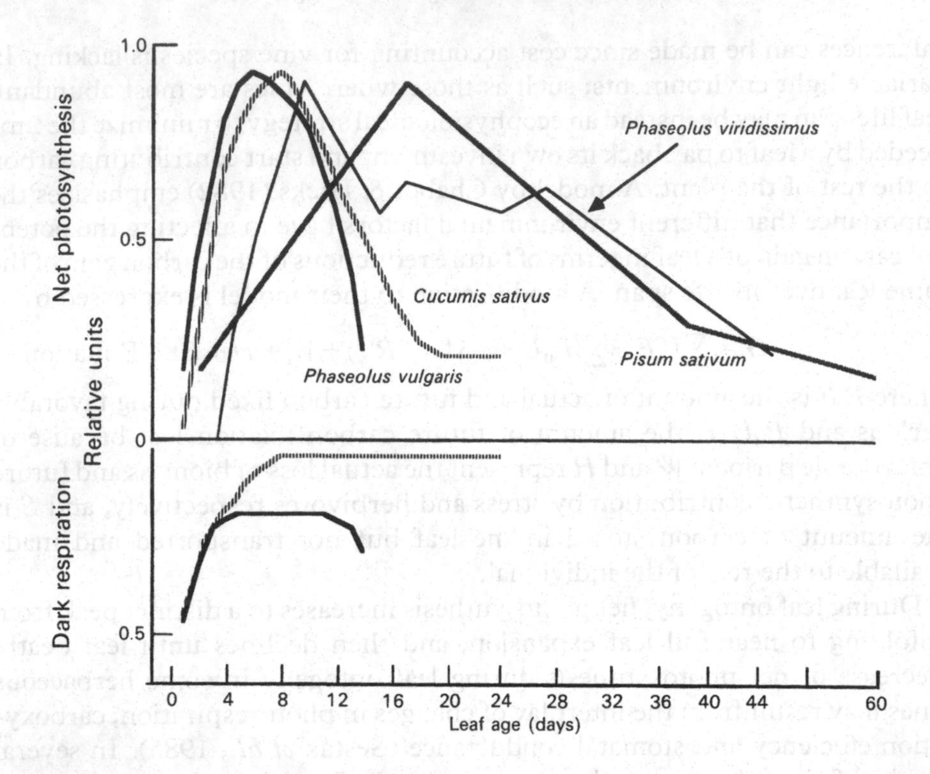

Figure 7.5. Ontogeny of leaf gas exchange in vine species. Rates are expressed as the proportion of their maximum during the leaf life span. All original data were modified to relative rates. Data on *Pisum sativum* from Bethlenfalvay & Phillips (1977), *Phaseolus viridissimus* from Saeki (1959), *Cucumis sativus* from Hopkinson (1964) and *Phaseolus vulgaris* from Sesták *et al.* (1985).

compounds in vines may involve hidden costs in terms of future carbon contribution resulting from losses in photosynthetic surface (see Equation 3), although the presence of photosynthetic compensatory mechanisms cannot be ruled out.

Photosynthesis in other tissues

A number of other chlorophyllous tissues are present in vine species and their contribution to the carbon economy of the individual could be important by contributing a considerable amount of photosynthate to their own carbon needs. A number of species of vines and other growth forms within a tropical deciduous forest in Costa Rica have chlorophyllous developing embryos, suggesting that carbon contribution by embryos to self-supporting seed development may be considerable (Janzen, 1982). Photosynthetic response to light is similar in pods and leaves so that fruit carbon self-support will be larger under high light conditions (Littleton *et al.*, 1981). Pod photosynthetic contribution to developing seeds ranges from 67 to 80% in leguminous vine species but these figures vary with developmental stage and source–sink relationships between subtending leaves and fruits (Flinn & Pate, 1970; Flinn, 1974; Littleton *et al*, 1981). In peas (*Pisum sativum*), positive carbon exchange in pods is present only after they cease growing and seed weight begins to increase (Flinn, 1974). Important contributions to the carbon requirements of seeds are made by pods at the beginning of fruit growth but carbon contributions from other sources are needed at the seed maturation stage (Flinn & Pate, 1970; Flinn, 1974).

Other green structures such as peduncles and stems may also contribute some carbon (Littleton *et al.*, 1981). Some vine species, like *Marsdenia laxiflora*, many *Cissus* species, and most herbaceous vines produce some type of unspecialized green shoots utilized as 'exploratory' structures searching for support (Peñalosa, 1982). Photosynthesis by these shoots may be very limited and growth subsidized by other plant structures may represent important costs for the individual. Both the ecological and physiological importance of these 'exploratory' shoots in the carbon economy of the individual are only intuitive at present.

Concluding remarks

It is interesting to note that some of the best studied photosynthetic responses to light and their interaction with other resources are those for vines, such as *Phaseolus*, *Hedera helix* and *Solanum dulcamara*. There is still a lack of comprehensive studies on the photosynthetic and gas exchange responses of vines, in particular within their native ecological settings. Vine species do better within heterogeneous light environments and some of the responses discussed above may indicate their high degree of physiological plasticity to such type of conditions.

Low stomatal conductances in many vines do not seem to correlate well with their moderate photosynthetic rates. Differential sensitivity of stomata to water vapor pressure deficit seems to override any biochemically imposed control. One may expect that vine species living in heterogeneous light environments with frequent high intensity lightflecks should have higher stomatal conductances to avoid limiting photosynthetic capacity; however, this does not seem to be the case. We need more data to understand further the vine strategies in regulating carbon uptake and water loss.

Habitat and growth dynamics of vines require a more integrated approach to carbon uptake and loss, i.e. their temporal and spatial carbon balance, to understand major selective pressures in photosynthetic and gas exchange traits. That importance is strongly suggested by changes in the acclimation potential of vines depending on resource availability and the fast pay-back period for the amount of carbon invested in a leaf. Rapid turnover means that costs must be paid soon to ensure a leaf's contribution within a potentially fast changing environment.

At present, photosynthetic and gas exchange responses of vines can only be roughly sketched and mostly by borrowing pieces from many and dissimilar species living sometimes in contrasting environments. It is hoped that this paper serves to stimulate the exploration of the many unanswered aspects of vine photosynthesis, its dynamics and integration.

Summary

Vines are characterized by allocating a large percentage of their biomass to photosynthetic surface that turns over rapidly. Biochemical, biophysical and physiological aspects of vine leaf photosynthesis and respiration and their leaf life-span integrated carbon gains and costs are practically unknown. Optimization models of leaf size and shape based on the temperature effects on photosynthesis are not satisfactory to explain vine species leaf characteristics while light and water availability seem to be more important constraints in their leaf physiology. A comparison of temperate and tropical vine species suggests a large amount of plastic adjustments in photosynthesis to light conditions both during leaf development and after maturation. Some interesting but not fully understood biochemical and physiological mechanisms seem to be involved in the adaptation of vine photosynthesis to heterogeneous light environments, such as rapid photosynthetic acclimation responses in C_3 plants, and the presence of CAM (Crassulacean acid metabolism) in vine species growing under deep-shade conditions within tropical rainforests. Leaf dimorphism in vines and reversible leaflet movements in several leguminous vines are important mechanisms to maximize carbon gain while reducing transpiration, excessive heat load and photoinhibitory conditions. Fragmentary data on photosynthesis and gas exchange of vines and their different components (photosynthetic capacity, cost of construction and maintenance of leaf biomass, leaf life-span and environmental constraints on photosynthesis) are insufficient to construct a leaf (and individual) carbon balance. Further studies integrating photosynthetic and ecophysiological characteristics are needed in order to understand the extensive three-dimensional 'foraging' surface developed in vines apparently as an adaptation to spatially heterogeneous light environments.

References

Adams, W. W. III, Terashima, I., Brugnoli, E. & Demmig, B. (1988). Comparisons of photosynthesis and photoinhibition in the CAM vine *Hoya australis* and several C_3 vines growing on the coast of eastern

Australia. *Plant, Cell and Environment* **11**: 173–81.

Bauer, H. & Thöni, W. (1988). Photosynthetic light acclimation in fully developed leaves of the juvenile and adult life phases of *Hedera helix*. *Physiologia Plantarum* **73**: 31–7.

Bell, A. (1984). Dynamic morphology: A contribution to plant population ecology. In *Perspectives on Plant Population Ecology*, ed. R. Dirzo and J. Sarukhán, pp. 48–65. Sinauer, Sunderland, Mass.

Benecke, U., Schulze, E.-D., Matyssek, R. & Havranek, W. M. (1981). Environmental control of CO_2-assimilation and leaf conductance in *Larix decidua* Mill. 1. A comparison of contrasting natural environments. *Oecologia* **50**: 54–61.

Bethlenfalvay, G. J. & Phillips, D. A. (1977). Ontogenic interactions between photosynthesis and symbiotic nitrogen fixation in legumes. *Plant Physiology* **60**: 419–21.

Björkman, O. (1981). Responses to different quantum flux densities. In *Physiological Plant Ecology I, Encyclopedia of Plant Physiology, n.s., Vol. 12A*, ed. O. L. Lange, P. S. Nobel, C. B. Osmond and H. Ziegler, pp. 57–107. Springer-Verlag, Berlin.

Bloom, A., Chapin, F. S. & Mooney, H. A. (1985). Resource limitation in plants- An economic analogy. *Annual Review of Ecology and Systematics* **16**: 363–92.

Boardmann, N. K. (1977). Comparative photosynthesis of sun and shade plants. *Annual Review of Plant Physiology* **28**: 355–77.

Burnside, C. A. & Böhning, R. H. (1957). The effect of prolonged shading on the light saturation curves of apparent photosynthesis in sun plants. *Plant Physiology* **32**: 61–3.

Caballé, G. (1984). Essai sur la dynamique des peuplements de lianes ligneuses d'une forêt du Nord-Est du Gabon. *Revue d'Ecologie* **39**: 3–36.

Carter, G. A. & Teramura, A. H. (1988). Vine photosynthesis and relationships to climbing mechanics in a forest understory. *American Journal of Botany* **75**: 1011–18.

Castellanos, A., Mooney, H. A., Bullock, S. H., Jones, C. & Robichaux, R. (1989). Leaf, stem and metamer characteristics of vines in a tropical deciduous forest in Jalisco, Mexico. *Biotropica* **21**: 41–9.

Chabot, B. F. & Hicks, D. J. (1982). The ecology of leaf life spans. *Annual Review of Ecology and Systematics* **13**: 229–59.

Chaves, M. M., Harley, P. C., Tenhunen, J. D. & Lange, O. L. (1987). Gas exchange studies in two Portuguese grapevine cultivars. *Physiologia Plantarum* **70**: 639–47.

Chazdon, R. L. & Pearcy, R. W. (1986). Photosynthetic responses to light variation in rainforest species. II. Carbon gain and photosynthetic

efficiency during lightflecks. *Oecologia* **69**: 524–31.
Chow, W. S. & Anderson, J. M. (1987). Photosynthetic responses of *Pisum sativum* to an increase in irradiance during growth. 1. Photosynthetic activities. *Australian Journal of Plant Physiology* **14**: 1–8.
de Luca, P., Alfani, A. & Virzo de Santo, A. (1977). CAM, transpiration, and adaptive mechanisms to xeric environments in the succulent Cucurbitaceae. *Botanical Gazette* **138**: 474–8.
Fetcher, N. (1981). Leaf size and leaf temperature in tropical vines. *American Naturalist* **117**: 1011–14.
Field, C. & Mooney, H. A. (1986). The photosynthesis-nitrogen relationship in wild plants. In *On the Economy of Plant Form and Function*, ed. T. J. Givnish, pp. 25–55. Cambridge University Press, Cambridge.
Flinn, A. M. (1974). Regulation of leaflet photosynthesis by developing fruit in the pea. *Physiologia Plantarum* **31**: 275–8.
Flinn, A. M. & Pate, J. S. (1970). A quantitative study of carbon transfer from pod and subtending leaf to the ripening seeds of the field pea (*Pisum arvense* L.). *Journal of Experimental Botany* **21**: 71–82.
Forseth, I. N. & Teramura, A. H. (1986). Kudzu leaf energy budget and calculated transpiration: The influence of leaflet orientation. *Ecology* **67**: 564–71.
Forseth, I. N. & Teramura, A. H. (1987). Field photosynthesis, microclimate and water relations of an exotic temperate liana, *Pueraria lobata*, kudzu. *Oecologia* **71**: 262–7.
Gauhl, E. (1976). Photosynthetic response to varying light intensity in ecotypes of *Solanum dulcamara* L. from shaded and exposed habitats. *Oecologia* **22**: 275–86.
Gentry, A. H. (1983). Lianas and the 'paradox' of contrasting latitudinal gradients in wood and litter production. *Tropical Ecology* **24**: 63–7.
Givnish, T. J. & Vermeij, G. J. (1976). Sizes and shapes of liane leaves. *American Naturalist* **110**: 743–78.
Greer, D. H. & Laing, W. A. (1988). Photoinhibition of photosynthesis in intact kiwifruit (*Actinidia deliciosa*) leaves: Effect of light during growth on photoinhibition and recovery. *Planta* **175**: 355–63.
Gross, L. J. (1982). Photosynthetic dynamics in varying light environments: A model and its application to whole leaf carbon gain. *Ecology* **63**: 84–93.
Gross, L. J. & Chabot, B. F. (1979). Time course of photosynthetic response to changes in incident light energy. *Plant Physiology* **63**: 1033–8.
Hall, A. E. & Schulze, E.-D. (1980). Stomatal response to environment and a possible interrelation between stomatal effects on transpiration

and CO_2 assimilation. *Plant, Cell and Environment* 3: 467–74.

Harper, J. L. & Bell, A. (1979). The population dynamics of growth form in organisms with modular construction. In *Population dynamics, the 20th Symposium of the British Ecological Society*, ed. R. M. Anderson *et al.*, pp. 29–52. Blackwell, Oxford.

Hladik, A. (1974). Importance des lianes dans la production foliaire de la forêt équatoriale du Nord-Est du Gabon. *Comptes Rendus, Académie des Sciences (Paris), Sér. D* 278: 2527–30.

Hoflacher, H. & Bauer, H. (1982). Light acclimation in leaves of the juvenile and adult life phases of ivy (*Hedera helix*). *Physiologia Plantarum* **56**: 177–82.

Hopkinson, J. M. (1964). Studies on the expansion of the leaf surface. IV. The carbon and phosphorous economy of a leaf. *Journal of Experimental Botany* **15**: 125–37.

Janzen, D. H. (1982). Ecological distribution of chlorophyllous developing embryos among perennial plants in a tropical deciduous forest. *Biotropica* **14**: 232–6.

Jones, H. G. (1983). *Plants and Microclimate.* Cambridge University Press, Cambridge.

Jordan, C. F. & Murphy, P. G. (1978). A latitudinal gradient of wood and litter production, and its implication regarding competition and species diversity in trees. *American Midland Naturalist* **99**: 451–34.

Jurik, T. W., Chabot, J. F. & Chabot, B. F. (1979). Ontogeny of photosynthetic performance in *Fragaria virginiana. Plant Physiology* **63**: 542–7.

Jurik, T. W. & Chabot, B. F. (1986). Leaf dynamics and profitability in wild strawberries. *Oecologia* **69**: 296–304.

Keuneman, E. A., Wallace, D. H. & Ludford, P. M. (1979). Photosynthetic measurements of field-grown dry beans and their relation to selection for yield. *Journal of American Society for Horticultural Science* **104**: 480–2.

Küppers, M. (1984). Carbon relations and competition between woody species in a central European hedgerow. III. Carbon and water balance on the leaf level. *Oecologia* **65**: 94–100.

Küppers, M., Wheeler, A. M., Küppers, B. I., Kirschbaum, M. U. F. & Farquhar, G. D. (1986). Carbon fixation in eucalypts in the field. Analysis of diurnal variations in photosynthetic capacity. *Oecologia* **70**: 273–82.

Larcher, W. (1980). *Physiological Plant Ecology*, 2nd edition. Springer-Verlag, Berlin.

Littleton, E. J., Dennett, M. D., Elston, J. & Monteith, J. L. (1981). The growth and development of cowpeas (*Vigna unguiculata*) under tropical field conditions. 3. Photosynthesis of leaves and pods. *Journal of Agricultural Science, Cambridge* **97**: 539–50.

Loveys, B. R. & Kriedemann, P. E. (1973). Rapid changes in abscisic acid-like inhibitors following alterations in vine leaf water potential. *Physiologia Plantarum* **28**: 476–9.

Mooney, H. A. (1972). The carbon balance of plants. *Annual Review of Ecology and Systematics* **3**: 315–46.

Mooney, H. A. (1981). Primary production in Mediterranean-climate regions. In *Mediterranean-type Shrublands*, ed. F. di Castri, D. W. Goodall and R. L. Specht, pp. 249–55. Elsevier, The Netherlands.

Mooney, H. A., Troughton, J. H. & Berry, J. A. (1977). Carbon isotope ratio measurements of succulent plants in Southern Africa. *Oecologia* **30**: 295–305.

Nobel, P. (1974). *An Introduction to Biophysical Plant Physiology*. W. H. Freeman, San Francisco.

Olivares, E., Urich, R., Montes, G., Coronel, I. & Herrera, A. (1984). Occurrence of crassulacean acid metabolism in *Cissus trifoliata* L. (Vitaceae). *Oecologia* **61**: 358–62.

Osmond, B., Björkman, O. & Anderson, D. J. (1980). Physiological processes in plant ecology. Toward a synthesis with *Atriplex*. *Ecological Studies 36*. Springer-Verlag, Berlin.

Parkhurst, D. F. & Loucks, O. L. (1972). Optimal leaf size in relation to environment. *Journal of Ecology* **60**: 505–37.

Patterson, D. T. (1975). Photosynthetic acclimation to irradiance in *Celastrus orbiculatus* Thunb. *Photosynthetica* **9**: 140–4.

Pearcy, R. W., Osteryoung, K. & Calkin, H. W. (1985). Photosynthetic response to dynamic light environments by Hawaiian trees. *Plant Physiology* **79**: 896–902.

Pearson, C. (1974). Daily changes in carbon-dioxide exchange and photosynthate translocation of leaves of *Vicia faba*. *Planta* **119**: 59–70.

Peñalosa, J. (1982). Morphological specialization and attachment success in two twining lianas. *American Journal of Botany* **69**: 1043–5.

Peñalosa, J. (1983). Shoot dynamics and adaptive morphology of *Ipomoea phillomega* (Vell.) House (Convolvulaceae), a tropical rainforest liana. *Annals of Botany* **52**: 737–54.

Peñalosa, J. (1985). Dinámica de crecimiento de lianas. In *Investigaciones sobre la regeneración de selvas altas en Veracruz, México, Vol. II*, ed. A. Gomez-Pompa and S. del Amo, pp. 147–69. Alhambra, México.

Putz, F. E. (1983). Liana biomass and leaf area of a 'Tierra firme' forest in the Rio Negro Basin, Venezuela. *Biotropica* **15**: 185–9.

Putz, F. E. (1984). The natural history of lianas on Barro Colorado Island, Panama. *Ecology* **65**: 1713–24.

Putz, F. E. & Windsor, D. M. (1987). Liana phenology on Barro Colorado Island, Panama. *Biotropica* **19**: 334–41.

Saeki, T. (1959). Variation of photosynthetic activity with aging of leaves and total photosynthesis in a plant community. *Botanical Magazine (Tokyo)* 72: 404–8.

Schenck, H. (1892). Beiträge zur Biologie und Anatomie der Lianen, im Besonderen der Brasilien einheimischen Arten. 1. Beiträge zur Biologie der Lianen. In *Botanische Mittheilungen aus den Tropen 4*, ed. A. F. W. Schimper, pp. 1–253. G. Fischer, Jena.

Sesták, Z., Tichá, I., Catsky, J., Solarová, J., Pospisilová, J. & Hodanová, D. (1985). Integration of photosynthetic characteristics during leaf development. In *Photosynthesis during Leaf Development*, ed. Z Sesták, pp. 263–86. Dr W. Junk, The Netherlands.

Sheriff, D. W. & Ludlow, M. M. (1984). Physiological reactions to an imposed drought by *Macroptilium atropurpureum* and *Cenchrus ciliaris* in a mixed sward. *Australian Journal of Plant Physiology* **11**: 23–34.

Shim, K. K., Ahn, Y. H. & Yoo, M. S. (1985). Studies on photosynthesis of perennial vines (*Actinidia arguta, Celastrus orbiculatus, Lonicera japonica* var. *aureoreticulata, Lonicera sempervirens*). *J. Korean Soc. Hort. Sci.* **26**: 44–50.

Slade, A. J. & Hutchings, M. J. (1987). Clonal integration and plasticity in foraging behaviour in *Glechoma hederacea*. *Journal of Ecology* 75: 1023–36.

Suzuki, W. (1987). Comparative ecology of *Rubus* species (Rosaceae). I. Ecological distribution and life history characteristics of three species, *R. palmatus* var. *coptophyllus, R. microhyllus* and *R. crataegifolius*. *Plant Species Biology* **2**: 85–100.

Swanson, C. A. & Hoddinott, J. (1978). Effect of light and ontogenetic stage on sink strength in bean leaves. *Plant Physiology* **62**: 454–7.

Syvertsen, J. P. (1984). Light acclimation in citrus leaves. II. CO_2 assimilation and light, water, and nitrogen use efficiency. *Journal of the American Society for Horticultural Science* **109**: 812–17.

Taylor, S. E. (1975). Optimal leaf form. In *Perspectives of Biophysical Ecology*, ed. D. M. Gates and R. B. Schmere, pp. 73–86. Springer-Verlag, Berlin.

Tenhunen, J. D., Lange, O. L., Gebel, J., Beyschlag, W. & Weber, J. A. (1984). Changes in photosynthetic capacity, carboxylation efficiency, and CO_2 compensation point associated with midday stomatal closure and midday depression of net CO_2 exchange of leaves of *Quercus suber*. *Planta* **162**: 193–203.

Ting, I. P., Hann, J., Holbrook, N. M., Putz, F. E., Sternberg, L. da S. L., Price, D. & Goldstein, G. (1987). Photosynthesis in hemiepiphytic

species of *Clusia* and *Ficus*. *Oecologia* 74: 339–46.
Turgeon, R. & Webb, J. A. (1975). Leaf development and phloem transport in *Cucurbita pepo*: Carbon economy. *Planta* **123**: 53–62.
Virzo de Santo, A., Alfani, A., Greco, L. & Fioretto, A. (1980). Environmental influences on CAM activity of *Cissus quadrangularis*. *Journal of Experimental Botany* **31**: 75–82.
Weber, J. A., Tenhunen, J. D. & Lange, O. L. (1985). Effects of temperature at constant air dew point on leaf carboxylation efficiency and CO_2 compensation point of different leaf types. *Planta* **167**: 81–8.
Winter, K., Wallace, B. J., Stocker, G.C. & Roksandic, Z. (1983). Crassulacean acid metabolism in Australian vascular epiphytes and some related species. *Oecologia* 57: 129–41.
Winter, K., Osmond, C. B. & Hubick, K. T. (1986). Crassulacean acid metabolism in the shade. Studies on an epiphytic fern, *Pyrrosia longifolia*, and other rainforest species from Australia. *Oecologia* **68**: 224–30.
Ziegler, H., Batanouny, K. H., Sankhla, N., Vyas, O. P. & Stichler, W. (1981). The photosynthetic pathway types of some desert plants from India, Saudi Arabia, Egypt and Iraq. *Oecologia* **48**: 93–9.

8

Heteroblastic development in vines

DAVID W. LEE AND JENNIFER H. RICHARDS

Introduction

All plants pass through a series of developmental changes from seedling to reproductive age. In many plants the shifts in development are subtle, involving internode distance or small changes in leaf size and shape. In others the changes are profound, so much so that early and late stages have been identified as separate species. Goebel (1900) recognized these differences in the degree of developmental change, designating the former as **homoblastic** and the latter as **heteroblastic**. Since Goebel's treatment, the concept of heteroblastic development has been extended to species in which the juvenile to adult transition is more gradual, which includes the majority of plants (Allsopp, 1965), and the term has been applied to developmental changes in floral form in the reproductive phase (Lord, 1979).

The following traits can change during heteroblastic development: (i) leaf size and shape; (ii) leaf anatomy; (iii) phyllotaxis; (iv) internode length; (v) stem thickness; (vi) shoot apex structure and zonation; (vii) tropic response; (viii) regenerative capacity; (ix) physiology; (x) reproductive status (Goebel, 1900; Troll, 1937, 1939; Allsopp, 1965, 1967; Doorenbos, 1965; Richards, 1983). The change from juvenile to adult character states can occur at different times for each of these characters, e.g. changes in leaf shape can occur before changes in reproductive status. This temporal variability indicates that the concept of heteroblasty probably includes a number of indirectly related developmental changes (Allsopp, 1965; Borchert, 1976).

The complexity of the interrelationships of heteroblastic characters can be seen when considering the relationship of heteroblastic development to apical meristem size. Apical meristem size commonly increases during plant ontogeny. Heteroblastic changes, such as an increase in leaf size, have thus been shown to be accompanied by an increase in apical meristem size (Crotty,

1955; Franck, 1976; Kaplan, 1980; Mueller, 1982a). This increase in apical meristem size is also found in lateral branch reiterations, which often resemble heteroblastic sequences (Kaplan, 1973). Changes in leaf *shape* during heteroblastic development, however, are not necessarily a direct result of increased leaf size (Hammond, 1941; Kaplan, 1980). Thus, the changes in leaf morphology associated with heteroblastic development cannot be attributed solely to increases in meristem size, although this increase may contribute to or allow such changes.

Although heteroblastic development is considered to be under intrinsic control, environmental factors can affect the course of the developmental changes. To Goebel (1900) the differences between the juvenile (non-reproductive) and adult (reproductive) stages showed that '**the adaptation of the juvenile form to external relationships is different from that of the adult form** . . .'. Emergent aquatics, some xerophytes, and vines have some of the most extreme heteroblastic changes in the plant kingdom. Emergent aquatics and vines also show extensive environmentally induced variation in form (plasticity) that can be superimposed on heteroblastic development. The term heterophylly is commonly used to describe the marked variation in leaf form seen in aquatics and vines. Although controls on environmental plasticity have been investigated experimentally in aquatics (e.g. Bodkin, Spence & Weeks, 1980; Deschamps & Cooke, 1983; Richards & Lee, 1986; Goliber & Feldman, 1990), such data are non-existent for vines. In addition, the relationship of heterophylly to heteroblastic development has not been well defined for either of these types of plants.

The problems of understanding the role of intrinsic vs. extrinsic (i.e. environmental) controls on heteroblastic transitions are especially apparent in the literature on vine development. Juvenile vines appear adapted to the extreme shade conditions to which the plants are exposed in nature, but research defining the microenvironment in which vines develop is lacking, as are experimental data on effects of different environmental parameters on heteroblastic development and heterophylly in vines. The purpose of this chapter is to describe heteroblastic development in a variety of vines, to present a model for vine development, and to suggest approaches to research on the adaptive significance and environmental control of vine heteroblasty.

Heteroblasty in vines: examples

Most vines exhibit marked changes in development from juvenile to adult stages – they are heteroblastic in Goebel's original sense of the term. The vine growth habit allows plants to begin life in the profound shade under vegetation, then ascend to more direct exposure to sunlight.

The challenge that a vine faces during development is to survive in a

Figure 8.1. Juvenile and adult stages of *Hedera helix*. Drawing from Sinott (1960).

particular microenvironment and to accumulate enough resources in that environment to allow directed growth into a new microenvironment. In order to understand the types of variations seen in heteroblastic development, we describe below the developmental ecology of some of the best studied or most remarkable vine taxa.

Temperate dicotyledons

The English ivy (*Hedera helix* L., Araliaceae) is a woody vine native to European forests and widely introduced into temperate areas of the world. In its juvenile stage the vine develops palmately lobed leaves, an alternate phyllotaxis, pubescent stems, and a plageotropic, climbing growth habit (Figure 8.1). The climbing stems are anchored by adventitious roots. The adult form, in addition to the development of flowers, produces ovate leaves spirally arranged on orthotropically growing branches (Figure 8.1; Wareing & Frydman, 1976). These branches extend away from the juvenile form's support and do not produce roots. The juvenile stage is associated with high endogenous gibberellin levels, and the transition to the adult stage may result from reduced gibberellic acid (GA) production in the absence of roots. Rogler & Hackett (1975) showed that applications of GA_3 caused a reversion of the mature phase to a juvenile-like plant.

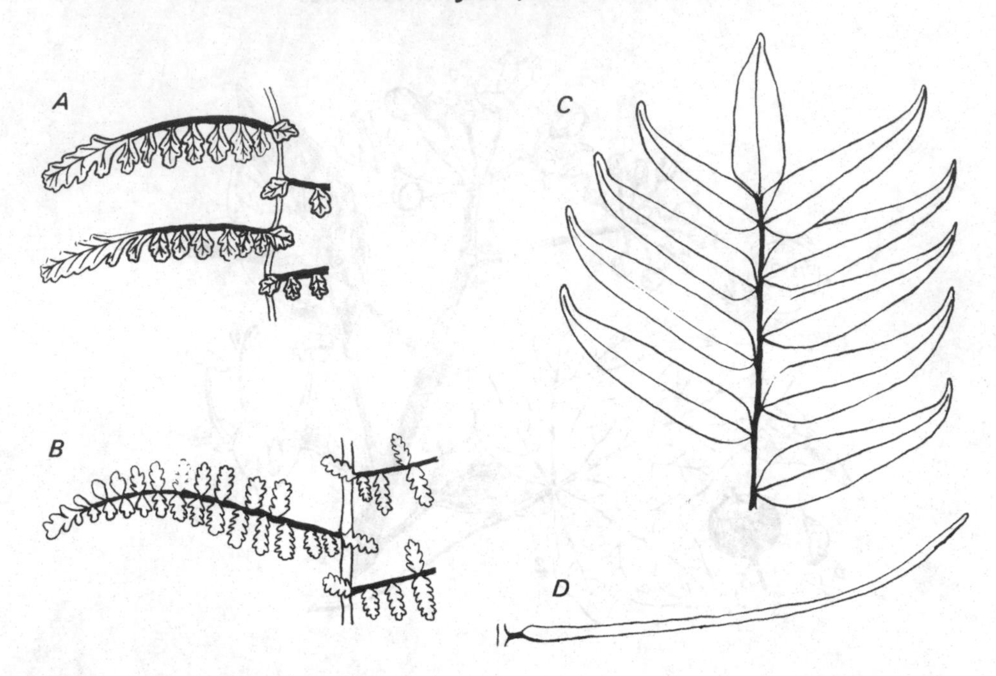

Figure 8.2. A diagram of leaf morphology in *Teratophyllum aculeatum* (Aspleniaceae). Four markedly different leaf types are produced at different stages in the life cycle. *A*, Leaves on young plants; *B*, intermediate leaf in heteroblastic series; *C*, tip of adult leaf; *D*, pinna of fertile frond. Redrawn from Holttum (1968).

Ferns

Many vining ferns exhibit marked differences in leaf morphology during their life, particularly between sterile and fertile leaves. The most extreme examples are found in the genus *Teratophyllum* (Aspleniaceae), which has 12 species in Southeast Asia. These ferns are root climbers that grow high in rainforest trees. In *T. aculeatum*, native to lowland rainforest in the region, four markedly different leaf types are produced at different stages in the life cycle (Holttum, 1968; Figure 8.2). Young plants produce very small pinnate leaves that bear pinnae only on the lower side of the midrib. Successively produced leaves are larger and develop pinnae on both sides. Transitional leaves with larger, lobed pinnae jointed to the rachis may occur. Adult plants produce fronds up to 40 cm long with 16 or more pairs of simple pinnae. Fertile fronds are similar, but the pinnae are less than 2 mm wide. A second climbing genus in the region, *Lomagramma*, also has a striking heteroblastic leaf sequence (Holttum, 1968).

The best-studied vining fern is undoubtedly the genus *Lygodium* (Schizaeaceae), which is widely distributed in disturbed vegetation throughout the tropics. Mueller (1982a, b; Figure 8.3) has studied the shoot morphology and

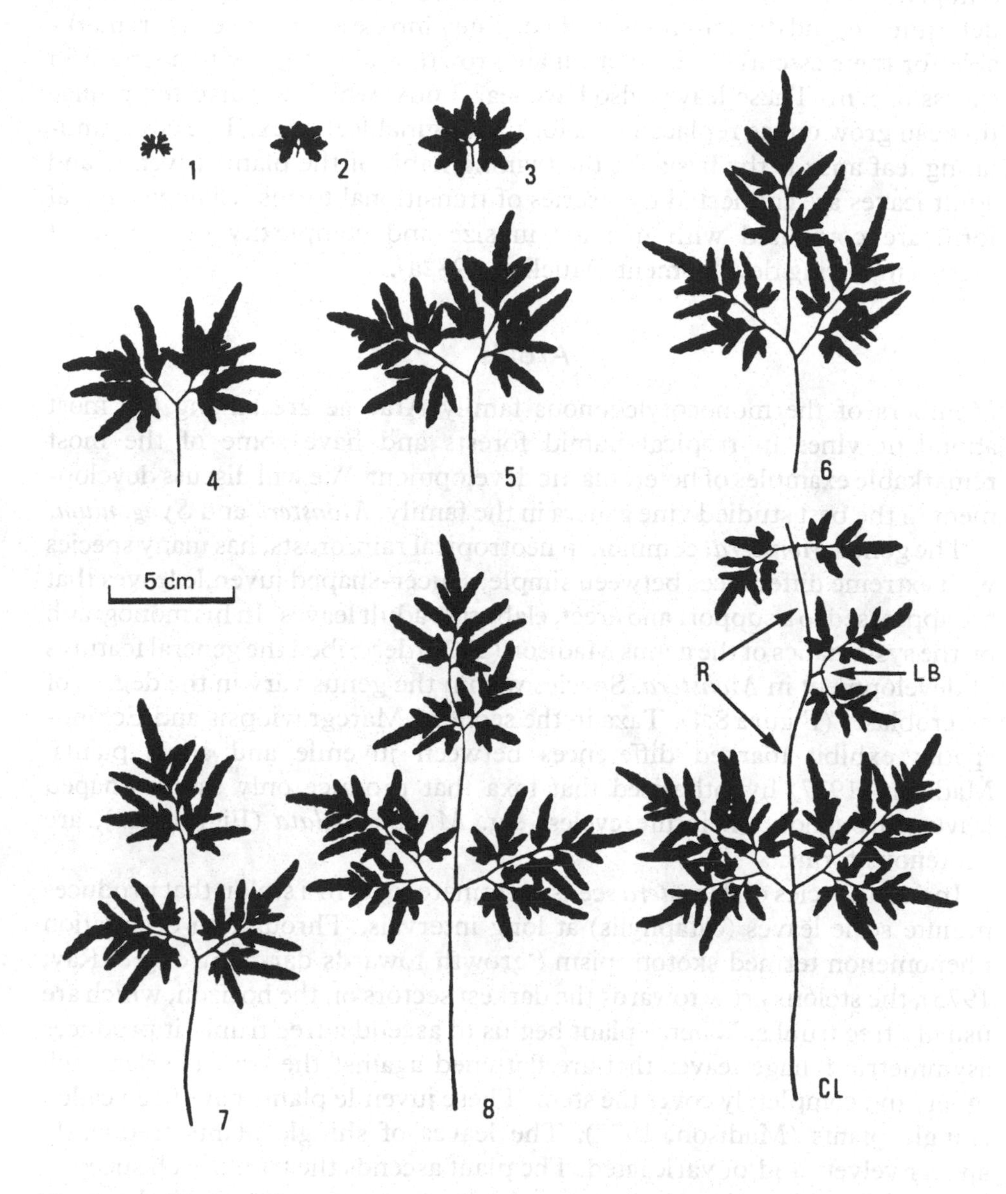

Figure 8.3. Diagram of the foliar heteroblastic sequence in *Lygodium japonicum* showing the first 8 leaves on the shoot (1–8) and part of the adult climbing leaf (CL). The juvenile (1–3) and adult (CL) leaves are connected by a series of transitional forms (4–8) that gradually increase in size and asymmetry of pinnae outgrowth. Diagram from Mueller (1982a).

ontogeny of *L. japonicum*. The first leaves on the sporophyte are small, determinate, and dichotomously lobed. The pinnate adult leaves are remarkable for their essentially indeterminate growth and can grow to a length in excess of 5 m. These leaves also have leaf 'buds' which are arrested pinnae that can grow out as replacements for the original leaf apex. The circumnutating leaf apex is the basis for the twining habit of the plant. Juvenile and adult leaves are connected by a series of transitional forms. Changes in leaf form are correlated with increase in size and complexity of the shoot meristem during development (Mueller, 1982a).

Aroids

Members of the monocotyledonous family Araceae are among the most abundant vines in tropical humid forests and have some of the most remarkable examples of heteroblastic development. We will discuss development in the best studied vine genera in the family, *Monstera* and *Syngonium*.

The genus *Monstera*, common in neotropical rainforests, has many species with extreme differences between simple, saucer-shaped juvenile leaves that are appressed to a support and erect, elaborate adult leaves. In his monograph on the systematics of the genus Madison (1977) described the general features of development in *Monstera*. Species within the genus vary in the degree of heteroblasty (Figure 8.4). Taxa in the sections Marcgraviopsis and Echinospadix exhibit marked differences between juvenile and adult plants. Madison (1977) hypothesized that taxa that produce only saucer-shaped leaves throughout their life cycles, e.g. *M. tuberculata* (Figure 8.4), are neotenous forms.

In most species of *Monstera* seeds germinate to form a stolon that produces minute scale leaves (cataphylls) at long intervals. Through an orientation phenomenon termed skototropism ('growth towards dark': Strong & Ray, 1975), the stolons grow towards the darkest sectors on the horizon, which are usually tree trunks. When a plant begins to ascend a tree trunk, it produces asymmetric foliage leaves that are flattened against the tree, overlap each other, and completely cover the stem. These juvenile plants have been called shingle plants (Madison, 1977). The leaves of shingle plants frequently appear velvety and/or variegated. The plant ascends the trunk, each successive leaf larger than the previous one. Some distance up the trunk there is a transition to the adult form, which produces large erect lobed leaves with holes in the laminae (Figure 8.5) as well as inflorescences. Should the vine outgrow its support, the stem hangs free and grows down to the forest floor. These descending axes, described as 'flagelliform' shoots (Madison, 1977; Blanc, 1980), produce long internodes and reduced leaves.

This description indicates that shoot growth in *Monstera* species displays

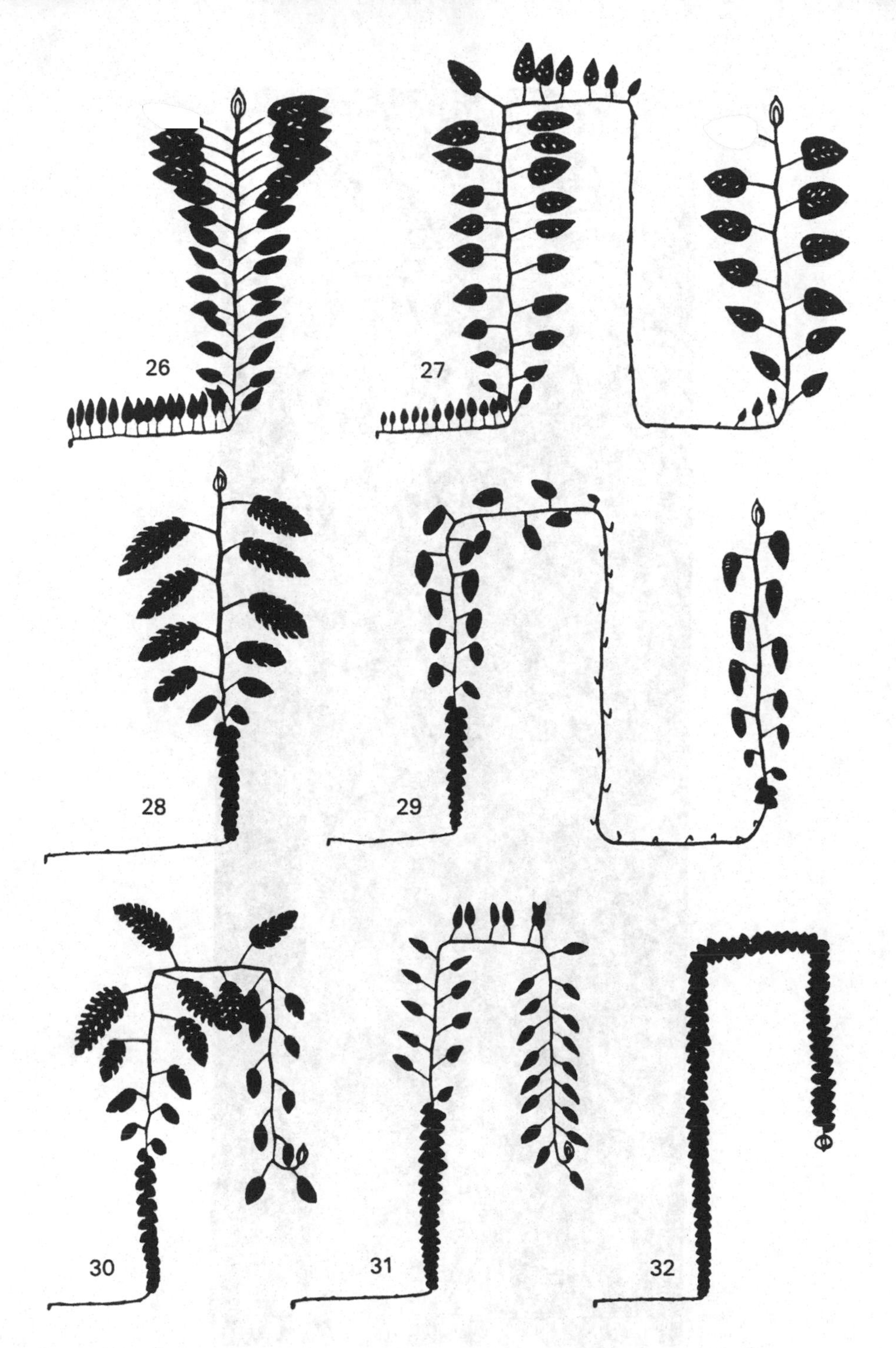

Figure 8.4. Diagram showing the general features of shoot development in the genus *Monstera*. Species differ from those having no leaf heteromorphy (e.g. *M. tuberculata*) to those with very marked heteromorphy (e.g. *M. siltepecana*). In each diagram the shoot begins at the seed on the left. 26. *M. lechleriana*; 27. *M. siltepecana*; 28. *M. punctata*; 29. *M. acuminata*; 30. *M. dubia*; 31. *M. pittieri*; 32. *M. tuberculata*. Diagram from Madison (1977).

Figure 8.5. Climbing vine of *Monstera tenuis* in lowland tropical forest at La Selva, Costa Rica. S, shingle leaf stage; T, transitional leaves; A, adult leaves.

both marked heteroblasty and great plasticity. The exact nature and interrelationship of these two phenomena need to be defined experimentally and in relation to the environments in which these plants grow (see section on developmental ecology, below).

The shifts between heteroblastic forms in *Monstera* are correlated with changes in the vine's light environment. Other observations, however, indicate the importance of endogenous growth patterns or other environmental controls on heteroblastic transitions (Ray, 1983a). For example, individuals of *M. tenuis* growing on exposed tree trunks still showed the progression of gradually increasing saucer leaf sizes, and the switch from juvenile to adult form occurred once these leaves reached a maximum diameter (Oberbauer *et al.*, 1980). Light also appears to have little influence on the transition from adult to flagelliform stages.

The genus *Syngonium* is abundant in the same neotropical forests as *Monstera*. It exhibits a different and much more uniform developmental strategy than *Monstera*. Ray (1983b, 1987) has described the life history patterns of taxa native to Costa Rica, particularly *S. triphyllum*. Seeds germinate on the forest floor and initially develop into small plants with a rosette of about 10 small leaves (Figure 8.6). The rosette stem tip then produces a slender prostrate stem with very small leaves at 8 cm intervals. These stems are skototropic (Ray, 1983b). If a tree is not encountered after 30 internodes, the plant reverts to its rosette form, produces 10–15 leaves, then switches back to the prostrate form. This alternation continues until a tree trunk is encountered. The stem then begins to climb the trunk and produces successively thicker stems and larger leaves. Each leaf has deeper lobes than the previous leaves. When the stem attains a diameter of approximately 14 mm, terminal inflorescences can be produced, in which case axillary branches continue the vegetative growth. The axillary branches repeat the trunk-based portion of the growth cycle.

Climbing stems that reach the top of a tree become detached and hang down. Successive internodes then decrease in diameter and increase in length, while the leaves decrease in size. When this hanging stem reaches the ground, it elongates rapidly and produces extremely small leaves, until it contacts another tree trunk and repeats the trunk-based cycle.

Tropical dicotyledons

Considerable diversity in growth strategies occurs among the dicotyledonous tropical vines. This variability probably results in part from diversity among species in the method of climbing (adventitious roots, hooks or tendrils, twining, etc.). The taxa described here are representative of this diversity.

Morning glory vines (*Ipomoea* spp., Convolvulaceae) climb by twining.

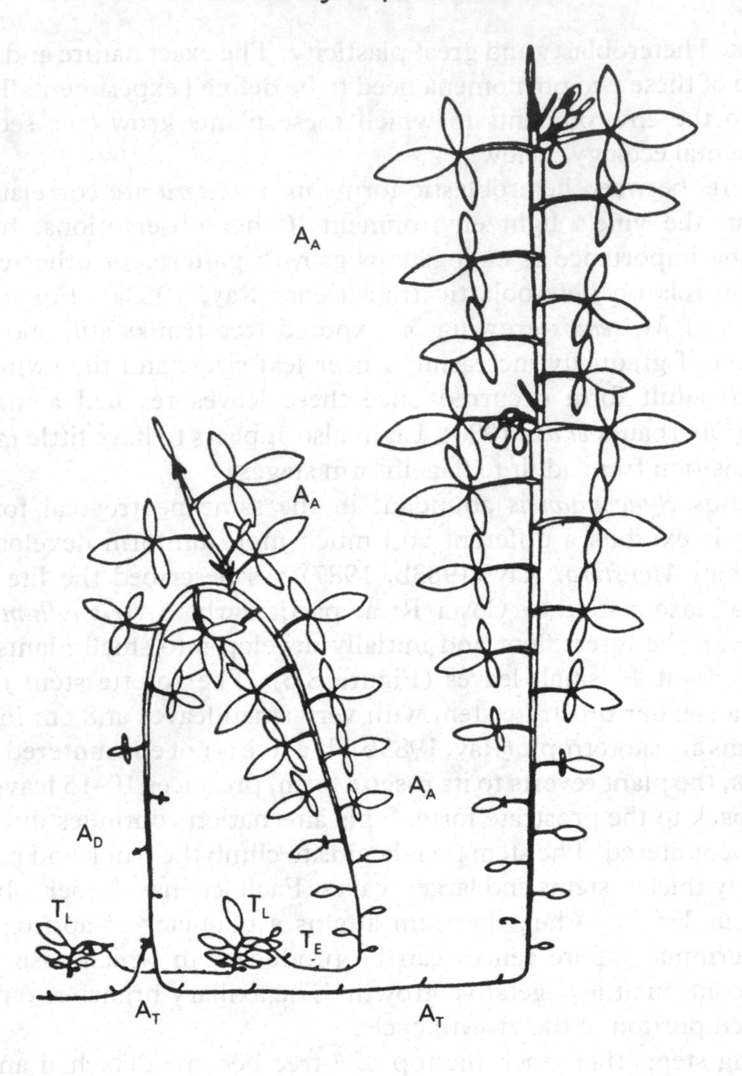

Figure 8.6. Diagram of shoot growth in *Syngonium* as exemplified by *S. triphyllum*. Growth begins with a terrestrial leafy stage (T_L) on left. T_E, terrestrial elongate; A_A, aerial ascending; A_D, aerial descending; A_T, aerial terrestrial. Diagram from Ray (1987).

Many morning glory vines exhibit a transition with age from simple cordate to deeply lobed leaves. The influence of environmental factors on this transition was first investigated by Ashby (1948a, b), Ashby & Wangerman (1950), and later by Njoku (1956, 1957, 1958). Temperature and mineral nutrition affected the shift in leaf development but were not as important as light intensity. Njoku (1956) showed that reduced light intensity retarded the rate of change in leaf morphology from juvenile to adult, and that plants grown in 72% and 78% shade did not produce adult leaves during the course

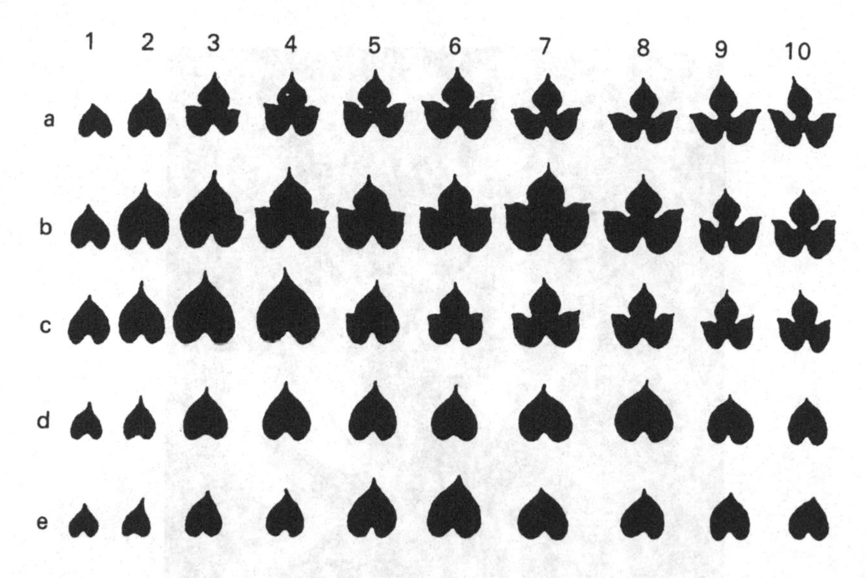

Figure 8.7. Diagram of the first ten leaves produced by seedlings of *Ipomoea caerulea* under different light intensities. a, full daylight; b, 26% shade; c, 44% shade; d, 72% shade; e, 78% shade. Diagram from Njoku (1956).

of the experiment (Figure 8.7). Njoku (1958) also showed that application of gibberellin to mature plants induced a shift to juvenile leaf shape.

Peñalosa (1983) investigated the demography of shoot types produced by *Ipomoea phillomega*, which grows in primary and secondary forest in Central America. This species produces two primary shoot types: (i) rapidly extending stolons with early deciduous leaves; (ii) erect twiners with longer-lived leaves and slower elongation rates. In studying the demography of these shoot types Peñalosa showed that high light intensity favors the development of twining shoots. Primary twining shoots originate from buds on stolons, and secondary twining shoots originate from buds on primary twining shoots.

The passion fruits (*Passiflora* spp., Passifloraceae) are important vines in neotropical forests (Killip, 1938). *Passiflora vitifolia*, which is native to tropical forests in Central America, initially develops as a compact plant with short internodes (Smiley, 1983). Juvenile plants are found in shady understory conditions and can persist for an undetermined amount of time. In bright light the internode distance lengthens, the plant develops a vining growth form, and the leaves progressively become more lobed. Larger vines grow into the canopy, but the scarlet flowers and mottled fruits generally occur on lower, leafless branches in the understory.

Cat-claw (*Macfadyena unguis-cati* (L) A. Gentry, Bignoniaceae) is a common forest liana that occurs from Mexico to Argentina and has been

Figure 8.8. Juvenile and adult foliage of *Macfadyena unguis-cati* in Miami, FL. Sd, seedling stage; J, juvenile vining stage; A, a single node from the adult vine.

introduced into the USA around the Gulf of Mexico. It has two juvenile forms (Figure 8.8; Gentry, 1983). The seedlings are erect with opposite and simple leaves. Seedlings develop into juvenile vines with bifoliate leaves. Each leaf develops a tendril with hooked 'cat-claw' tips. This stage is apparently skototropic and grows toward tree trunks or other dark silhouettes. After ascending the trunk, the vine produces free-hanging shoots with large bifoliate leaves that lack cat-claws (Figure 8.8). These adult shoots flower and fruit profusely.

Some insectivorous plants, which are often strongly heteroblastic (Franck, 1976), are also vines (e.g. *Nepenthes*). A spectacular example of extreme leaf dimorphism in an insectivorous plant is the tropical African liana *Triphyophyllum peltatum* (Dioncophyllaceae: Cremers, 1973; Green, Green & Heslop-Harrison, 1979). The seedling of *T. peltatum* produces lanceolate leaves up to 35 cm long and arranged spirally on a stout stem (Figure 8.9). Groups of these leaves alternate in an annual cycle with groups of filiform glandular

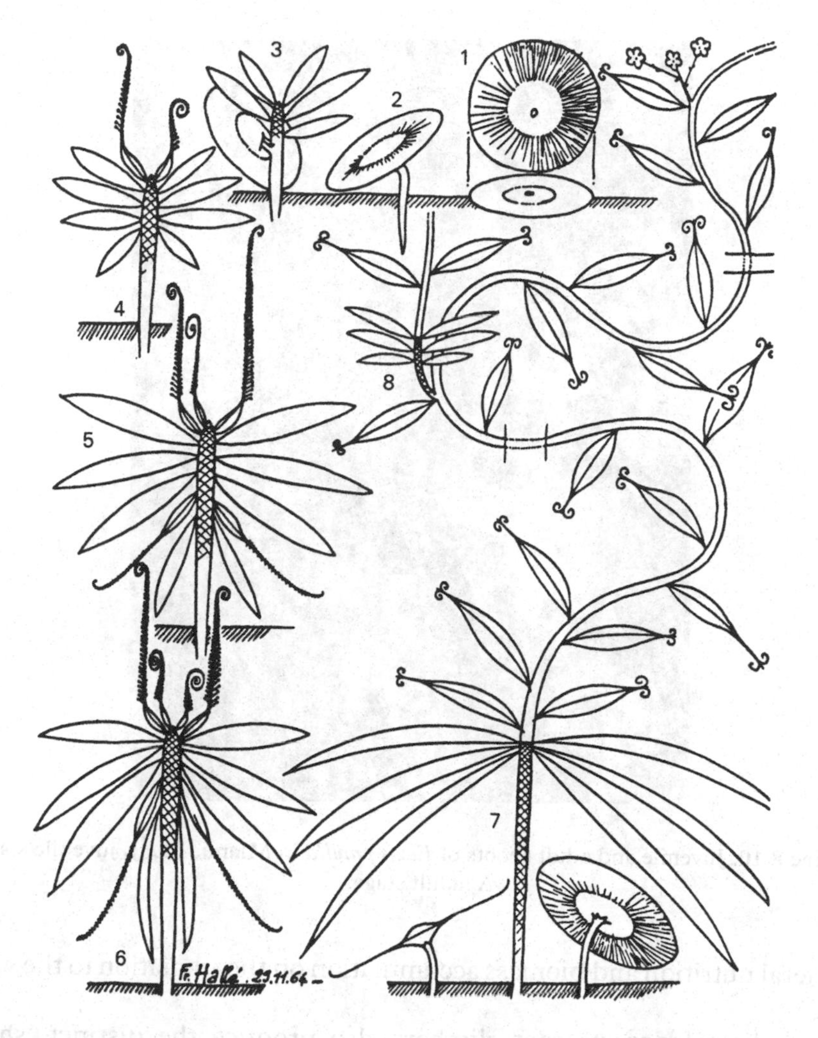

Figure 8.9. Diagram of developmental stages in the tropical African liana *Triphyophyllum peltatum* (Dioncophyllceae), which combines the vining habit with insectivory. Seedling stage with lanceolate leaves (1–3); juvenile stage with lanceolate and insectivorous leaves (4–6); adult vine produced by juvenile plant (7) and showing reiteration from axillary buds (8). Drawing by Francis Hallé.

leaves up to 30 cm long. The glandular leaves trap and digest insects. Adult vines, which can be up to 50 m long, are produced from the terminal or axillary buds of juvenile plants. Adult leaves are about half the length of juvenile ones, leathery, and oblong-elliptic, with two short hooks produced at the tip of each leaf (Figure 8.9). Thus, insectivory is limited to the juvenile, non-vining form. This species might lend itself readily to a study of the effect

Figure 8.10. Juvenile and adult shoots of *Ficus pumila* in Miami, FL. J, juvenile stage; A, adult stage.

of mineral nutrition and biomass accumulation on the transition to the vining habit.

Many dicotyledonous root climbers also produce the distinct 'shingle plant' juvenile morphology that was described for *Monstera*. This heteroblastic stage has been observed in the genera *Piper* (Blanc & Andraos, 1983), *Marcgravia*, *Metrosideros*, *Hoya*, *Conchophyllum*, and *Ficus* (Madison, 1977). Taxa in *Ficus* section Rhizocladus (76 species in Asia and Australasia: Corner, 1965) grow in this manner. *F. pumila* L., which belongs to this section, is an important ornamental vine in tropical and subtropical regions (Figure 8.10). Seedlings of this species establish on walls or tree trunks, grow vertically, and produce velvet-surfaced leaves that are appressed to the support. Older shoots produce lateral branches that grow away from the surface and develop leathery, oval leaves. These branches can bear flowers and fruits.

This survey shows that vines employ a variety of developmental patterns. Upon germination, seedlings can grow erect or can immediately begin horizontal growth in search of a support. Juvenile stages can be erect and self-supporting, stoloniferous and searching, or climbing. Adult stages can be climbing or free-hanging. What controls the sequence of growth forms that characterizes a particular species?

Development ecology of vine heteroblasty

Vine heteroblasty and environmental plasticity: models

Plant developmental patterns cover a range from those that are internally controlled, pre-programmed, and not sensitive to environmental factors to those that are very responsive to environmental changes. Heteroblastic development is generally considered to be an internally controlled, relatively predictable progression. Heteroblastic development in vines is interesting because control of this progression from juvenile to adult has been subjected to substantial environmental regulation. Figure 8.11 presents models for heteroblastic development in herbs or trees and in vines. In the former, the progression through developmental stages proceeds predictably from the seedling stage (1) to the adult (5). Change from one stage to the next is controlled primarily by some internal regulatory mechanism that, presumably, assesses the developmental status of the plant and causes the appropriate phase shifts. The mechanism for this control is unknown.

In vines (Figure 8.11*B*), in contrast, particular stages in the developmental progression can be prolonged (e.g. the increased number of stages 2, 3 and 4, compared with Figure 8.11*A*) or even re-entered (e.g. Figure 8.11*B*, arrows from stages 3 and 4 to stage 2), depending on the plant's environment. Figure 8.11*B* shows diagramatically the prolonged stages as similar throughout with sudden shifts between stages. Alternatively, gradual changes may occur within each stage, as in Figure 8.11*C*. Progression from one stage to the next probably still depends on an internal control, but sensitivity to external conditions has been superimposed on this internal control. These models emphasize that what is striking about heteroblastic development in vines is not the phenomenon *per se* but the environmental regulation of the phenomenon, which allows for the prolongation or repetition of certain stages.

The models presented here for vine heteroblastic development make testable predictions. For example, in Figure 8.11*B*, *C* a step-wise progression is shown in going from stage 1 to 5, but reversions can skip steps. Is this realistic? Madison's (1977) and Ray's (1987) studies of aroids imply that unique stages are involved in reversions (e.g. descending shoots, Figure 8.4;

A Heteroblastic development in herbs or trees

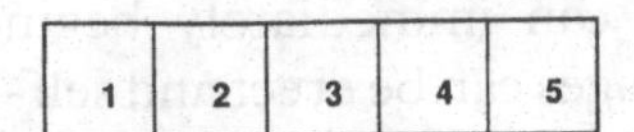

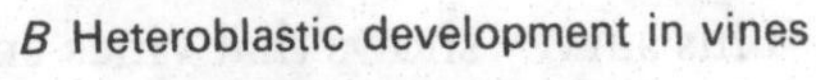

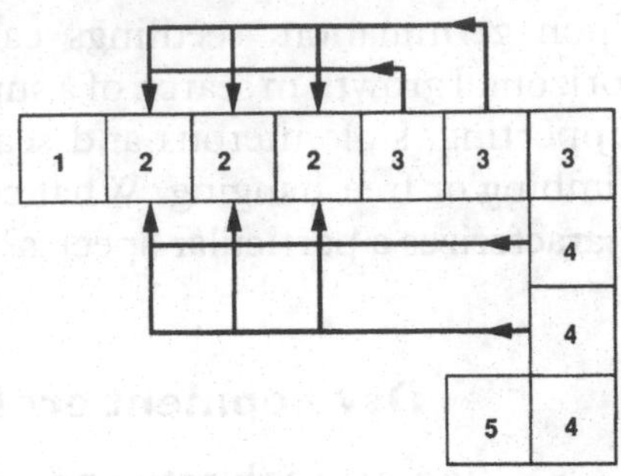

C Gradual heteroblastic development in vines

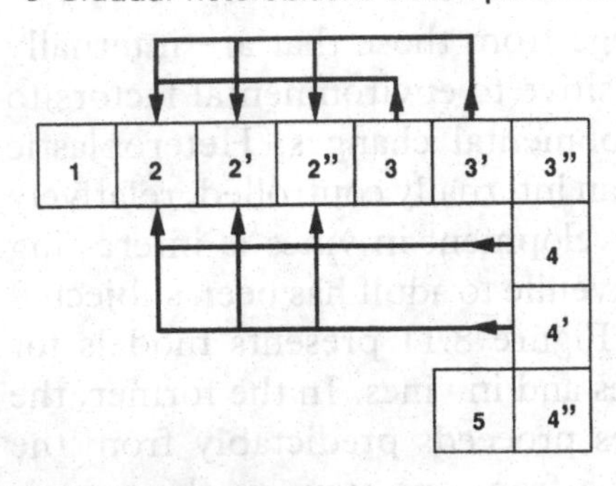

D Unique reversion stages in vine heteroblasty

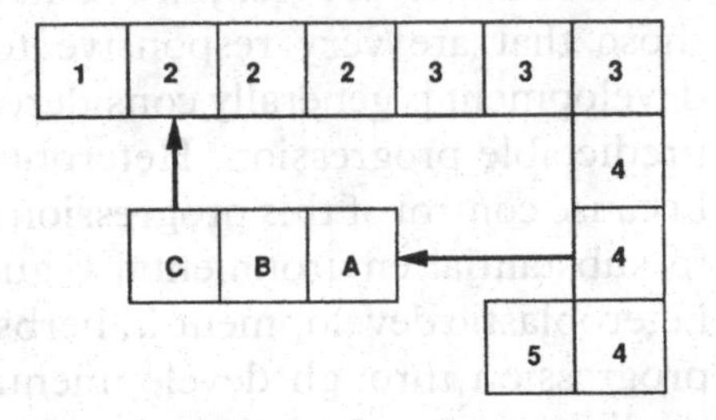

Figure 8.11. Models of heteroblastic development in herbs or trees (*A*) and vines (*B*, *C*, *D*). Numbers 1–5 represent particular stages in a heteroblastic sequence. The arrows show reversions. *B* and *C* differ in whether a particular stage is stable until it switches to the next stage (*B*) or whether a gradual change occurs within each stage (*C*). *D* illustrates the possibility that reversions involve unique developmental stages (*A*, *B*, *C*).

A_D and A_T, Figure 8.6). Such stages could be modeled as in Figure 8.11*D*. Experiments that attempt to induce reversions and that carefully analyze the nature of reversions in a number of vines are needed to evaluate this aspect of these models. Such experiments would also show whether reversions can occur at any stage or are limited to only some stages (e.g. stages 3 and 4 can revert, but not 5).

Consideration of these models raises questions about the architecture of vine heteroblastic development and how it differs among species. For example, when reversions occur, are they transformations of the original shoot apex, or are they reiterations of the heteroblastic sequence by lateral branches? Answers to this latter question have basic implications for how heteroblastic development is controlled in a given species, i.e. can a single apex be returned to a prior developmental stage? More detailed descriptions of the architecture and development of a number of vine species are needed before we will be able to generalize about vine architecture.

These models of heteroblasty in vines emphasize that vines are subjected to drastic changes in environmental factors as they grow. This observation raises two questions. The first is, why do canopy trees, which pass through the same environmental gradient, not show the same marked heteroblastic development; while the second is, what are the significant environmental factors that regulate vine heteroblasty?

Heteroblasty in vines vs. canopy trees

Although both canopy trees and vines germinate and grow in the forest understory but expand into the very different environment of the canopy, canopy trees do not have as marked heteroblasty as vines. Several reasons for this difference seem possible.

The first reason focuses on the basic difference between trees and vines: trees support themselves, whereas vines rely on external support. The two habits represent different resource allocation responses to the problem of growing from shade into sunlight. Trees invest in support, pushing their apical meristems to higher light levels. Vines invest in mobility. They have to find a support that will enable them to grow into the canopy. To do this they must have a low-cost developmental stage that allows them to seek a support in an energy-limited environment. Juvenile stages are generally smaller and simpler than adult stages. Thus, juvenile stages of vines may have been prolonged and further modified through selection for horizontal, searching growth.

A second possible reason for differences between trees and vines in their heteroblastic development relates to environmental differences that are an outcome of their different growth habits. Trees occupy a more predictable environment than vines. They are fixed in place and must either survive in place until a gap occurs in their vicinity or push themselves up through the canopy. Although the changes that they perceive in their environment can be gradual or sudden, those changes are unidirectional (e.g. increased light, decreased humidity, increased temperature).

Vines, in contrast, are more likely to encounter alternating changes in environmental parameters. If they overgrow their support or if it breaks, they will re-enter an environment that they have previously outgrown. Thus, whereas the tree's environment is either radically or gradually changing in one direction along a gradient, the vine's environment has cyclical changes that involve recurring increases and decreases in environmental factors. Vines must be able to respond to these changes and, therefore, plasticity must have greater selective advantage in heteroblastic development of vines than of trees.

A third possible reason for differences in vine and tree heteroblasty again derives from their differences in habit. Because trees have internal mechanical support, displaying flowers and bearing fruit do not represent new structural demands on plant resources. For vines, however, the need to expose flowers to pollinators and then to support the weight of developing fruit may present a new mechanical demand, which requires a new architecture. Thus, many root climbers, such as *Hedera* spp. and *Ficus pumila*, produce a rootless adult phase that grows away from the external support.

These hypotheses generate testable predictions for studies of the development and architecture of particular vine species. If support for reproductive parts is a significant factor in the evolution of adult vine morphology, then relative investment in support structures should be greater in the adult than in the juvenile phases. How do support costs change between juvenile and adult phases of species like *Triphyophyllum peltatum*, that are erect as juveniles and vining as adults? How does resource allocation in this type of vine compare with that of a species such as *Ficus pumila*, which is vining in the juvenile phase, but free-hanging in the adult?

If juvenile vines are both energy-limited and environmentally plastic, then their developmental stage could be controlled experimentally. For example, many vines begin growth with an upright stage similar to tree seedlings, then transform to a horizontal searching stage (cf. *Syngonium*, *Passiflora*, *Macfadyena*, above). If biomass accumulation regulates transformation to the searching stage, then experimental removal of leaf area from the upright stage should delay the change.

If developmental plasticity and environmental responsiveness have been selected for in vines, then vines should show a greater ability to respond to experimental manipulations of the environment than related tree or herb species. Studies of environmental plasticity in vine and tree species of *Bauhinia* could be used to test this hypothesis.

Environmental influences on heteroblasty in vines

Possible environmental influences on vine heteroblasty include gravity, tactile stimulation, temperature, water availability and light. Vines that overtop the trees on which they grow hang free and frequently revert to a more juvenile condition. As they hang down, they are re-oriented in the gravitational field, so gravity could stimulate the reversion. Movement of the free-swinging vine (similar to touch) could also be a stimulus. Tactile stimulation is important in the transformation of young shoots or tendrils when they come into contact with a vertical trunk, but because climbing a support involves a re-orientation, gravity could also play a role. Changes in relative humidity are likely to be correlated with other variables, such as temperature and light. The higher temperatures of more exposed environments could stimulate heteroblastic transitions in vines, and temperature affected the heteroblastic series in leaf form reported by Njoku (1957).

Although all of these environmental parameters may affect vine heteroblasty, light is likely to be the major environmental factor influencing phase shifts in vines. Mechanisms are known that regulate plant responses to changes in both light quantity and quality (Smith, 1981; Kendrick & Kronenberg, 1986), and both aspects of light change as vines climb. In order to study light effects on vine heteroblastic development, we must first know how the light climate differs between the forest floor and canopy.

Light climates. Light available for plant photosynthesis (quanta at 400–700 nm or photosynthetic photon flux [PPF]) decreases as it passes through a canopy. In tropical humid forests light levels are often less than 0.5% of light above the canopy, with average values between 1 and 2% (Chazdon & Fetcher, 1984a, b; Pearcy, 1983; Lee & Paliwal, 1988; Oberbauer *et al.*, 1988). In a tropical deciduous forest understory after the rainy season the light level was about 10% of that above the canopy, although PPF during the rainy seasons may be similar to levels in the understory of humid tropical forests (Lee, 1989).

The understory spectrum results from radiation transmitted through and reflected by leaves, penumbral radiation, skylight and direct sunlight. Radiation passing through the canopy is strongly altered in spectral quality. The most significant change is in the ratio of red to far-red quanta (quanta in a 10 nm band width centering on 660 and 730 nm, or R:FR), which affects the equilibrium between phytochrome P_r and P_{fr} (Smith, 1982). The change in R:FR beneath the canopy results from the optical properties of leaves, which absorb in the visible and reflect or transmit in the far-red (Gates *et al.*, 1965). Typical R:FR values for rainforest understory are 0.30–0.40, compared with 1.25 for sunlight. Extreme-shade R:FR values may be less than 0.25 (Lee, 1987). Differences in spectral quality at a rainforest site on Barro Colorado Island in Panama are illustrated in Figure 8.12, which shows that the greatest shift in spectral quality occurs at the lowest PPF, as in going from shade to a light fleck.

Even under the shady conditions in a forest, light quality is heterogenous. Spectral alteration is greater toward the horizon because solar radiation passes through more foliage layers, and there are fewer gaps in the canopy. Horizontal measurements in humid tropical forest at La Selva, Costa Rica, were substantially lower in PPF (less than 0.2%) and in R:FR (approximately 0.10) than vertical ones (see Lee, 1987, for methods).

Tree trunks also contribute to the understory light climate. Tree barks measured at Barro Colorado Island (five species; see Lee & Graham, 1986, for methods) reflected approximately 10% of incident light, with a range of

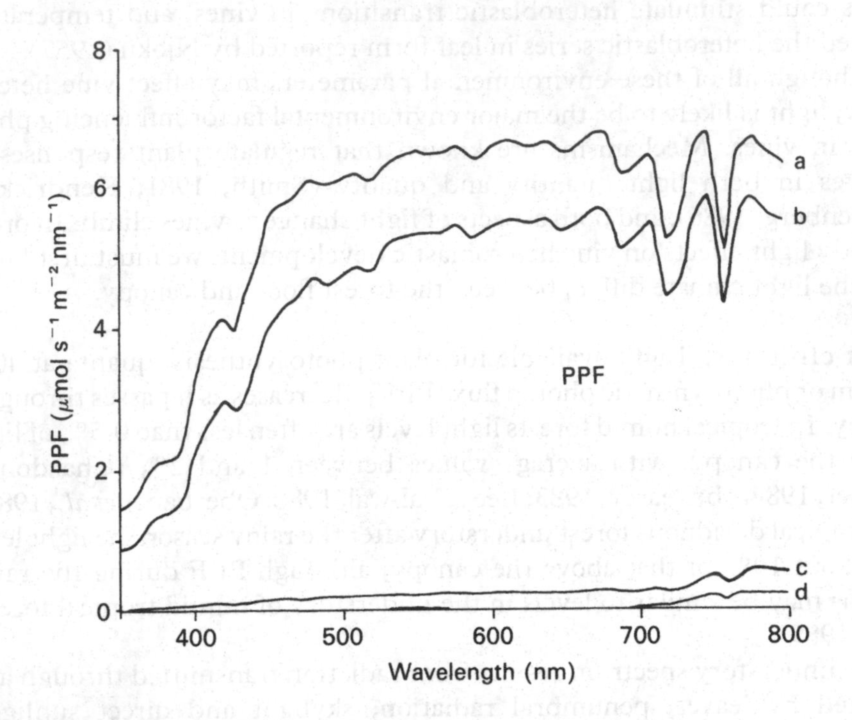

Figure 8.12. Graph of spectral radiation from 300 to 800 nm in a tropical rainforest on Barro Colorado Island, Panama. Measurements were taken in full sun (a), a large gap (b), a sunfleck in the understory (c), and deep shade in the understory (d). A comparison of radiation at 660 nm and 730 nm shows that a large shift in R:FR occurs at low PPF in going from shade (d) to a light fleck (c). Data from Lee (1987).

4–25%. More interestingly, the trunks reflected light that was even more strongly altered in R:FR than the diffuse background radiation, since bark reflects much more strongly in the far-red region (Figure 8.13). The mean R:FR of radiation reflected by the bark of five species was reduced to 0.047, compared with R:FR = 0.103 for understory radiation.

This discussion of light climates shows that since vines begin their growth on the rainforest floor, then ascend to the canopy, they experience extreme differences in radiation quantity (PPF) and spectral quality (R:FR). Vines undergo developmental changes, however, at intermediate elevations in the forest, such as on trunks or upper branches. To know if light climates are important in regulating developmental states in a heteroblastic series, we must know what the light conditions are at these intermediate elevations. Yoda (1974) has measured gradients in intensity with elevation in the forest. By extrapolating from a correlation between percentage of solar PPF and

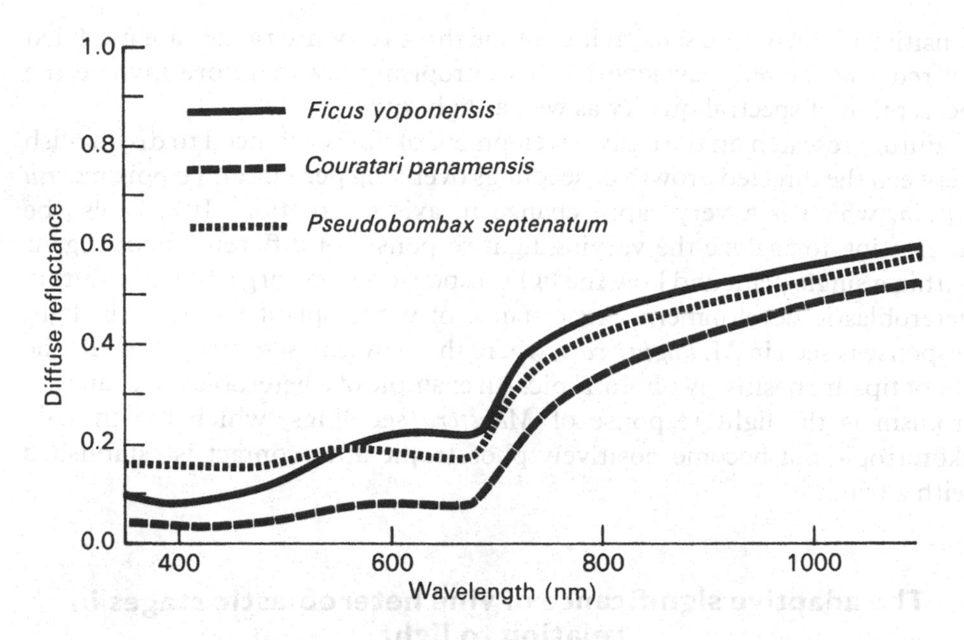

Figure 8.13. A graph of the radiation from 300–1100 nm reflected from the bark of three trees in the lowland tropical rainforest at Barro Colorado Island, Panama.

R:FR for measurements on the rainforest floor (Lee, 1987), we can predict R:FR for light at different elevations from Yoda's (1974) study. Such predictions suggest that light levels at 10 m above the forest floor could have R:FR as high as 0.70, which represents a rapid change in light quality between the forest floor and intermediate levels in the forest. If we are to learn more about the effects of light climates on vine development, however, we need actual measurements of the light microenvironments in which developmental changes occur, as well as evidence that environmentally significant levels of PPF and R:FR can influence vine development.

Skototropism. Light climates *can* exert a profound effect on early vine development, as illustrated by skototropism. Responses similar to skototropism have also been observed in the tendrils of vines in the Bignoniaceae, as in *Macfadyena unguis-cati* and *Bignonia capreolata*, studied by Darwin (1876).

The mechanism for the skototropic response is not known and merits further study. Strong & Ray (1975) suggested that skototropism may be the classic negative phototropic response, where seedlings are positively phototropic at very low quantum flux densities, then negatively phototropic at intermediate levels, then positively phototropic at still higher levels (Briggs, 1963). Although this phototropic response is mediated by a blue-light sensitive pigment, the sensitivities of the three response ranges are modified by red and far-red wavelengths. Skototropism may therefore involve the perception of spectral quality as well as light quantity.

Future research on the early development of vines will need to distinguish between the directed growth of seedlings over long periods and tropism *sensu stricto*, which is a very rapid change in axis orientation. It will also be interesting to analyze the varying light responses of different shoot organs within a single plant and how the light responses of one organ change during heteroblastic development. An instance of within-plant variation in light response is seen in *M. unguis-cati*, where the claws are skototropic, while the shoot tips are positively phototropic. An example of a heteroblastic change in tropism is the light response of *Monstera* seedlings, which are initially skototropic but become positively phototropic after contact is established with a trunk.

The adaptive significance of vine heteroblastic stages in relation to light

If light is the environmental factor to which vine heteroblastic development is responding, then vine morphology at different stages in development must represent adaptations to the light environments encountered at those stages. The vine characteristics most obviously correlated with light climate involve plant architecture, leaf morphology, and leaf anatomy. Here we will discuss research on (i) the relationship of vine architecture and leaf form to light environment and (ii) the significance of changes in leaf anatomy on the efficiency of light capture.

Architecture

An important strategy in plant survival is the optimization of energy capture through leaf surface display (Givnish, 1987). For xeric and very sunny environments, plants may minimize radiation absorption through pubescence or surface waxes and diurnal leaf movements. Extreme-shade adapted plants, however, expose foliar surfaces so as to maximize absorption of radiation. Strategies of leaf display are in part dependent on intrinsic patterns of branching and phyllotaxy, which have been called a plant's architecture (Hallé, Oldeman & Tomlinson, 1978).

Plants in extreme shade must minimize self-shading or the lower leaves may not have sufficient radiation to support photosynthesis. Branching patterns in shade-adapted plants typically produce monolayers of foliage

(Horn, 1971). Monolayers are formed by (i) branching horizontally; (ii) altering phyllotaxy or bending petioles to display leaves plagiotropically; (iii) combining leaf size, shape and display to minimize overlapping.

The architecture of vines has been neglected compared to that of tropical trees (Hallé *et al.*, 1978). Cremers (1973, 1974) described the architecture of 20 African vines. Many of these conformed to the architectural models characteristic of trees, but some vines developed by models that differed from the 24 tree models described by Hallé *et al.*, (1978).

In most of the vines that Cremers (1973, 1974) described the initial axis growth was orthotropic with leaves produced in a spiral. In the juvenile vining stage leaves were typically smaller than seedling leaves and were displayed plagiotropically, while axis extension varied from monopodial to sympodial. Some vines produced lateral branches that formed remarkably flat foliar surfaces at long intervals, as in *Raphiostylis beninensis* (Icacinaceae). For vines that were tendril-climbers, the production of tendrils was often one of the modifications in growth patterns that distinguished seedling and climbing stages.

Additional careful studies of vine architecture in relation to specific habitats would enable us to classify the diversity in developmental patterns responsible for the vining habit and to relate these patterns to the light environments of different species.

Leaf morphology

Building on the work of Parkhurst & Loucks (1972) and Taylor (1975), Givnish & Vermeij (1976) postulated that leaf form represents a compromise that maximizes the difference between photosynthetic profits and transpirational costs. A significant variable affecting gas exchange, water loss and leaf temperature is boundary layer resistance, which varies as a function of the effective leaf diameter. Givnish & Vermeij (1976) predicted that leaves along a vertical forest profile should be small in the shady and mesic understory, increase in size with elevation, then decrease in size in the more xeric and sunny conditions of the canopy. Superficial field observations in Venezuela and Costa Rica seemed consistent with their predictions (Givnish & Vermeij, 1976).

Many vines develop extensive lobing or holes in the lamina as they climb. These changes in form may decrease the effective leaf size and reduce boundary layer resistance, resulting in a functionally smaller leaf. Fetcher (1981) measured leaf temperatures and water vapor conductance of vine leaves of different sizes and solar exposures. He observed no significant correlations in leaf temperature with leaf diameter for the five taxa studied, although measurement was limited to a brief period during the rainy season (Fetcher, 1981). Both experimental studies and field analyses are needed to understand the interaction of boundary-layer effects and light climate effects on leaf form in vine heteroblastic development.

'Shingle plants' are found in the juvenile stages of phyletically diverse taxa, such as *Teratophyllum*, *Monstera*, and *Syngonium* (cf. Figures 8.4 and 8.5). This convergence in form suggests that shingle plant form provides a solution to some environmental problem(s) (Givnish, 1987). A number of hypotheses have been advanced to explain the adaptive significance of shingle leaves in juvenile vines. Goebel (1900) suggested that shingle leaves protect adventitious roots. Givnish & Vermeij (1976) speculated that these appressed leaves have lower support costs and are better oriented for light capture, since they face away from the trunk silhouette. Madison (1977) suggested that transpiration is reduced in such appressed leaves. Shingle leaves may also trap CO_2 from trunk respiration and so increase photosynthesis.

Measurements of microenvironmental parameters such as temperature, CO_2 concentration, humidity, and light climates, could eliminate some of these hypotheses. For example, is CO_2 concentration higher beneath shingle leaves? Oriented measurements of light microenvironments immediately adjacent to trunks are needed to substantiate Givnish & Vermeij's (1976) speculations, especially since light from the horizon in a forest is generally reduced in both quantity and quality, compared with vertical illumination (see section on environmental influences, above). An analysis of differences in allocation to support between appressed and non-appressed leaves in a heteroblastic series could begin to test the importance of support costs. Experimental manipulations could also contribute to understanding the adaptive significance of shingle plant morphology. For example, what is the effect on plant survival if shingle leaves are lifted away from the trunk to expose adventitious roots?

In twiners and tendril- or hook-climbers leaf form does not so closely correspond with Givnish & Vermeij's (1976) predictions. The climbing stage of *Passiflora vitifolia* produces progressively larger and more lobed leaves, and leaves at maturity are not smaller in size. In *Macfadyena unguis-cati* the seedling leaves are large and simple, the vine stage produces very small leaves with hooks, and the adult stage develops large bifoliate leaves. In *Ipomoea* spp. leaf size and lobing increase with age, and presumably, exposure to sunlight. Reduction in light quantity retards this transition (Njoku, 1956). Although lobing would tend to decrease the effective leaf diameter, it is difficult to discuss the functional significance of these changes without a detailed examination of the separate effects of size and lobing and studies of the ecological requirements of individual taxa.

Leaf anatomy and optical properties

Leaves of vine heteroblastic stages frequently differ in anatomy, particularly among the root climbers. Juvenile leaves appear to be adapted to shade conditions (D. W. Lee, personal observation). Shade-adapted leaves are generally thinner and have lower specific weights than sun-adapted leaves. The thinner leaves have fewer cell layers and fewer columnar palisade parenchyma cells. In order to determine whether these are features characteristic of a particular heteroblastic stage or whether these are anatomical adaptations to light conditions, plants from different stages must be grown under experimental light conditions. Here we will discuss some implications for physiological function of leaf anatomical characteristics and then present preliminary data on differences between juvenile and adult leaves of two root-climbing vines.

Variations in leaf anatomy can be understood as adaptations to the photosynthesis/transpiration compromise discussed previously. Adult vine leaves typically possess the anatomical features seen in Figures 8.14 and 8.16. The columnar palisade cells contain most of the chloroplasts but allow a portion of absorbed radiation to reach chloroplasts in the spongy mesophyll (Sharkey, 1985; Lee *et al.*, 1990). Intercellular spaces allow gas exchange, and the amount of cell surface exposed to air is a major determinant in photosynthesis (Nobel, Zaragoza & Smith, 1975). Intercellular spaces also contribute to light scatter within leaves, and their extent and distribution partially determine the degree of absorption of light by leaves (Willstatter & Stoll, 1918; Gausman *et al.*, 1969; Allen, Gausman & Richardson, 1973).

Juvenile leaves of vines often differ from the above description. Juvenile vine leaves can have (i) a satiny sheen due to convexly curved epidermal cells; (ii) anthocyanic coloration on the abaxial surface; (iii) variegation; (iv) lack of differentiation into a palisade and spongy mesophyll. In the root climbers *Monstera*, *Marcgravia* and *Ficus*, for example, most juvenile saucer leaves have the first and fourth characteristics, and a few *Monstera* taxa are variegated. Similar characteristics are found in extreme-shade plants of tropical rainforests (Haberlandt, 1914; Richards, 1952; Burtt, 1978; Roth, 1984; Lee, 1986).

The adaptive significance of these features in juvenile vines and extreme-shade plants are probably similar. Convexly curved epidermal cells can refract light onto specially oriented chloroplasts and increase photosynthetic efficiency (Bone, Lee & Norman, 1985). Such cells can also increase light capture efficiency at oblique angles under diffuse light conditions. Anthocyanic undersurface coloration is associated with greater absorption above 650 nm (Lee, Lowry & Stone, 1979; Lee & Graham, 1986; Lee, 1986).

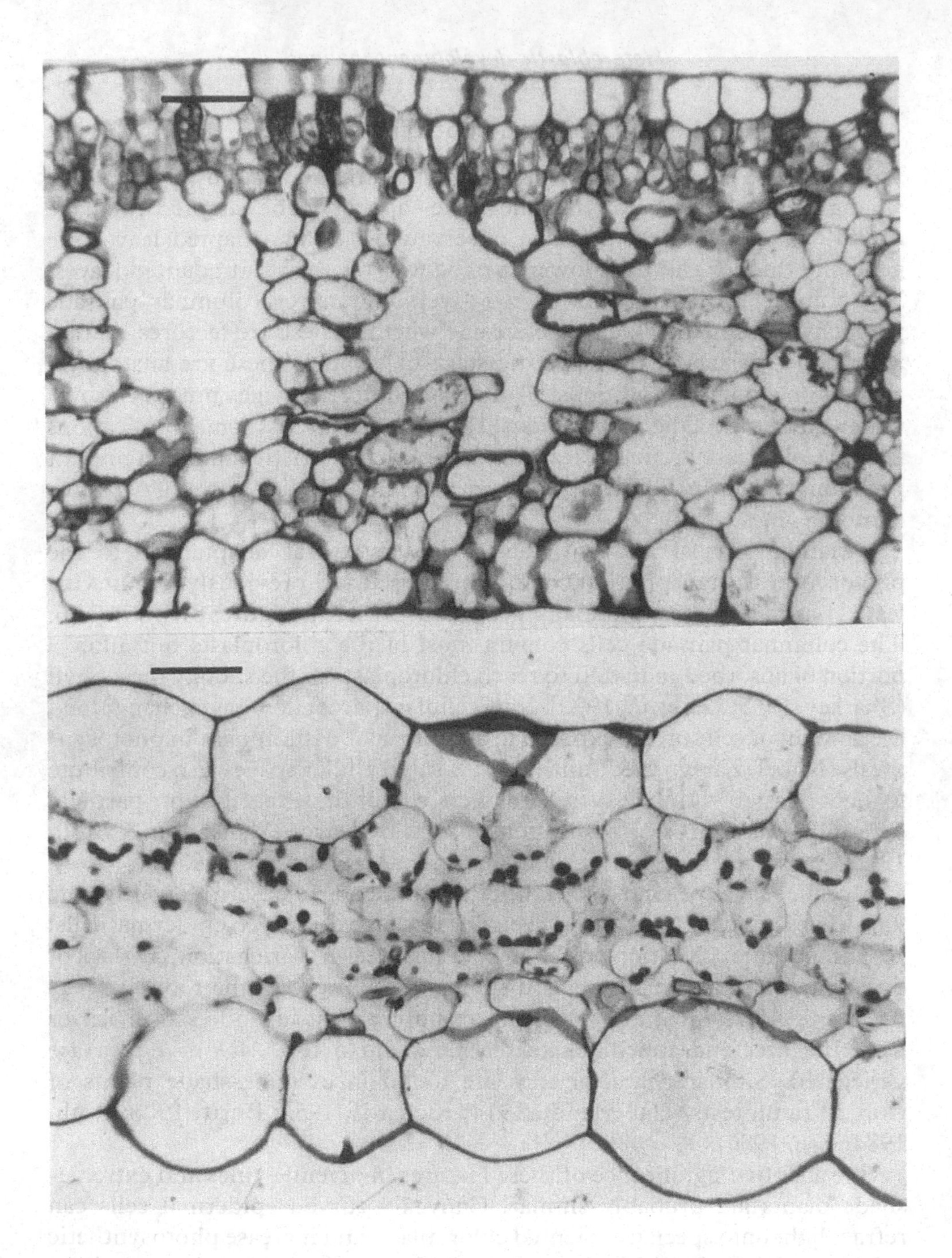

Figures 8.14 (above), **8.15** (below). Cross-sections of adult (Fig. 8.14) and juvenile (Fig. 8.15) leaves of *Monstera adansonii*. Bars = 50 μm. The juvenile leaves are thinner, lack a distinct palisade mesophyll layer, and have much larger, convexly curved, epidermal cells. Palisade cells in the thicker adult leaves are columnar, epidermal cells are flat, and chloroplasts are distributed throughout the mesophyll. In the juvenile leaves the chloroplasts are concentrated on the lower sides of the top two mesophyll layers.

Variegation has been interpreted as aposematic coloration and as a means of reducing leaf temperature (Smith, 1986). The presence of equi-diameter palisade cells in shade leaves is associated with the distribution of chloroplasts either in a dense monolayer on the adaxial surface or in smaller cells packed with large chloroplasts. Such a monolayer increases the efficiency of leaf absorption by minimizing the 'sieve' effect (Duysens, 1956). In leaves of sun-adapted plants, the columnar palisade cells allow light to pass through, making the leaf less efficient in energy capture but exposing more chloroplasts to light. Leaves of extreme-shade plants also have fewer cell layers, less lignification, lower specific weights, lower chlorophyll content, and lower protein content (Björkman, 1981; Lee *et al.*, 1990). Such leaves represent a less metabolically costly production of surface for energy capture.

As an example of the differences between juvenile and adult vine leaf anatomy and of the similarities between juvenile vine leaves and the leaves of extreme-shade plants, we present below the results of an anatomical analysis of leaves of two root climbers. The species are *Monstera adansonii* Schott. and *Syngonium rayii* Croat & Grayum ined., both aroids common at La Selva, Costa Rica (Figures 8.14–8.17; Table 8.1).

Monstera adansonii grows in secondary forest and disturbed sites. Its shingle leaves are typical of those in the genus, and its adult leaves are not lobed but have numerous perforations (cf. Figure 8.4). The most pronounced anatomical difference between juvenile shingle leaves and adult leaves is the difference in size of epidermal cells, for the juvenile form has much larger cells. Palisade cells in the adult leaves are more columnar than those of the shingle leaves, but they are reduced in size. The chloroplasts in adult leaves are distributed throughout the mesophyll. In the juvenile leaves the chloroplasts are concentrated on the abaxial surfaces of the top two mesophyll cell layers (Figure 8.15).

The juvenile leaves have (i) a lower percentage dry weight; (ii) a lower specific weight; (iii) thinner cross-sections; (iv) equi-diametric palisade cells; (v) more air spaces; (vi) less chlorophyll per unit area (Table 8.1).

The juvenile leaves absorb less radiation than the adult leaves (Figure 8.18), but their efficiency of light absorption by chlorophyll is greater. The ratio of attenuance (*in vivo* absorbance by chlorophyll) to absorbance (absorbance by chlorophyll extract at an equivalent concentration) is higher than in the adult leaf. This implies reduced sieve effects in the juvenile leaves and a more efficient distribution of chloroplasts for absorption (Lee *et al.*, 1990).

The convexly curved cells of the juvenile leaves focus light (Bone *et al.*, 1985). Focusing increases light intensity at the upper level of the choroplasts (57 μm) by 11.7 times. The adult leaves have flat epidermal cell walls with no focusing effects. Thus, the epidermal cells of juvenile leaves increase the

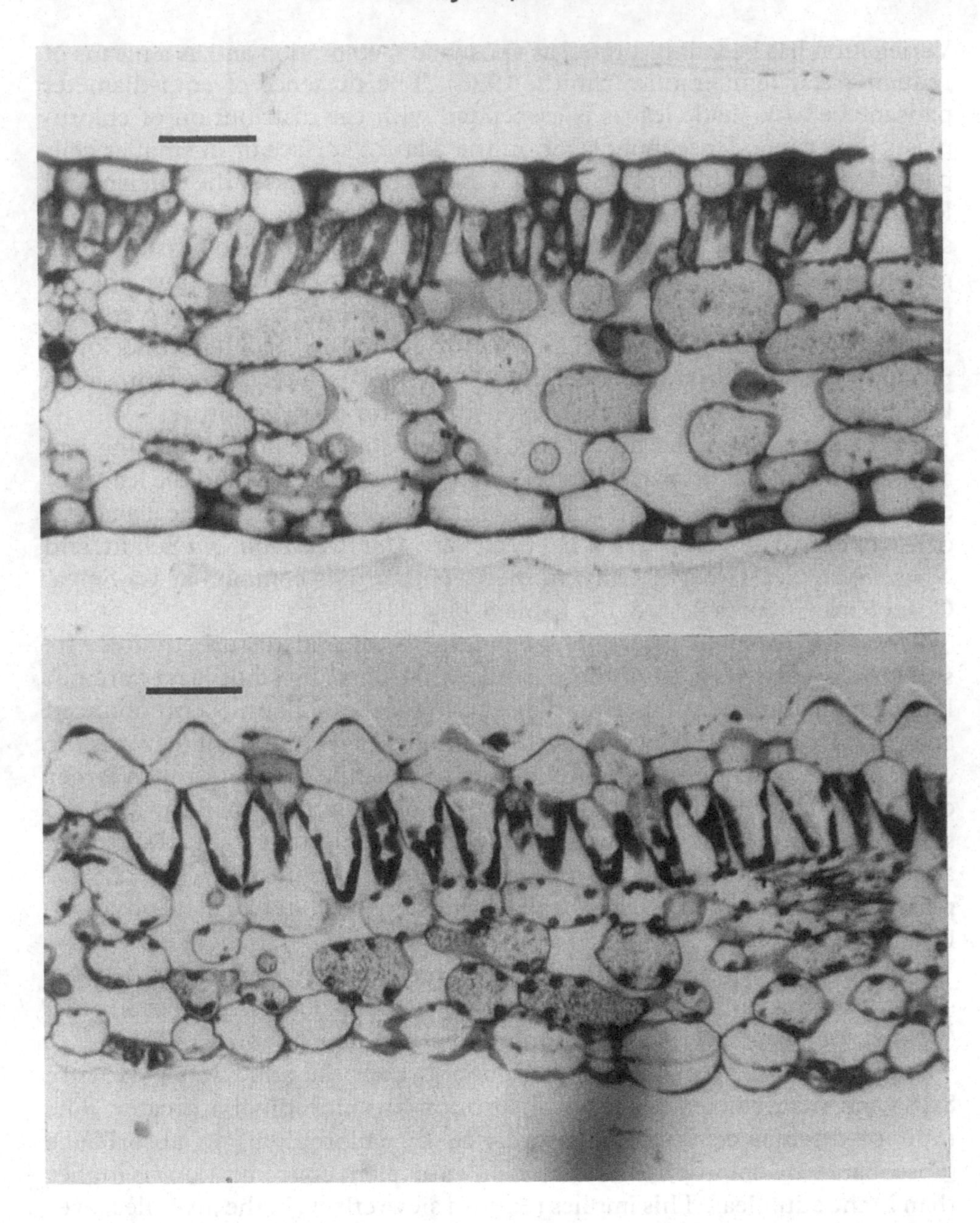

Figures 8.16 (above), **8.17** (below). Cross-sections of the adult (Fig. 8.16) and juvenile (Fig. 8.17) leaves of *Syngonium rayii*. Bars = 50 μm. The juvenile and adult leaves do not show the marked differences seen in *Monstera adansonii*, although the juvenile leaves have convexly curved epidermal cells.

Table 8.1. *Comparisons of anatomical and optical characteristics of juvenile and adult leaves of two vines*

Character	*Monstera adansonii*		*Syngonium rayii*	
	Juvenile	Adult	Juvenile	Adult
Percentage dry weight	10.0 ± 0.8	14.7 ± 0.7	14.4 ± 1.0	17.7 ± 2.7
Specific weight (mg cm^{-2})	2.78 ± 0.58	4.98 ± 0.21	2.45 ± 0.19	3.50 ± 0.68
Thickness (μm)	206 ± 44	360 ± 46	169 ± 18	188 ± 19
Palisade height/width	0.91 ± 0.22	2.17 ± 0.35	1.98 ± 0.35	2.66 ± 0.32
Number of cells thick	5.9 ± 0.3	13.2 ± 1.0	5.8 ± 0.4	6.8 ± 0.4
Portion of air space	0.25 ± 0.06	0.12 ± 0.02	0.22 ± 0.05	0.27 ± 0.04
Chlorophyll/area (μg cm^{-2})	27 ± 3	43 ± 6	63 ± 11	45 ± 7
Percentage of PPFD absorptance	84.9 ± 1.5	88.4 ± 1.9	94.9 ± 0.6	89.4 ± 1.6
Attenuance/absorbance at 652 nm	1.06 ± 0.05	0.80 ± 0.09	0.57 ± 0.02	0.76 ± 0.04
Concentration factors for direct sunlight/depth				
chloroplast	11.7 × /57 μm	–	6.00 × /22 μm	–
optimal	13.9 × /70 μm	–	7.55 × /33 μm	–

efficiency of light capture by refracting light onto the chloroplasts, although the rounded cells are not likely to increase the absorption of more oblique light rays (Bone *et al.*, 1985). This comparison of the structure of adult and juvenile leaves of *M. adansonii* shows that the shade-adapted juvenile leaves are less costly in construction and more efficient in energy capture than the adult leaves.

Syngonium rayii is a root climber that ascends to 2–3 m in the forest at La Selva, especially in disturbed areas. The juvenile form is a small rosette of almost black leaves found on the forest floor in deep shade. Juvenile leaves are velvety and about 3 cm long. Adult leaves are a glossy green, cordate, and up to 10 cm long.

Juvenile leaves of *S. rayii* have a significantly lower percentage dry weight and specific weight than adult leaves (Table 8.1). They are slightly thinner and lack approximately one cell layer present in adult leaves (Figure 8.16 vs. Figure 8.17). The proportion of air space does not differ significantly between juvenile and adult leaves, but juvenile leaves have more chlorophyll per unit area. The juvenile leaves absorb significantly more photosynthetic radiation than adult leaves (Table 8.1, Figure 8.19). Part of this increase, however, results from absorbance by anthocyanin at 525 nm and does not contribute to photosynthesis (Lee & Graham, 1986). The efficiency of energy capture (the attenuance/absorbance ratio at 652 nm) is lower in juvenile compared with adult leaves. Juvenile leaf epidermal cells have a convex

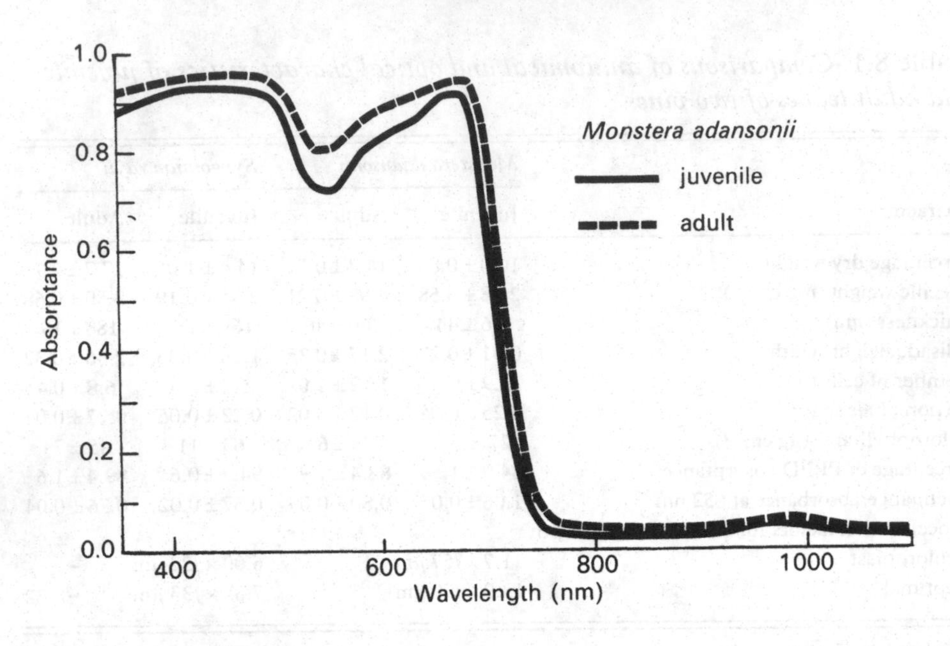

Figure 8.18. The absorptance of radiation between 300 and 1100 nm by juvenile and adult leaves of *Monstera adansonii*.

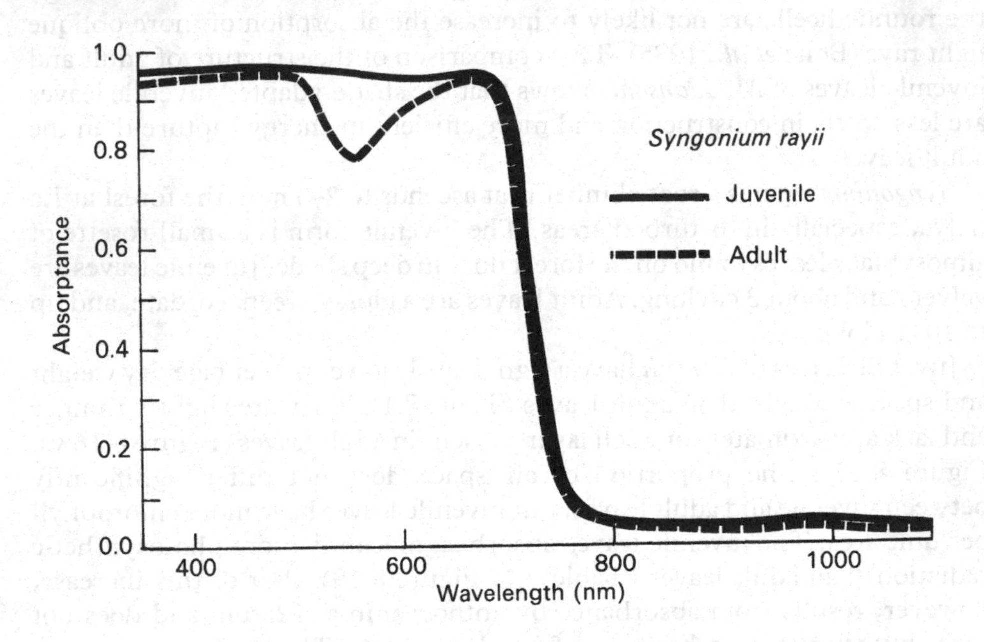

Figure 8.19. The absorptance of radiation between 300 and 1100 nm by juvenile and adult leaves of *Syngonium rayii*.

curvature that focuses light, increasing intensity 6.0 times at the level of the chloroplasts.

In *S. rayii* both juvenile and adult leaves appear shade-adapted, but the two stages have somewhat different adaptations. The adult leaves have a more efficient distribution of chlorophyll, but the juvenile leaves can concentrate light onto chloroplasts.

There are few analyses for leaves at different heteroblastic stages comparable with those presented here. Nilleson & Karetens (1955) showed similar anatomical differences between juvenile and adult leaves of *Marcgravia umbellata*. Analyses of anatomical and leaf optical properties in other vines would allow comparison of adaptive strategies among leaves of heteroblastic stages both within and between vine species.

Experimental approaches to the effects of light on vine development

As Figure 8.11*B* and *C* illustrate, a major challenge in understanding the control of heteroblastic development in vines is to distinguish between intrinsically determined and environmentally controlled components of development. A second challenge is to determine which environmental factors are significant. Here we will discuss experimental ways to study the possible effects of a major environmental factor, light, on vine heteroblastic development.

Light could affect vine development in several ways. Within a morphologically distinct phase, such as the shingle plant of root-climbers, different light regimes could alter the degree of shade adaptation by individual leaves. Light could also act as a cue to switch from one developmental stage to another, or, if some other factor controls the switch, light could change the timing of the switch.

If light affects heteroblastic development, either light quantity (PPF or blue wavelengths) or light quality (R:FR) could be the significant factor. Most research on light effects on plant, and vine, development has been limited to alterations in light quantity and has not considered the effects of light quality at all. Light quality has been shown to affect the developmental responses of some plants, particularly European woodland herbs, to shading (Smith, 1982). Because light quantity and quality tend to covary in natural environments, such as under a forest canopy, experimental approaches are needed to separate their effects.

Lee (1988) distinguished between the effects of reduced PPF and lowered R:FR on the juvenile development of three tropical leguminous vines. Shadehouses were constructed which allowed the penetration of 3% of sunlight. One house did not reduce R:FR, while the other decreased R:FR to

0.35, which was similar to the R:FR in nearby forest understory (Lee, 1989). The vines were twiners, and none were strongly heteroblastic, but the research provides an example of the approach needed to understand the effects of light quantity and quality on vine heteroblastic development.

Two of the three vines, *Mucuna pruriens* and *Caesalpinia bondicella*, were not strongly heteromorphic in their development. Reduced PPF and R:FR affected anatomy and morphology in different, often additive, ways. For example, internodes produced under reduced light quantity were longer than internodes developed in full sunlight for both species. Internodes grown in reduced light quantity and reduced R:FR were not significantly different from internodes in reduced light quantity for *C. bondicella* but were even longer for *M. pruriens* (Lee, 1988).

The third species, *Abrus precatorius*, produced a compact rosette of small compound leaves as a juvenile. After expanding 14 or 15 leaves, the internode distance suddenly increased and the plant adopted its twining habit. Reduced PPF and R:FR profoundly affected the morphology and anatomy of juvenile plants (Table 8.2). Although shade increased internodal lengths, the plants still grew erect. None of the treatments accelerated the heteroblastic change from erect to twining habit. Thus, the intrinsic pattern of heteroblastic development in *A. precatorius* was not affected by light level or quality, although the morphology and anatomy within one heteroblastic stage were altered.

Similar research has been conducted on varieties of beans (*Phaseolus vulgaris* L.) with different growth habits (Kretchmer *et al.*, 1977). Vining varieties of beans typically begin their development with a bushy growth form and later change to a climbing habit. Increased day length can moderately increase the rate of transition to a vining habit in some varieties. Plants exposed to red light in the middle of the dark period are most likely to convert to a vining habit, and this effect can be reversed by far-red irradiation (Kretchmer *et al.*, 1977). Although such a treatment is not equivalent to continuous exposure to altered spectral quality, the results strongly suggest that phytochrome is involved in developmental control of the heteroblastic transition to a vining habit in *Phaseolus*.

Future research on vine heteroblastic development should make comparisons between species and between different developmental stages within a species. These studies should include careful architectural analyses and detailed measurements of variables in the microenvironment associated with different developmental stages. Such information can be used to design experimental studies that will enable us to distinguish the internal and external factors that control heteroblastic development and to understand the adaptive nature of heteroblasty in vines. These are old but unresolved problems in plant biology.

Table 8.2. *Effect of light treatments on the morphology and anatomy of juveniles of* Abrus precatorius

Trait	Treatment				
	100% PPF R:FR 1.10	3% PPF R:FR 0.33	3% PPF R:FR 1.10	*F*	shade
Percentage dry weight allocation					
Leaves	54.0±2.7	57.6±1.9	51.0±1.9	***	***
Stems	33.6±2.5	32.7±3.3	39.6±1.6	***	***
Roots	12.4±4.6	9.7±1.7	9.5±2.4	*	NS
Internode distance (cm)	0.3±0.1	1.0±0.3	1.5±0.2	***	**
Leaf					
Area (cm^2)	5.46±0.80	7.97±1.63	5.20±0.49	***	***
Specific wt (mg cm^{-2})	5.49±1.33	1.86±0.24	1.34±0.21	***	***
Chlorophyll					
mg/g dry wt	4.56±0.72	14.90±2.72	17.59±3.96	***	NS
$\mu g/cm^2$ area	24 ±4	28±5	28±5	NS	NS
a/b	2.35±0.13	2.11±0.09	2.10±0.01	***	NS
Thickness					
Leaf	120 ±15	76±6	100±13	***	***
Palisade	55±10	19±3	14±3	***	**
Stomatal density (10^3 cm^{-2})	12.11±1.47	4.88±1.07	8.39±0.54	***	***

Sample sizes for plants were $N=5$, and $N=10$ for leaf characteristics. Statistical comparisons show the levels of significance for ANOVA between all samples (*F*), and student's T test for the shade treatments (shade): <0.01 = ***; <0.02 = **; <0.05 = *; and >0.05 = NS.
From Lee (1988).

Summary

Vines display dramatic developmental changes between juvenile and adult stages and have frequently been used as striking examples of the developmental phenomenon of heteroblasty. In vines heteroblastic stages are strongly correlated with the microclimatic conditions in which each stage occurs, from the humid and shady conditions of juveniles to the sunny and drier environments of adults. We propose that the unique aspect of vine development is not heteroblasty *per se*, but the environmental control and consequent plasticity of heteroblastic development in the vining habit. Traits that vary between heteroblastic stages in vines include leaf morphology and anatomy, phyllotaxy, internode length, and physiology. Evidence suggests that these variations are adapted to the environmental conditions encountered by each stage, but experimental evidence on how the developmental changes are controlled is lacking. Which aspects of these shifts are genetically preprogrammed and which are induced or modified by changes in environmental conditions? Light quantity and quality vary among vine microhabitats and have been shown to affect heteroblastic development in vines, as well as other plants. Research on the role of such environmental cues in regulating heteroblastic development in vines will make an important contribution to our understanding of vine biology and, more generally, of epigenetic effects on plant development.

References

Allen, W. A., Gausman, H. W. & Richardson, A. J. (1973). Willstatter–Stöll theory of leaf reflectance evaluated by ray tracing. *Applied Optics* 12: 2448–53.

Allsopp, A. (1965). Heteroblastic development in cormophytes. In *Handbuch der Pflanzenphysiologie, Vol. 15, No. 1*, ed. W. Ruhland *et al.*, pp. 1172–221. Springer-Verlag, Berlin.

Allsopp, A. (1967). Heteroblastic development in vascular plants. *Advances in Morphogenesis* 6: 127–71.

Ashby, E. (1948a). Studies in the morphogenesis of leaves. I. An essay on leaf shape. *New Phytologist* 47: 153–76.

Ashby, E. (1948b). Studies in the morphogenesis of leaves. II. The area, cell size and cell number of leaves of *Ipomoea* in relation to their position on the shoot. *New Phytologist* 47: 177–95.

Ashby, E. & Wangermann, E. (1950). A note on the origin of differences in cell size among leaves at different levels of insertion on the stem. *New Phytologist* 49: 189–92.

Björkman, O. (1981). Responses to different quantum flux densities. In *Physiological Plant Ecology I, Encyclopedia of Plant Physiology, n.s., Vol. 12A*, pp. 57–107. Springer-Verlag, Heidelberg.

Blanc, P. (1980). Observations sur les flagelles des Araceae. *Adansonia, Sér. 2* 20: 325–38.

Blanc, P. & Andraos, K. (1983). Remarques sur la dynamique de croissance dans le genre *Piper* L. (Piperaceae) et les genres affinés. *Bulletin du Museum National d'Histoire Naturelle, 4ᵉ Sér.* 5: 259–82.

Bodkin, P. C., Spence, D. H. N. & Weeks, D. C. (1980). Photoreversible control of heterophylly in *Hippuris vulgaris* L. *New Phytologist* 84: 533–42.

Bone, R. E., Lee, D. W. & Norman, J. M. (1985). Epidermal cells functioning as lenses in leaves of tropical rain-forest shade plants. *Applied Optics* 24: 1408–12.

Borchert, R. (1976). The concept of juvenility in woody plants. *Acta*

Horticulturae **56**: 21–36.
Briggs, W. R. (1963). The phototropic response of higher plants. *Annual Review of Plant Physiology* **14**: 311–52.
Burtt, B. L. (1978). Notes on rain-forest herbs. *The Gardens Bulletin, Singapore* **29**: 37–49.
Chazdon, R. L. & Fetcher, N. (1984a). Photosynthetic light environments in a lowland tropical rain forest in Costa Rica. *Journal of Ecology* **72**: 553–64.
Chazdon, R. L. & Fetcher, N. (1984b). Light environments of tropical forests. In *Physiological Ecology of Plants of the Wet Tropics*, ed. E. Medina, H. A. Mooney and C. Vasquez-Yanes, pp. 27–36. W. Junk, The Hague.
Corner, E. J. H. (1965). Checklist of *Ficus* in Asia and Australasia with keys to identification. *The Gardens Bulletin, Singapore* **21**: 1–186.
Cremers, G. (1973). Architecture de quelques lianes d'Afrique tropicale, 1. *Candollea* **28**: 249–80.
Cremers, G. (1974). Architecture de quelques lianes d'Afrique tropicale, 2. *Candollea* **29**: 57–110.
Crotty, W. J. (1955). Trends in the pattern of primordial development with age in the fern *Acrostichum danaefolium*. *American Journal of Botany* **42**: 627–36.
Darwin, C. R. (1876). *The Movements and Habits of Climbing Plants*, 2nd edition. D. Appleton and Co., New York.
Deschamps, P. & Cooke, T. J. (1983). Leaf dimorphism in aquatic angiosperms: significance of turgor pressure and cell expansion. *Science* **219**: 505–7.
Doorenbos, J. (1965). Juvenile and adult phases in woody plants. *Handbuch der Pflanzenphysiologie, Vol. 15, No. 1*, ed. W. Ruhland *et al.*, pp. 1222–35. Springer-Verlag, Berlin.
Duysens, L. N. M. (1956). The flattening of the absorption spectrum of suspensions, as compared to that of solutions. *Biochimica et Biophysica Acta* **19**: 1–12.
Fetcher, N. (1981). Leaf size and leaf temperature in tropical vines. *American Naturalist* **117**: 1011–14.
Franck, D. H. (1976). Comparative morphological and early leaf histogenesis of adult and juvenile leaves of *Darlingtonia californica* and their bearing on the concept of heterophylly. *Botanical Gazette* **137**: 20–34.
Gates, D. M., Keegan, H. J., Schleter, J. L. & Weidner, V. R. (1965). Spectral qualities of plants. *Applied Optics* **4**: 11–20.
Gausman, H. W., Allen, W. A., Myers, V. I. & Cardenas, R. (1969). Reflectance and internal structure of cotton leaves, *Gossypium*

hirsutum L. *Agronomy Journal* **61**: 374–76.

Gentry, A. H. (1983). *Macfadyena unguis-cati*. In *Costa Rican Natural History*, ed. D. H. Janzen, pp. 272–3. University of Chicago Press, Chicago, Ill.

Givnish, T. J. (1987). Comparative studies of leaf form: assessing the relative roles of selective pressures and phylogenetic constraints. *New Phytologist* **106** (Suppl.): 131–60.

Givnish, T. J. & Vermeij, G. J. (1976). Sizes and shapes of liane leaves. *American Naturalist* **110**: 743–78.

Goebel, K. (1900). *Organography of plants, Part I. General Organography*. Oxford University Press (facsimile of the 1900 edition, Hafner Publ. Co., New York, 1969).

Goliber, T. E. & Feldman, L. J. (1990). Developmental analysis of leaf plasticity in the heterophyllous aquatic plant *Hippuris vulgaris*. *American Journal of Botany* 77: 399–412.

Green, S., Green, T. L. & Heslop-Harrison, Y. (1979). Seasonal heterophylly and leaf gland features in *Triphyophyllum* (Dioncophyllaceae), a new carnivorous plant genus. *Botanical Journal of the Linnean Society* 78: 99–116.

Haberlandt, G. (1914). *Physiological Plant Anatomy*. Macmillan, London (reprint, Today and Tomorrow's Printers and Publishers, Delhi, India, 1965).

Hallé, F., Oldeman, R. A. A. & Tomlinson, P. B. (1978). *Tropical Trees and Forests*. Springer-Verlag, New York.

Hammond, D. (1941). The expression of genes for leaf shape in *Gossypium hirsutum* L. and *Gossypium arboreum* L. II. The expression of genes for leaf shape in *Gossypium arboreum* L. *American Journal of Botany* **28**: 138–50.

Holttum, R. E. (1968). *A Revised Flora of Malaya, Vol. II, Ferns of Malaya*, 2nd edition. Government Printing Office, Singapore.

Horn, H. (1971). *The Adaptive Geometry of Trees*. Princeton University Press, Princeton, NJ.

Kaplan, D. R. (1973). Comparative developmental analysis of the heteroblastic leaf series of axillary shoots of *Acorus calamus* L. (Araceae). *La Cellule* **69**: 235–90.

Kaplan, D. R. (1980). Heteroblastic leaf development in *Acacia*. *La Céllule* 73: 136–203.

Kendrick, R. E. & Kronenberg, G. H. M. (ed.) (1986). *Photomorphogenesis in Plants*. Martinus Nijhoff, Dordrecht.

Killip, E. P. (1938). The American species of Passifloraceae. *Field Museum of Natural History, Botanical Series* **19**: 1–613.

Kretchmer, P. J., Ozbun, J. L., Kaplan, S. L., Laing, D. R. & Wallace, D.

H. (1977). Red and far-red light effects on climbing in *Phaseolus vulgaris* L. *Crop Science* **17**: 797–9.

Lee, D. W. (1986). Unusual strategies of light absorption in rainforest herbs. In *On the Economy of Plant Form and Function*, ed. T. J. Givnish, pp. 105–31. Cambridge University Press, Cambridge.

Lee, D. W. (1987). The spectral distribution of radiation in two neotropical forests. *Biotropica* **19**: 161–6.

Lee, D. W. (1988). Simulating forest shade to study the developmental ecology of tropical plants: juvenile growth in three vines. *Journal of Tropical Ecology* **4**: 281–92.

Lee, D. W. (1989). Canopy dynamics and light climates in a tropical moist deciduous forest in India. *Journal of Tropical Ecology* **5**: 65–79.

Lee, D. W., Lowry, J. B. & Stone, B. C. (1979). Abaxial anthocyanin layer in leaves of tropical rainforest plants: enhancer of light capture in deep shade. *Biotropica* **11**: 70–7.

Lee, D. W. & Graham, R. (1986). Leaf optical properties of rainforest sun and extreme shade plants. *American Journal of Botany* **73**: 1100–8.

Lee, D. W. & Paliwal, K. (1988). The light climate of a South Indian tropical evergreen forest. *GeoBios* **15**: 3–6.

Lee, D. W., Bone, R. A., Tarsis, S. & Storch, D. (1990). Correlates of leaf optical properties in tropical forest extreme shade and sun plants. *American Journal of Botany* **77**: 370–80.

Lord, E. (1979). The development of cleistogamous and chasmogamous flowers in *Lamium amplexicaule* (Labiatae): an example of heteroblastic inflorescence development. *Botanical Gazette* **140**: 39–50.

Madison, M. (1977). A revision of *Monstera* (Araceae). *Contributions from the Gray Herbarium of Harvard University* **207**: 3–100.

Mueller, R. J. (1982a). Shoot ontogeny and the comparative development of the heteroblastic leaf series in *Lygodium japonicum* (Thunb.) Sw. *Botanical Gazette* **143**: 424–38.

Mueller, R. J. (1982b). Shoot morphology of the climbing fern *Lygodium* (Schizaeaceae): general organography, leaf initiation, and branching. *Botanical Gazette* **143**: 319–30.

Nilleson, G. A. & Karetens, W. K. H. (1955). Remarks on the morphology and anatomy of the dimorphous leaves of *Marcgravia umbellata* Jacq. *Proceedings, Koninklijke Nederlandse Akademie van Wetenschappen, Series C* **58**: 554–66.

Njoku, E. (1956). Studies in the morphogenesis of leaves. XI. The effect of light intensity on leaf shape in *Ipomoea caerulea*. *New Phytologist* **55**: 91–110.

Njoku, E. (1957). The effect of mineral nutrition and temperature on leaf

shape in *Ipomoea caerulea*. *New Phytologist* **56**: 154–71.
Njoku, E. (1958). The effect of gibberellic acid on leaf form. *Nature (London)* **182**: 1097–8.
Nobel, P. S., Zaragoza, L. J. & Smith, W. K. (1975). Relation between mesophyll surface area, photosynthetic rate, and illumination level during development for leaves of *Plectranthus parviflorus* Henckel. *Plant Physiology* 55: 1067–70.
Oberbauer, S. F., Boring, L., Herman, K., Lodge, D., Ray, T. & Trombulak, S. (1980). Leaf morphology of *Monstera tenuis*. In *Tropical Biology: an ecological approach, No. 79.1*: 24–8. Organization of Tropical Studies, Ciudad Universitaria, Costa Rica.
Oberbauer, S. F., Clark, D. B., Clark, D. A. & Quesada, M. (1988). Crown light environments of saplings of two species of rainforest emergent trees. *Oecologia* 75: 207–12.
Parkhurst, D. L. & Loucks, O. L. (1972). Optimal leaf size in relation to environment. *Journal of Ecology* **60**: 505–37.
Pearcy, R. W. (1983). The light environment and growth of C_3 and C_4 tree species in the understory of a Hawaiian forest. *Oecologia* 58: 19–25.
Peñalosa, J. (1983). Shoot dynamics and adaptive morphology of *Ipomoea phillomega* (Vell.) House (Convolvulaceae), a tropical rainforest liana. *Annals of Botany* 52: 737–54.
Ray, T. S. (1983a). *Monstera tenuis*. In *Costa Rican Natural History*, ed. D. H. Janzen, pp. 278–80. University of Chicago Press, Chicago, Ill.
Ray, T. S. (1983b). *Syngonium triphyllum*. In *Costa Rican Natural History*, ed. D. H. Janzen, pp. 333–5. University of Chicago Press, Chicago, Ill.
Ray, T. S. (1987). Cyclic heterophylly in *Syngonium* (Araceae). *American Journal of Botany* 74: 16–26.
Richards, J. H. (1983). Heteroblastic development in the water hyacinth, *Eichhornia crassipes* Solms. *Botanical Gazette* **144**: 247–59.
Richards, J. H. & Lee, D. W. (1986). Light environments and developmental control of leaf morphology in water hyacinth (*Eichhornia crassipes* Solms.). *American Journal of Botany* 73: 1741–7.
Richards, P. (1952). *The Tropical Rainforest*. Cambridge University Press, London.
Rogler, C. E. & Hackett, W. P. (1975). Phase change in *Hedera helix*: Induction of mature to juvenile phase change by gibberellin A_3. *Physiologia Plantarum* **34**: 141–7.
Roth, I. (1984). *Stratification of Tropical Forests as Seen in Leaf Structure*. W. Junk, The Hague.
Sharkey, T. D. (1985). Photosynthesis in intact leaves of C_3 plants: physics, physiology and rate limitations. *Botanical Review* **51**: 53–105.

Sinott, E. W. (1960). *Plant morphogenesis*. McGraw Hill, New York.
Smiley, J. T. (1983). *Passiflora vitifolia*. In *Costa Rican Natural History*, ed. D. H. Janzen, pp. 299–301. University of Chicago Press, Chicago, Ill.
Smith, A. P. (1986). Ecology of a leaf color polymorphism in a tropical forest species: habitat segregation and herbivory. *Oecologia* **69**: 283–7.
Smith, H. (ed.) (1981). *Plants and the Daylight Spectrum*. Academic Press, London.
Smith, H. (1982). Light quality, photoreception and plant strategy. *Annual Review of Plant Physiology* **33**: 481–518.
Strong, D. R. & Ray, T. S. (1975). Host tree location behavior of a tropical vine (*Monstera gigantea*) by skototropism. *Science* **190**: 804–6.
Taylor, S. E. (1975). Optimal leaf form. In *Perspectives of Biophysical Ecology*, ed. D. M. Gates and R. B. Schmere, pp. 73–86. Springer-Verlag, New York.
Troll, W. (1937). *Vergleichende Morphologie der Höheren Pflanzen. Band 1. Vegetationsorgane. Teil 1*. Gebrüder Borntraeger, Berlin (Reprint, 1967, O. Koeltz, Koeningstein-Taunus).
Troll, W. (1939). *Vergleichende Morphologie der Höheren Pflanzen. Band I. Vegetationsorgane. Teil 2*. Gebrüder Borntraeger, Berlin (Reprint, 1967, O. Koeltz, Koeningstein-Taunus).
Wareing, P. F. & Frydman, V. M. (1976). General aspects of phase change, with special reference to *Hedera helix*. *Acta Horticulturae* **56**: 57–69.
Willstatter, R. & Stoll, A. (1918). *Untersuchungen über die Assimilation der Kohlensäure*. Springer-Verlag, Berlin.
Yoda, K. (1974). Three-dimensional distribution of light intensity in a tropical rainforest of West Malaysia. *Japanese Journal of Ecology* **24**: 247–54.

9
Physiological ecology of mesic, temperate woody vines

ALAN H. TERAMURA, WARREN G. GOLD
AND IRWIN N. FORSETH

Introduction

Distribution, diversity and abundance

In North America, woody vines (lianas) of mesic, temperate environments can be found across a wide range of physiognomic regions, from open fields to climax hardwood forests. Within this range, lianas are predominantly associated with the arborescent vegetation of mid- to late-successional sites in eastern deciduous forests. The deciduous forest of eastern North America exhibits a pronounced gradient from south to north of decreasing mean daily temperature and length of growing season, while there is an east-to-west gradient of decreasing precipitation and increasing temperature variation (Figure 9.1). Within these gradients, lianas appear to be most abundant in the southeastern (warmest, highest precipitation) portion of the biome.

The woody vines native to the eastern deciduous forests are taxonomically diverse (Braun, 1950). Duncan (1975) lists a total of 30 genera in 21 families occurring in the southeastern USA (Table 9.1). This diversity for temperate vines is much less than that for vines of tropical ecosystems (Putz, 1984; Putz & Chai, 1987). The pattern of localized dominance for lianas in eastern North America has dramatically changed over the last 100 years with the rapid spread of exotic species such as *Lonicera japonica* Thunb., *Pueraria lobata* (Willd.) Owhi, and *Celastrus orbiculatus* Thunb. (Sasek, 1985; McNab & Meeker, 1987; Patterson, 1973; Wechsler, 1977) and with increasing disturbance resulting from human activities.

The proliferation of vines in a temperate forest involves a variety of factors, many of which are probably species-specific. The increase of lianas in forest patches may be associated with the presence of nutrient-rich soils in tropical communities (Janzen, 1971; Putz, 1983, 1985; Putz & Chai, 1987) but data for temperate areas are still lacking. Disturbance may also promote vine

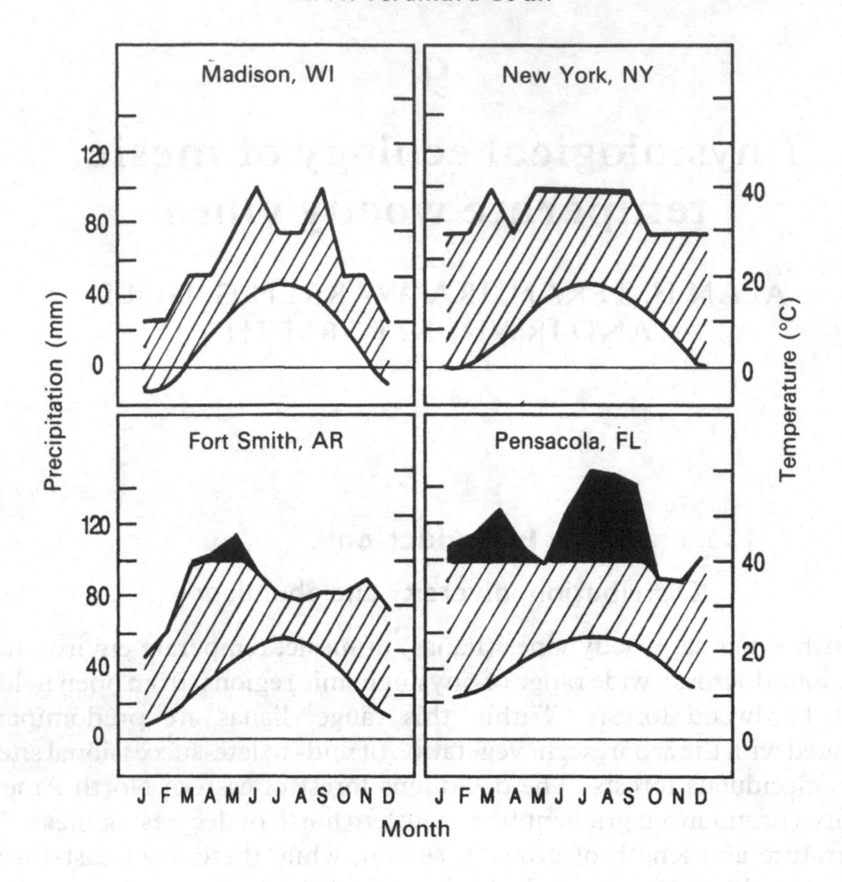

Figure 9.1. Climatic patterns for stations in the northwestern (Madison), northeastern (New York), southwestern (Fort Smith) and southeastern (Pensacola) portions of temperate deciduous forest in North America. Top curve is mean monthly precipitation, bottom curve is mean monthly temperature. Data from Ruffner (1985).

abundance, with concomitant increases in light penetration and an alteration of forest floor microenvironment (Lutz, 1943; Gysel, 1951; Richards, 1952; Boring, Monk & Swank, 1981). The association of greater liana density with forest edges and disturbed areas (Lutz, 1943; Slezak, 1976; Thomas 1980; Putz, 1984), and increases in liana density following disturbance (Slezak, 1976; Boring *et al.*, 1981; McNab & Meeker, 1987) are well documented. Dominance of lianas in forest patches may cause, as well as result from, disturbance. For instance, *L. japonica* may alter species composition, vegetation structure and foliage production (especially in the shrub layer) of understory vegetation in mature deciduous forests, resulting in greater light penetration and a long-term dominance of *L. japonica* (Thomas, 1980).

Table 9.1. *Compendium of plant families and genera with species exhibiting liana growth forms in the southeastern USA. Also listed is the climbing method and leaf longevity (*D*, deciduous;* SE*, semi-evergreen;* E*, evergreen) of species representatives*

Family	Twining	Tendrils	Tendrils with adhesive disks	Twisting leaf stalks	Aerial roots	Leaf longevity
Liliaceae						
Smilax		×				D,SE,E
Aristolochiaceae						
Aristolochia	×					D
Polygonaceae						
Brunnichia		×				D
Ranunculaceae						
Clematis				×		D
Lardizabalaceae						
*Akebia**	×					D
Menispermaceae						
Menispermum	×					D
Cocculus	×					SE
Calycocarpon	×					D
Schisandraceae						
Schisandra	×					D
Saxifragaceae						
Decumaria					×	D
Fabaceae						
*Wisteria***	×					D
*Pueraria**	×					D
Anacardiaceae						
Toxicodendron					×	D
Celastraceae						
*Celastrus***	×					D
*Euonymus**					×	E
Rhamnaceae						
Berchemia	×					D
Vitaceae						
Vitis		×				D
Parthenocissus			×			D
*Ampelopsis***		×				D
Cissus		×				D
Passifloraceae						
Passiflora		×				D

Table 9.1. (*cont.*)

Family	Twining	Tendrils	Tendrils with adhesive disks	Twisting leaf stalks	Aerial roots	Leaf longevity
Araliaceae						
*Hedera**					×	E
Ericaceae						
Pieris				×		E
Loganiaceae						
Gelsemium	×					SE,E
Apocynaceae						
Trachelospermum	×					D
Asclepiadaceae						
*Periploca**	×					D
Cynanchum	×					D
Bignoniaceae						
Bignonia			×			E
Campsis					×	D
Caprifoliaceae						
*Lonicera***	×					D,SE,E

*, Exotic species; **, native and exotic species.
Data compiled from Brown & Brown (1972) and Duncan (1975).

Lonicera japonica can also arrest succession of native vegetation in open fields by maintaining a ground cover which suppresses growth of both early and late successional species (Little, 1961; Thomas, 1980; Friedland and Smith, 1982).

Heterogeneity of the physical environment

The microhabitats which temperate, mesic vines occupy are highly heterogeneous both within and among different habitat types (e.g. forest edges, understories, or gaps). The success of a vine in a particular microhabitat will depend largely on its ability to adjust physiologically to its environment. Thus, a major portion of this chapter is devoted to an examination of physiological responses. However, the specific effects of the physical environment on vine foliage can be moderated by plant attributes such as rapid leaf movements or, over a longer term, by canopy development. Substantial differences in such characteristics among mesic, temperate lianas

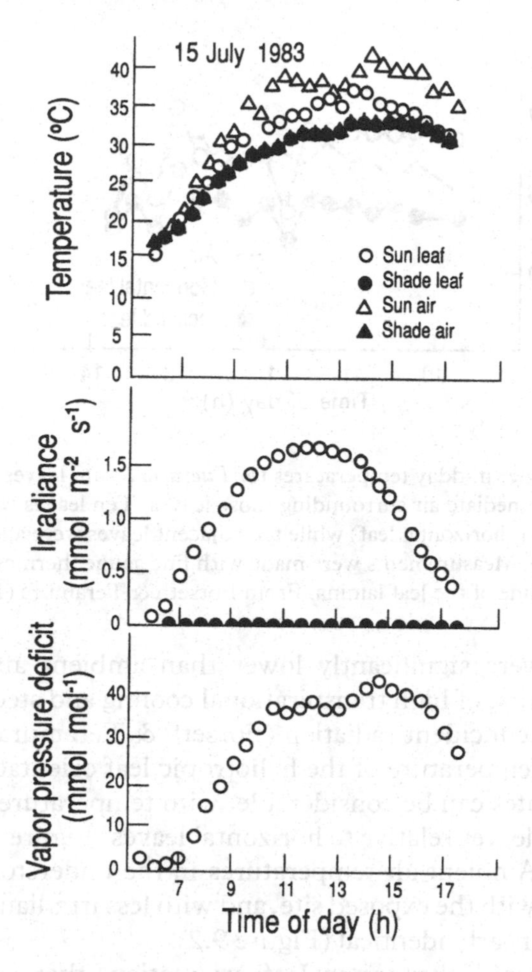

Figure 9.2. Microclimatic variables measured for leaves of *Pueraria lobata* in an exposed, open site and an understory site in College Park, MD, USA on 15 July 1983. From Forseth & Teramura (1987).

may lead to significant differences among species in the foliar microenvironment.

For example, representative mid-season microclimatic conditions experienced by *P. lobata* in central Maryland, USA on an exposed site and an adjacent understory site are shown in Figure 9.2 (Forseth & Teramura, 1987). Horizontal irradiance may exceed 1800–2000 μmol m^{-2} s^{-1} in the exposed site, while it may be only 1–2% of this value in the shade. Concomitant with high radiation in the exposed site are ambient air temperatures exceeding 35 °C and leaf-to-air water vapor concentration differences (Δw) greater than 40 mmol mol^{-1}. However, leaf temperatures for *P. lobata* in

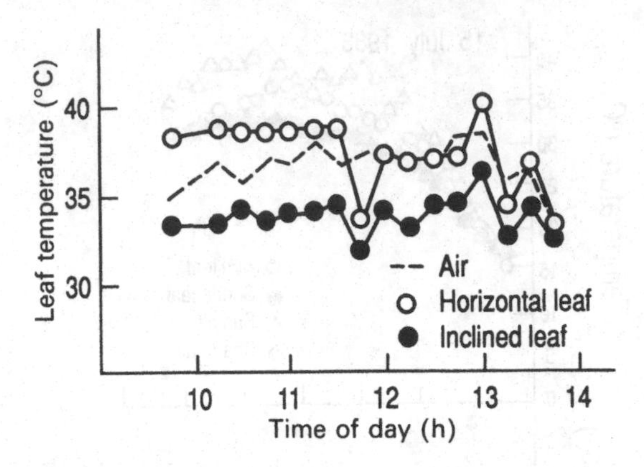

Figure 9.3. Summer midday temperatures for *Pueraria lobata* leaves of different orientations and the immediate air surrounding those leaves. Ten leaves were restrained in an horizontal position (horizontal leaf) while ten adjacent leaves were allowed to orient naturally (inclined leaf). Measurements were made with fine gauge thermocouples affixed to the shaded side of the leaf lamina. From Forseth & Teramura (1986).

the exposed site were significantly lower than ambient air temperature through a combination of high transpirational cooling and steep midday leaf angles which reduce incident radiation (Forseth & Teramura, 1986, 1987). The effect on leaf temperature of the heliotropic leaf orientation seen in *P. lobata* in exposed sites can be considerable, with temperature reductions of 5–6 °C for inclined leaves relative to horizontal leaves (Figure 9.3; Forseth & Teramura, 1986). Ambient air temperatures in the understory were much reduced compared with the exposed site, and with less irradiance, leaf and air temperatures were nearly identical (Figure 9.2).

In addition to rapid changes in leaf orientation, there are significant differences in leaf arrangement within the canopy which may lead to pronounced differences in temperature and Δw of the leaf microenvironment. For example, a study of three sympatric woody vines (Bell, Forseth & Teramura, 1988; Bell, 1986) showed marked differences among species in midday leaf temperatures associated with species differences in leaf angle and arrangement. Mean midday leaf-to-air (leaf minus air) temperature differences in exposed sites pooled from four sampling dates in 1985 were 1.4 °C for *L. japonica* and 0.5 °C for *Vitis vulpina* L. and *Parthenocissus quinquefolia* Planch. *Lonicera japonica* arrays its leaves nearly horizontally, resulting in an average midday incident irradiance of 1521 ± 51 μmol m^{-2} s^{-1} (mean ± standard error) and mean ambient air temperature near the leaves of 33.2 ± 0.5 °C. Its extensive canopy development forms a large radiation-absorbing surface low to the ground, resulting in a steep temperature gradient just above its canopy. *Vitis vulpina* had nearly vertical leaves in the exposed site, with mean midday incident irradiance of 949 ± 87 μmol m^{-2} s^{-1} and a substantially lower mean ambient air temperature of 31.1 ± 1.1 °C. *Parthenocissus quinquefolia* has a five-lobed, dissected leaf which curls its edges under high irradiance. It received a mean midday irradiance of 1267 ± 107 μmol m^{-2} s^{-1} and was in the coolest part of the temperature gradient with mean ambient air temperatures of 30.6 ± 0.9 °C. A similar study of *L. japonica* and *Smilax rotundifolia* L. was conducted in the center of a canopy gap where both species tended to display their leaves horizontally and received similar irradiance in the same part of the air temperature gradient (E. Whereat and A. Teramura, unpublished). From May through October, when irradiance was greater than 1000 μmol m^{-2} s^{-1}, mean leaf-to-air temperature differences were 3.2 °C and 2.3 °C for *L. japonica* and *S. rotundifolia*, respectively. Among the five species, two maintain leaf temperatures consistently above ambient (*L. japonica* and *S. rotundifolia*), one species consistently below ambient (*P. lobata*) and two maintain leaf temperature consistently near ambient (*P. quinquefolia* and *V. vulpina*). These differences might have significant implications for the comparative physiological activity and water loss of these co-occurring lianas. However, differences among these species in their occurrence on the forest floor vs. upper canopy layer have not been as clearly defined.

Liana–host interaction

Woody vines play a major ecological role through their impacts on hosts for support. However, most assessments of these impacts are anecdotal, based on the observation that non-infested hosts grow faster or reproduce more than infested hosts. For example, Lutz (1943) reported that wild grapevines caused crown deformation by constriction and shading in saplings and middle-aged trees in eastern deciduous forests of North America. Grapevine infestation has also been positively associated with ice damage in mixed temperate hardwood forests (Siccama, Weir & Wallace, 1976). Commonly observed damage and death of trees infested by *Celastrus* spp. in deciduous forests of the eastern USA has been attributed to constriction of host vascular tissues by this twining liana, and the subsequent interruption of photoassimilate translocation (Lutz, 1943). This often results in the spiral growth of vascular tree cambium around the twining liana.

Kudzu, *P. lobata*, is a weedy, leguminous vine introduced into North America from Asia in 1876. Although it was initially planted widely as cattle fodder and for erosion control, it has since escaped cultivation and is considered a major pest in the southeastern USA (Tanner *et al.*, 1979). The basis for its ability to suppress growth in host trees has not been formally examined, but is thought primarily to involve competition with its hosts for light. Kudzu is capable of prolific stem elongation and thereby is efficient at climbing and overtopping trees of 20 m or greater in height (Brender, 1960). It also allocates a large proportion of its biomass to leaves, forming dense canopies with leaf area indices exceeding 7 (Wechsler, 1977).

Because of its weedy nature and rapid spread, *L. japonica*, an east Asian vine introduced into North America as an ornamental, has been studied extensively. Friedland & Smith (1982) reported that infestation by *L. japonica* was associated with increased allocation to leaf biomass in *Solidago rugosa*, a common early successional herb in temperate old-field sites. Growth suppression and crown deformation in shrubs and small trees in forest gaps have also been associated with infestation by *L. japonica* (Little, 1961). In a study of the impact of exotic species on the vegetation of Potomac Island, USA, the presence of *L. japonica* was always associated with reduced tree growth, particularly in the dominant species (Thomas, 1980). Thomas (1980) attributed this to host stem constriction by the twining stem of *L. japonica*.

In most of the studies mentioned above, direct effects of vines on their hosts were not separated from potential differences in habitat preferences. Stevens (1987) points out that an alternative explanation for positive associations between vine infestation and reduced growth for hosts may be that vines predominate in habitats where host trees do not perform well, or vine spread may be more rapid in host trees that have low vigor or that are already competitively suppressed. Therefore, studies correlating vine infestation with host growth may be misleading, and manipulative experiments involving vine removal are needed to separate direct vine effects on their hosts from habitat-related effects (Stevens, 1987).

Such removal experiments in mesic, temperate forests have been few, and their results are far from definitive. Bruner (1967) removed *L. japonica* vines from a group of yellow poplar saplings four times over a 5-year period with herbicide applications at a moist, early-successional site in South Carolina, USA. This treatment did not reduce the number of saplings with heavy vine infestations (100% in both treated and untreated groups), although it did lower mortality compared with non-treated saplings. Although the herbicide removed *L. japonica* and its roots attached to the hosts, seeds of *L. japonica* in the soil, which were unaffected by spraying, initiated rapid reinfestation of the treated saplings (Bruner 1967). Elimination of vines and other weeds by disking the soil around hardwood trees was found to improve tree growth and nutrient uptake (Kennedy, 1981). A complication of interpreting this treatment, however, was that the disking also may have improved soil structure, water infiltration and storage, soil–atmosphere gas exchange, organic matter content, nutrient availability and the ease of root elongation (Kennedy, 1981). Whigham (1984) examined the effects of *L. japonica* on the growth of *Liquidambar styraciflua* trees growing in a mid-successional field in Maryland, USA. The experimental trees were free of vines in the upper crown, but were subjected to heavy vine infestations through the lower- and mid-canopy zones. Complete vine removal resulted in greater increases in host tree growth than did removal of vines from only the aboveground portions of the tree. Whigham (1984) suggested that belowground competition for water and/or nutrients was therefore more important in this specific case than were aboveground interactions.

In general, the nature of liana–host interactions in temperate forests remains largely descriptive and speculative. Quantitative information on liana–host competition, both above and belowground, and the mechanisms of such competition, is sorely lacking. Other aspects of liana–host interactions in temperate, mesic areas, such as allelopathy or any positive aspects of these interactions, are even less explored. Trémolières *et al.* (1988) demonstrated summertime enhancement of nitrogen concentrations in the litter layer of a temperate hardwood forest resulting from woody vine (*Hedera helix* L.) litter. Such changes in nutrient input could potentially have positive influences on associated trees, but this has not been examined directly.

Vine growth

Biomass allocation

Aboveground allocation. The liana life form utilizes other woody plants as support, which should result in the increased allocation of leaf tissue over that of woody support tissue. Putz (1983) has shown that tropical lianas support more foliar biomass than trees of equal stem basal diameters. However, no substantial differences existed for small individuals (<5 cm diameter). Unfortunately, few data exist on the aboveground biomass partitioning of mature temperate lianas. Wechsler (1977) found that 21–28% of total shoot dry mass was in leaf biomass for mature *P. lobata* in an open site in Georgia, USA. This is much greater than the typical shoot allocation to photosynthetic biomass for mature deciduous trees of temperature zones, which have 1–2% of their dry biomass as leaves (Larcher, 1980; Vitousek *et al.*, 1988).

Studies of shoot biomass allocation in 2–5-month-old potted seedlings of *P. lobata* and *L. japonica* found that 60% of total shoot dry weight was allocated to leaves in both species (Tsugawa, Tange & Masui, 1979; Sasek, 1983). In a field study, Tsugawa & Kayama (1981) found a similar allocation to leaves within the current year's growth of mature *P. lobata* vines during most of the growing season, although this value varied seasonally, decreasing to 40% by the end of the growing season. Allocation patterns can also be affected by the directional orientation of liana growth. Frey & Frick (1987) found a higher allocation to leaves in vertically growing shoots (58.1%)

compared with horizontal shoots (54.8%) of 3-month-old rooted cuttings of *H. helix*. Greater allocation to stems for horizontal growth might enhance exploration for a suitable climbing support, but the ecological significance of these rather small differences reported by Frey & Frick (1987) is uncertain. Sasek (1983) demonstrated differences in aboveground allocation between similar, congeneric temperate liana seedlings. The proportion of shoot dry weight in leaves of *Lonicera sempervirens* L. seedlings was only 46%, compared with 60% for *L. japonica*, following 70 days of growth in a greenhouse. Sasek (1983) proposed that the greater allocation to leaves in the exotic *L. japonica* may contribute to its competitive superiority and rapid proliferation in the deciduous forests of eastern North America. However, extrapolation of allocation information for seedlings or current annual growth to mature plants in temperate climates is difficult without quantitative information on seasonal changes in allocation patterns and life history characteristics such as stem persistence and foliage longevity. Such information is scant for temperate lianas, with the exception of the ongoing studies of *P. lobata* by Tsugawa and co-workers (e.g. Tsugawa *et al.*, 1979, 1980; Tsugawa & Kayama, 1980, 1981).

Because of variations in specific leaf weight (leaf mass per leaf area), the proportion of dry mass allocated to leaves may not always reflect the actual investment in photosynthetic surface area. To address this problem, Wechsler (1977) developed 'collection/support ratios' (CSR), which represent a measure of the amount of leaf area relative to the shoot biomass supporting it (leaf area/shoot biomass). The average midsummer CSR for mature *P. lobata*, from quadrats in heavily infested areas, was 110 $cm^2\ g^{-1}$ (Wechsler, 1977). Abramovitz (1983) found even higher CSR values (144–153 $cm^2\ g^{-1}$) for *P. lobata* at a Maryland, USA site. Although values of CSR will undoubtedly vary with plant age and size, CSRs for temperate trees are an order of magnitude lower, ranging from 3 to 9 for mature trees and from 9 to 17 for saplings (Wechsler, 1977). This greater investment in photosynthetic surface area vs. support tissue by temperate lianas compared with trees from the same community appears to provide at least a partial basis for the higher growth rates of lianas.

Below vs. aboveground biomass allocation. To our knowledge, no detailed studies of root growth in field populations of temperate woody vines have been attempted. Wechsler (1977) reported that root mass accounted for 52% of total biomass for *P. lobata* in the spring at a site in Georgia, USA. This proportion steadily increased through the summer to 65% in the autumn (Wechsler, 1977), reflecting the partial die-back of above-ground stems and the translocation of assimilates to its massive root system for storage during the winter months. These values for root biomass allocation are much greater than those for *Liriodendron tulipifera*, a common deciduous tree in the eastern USA, which allocates approximately 10% of its biomass to roots (Caldwell, 1987), or even greater than a general average of 19.5% for dominant woody species of temperate, deciduous forests (Vitousek *et al.*, 1988). Other temperate liana species may have root biomass allocation patterns much more comparable to deciduous forest trees, but only data for liana seedlings are available. Allocation to root systems in potted seedlings of *L. japonica* and *L. sempervirens* (9.4 and 11.6% of total biomass, respectively: Sasek, 1983) were much lower than values for *P. lobata* seedlings (71%: Tsugawa *et al.*, 1979).

Despite the relatively modest root biomass allocation reported for some of these liana seedlings and many mature temperate forest woody plants, the total carbon cost of roots may be considerable owing to high rates of root turnover. Although roots may comprise only 10% of *L. tulipifera's* biomass at any one time, 60% of annual carbon gain is allocated belowground (Caldwell, 1987). In a European hedgerow, where there was considerable competition for light, Kuppers (1985) found that four of the five woody species allocated about 30% of assimilates to the root system annually. The fifth species, *Rubus corylifolius*, a liana, allocated over 70% of assimilates to roots (Kuppers, 1985). In spite of the considerable cost and importance of maintaining root systems (Caldwell & Richards, 1986) and the apparent wide range of root biomass allocation in temperate lianas, our knowledge of root systems and carbon allocation patterns in these plants remains limited.

Vegetative vs. reproductive biomass allocation. Carbon and nutrient allocation to reproduction have important ecological and evolutionary implications for plants (Harper, 1977). Horizontal growth can vegetatively propagate a species in adjacent, new areas. However, successful sexual reproduction, dispersal and germination of seeds are crucial for the colonization of broad geographic areas and relatively distant habitats. The proportion of aboveground biomass may be a crude measure of actual resource allocation to reproduction (Harper, 1977); however, the only estimate available of resource allocation to sexual reproduction in temperate lianas is the aboveground biomass of reproductive structures. Typical values for the proportion of shoot biomass allocated to seed production in woody perennials from temperate areas range from 5 to 20% (Abrahamson, 1975; Harper, 1977). In contrast, the proportion of shoot dry weight invested in seeds for field populations of *P. lobata* ranged from 1 to 2% at sites in central Maryland, USA (Abramovitz, 1983) and 2 to 3% at a site in Japan (Tsugawa & Kayama, 1981). This supports the general observation that *P. lobata* dominates localized areas through prodigious growth of vegetative tissue, while its natural rate of long-distance colonization is not as great as that of other

Table 9.2. *Shoot elongation rates for various vine species*

	Shoot elongation (mm day^{-1})	Range (mm day^{-1})
Twiners		
Monocots ($N=5$)	72±24	15–143
Dicots ($N=15$)	73±9	1–142
All species	73±9	1–143
Tendril climbers		
Monocots ($N=2$)	47±12	35–59
Dicots ($N=11$)	51±9	20–110
All species	50±7	20–110
Root climbers		
Monocots ($N=5$)	13±4	3–22
Dicots ($N=3$)	13±3	8–16
All species	13±2	3–22

Data compiled from French (1977), Wechsler (1977), Abramovitz (1983) and Bell *et al.* (1988).

temperate vines (e.g. *L. japonica*: Leatherman, 1955; Wechsler, 1977; Abramovitz, 1983). Its apparent wide distribution throughout the south-eastern USA is a result of intentional planting rather than effective seed dispersal (Abramovitz, 1983).

Growth rates

The rate of stem elongation in lianas plays a central role in the colonization and exploitation of suitable climbing substrates. The rate of stem elongation in mesic, temperate lianas varies tremendously among species, habitat and climbing habit (Table 9.2). For example, *P. lobata* is known for its prolific colonization of open, disturbed sites and displays stem elongation rates ranging from 62 to 127 mm day^{-1} (Wechsler, 1977; Abramovitz, 1983). Stem elongation rates from a population of *P. lobata* in its native Japan were considerably lower, 43 mm day^{-1}, despite growing in a higher precipitation regime than the North American populations (Tsugawa & Kayama, 1981). However, mean maximum air temperatures throughout the growing season were about 3–5 °C lower at the site in Japan. A notable exception among twining vines is the low stem elongation rate found in the evergreen, *L. japonica*, with seasonal maximum elongation rates ranging from 4.6 to 8.9 mm day^{-1} in central Maryland (Bell *et al.*, 1988; E. Whereat and A. Teramura, unpublished data). The exclusion of *L. japonica* from Table 9.2

would increase the minimum value reported for dicotyledonous twiners to 39 mm day^{-1}. On an annual basis, shoot elongation of *L. japonica* (0.5–10 m) was much less than that of two sympatric lianas, *P. quinquefolia* (2.3 m) and *V. vulpina* (3.0 m) in central Maryland (Bell *et al.*, 1988; see Figure 9.5). This relatively low rate of stem elongation may be related to *L. japonica*'s evergreen habit (therefore longer growing season) and an extensively branched growth pattern with short internode lengths, emphasizing horizontal spread over vertical growth (Sasek, 1983; Bell *et al.*, 1988). This results in considerable secondary and tertiary branching during the growing season.

Climbing mechanics

Liana occurrence in a range of environments is often restricted by its climbing mechanics (Putz, 1984). The principal climbing mechanisms of temperate lianas and their taxonomic distributions are illustrated in Figure 9.4 and Table 9.1. Stem twiners (e.g. *P. lobata* and *L. japonica*) are restricted to climbing small to medium diameter objects, generally less than 10 cm. Hence, simple twiners are most commonly found in early to mid-successional, high-irradiance environments, where hosts of relatively small to medium diameter are available for climbing (Darwin, 1865; Carter & Teramura, 1988a). Vines with more specialized climbing mechanisms, such as modified leaves and tendrils, may attach to a variety of protruding objects (e.g. flakes of bark, leaf petioles and other liana stems) as well as smaller diameter branches. Petiole twining and tendriled lianas may allocate less biomass in climbing structures than stem twiners since these smaller, specialized structures circumvent the need for the main stem to coil (Darwin, 1865; Carter & Teramura, 1988a). This may restrict these specialized vines to smaller diameter supports than stem twiners (Putz, 1984). Therefore, they are more successful in the understory and gap edges of mid- to late-successional forests. Finally, vines that climb by roots and tendrils with adhesive disks are effective at ascending broad, smooth trunks of large trees. Hence, they are capable of exploiting habitats within the understory of late-successional forests. Although these general patterns are readily observable, exceptions are common. For instance, the temperate root-climber, *Toxicodendron radicans* L., is also quite successful at colonizing the open, high-irradiance environment of steep, rocky cliffs because of its root-climbing mechanism.

Interactions with the environment

Phenology

In temperate environments, seasonal patterns of plant activity can have important implications for resource capture and growth. The majority of

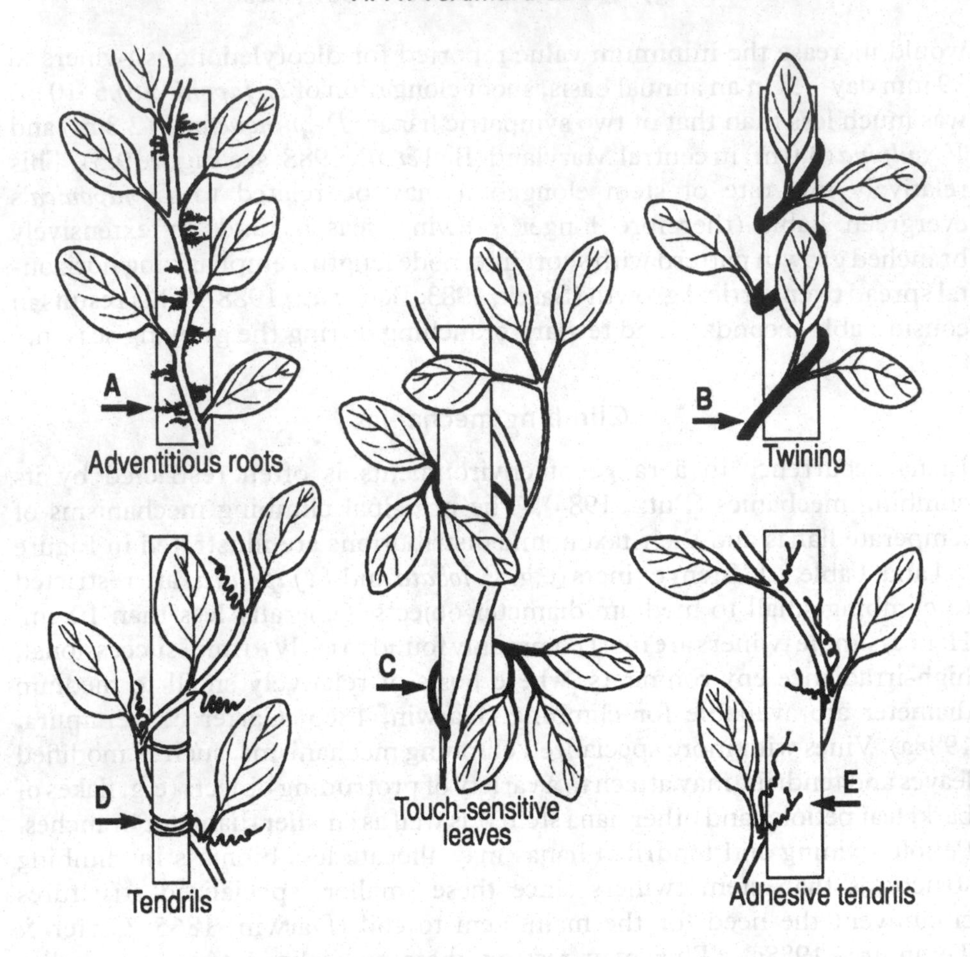

Figure 9.4. Climbing mechanisms used by temperate vines. Habitat associations of these climbing methods are discussed further in the text. Taxonomic distribution is listed in Table 9.1. From Carter & Teramura (1988a).

plant activity in mesic, temperate deciduous forests is confined to non-winter months, when temperatures are sufficient for physiological activity of the dominant species. Unlike seasonally dry regions, there are no consistent trends in the seasonal availability of soil resources, such as water and mineral nutrients, in the eastern deciduous forests of North America (Figure 9.1; Bell *et al.*, 1988). There are, however, droughts of relatively short duration that occur on a year-to-year basis that may alter the seasonal peak in physiological activity. For example, Bell *et al.* (1988) showed that seasonal maxima of stomatal conductance (g_s) in three woody vines in Maryland occurred in the late summer of 1985, a year in which precipitation was evenly distributed from May to September. But, in 1984, mid- to late-summer dry periods

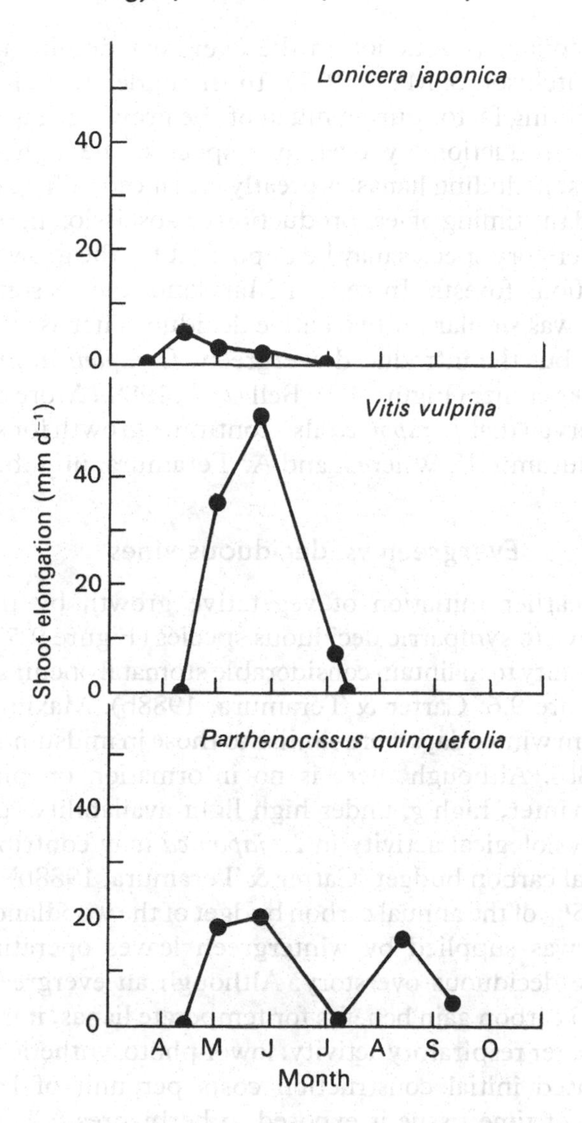

Figure 9.5. Seasonal course of shoot elongation in 1984–5 growing seasons for three sympatric vine species in College Park, MD, USA. Data represent the mean of 10–15 tagged shoots per species. From Bell *et al.* (1988).

resulted in seasonal maxima in g_s during late spring and early summer. A recent study of two temperate lianas also found that conductance was low during late summer drought; however, drought was coincident with the seasonal peak in evaporative demand (Whereat, 1990).

In contrast to water availability, there are highly predictable seasonal variations in the availability of light within mesic, temperate forests coincident with foliage production in the overstory dominants (Taylor & Pearcy, 1976; Hutchison & Matt, 1977). In the understory, low irradiance can often be a limiting factor during much of the growing season. However, prior to foliage production by overstory species, the light available to understory plants, including lianas, is greatly enhanced. Thus, the longevity of vine leaves and the timing of leaf production or abscission in relation to leaf production in overstory species may be important for the growth of lianas in temperate deciduous forests. In central Maryland, the seasonal pattern of shoot elongation was similar for two native deciduous lianas (*P. quinquefolia* and *V. vulpina*), but the introduced evergreen, *L. japonica*, initiated shoot elongation 3 weeks earlier (Figure 9.5) (Bell *et al.*, 1988). More detailed work has recently observed that *L. japonica* also continues growth for several weeks longer through autumn (E. Whereat and A. Teramura, unpublished).

Evergreen vs. deciduous vines

In addition to earlier initiation of vegetative growth by the evergreen *L. japonica* relative to sympatric deciduous species (Figure 9.5), this species also showed an ability to maintain considerable stomatal opening during mild winter days (Figure 9.6; Carter & Teramura, 1988b). Maximum g_s values during these warm winter days were similar to those in midsummer (Carter & Teramura, 1988b). Although there is no information on photosynthetic activity during winter, high g_s under high light availability suggested that non-summer physiological activity in *L. japonica* may contribute substantially to its annual carbon budget (Carter & Teramura, 1988b). Jurik (1980) calculated that 45% of the annual carbon budget of the woodland strawberry, *Fragaria vesca*, was supplied by wintergreen leaves operating after leaf abscission of the deciduous overstory. Although an evergreen habit may provide extended carbon gain benefits for temperate lianas, it may also incur costs, such as greater respiratory activity, lower photosynthetic rates per unit leaf area, increased initial construction costs per unit of leaf area, and increased length of time tissue is exposed to herbivores (Chabot & Hicks, 1982).

There are other problems faced by evergreen vines as well. The winter survival of *L. japonica* leaves in Maryland appears to be habitat-specific, with plants in exposed sites experiencing more leaf damage and loss than ones in understory or protected sites. Visual symptoms of leaf damage include leaf curling and rolling and anthocyanin accumulation. Evergreen vines must also prevent disruption of the vascular system and provide xylem flow to metabolically active leaves during periods of low temperatures. The large vessel elements of most vines are particularly susceptible to cavitation and the spread of embolisms formed at low temperatures (Scholander, 1957). Hence,

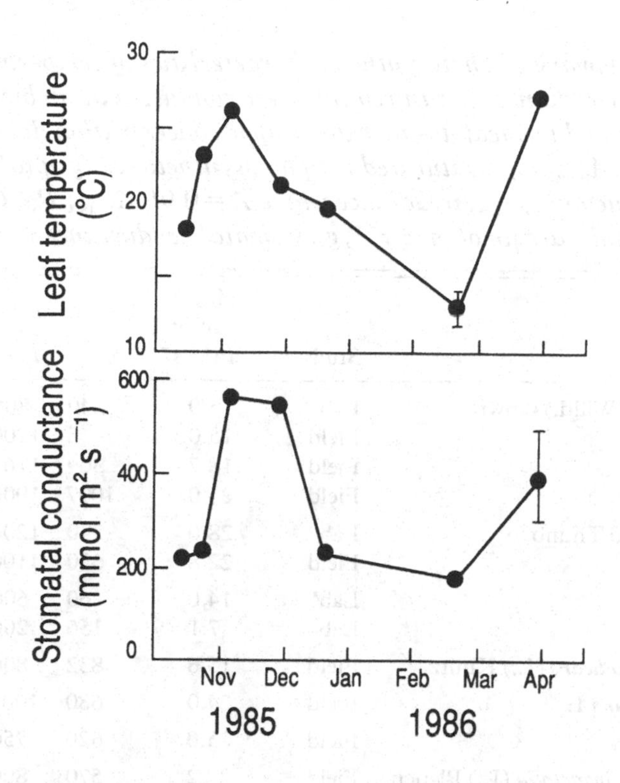

Figure 9.6. Midday values of ambient temperature and stomatal conductance collected several times during 1985–6 for a population of *Lonicera japonica* in College Park, MD, USA. From Carter & Teramura (1988b).

it is not surprising that the xylem vessel elements of the evergreen *L. japonica* are much shorter and smaller in diameter than those of associated deciduous lianas (Bell, 1986). Perhaps as a result of these and other constraints associated with the evergreen habit, the occurrence of evergreen foliage on woody temperate vines is relatively low (Table 9.1). Only 20% of lianas which occur in Maryland have evergreen foliage (Brown & Brown, 1972; Duncan, 1975).

Physiological features

Photosynthetic responses

Light-saturated rates of net photosynthesis (A_{max}) for temperate lianas grown under high irradiance range from 16 to 28 μmol CO_2 m^{-2} s^{-1} (Table 9.3). These values are at the high end of the range reported for early-successional species from mesic, temperate environments (Bazzaz, 1979). They are also considerably higher than the range of A_{max} reported for deciduous overstory

Table 9.3. *Summary of photosynthetic characteristics of temperate lianas grown in high irradiance. Measurements were conducted at ambient CO_2 concentrations and low leaf-to-air water vapor concentration deficits (6–10 mmol mol^{-1}). A_{max}, light saturated net photosynthetic CO_2 uptake; g_s, stomatal conductance; I_{sat}, irradiance where $A = 0.9^* A_{max}$; R_d, CO_2 efflux in the dark: all units are μmol m^{-2} s^{-1}, g_s, stomatal conductance (μmol m^{-2} s^{-1})*

Species	Study site	A_{max} (μmol m^{-2} s^{-1})	g_s	I_{sat}	R_d	Source
Pueraria lobata (Willd.) Ohwi	Lab	15.0	240	800	1.0	1
	Field	15.0	–	1200	0.5	2
	Field	18.7	309	1167	0.9	3
	Field	25.0	1097	1000	2.0	4
Lonicera japonica Thunb.	Lab	28.0	340	1200	–	1
	Field	22.8	620	1100	3.0	4
Hedera helix L.[a]	Lab[b]	14.0	260	600	–	5
	Lab[c]	7.1	156	200	–	5
Toxicodendron radicans (L.) Kuntz.	Field	15.6	832	800	2.5	4
Clematis virginiana L.	Field	24.0	680	1000	4.0	4
Vitis vulpina L.	Field	25.0	620	750	2.0	4
Parthenocissus quinquefolia (L.) Planch.	Field	27.2	570	800	2.0	4

[a]Growth regime was only 300–500 μmol m^{-2} s^{-1} PPFD.
[b]Adult leaves produced during flowering.
[c]Juvenile leaves produced during vegetative growth.
Sources: 1. Sasek (1985); 2. Wechsler (1977); 3. Forseth & Teramura (1987); 4. Carter *et al.* (1989); 5. Hoflacher & Bauer (1982).

tree species in moist temperate areas (3–16 μmol m^{-2} s^{-1}) with which these lianas often compete (Hicks & Chabot, 1985). Consistent with these high rates of A_{max}, dark respiration (R_d) rates (Table 9.3) are also at the high end of the range reported for most early-successional woody species (0.7–2.0: Hicks & Chabot, 1985) and early-successional weeds (0.6–1.3: Bazzaz & Carlson, 1982).

Although all of the lianas discussed in this chapter utilize the C_3 photosynthetic pathway and would consequently be expected to show considerable photorespiration, this has generally not been examined in temperate woody vines. In two cultivars of the Mediterranean vine, *Vitis vinifera*, A_{max} decreased about 50% in normal air compared with measurements in air with only 1% O_2 (Chaves *et al.*, 1987). Although this method of determining photorespiration is thought to be an overestimation (Jones, 1983), it still falls

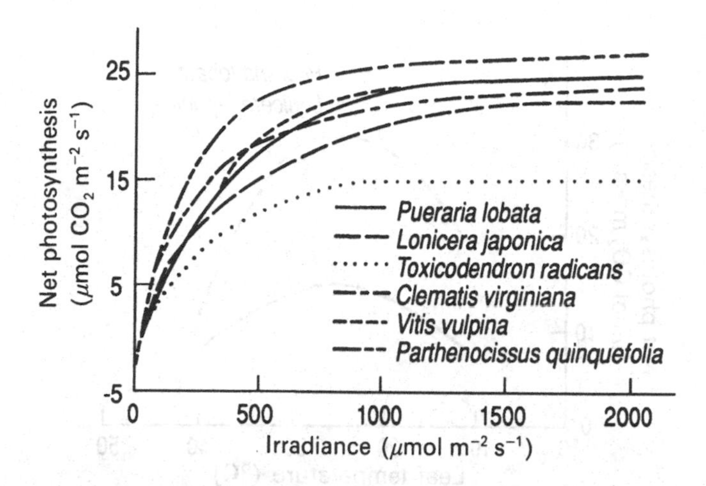

Figure 9.7. Response of net photosynthetic CO_2 uptake to incident photosynthetic photon flux density for six species of vines planted in an open field in central Maryland, USA. Curves represent the mean from three leaves of each species. From Carter *et al.* (1989).

within the range for most C_3 plants, which typically show a photorespiratory reduction of A_{max} of 20–50% (Larcher, 1980).

Although rates of A_{max} are generally higher than for other mesic temperate woody species, the photosynthetic response to irradiance in mesic temperate lianas grown under high irradiance is similar for other C_3 species in this environment (Figure 9.7). The PPFD at which photosynthesis reaches 90% of A_{max} (I_{sat}) ranges from 750 to 1200 μmol m^{-2} s^{-1}. This is within the range of 35–60% of maximum solar flux reported for I_{sat} values in most temperate deciduous species (Larcher, 1980).

Most temperate woody species exhibit broad photosynthetic temperature optima, and considerable acclimation potential (Berry & Björkman, 1980). Although data on mesic temperate lianas is scant, both *P. lobata* and *L. japonica* show rather wide temperature responses (Figure 9.8), with optima ranging from 20 to 30 °C (Wechsler, 1977; Sasek, 1985). Because of these broad photosynthetic temperature responses and the potential for thermal acclimation, seasonal reductions in carbon income due to unfavorable temperatures for a number of temperate woody species are relatively small, having been estimated to be 3–9% (Jurik, 1980; Kuppers, 1984c; Hicks & Chabot, 1985). We hypothesize that temperature extremes probably place little limitation on the carbon gain of temperate, deciduous lianas, although there is little direct experimental information. Photosynthesis in *P. lobata* appears to be well adapted to the temperatures typical of the field. Leaf photosynthesis does not undergo appreciable decline until temperatures

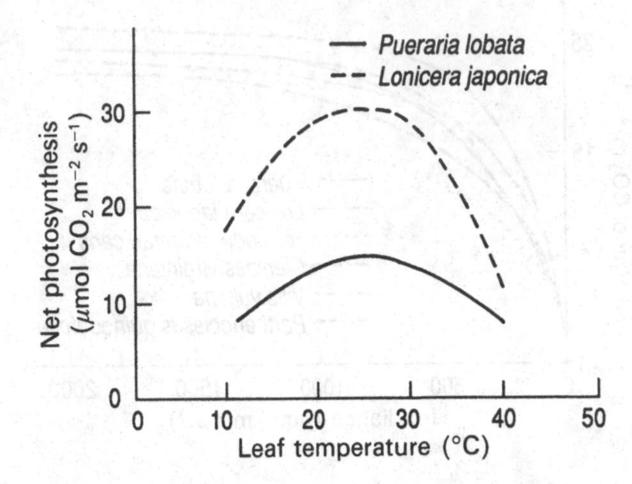

Figure 9.8. Response of net photosynthetic CO_2 uptake to leaf temperature for *Pueraria lobata* and *Lonicera japonica*. Figure was drawn from data in Sasek (1985). Plants were grown in pots in the phytotron at Duke University, NC, USA.

exceed 35 °C, which happens only rarely in the field, even on warm, sunny days, because of midday leaf orientation and relatively high rates of transpirational cooling (Forseth & Teramura, 1987). Although no information exists on photosynthetic temperature acclimation in non-cultivated, temperate lianas, photosynthetic temperature acclimation in two Mediterranean cultivars of *V. vinifera* from contrasting temperature regimes was demonstrated by Chaves *et al.* (1987).

Photosynthesis in temperate lianas has been found to be sensitive to changes in leaf-to-air water vapour concentration (Δw). Carter, Teramura & Forseth (1989) showed A_{max} decreases of 24–56% in six species of liana when Δw was increased from 10 to 40 mmol mol^{-1}. Forseth & Teramura (1987) similarly measured a 56% decline in A_{max} in *P. lobata* when Δw ranged from 10 to 35 mmol mol^{-1}. In most cases, the decline in A_{max} was accompanied by a substantial decrease in both g_s and intercellular CO_2 concentration (c_i). This implies that at least a portion of the decrease in A_{max} was a result of stomatal closure. In *L. japonica* and *P. quinquefolia*, however, c_i remained relatively constant, while both g_s and A_{max} declined (Carter *et al.*, 1989). However, without direct information on the behavior of stomata, it is impossible to conclude whether this response was due to a parallel decrease in mesophyll photosynthetic activity and stomatal aperture, or if it was due to patchy stomatal closure in localized areas of the epidermis. Although Carter *et al.* (1989) included exotic and native liana species representing a variety of climbing mechanisms, no associations between climbing mechanics and response to Δw were immediately apparent.

Although photosynthetic sensitivity to Δw is widespread in a variety of plant life forms and habitats (Tibbits, 1979), there is variability in the degree of response to Δw. Decreases in A_{max} from only slight reductions to up to 70% have been documented over a 30 mmol mol^{-1} range of Δw. For example, Bunce (1984, 1986) reported declines in A_{max} of 20–40% over a Δw range of 10–35 mmol mol^{-1} for two weedy annuals, *Abutilon theophrasti* and *Chenopodium album*. Schulze & Kuppers (1979) reported 50–70% reductions in A_{max} for the temperate woody shrub *Corylus avellana* from a Δw of 10–30 mmol mol^{-1}. The magnitudes of the reductions in A_{max} with increasing Δw found by Carter *et al.* (1989) and Forseth & Teramura (1987) for temperate mesic lianas are comparable to other temperate plant species.

Sensitivity of A and g_s to change in Δw might be particularly important for vines which occupy open, early-successional sites where soil moisture can be seasonally limiting and evaporative demand can change rapidly over a large range (Figure 9.2). Forseth & Teramura (1987) documented diurnal changes in Δw which were commonly in excess of 50 mmol mol^{-1} at a site in central Maryland dominated by *P. lobata*. A feedforward stomatal closure mechanism (Cowan & Farquhar, 1977) accompanying the reduction in A_{max} with increasing Δw would allow these vines to curtail water loss prior to substantial reduction in plant water status. As will be discussed in the water relations section of this chapter, regression models and boundary line analyses of stomatal response to environmental variables for three sympatric temperate vine species have shown irradiance and Δw to be the variables most highly correlated to g_s in the field (Bell, 1986). While leaf water potential (ψ_l) was seldom a limiting factor, boundary line analysis indicated that the range of ψ_l experienced by these three vines during the growing season was much more restricted than the range reported for most overstory tree species (Roberts, Knoerr & Strain, 1979; Hinckley *et al.*, 1981; Bell *et al.*, 1988).

The response of A to changes in c_i has been used to examine the relative importance of stomatal vs. mesophyll limitations to A (Farquhar & Sharkey, 1982). Sasek (1985) used the response of A to c_i to demonstrate that decreases in A of two temperate vines grown in a high CO_2 atmosphere were primarily the result of reductions in mesophyll photosynthetic activity, rather than changes in g_s. Under normal atmospheric conditions, diffusional limitations reduced potential photosynthesis of potted plants of *P. lobata* and *L. japonica* in a glasshouse by about 10–20% (Sasek, 1985). This falls in the range of a variety of C_3 species (Wong, Cowan & Farquhar, 1979; Jones, 1985). Also, Kuppers (1984a) reported a consistent 10% reduction in A due to stomatal limitations in five woody species of a European hedgerow, including the liana *Rubus corylifolius*.

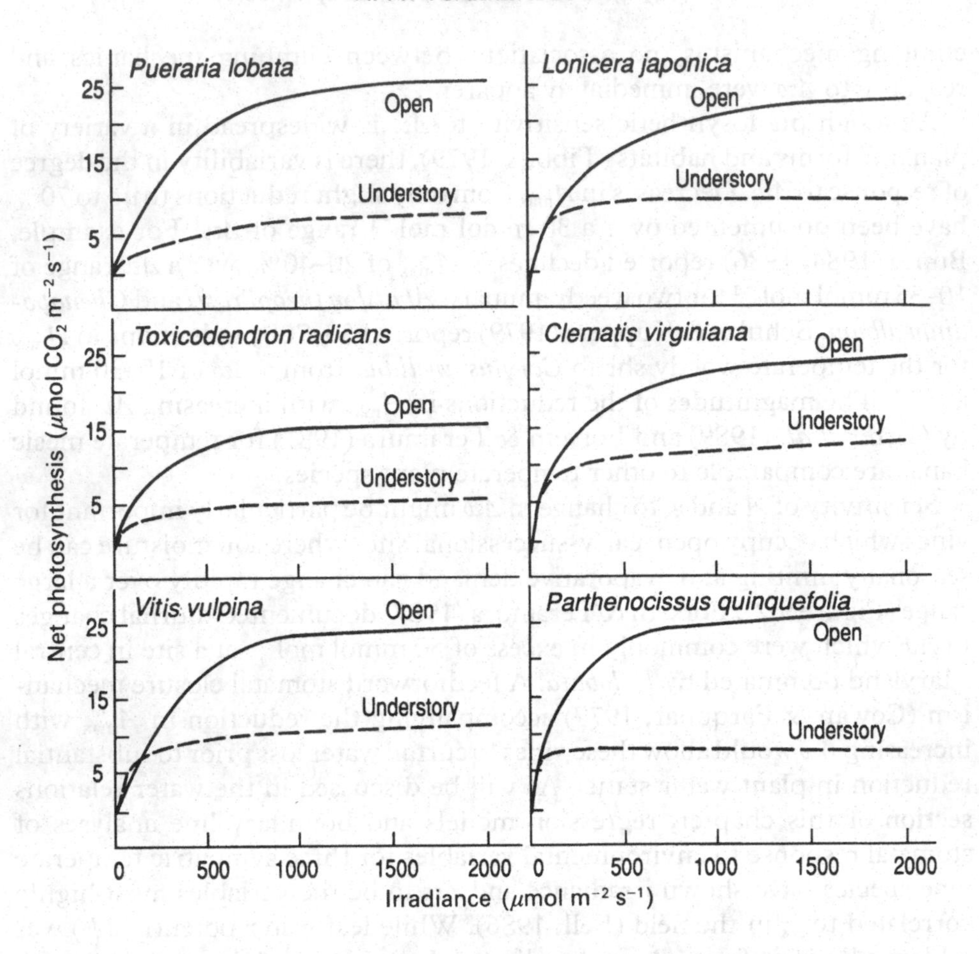

Figure 9.9. Response of net photosynthetic CO_2 uptake to incident photosynthetic photon flux density for six species of vines planted in an open field and an understory site. Plants were established from cuttings collected on naturally occurring individuals in the field. Measurements were made 1–3 months after transplantation and establishment and represent the means of 3–5 leaves for each species. From Carter & Teramura (1988a) and Carter *et al.* (1989).

Acclimation to light

The capacity to acclimate physiologically to a variety of irradiances is important for lianas, since different leaves on the same stem may be exposed to light environments ranging from 1 to 100% of full noon sunlight. All six temperate vines species studied by Carter & Teramura (1988a) and Carter *et al.* (1989) displayed photosynthetic light acclimation when grown in a deeply shaded understory site compared with plants grown in an open field (Figure 9.9; Tables 9.3, 9.4). Rates of A_{max}, I_{sat}, g_s and R_d were lower in individuals

Table 9.4. *Summary of photosynthetic characteristics of temperate lianas grown in low irradiance. Measurements were conducted at ambient CO_2 concentrations and low leaf-to-air water vapor concentration deficits (6–10 mmol mol^{-1}). PPFD$_c$, photosynthetic photon flux density where net photosynthetic CO_2 uptake is zero; A_l, net photosynthetic CO_2 uptake at an irradiance of 50 μmol m^{-2} s^{-1}; A_{max}, light saturated net photosynthetic CO_2 uptake; I_{sat}, irradiance where $A = 0.9 \times A_{max}$; R_d, CO_2 efflux in the dark: all units are μmol m^{-2} s^{-1}. g_s, stomatal conductance (mmol m^{-2} s^{-1})*

Species	Study site	PPFDd$_c$	A_l	A_{max}	g_s	I_{sat}	R_d	Source
Pueraria lobata	Field	43	0.5	7.2	105	860	1.5	1
	Field	–	–	14.0	234	590	0.3	2
Lonicera japonica	Field	35	1.3	9.0	137	280	2.7	1
Hedera helix	Field	30	1.1	6.4	113	360	2.0	1
	Lab[a]	–	–	5.9	132	250	–	3
	Lab[b]	–	–	4.1	76	100	–	3
Toxicodendron radicans	Field	32	1.0	5.5	105	260	2.0	1
Clematis virginiana	Field	32	2.4	12.9	362	520	3.0	1
Vitis vulpina	Field	25	1.9	11.1	125	320	2.0	1
Parthenocissus quinquefolia	Field	20	3.5	7.8	133	160	1.5	1

[a]Adult leaves produced during flowering.
[b]Juvenile leaves produced during vegetative growth.
Sources: 1. Carter & Teramura (1988a); 2. Forseth & Teramura (1987); 3. Hoflacher & Bauer (1982).

grown in the forest understory than in plants from the open site, but I_{sat} was near the mean daily maximum irradiance recorded in each environment.

Relative light acclimation potential may be related to climbing mechanics. Although most temperate lianas exhibit prolific growth or flowering in open areas (Brown & Brown, 1972; Slezak, 1976; Wechsler, 1977; Della-Bianca, 1978; Abramovitz, 1983), many species appear to possess broad physiological plasticity in terms of photosynthetic light acclimation (Carter & Teramura, 1988a; Carter *et al.*, 1989). Carter & Teramura (1988a) reported tendril climbers (*Vitis vulpina* and *S. rotundifolia*), especially ones with adhesive discs (*Parthenocissus quinquefolia*), displayed greater physiological acclimation to low light environments by possessing lower photosynthetic light compensation points, higher photosynthesis (A) and water use efficiency (WUE) under low light, and lower photosynthetic light saturation points. While *P. lobata* displayed poor adaptability to low light, the other twining liana, *L. japonica*, showed intermediate characteristics, as did the petiole

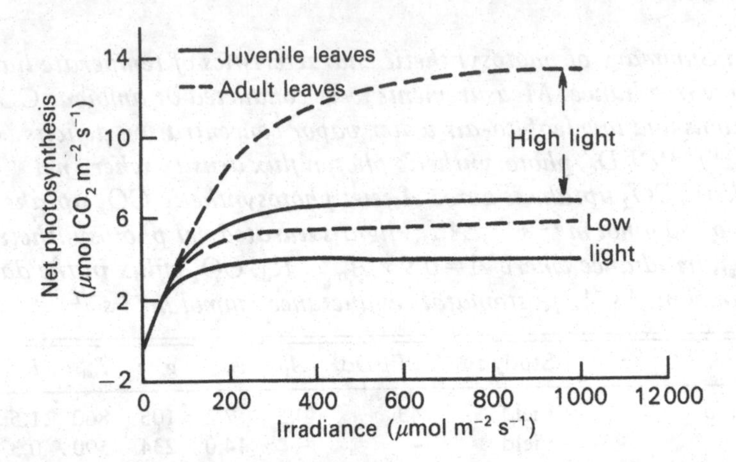

Figure 9.10. Response of net photosynthetic CO_2 uptake to incident photosynthetic photon flux density for leaves from the juvenile and adult life phases of *Hedera helix*. Low light leaves developed under a PPFD regime of 30–50 μmol m^{-2} s^{-1}, while high light leaves developed under a PPFD regime of 300–500 μmol m^{-2} s^{-1}. Redrawn from Hoflacher & Bauer (1982).

twiner, *Clematis virginiana* and the root climbers *Hedera helix* and *Toxicodendron radicans*.

Acclimation to high irradiance was also examined in the same lianas, with the exception of *S. rotundifolia* and *H. helix*, in an open field study (Carter *et al.*, 1989). Most species were capable of comparably high maximum A (23–27 μmol m^{-2} s^{-1}), except for the root climber *R. radicans* (16 μmol m^{-2} s^{-1}). Thus, tendriled climbers were found to have the broadest physiological plasticity, some twining vines were less suited to an understory existence, and root climbers were least suited to an open field environment.

Another exotic twining vine, *Celastrus orbiculatus*, has been reported to be able to acclimate to contrasting irradiances (Patterson, 1975). Although the three irradiance regimes used in that study were quite low (411, 137 and 21 μmol m^{-2} s^{-1}) seedlings showed higher A_{max} and I_{sat} under higher irradiances. In addition, acclimation from the lowest to highest irradiance regime was shown to occur after only 8 days.

The evergreen, temperate root climber *Hedera helix* also possesses a high degree of physiological plasticity to contrasting irradiances (Bauer & Thoni, 1988). *Hedera helix* produces two distinct types of leaves, juvenile leaves produced during the vegetative growth phase and adult leaves produced during flowering. The juvenile and adult leaves behave physiologically as shade and sun leaves, respectively (Hoflacher & Bauer, 1982). Under high irradiance, both leaf types increase A_{max} and I_{sat} (Tables 9.3, 9.4; Figure 9.10). In addition, other morphological and biochemical adjustments (e.g.

increases in specific leaf weight, number of chloroplasts per unit leaf area, chlorophyll a/b ratio) also occur (Hoflacher & Bauer, 1982). Although both types of leaves can acclimate to contrasting irradiances to some degree, juvenile leaves cannot adjust to high irradiance as well as adult leaves and adult leaves cannot adjust to low irradiance as well as juvenile leaves. Thus, the production of two physiologically distinct leaf types effectively extends the capability of *H. helix* to acclimate to the wide seasonal range of irradiances found in temperate forest understories.

Water relations

The availability of water during the growing season can be a principal determinant of species distribution and abundance across the landscape (Grime, 1979; Pallardy & Kozlowski, 1979; Hinckley *et al.*, 1981; Chambers *et al.*, 1985). Within temperate deciduous forests, water availability can be highly variable, depending upon the quantity, frequency and type of precipitation, evaporative demand, and substrate characteristics. On a large scale, the distribution of forest associations in Maryland has been correlated with water availability (Brush, Lenk & Smith, 1980) and lianas are listed as most common in mesic floodplain and adjacent forest habitats.

Response of stomata. Regulation of g_s is the principal means by which a plant controls water loss from its tissues. Stomatal response to changes in atmospheric humidity is a well-documented mechanism preventing high rates of water loss and subsequent water deficits under high evaporative demand (Schulze & Hall, 1982). Relative sensitivity of stomata to humidity differed among six temperate liana species examined by Forseth & Teramura (1987) and Carter *et al.* (1989) (Table 9.5). Stomatal conductance decreased by about 20–40% for most of the vines when Δw increased from 10 to 20 mmol mol^{-1}. However, at higher values of Δw, differences among species became more pronounced. The stomatal closure with increasing Δw resulted in little or no increase in the rate of transpirational water loss for most of these vines (Forseth & Teramura, 1987; Carter *et al.*, 1989).

Diurnal patterns of water use. In a field study with three temperate vine species, Bell (1986) and Bell *et al.* (1988) examined the sensitivity and range of stomatal responses to Δw, incident irradiance, leaf temperature (T_l), and bulk ψ_l over two growing seasons. They concluded that the major limiting factors to g_s for these vines in the field were irradiance and Δw. Stomatal sensitivity of the two deciduous species (*V. vulpina* and *P. quinquefolia*) to these factors was strong enough to result in mid-morning peaks in diurnal water loss, prior to diurnal peaks in evaporative demand (Figure 9.11; Bell *et*

Table 9.5. *Percentage of the stomatal conductance measured at a leaf-to-air water vapor concentration deficit (Δw) of 10 mmol mol^{-1} at values of Δw from 20 to 40 mmol mol^{-1}. Data were collected at saturating irradiance and ambient CO_2 concentrations on vines growing in exposed sites in the field*

	Δw (mmol mol^{-1})			
Species	20	30	40	Source
Pueraria lobata	44	29	–	Forseth & Teramura (1987)
	72	47	18	Carter *et al.* (1989)
Lonicera japonica	98	73	37	Carter *et al.* (1989)
Vitis vulpina	66	58	16	Carter *et al.* (1989)
Parthenocissus quinquefolia	73	51	28	Carter *et al.* (1989)
Toxicodendron radicans	80	47	23	Carter *et al.* (1989)
Clematis virginiana	88	56	42	Carter *et al.* (1989)

al., 1988). However, reduced stomatal sensitivity to Δw and higher midday leaf temperatures resulted in a midday peak in diurnal water loss for *L. japonica* (Figure 9.11; Bell, 1986; Bell *et al.*, 1988). Thus, differences in both stomatal sensitivity to environmental factors and variation in the physical environment play important roles in controlling the diurnal pattern of water loss in these temperate lianas.

Boundary line analysis of seasonal data (D. Bell and I. Forseth, unpublished data) showed that the range of ψ_l values over which these three vine species had g_s above zero was narrow compared with other deciduous forest species (Figure 9.12; Roberts, Strain & Knoerr, 1980; Hinckley *et al.*, 1981; Chambers *et al.*, 1985). However, maximum g_s was insensitive to high values of ψ_l. The narrow range of ψ_l measured in these plants suggests that soil water availability could limit the distribution of these vines. This is supported by observations of increased vine abundance in mesic deciduous forest associations (Brush *et al.*, 1980). Differences among the three vine species in the ranges of ψ_l over which gas exchange can be maintained indicate that *P. quinquefolia*, with the narrowest range, could be restricted to more mesic sites than the other two species (Figure 9.12). However, sensitivity analysis revealed that ψ_l was rarely the most limiting factor to g_s in the field for these vines. Direct stomatal responses to irradiance and Δw resulted in the maintenance of ψ_l above the threshold value for limitation to g_s for each species throughout most of the 2-year study (Bell, 1986). Thus, even though ψ_l may be important in restricting the occurrence of these lianas at the limits of their ranges, it does not appear to play a major role in limiting productivity once the vine is established (D. Bell and I. Forseth, unpublished data).

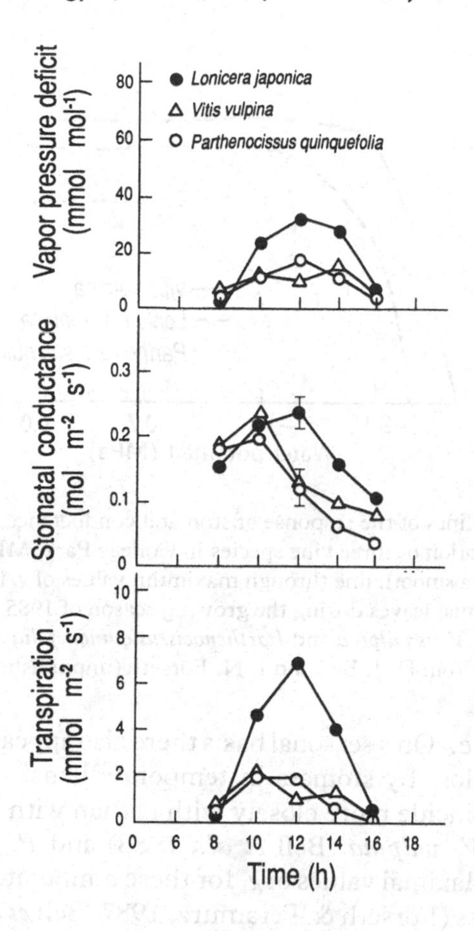

Figure 9.11. Diurnal courses of leaf-to-air water vapor concentration deficit, stomatal conductance and transpirational water flux on 13 September 1985 for populations of three sympatric vine species in College Park, MD, USA. Error bars for stomatal conductance represent ± 1 SE. Notice the correspondence of daily maxima in g_s, E and Δw in *Lonicera japonica*, while *Vitis vulpina* and *Parthenocissus quinquefolia* have daily maxima of g_s and E in morning hours, at lower Δw values. From Bell *et al.* (1988).

The year-round stomatal response of *L. japonica* and *S. rotundifolia* have recently been examined in the center of a canopy gap in Maryland, USA (E. Whereat and A. Teramura, unpublished data). During summer, leaf conductance was found to be most influenced by changes in irradiance and Δw, and predawn or diurnal ψ were relatively influential. It appeared that seasonal changes in the stomatal response to Δw may also occur. During winter, wintergreen leaves of *L. japonica* and evergreen stems of *S. rotundifolia* were found to be more influenced by leaf or stem temperature and the minimum temperature of the previous night. Evergreen lianas appear to possess the capacity to acclimate to a wide range of temperature regimes.

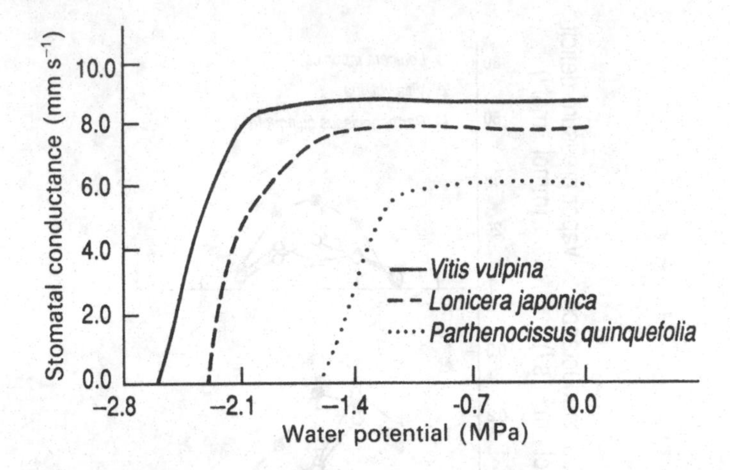

Figure 9.12. Boundary lines of the response of stomatal conductance to leaf water potential for a sympatric population of three vine species in College Park, MD, USA. Plots were constructed by fitting a smooth line through maximum values of g_s from scatter plots of data collected on individual leaves during the growing season of 1985. Total sample size for the season was 115 for *Vitis vulpina* and *Parthenocissus quinquefolia* and 180 for *Lonicera japonica*. From D. J. Bell and I. N. Forseth (unpublished data).

Seasonal water use. On a seasonal basis there also appears to be substantial control over water loss by stomata in temperate lianas. Seasonal peaks in transpiration (E) coincide more closely with g_s than with Δw for *P. quinquefolia, L. japonica, V. vulpina* (Bell *et al.*, 1988) and *P. lobata* (Forseth & Teramura, 1987). Maximal values of g_s for these temperate lianas (Table 9.3) growing in open sites (Forseth & Teramura, 1987; Bell *et al.*, 1988; Carter *et al.*, 1989) were at the high end of the range of values reported for other temperate woody species (Körner, Scheel & Bauer, 1979; Roberts *et al.*, 1979). Despite high g_s, and consequently high E, the diurnal ranges of ψ_l measured for most of these lianas were substantially narrower than for most other temperate woody species (e.g. Hinckley *et al.*, 1981; Hull & Wood, 1984; Bahari, Pallardy & Parker, 1985). The lowest daily minimum values of ψ_l in these lianas were reached in late summer and ranged from -1.4 to -2.4 MPa (Bell *et al.*, 1988; Whereat, 1990). Minimum diurnal values of ψ_l of most temperate deciduous shrubs and trees can range well below -2.0 MPa (Hinckley *et al.*, 1981, 1983). The maintenance of high minimum ψ_l despite high g_s and E suggests that temperate lianas have considerable quantities of stored water available (capacitance), or relatively high conductivity to water flow through the soil–plant–atmosphere continuum (Forseth & Teramura, 1987).

The maintenance of high leaf E with the development of only moderate ψ is dependent on supplying large volumes of water to individual leaves. However, on a whole plant level, this demand is more pronounced in lianas since they characteristically possess small diameter stems which supply relatively large leaf areas in comparison to trees. The hydraulic conductance of a cylindrical conducting element is proportional to the fourth power of the radius (Poiseuille's Law). Therefore, woody species can deliver large volumes of water either by high rates of flow through relatively few large vessels, or by reduced flow rates through many smaller vessels, or by some combination of the two (Carlquist, 1975; Hinckley *et al.*, 1981; Bell *et al.*, 1988).

The benefits of possessing larger vessel elements appear to be offset by their increased susceptibility to cavitation or embolism, induced by freezing or low xylem water potentials (Scholander, 1957; Zimmermann & Jeje, 1981; Bell *et al.*, 1988; Tyree & Sperry, 1989). In addition, since vessel diameter is inversely correlated with the number of vessels per unit of cross-sectional area, the conductivity of stems with large vessels can be reduced by relatively few cavitation events (Carlquist, 1975). There may be exceptions to this trend since xylem susceptibility may be more directly related to pit membrane pore diameters, which is often correlated with vessel size (Tyree & Sperry, 1989). Nevertheless, the vessel element diameters of lianas are typically 2–3 times greater than most trees, permitting high flow rates per cross-sectional area (Carlquist, 1975; Baas, 1986). Accordingly, the vulnerability of the water conducting system of lianas is thought to limit their distribution and productivity at higher latitudes and in drier areas (Carlquist, 1975; Baas, 1986; Bell *et al.*, 1988). There is evidence which suggests that some lianas are able to repair embolized vessels. Whereas the vessel elements of trees tend to retain their conducting capacity for a few years at most, some vessel elements of lianas may remain functional for 7–10 years or more (Zimmermann, 1983; Smart & Coombe, 1983; Putz, 1983). Sperry *et al.* (1987) reported that the stems of *Vitis* spp., which are air-filled during winters in New England, can be refilled and repaired by positive root pressure in spring. Although demonstration of xylem refilling is limited to the former example, root pressures have also been observed in a variety of temperate and tropical lianas (Putz, 1983). In addition, the rapid and vigorous regeneration of xylem and phloem in wounded or girdled liana stems may be important in overcoming vascular interruptions attributable to mechanical damage (Dobbins & Fisher, 1986).

It is possible that some lianas such as *P. lobata* avoid large xylem tensions (low ψ) while maintaining high E by utilizing the water capacitance of large tuberous roots (Forseth & Teramura, 1987). In addition, many temperate species may exert sufficient stomatal control under high Δw to avoid hazardous xylem tensions (Bell *et al.*, 1988; Carter *et al.*, 1989). Finally, the vascular anatomy of some sympatric temperate lianas may differ considerably from the typical liana pattern (Bell, 1986). *Lonicera japonica*, a ring porous species, possesses anatomical characteristics more akin to desert shrubs (small mean vessel diameter and large number of vessels/mm^2), whereas *V. vulpina* exhibits some of the largest vessel diameters, and hence the fewest elements per cross-sectional area, found among dicots (Carlquist, 1975; Bell, 1986). The ring porous anatomy of *L. japonica* may provide efficient, but vulnerable early-wood vessels in spring, and inefficient, but safe late-wood vessels in summer (Bell *et al.*, 1988; Baas, 1986).

Despite the high hydraulic conductivity of the vascular system in most lianas, the conductivity of liquid water from the soil to the leaf (G) in three temperate vines (*L. japonica, P. quinquefolia*, and *V. vulpina*) (Bell *et al.*, 1988) was comparable to that of some mesic chaparral shrubs (Calkin & Pearcy, 1984), temperate trees (Ginter-Whitehouse, Hinckley & Pallardy, 1983) and woody hedgerow species (Kuppers, 1984b). This may result from rhizospheric resistance to water movement rather than flow resistance in the vascular system (Passioura, 1982). In fact, seasonal patterns of G in temperate lianas (Bell *et al.*, 1988) and other temperate species (Ginter-Whitehouse *et al.*, 1983; Calkin & Pearcy, 1984) appeared to parallel changes in predawn values of ψ_l. Because predawn values of ψ_l are thought to reflect soil water potential in the rhizosphere, parallel changes in G and predawn ψ_l through the growing season support the idea of a relatively large rhizospheric resistance in the soil–leaf continuum.

Rooting patterns. No quantitative information is available on rooting patterns and root growth of temperate lianas in the field. Preliminary data on seasonal changes in predawn ψ_l for two species, *L. japonica* and *S. rotundifolia* L., suggest that these closely sympatric vines have substantially different functional rooting depths (Figure 9.13; E. Whereat and A. H. Teramura, unpublished data). Predawn ψ_l of *L. japonica* tracks fluctuations in soil water potentials at shallow depths (*c.* 5–15 cm), whereas predawn ψ_l values of *S. rotundifolia* reflect soil conditions deeper in the profile (15–40 cm). Such a separation in functional rooting distribution could have important implications for the coexistence of these two vines, which can be found growing together in dense aggregations within gaps of the deciduous forest canopy in central Maryland.

Osmotic adjustment. Alterations in cellular water volume and ψ_l relationships form the basis for several different mechanisms of plant response to decreased water availability (Parker *et al.*, 1982). These may take the form of changes in cell wall elastic modulus, solute content or size. Increases in solute concentration of cellular sap (osmotic adjustment) is an adaptation which allows the maintenance of favorable plant water status (e.g. ψ_l) and continued

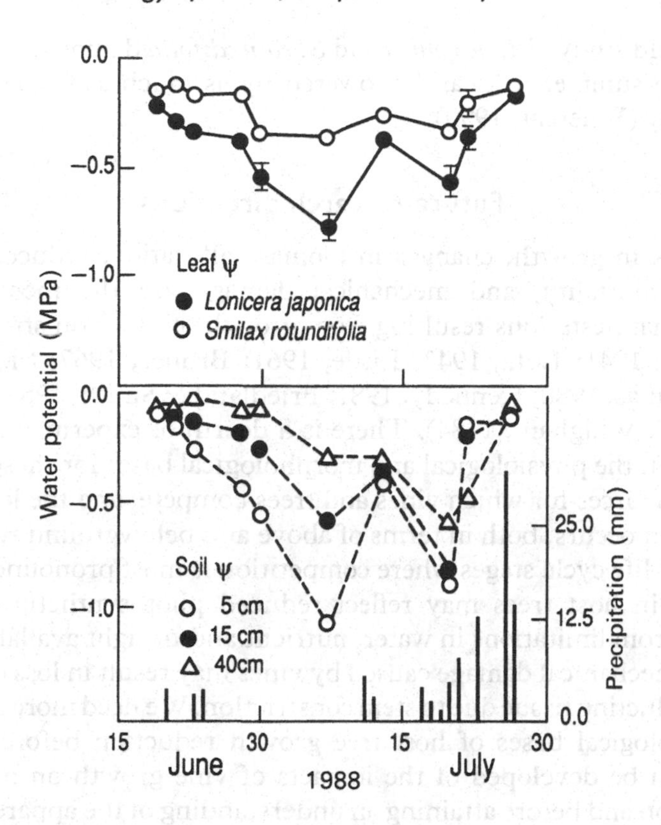

Figure 9.13. Predawn values of leaf water potential for sympatric populations of *Lonicera japonica* and *Smilax rotundifolia* collected on several dates during the summer of 1988 in comparison to adjacent soil water potentials at three different depths measured on the same days. Also plotted are precipitation events during this time period. From Whereat (1990).

water uptake under conditions of low soil water availability (Turner, 1986). Osmotic adjustment also results in higher levels of cell turgor pressure, which could be important for cell expansion, although the significance of leaf turgor maintenance to continued growth in drying soils has recently been questioned (Turner, 1986; Shackel, Matthews & Morrison, 1987). A moderate degree of osmotic adjustment was reported in the field for *V. vulpina* (lowering the osmotic potential at zero turgor by 0.45 MPa: Bell *et al.*, 1988). This degree of osmotic adjustment is comparable with that reported for some temperate mesic trees (Roberts *et al.*, 1980; Bahari *et al.*, 1985) and cultivated grapevines (Smart & Coombe, 1983), but smaller than that observed in other woody species (Roberts *et al.*, 1980; Bahari *et al.*, 1985; Abrams & Knapp, 1986). The other two temperate vines in the study (*P. quinquefolia* and *L. japonica*) did not show any lowering of osmotic potential as soil water availability decreased (Bell *et al.*, 1988). However, recent evidence from an ongoing field study of *L. japonica* and *S. rotundifolia* demonstrates that from mid to late summer, $\psi_{\pi o}$ can be lowered by as much as 0.7 and 0.4 MPa, respectively (Whereat, 1990).

Future research directions

Reductions in growth, changes in biomass allocation, reduced fecundity, increased mortality, and mechanical damage are the most commonly reported manifestations resulting from growth of vines on structural hosts (Featherly, 1941; Lutz, 1943; Little, 1961; Bruner, 1967; Siccama *et al.*, 1976; Thomas, 1980; Kennedy, 1981; Friedland & Smith, 1982; Putz, 1984, 1985, 1987; Whigham, 1984). There is a dearth of experimental evidence, however, on the physiological and morphological bases for these effects, the specific resources for which vines and trees compete, and the level at which competition occurs, both in terms of above and belowground resources and in terms of life cycle stages where competition is most pronounced. Growth reduction in host trees may reflect reduced photosynthetic carbon gain resulting from limitations in water, nutrient and/or light availability. Alternatively, mechanical damage caused by vines may result in loss of photosynthetic conducting tissue due to stem constriction. We need more research into the physiological bases of host tree growth reduction before a complete picture can be developed of the impacts of vine growth on mature forest composition and before attaining an understanding of the apparent competitive superiority of exotic over native vines.

Further research is needed on the comparative physiological ecology of liana species. We currently cannot answer questions about why vines like *V. vulpina* and *P. lobata* develop large canopies while others such as *P. quinquefolia* maintain small leaf areas and rarely become dominant. Future comparative work on relationships between climbing mechanics, physiological plasticity, hydraulic conductivity and rooting patterns should prove illuminating in this regard.

Temperate vines provide an ideal opportunity to study questions related to the evergreen versus deciduous growth habit since examples exist with differing degrees of evergreenness. For example, *Smilax rotundifolia* has evergreen stems which possess functional stomata but loses its deciduous leaves, while *Lonicera japonica* has evergreen leaves which persist throughout the winter months throughout most of its range. Wintergreen leaves of *L. japonica* and evergreen stems of *S. rotundifolia* may possess the capacity to acclimate to widely different temperature regimes. Other vines, such as *Pueraria lobata* and *Vitis vulpina*, have deciduous leaves only. Recently, some interest has been directed toward understanding the contribution of winter carbon gain by evergreen vines to the total annual carbon budget. Since the growth of evergreen species is largely restricted to non-winter months, storage and subsequent use of carbohydrates gained during the winter may be a particularly important aspect of their ecology. A liana growth habit may be especially well suited to taking advantage of early season growth based on stored reserves, thereby allowing lianas to climb unshaded structures and develop canopies which may shade later emergent deciduous competitors. In addition, further growth in the fall after deciduous species drop their leaves would place liana meristems closer to upper canopy layers.

Because vines occupy a diverse range of light environments and because some understory evergreen leaves may be exposed to markedly contrasting irradiances when the overstory vegetation is dormant, vines can provide opportunities to examine physiological and allometric acclimation to irradiance.

Some introduced vines, such as *P. lobata, L. japonica* and *Celastrus orbiculatus* have escaped cultivation and become aggressive weeds. Therefore vines also provide opportunities to examine fundamental questions related to their rapid growth rate and spread, particularly in relationship to less aggressive, co-occurring native vine species.

Summary

Vines can be found in a wide variety of habitats throughout temperate ecosystems and represent a large, taxonomically diverse group, including 21 families of plants. After escaping from cultivation, some introduced vines have become notably aggressive weeds and developed into a dominant part of the vegetation landscape, particularly in disturbed areas. Most aggressive vines have prodigious growth rates and play a major ecological role through their impact on host vegetation for support. Yet little is known of the specific mechanisms by which vines impact hosts nor whether this results primarily from above or belowground interactions.

Because vines use other woody plants for support, this generally results in an increased allocation of biomass into leaves compared with stems. Some vines, like *Pueraria lobata* (kudzu) allocate a large portion (>50%) of their total biomass into roots, yet very little (1 or 2%) into flowers and seeds. In these vines, vegetative growth and reproduction is much more important than sexual reproduction in determining their distribution. Rates of stem elongation are important to the colonization and exploitation of suitable habitats by vines and may be related to vine climbing mechanics. The fastest growing vines are stem twiners and tendril climbers and the slowest are root climbers. Climbing mechanics also seem to be related to the successional stage by determining trellis availability and light environment in which vines are commonly found. Simple stem twiners are commonly found in early to mid-successional, high irradiance environments because they are restricted to climbing small to medium diameter objects. Vines with tendrils or modified leaves that attach to leaf petioles and other smaller diameter objects and are more successful in understory and gap edges of mid to late successional forests. More specialized vines that climb by roots and tendrils with adhesive disks can climb large diameter objects and are found in the understory of late successional forests.

Vines have a large capacity to adjust physiologically to the spatial and temporal environmental heterogeneity found in temperate forests. Although maximum rates of photosynthesis are somewhat higher in vines than in other temperate woody species, the photosynthetic response to irradiance is similar to most C_3 species from high light environments. Some vines growing on the forest floor can acclimate to contrasting light environments such as is found in the tops of tree canopies. Others can osmotically adjust their internal water relations to the drought conditions which frequent the late summer months. Overall, vines have a tremendous ability to exploit the vast resource potential present in temperate environments.

References

Abrahamson, W. G. (1975). Reproductive strategies in dewberries. *Ecology* **56**: 721–6.

Abramovitz, J. N. (1983). *Pueraria lobata* Willd. (Ohwi) kudzu: limitations to sexual reproduction. Thesis, University of Maryland, College Park, Md.

Abrams, M. D. & Knapp, K. K. (1986). Seasonal water relations of three gallery forest hardwood species in northeast Kansas. *Forest Science* **32**: 687–96.

Baas, P. (1986). Ecological patterns of xylem anatomy. In *On the Economy of Plant Form and Function*, ed. T. J. Givnish, pp. 327–52. Cambridge University Press, Cambridge.

Bahari, Z. A., Pallardy, S. G. & Parker, W. C. (1985). Photosynthesis, water relations and drought adaptation in six woody species of Oak–Hickory forests in central Missouri. *Forest Science* **31**: 557–69.

Bauer, H. & Thoni, W. (1988). Photosynthetic light acclimation in fully developed leaves of the juvenile and adult life phases of *Hedera helix*. *Physiologia Plantarum* **73**: 31–7.

Bazzaz, F. A. (1979). The physiological ecology of succession. *Annual Review of Ecology and Systematics* **10**: 351–71.

Bazzaz, F. A. & Carlson, R. W. (1982). Photosynthetic acclimation to variability in the light environment of early and late successional plants. *Oecologia* **54**: 313–16.

Bell, D. J. (1986). Comparative water relations and resource partitioning in three temperate lianas. Thesis, University of Maryland, College Park, Md.

Bell, D. J., Forseth, I. N. & Teramura, A. H. (1988). Field water relations of three temperate vines. *Oecologia* 74: 537–45.

Berry, J. & Björkman, O. (1980). Photosynthetic response and adaptation to temperature in higher plants. *Annual Review of Plant Physiology* 311: 491–543.

Boring, L. R., Monk, C. D. & Swank, W. T. (1981). Early regeneration of a clear-cut southern Appalachian forest. *Ecology* 62: 1244–53.

Braun, E. L. (1950). *Deciduous Forests of Eastern North America.* The Blakiston Co., Philadelphia, PA.

Brender, E. V. (1960). Progress report on control of honeysuckle (*Lonicera japonica*) and kudzu. *Proceedings of the Southern Weed Science Society* 13: 187–93.

Brown, R. G. & Brown, M. L. (1972). *Woody Plants of Maryland.* Port City Press, Baltimore, Md.

Bruner, M. H. (1967). Honeysuckle – a bold competitor on bottomland hardwood sites. *Forest Farmer* 26: 9–17.

Brush, G. S., Lenk, C. & Smith, J. (1980). The natural forests of Maryland: An explanation of the vegetation map of Maryland. *Ecological Monographs* 50: 77–92.

Bunce, J. A. (1984). Effects of humidity on photosynthesis. *Journal of Experimental Botany* 35: 1245–51.

Bunce, J. A. (1986). Responses of gas exchange to humidity in populations of three herbs from environments differing in atmospheric water. *Oecologia* 71:117–20.

Caldwell, M. M. (1987). Root system competition in natural communities. In *Root Development and Function: Effects of the Physical Environment*, ed. P. J. Gregory. Cambridge University Press, Cambridge (in press).

Caldwell, M. M. & Richards, J. H. (1986). Competing root systems: morphology and models of absorption. In *On the Economy of Plant Form and Function*, ed. T. J. Givnish, pp. 251–73. Cambridge University Press, Cambridge.

Calkin, H. W. & Pearcy, R. W. (1984). Leaf conductance and transpiration, and water relations of evergreen and deciduous perennials co-occurring in a moist chaparral site. *Plant, Cell and Environment* 7: 339–46.

Carlquist, S. (1975). *Ecological Strategies of Xylem Evolution: A floristic approach.* Columbia University Press, New York.

Carter, G. A. & Teramura, A. H. (1988a). Vine photosynthesis and relationships to climbing mechanics in a forest understory. *American Journal of Botany* 75: 1011–18.

Carter, G. A. & Teramura, A. H. (1988b). Non-summer stomatal conductance for the invasive vines kudzu and Japanese honeysuckle. *Canadian Journal of Botany* **66**: 2392–5.

Carter, G. A., Teramura, A. H. & Forseth, I. N. (1989). Photosynthesis in an open field for exotic versus native vines of the southeastern USA. *Canadian Journal of Botany* **67**: 443–6.

Chabot, B. F. & Hicks, D. J. (1982). The ecology of leaf life spans. *Annual Review of Ecology and Systematics* **13**: 229–59.

Chambers, J. L., Hinckley, T. M., Cox, G. S., Metcalf, C. L. & Aslin, R. G. (1985). Boundary-line analysis and models of leaf conductance for four Oak-Hickory forest species. *Forest Science* **31**: 437–50.

Chaves, M. M., Harley, P. C., Tenhunen, J. D. & Lange, O. L. (1987). Gas exchange studies in two Portuguese grapevine cultivars. *Physiologia Plantarum* **70**: 639–47.

Cowan, I. R. & Farquhar, G. D. (1977). Stomatal function in relation to leaf metabolism and the environment. *Symposia of the Society for Experimental Biology* **31**: 471–505.

Darwin, C. (1865). The movements and habits of climbing plants. *Journal of the Linnean Society* **9**: 1–208.

Dobbins, D. R., & Fisher, J. B. (1986). Wound responses in girdled stems of lianas. *Botanical Gazette* **147**: 278–89.

Duncan, W. H. (1975). *Woody Vines of the Southeastern United States*. University of Georgia Press, Athens, GA.

Farquhar, G. D. & Sharkey, T. D. (1982). Stomatal conductance and photosynthesis. *Annual Review of Plant Physiology* **33**: 317–45.

Featherly, H. I. (1941). The effect of grapevines on trees. *Oklahoma Academy of Science Proceedings* **21**: 61–2.

Forseth, I. N. & Teramura, A. H. (1986). Kudzu leaf energy budget and calculated transpiration: The influence of leaflet orientation. *Ecology* **67**: 564–71.

Forseth, I. N. & Teramura, A. H. (1987). Field photosynthesis, microclimate and water relations of an exotic temperate liana, *Pueraria lobata*, kudzu. *Oecologia* **71**: 262–7.

French, J. C. (1977). Growth relationships of leaves and internodes in viny angiosperms with different modes of attachment. *American Journal of Botany* **64**: 292–304.

Frey, D. & Frick, H. (1987). Altered partitioning of new dry matter in *Hedera helix* L. (Araliaceae) induced by altered orientation. *Bulletin of the Torrey Botanical Club* **114**: 407–11.

Friedland, A. J. & Smith, A. P. (1982). Effects of vines on successional herbs. *American Midland Naturalist* **108**: 402–3.

Ginter-Whitehouse, D. L., Hinckley, T. M. & Pallardy, S. G. (1983). Spatial and temporal aspects of water relations of three tree species with different vascular anatomy. *Forest Science* **29**: 317–29.

Grime, J. P. (1979). *Plant Strategies and Vegetation Processes.* John Wiley and Sons, New York.

Gysel, L. W. (1951). Borders and openings of beech-maple woodlands in southern Michigan. *Journal of Forestry* **49**: 13–19.

Harper, J. L. (1977). *Population Biology of Plants.* Academic Press, New York.

Hicks, D. J. & Chabot, B. F. (1985). Deciduous forest. In *Physiological Ecology of North American Plant Communities*, ed. B. F. Chabot and H. A. Mooney, pp. 257–77. Chapman and Hall, New York.

Hinckley, T. M., Teskey, R. O., Duhme, F. & Richter, H. (1981). Temperate hardwood forests. In *Water Deficits and Plant Growth*, ed. T. T. Kozlowksi, pp. 154–208. Academic Press, New York.

Hinckley, T. M., Duhme, F., Hinckley, A. R. & Richter, H. (1983). Drought relations of shrub species: assessment of the mechanisms of drought resistance. *Oecologia* **59**: 344–50.

Hoflacher, H. & Bauer, H. (1982). Light acclimation in leaves of the juvenile and adult life phases of ivy (*Hedera helix*). *Physiologia Plantarum* **56**: 177–82.

Hull, J. C. & Wood, S. G. (1984). Water relations of oak species on and adjacent to a Maryland serpentine soil. *American Midland Naturalist* **112**: 224–34.

Hutchison, B. A. & Matt, D. R. (1977). The distribution of solar radiation within a deciduous forest. *Ecological Monographs* **47**: 185–207.

Janzen, D. H. (1971). Escape of juvenile *Dioclea megacarpa* vines from predators in a deciduous tropical forest. *American Naturalist* **105**: 97–112.

Jones, H. G. (1983). *Plants and Microclimate: A Quantitative Approach to Environmental Plant Physiology*. Cambridge University Press, Cambridge.

Jones, H. G. (1985). Partitioning stomatal and non-stomatal limitations to photosynthesis. *Plant, Cell and Environment* **8**: 95–104.

Jurik, T. W. (1980). Physiology, growth, and life-history characteristics of *Fragaria virginiana* Duchesne and *F. vesca* L. (Rosaceae). Dissertation, Cornell University, Ithaca, NY.

Kennedy, H. E. Jr. (1981). Foliar nutrient concentrations and hardwood growth influenced by cultural treatments. *Plant and Soil* **63**: 307–16.

Körner, C., Scheel, J. A. & Bauer, H. (1979). Maximum leaf diffusive conductance in vascular plants. *Photosynthetica* **13**: 45–82.

Kuppers, M. (1984a). Carbon relations and competition between woody species in a central European hedgerow. I. Photosynthetic characteristics. *Oecologia* **64**: 332–43.

Kuppers, M. (1984b). Carbon relations and competition between woody species in a central European hedgerow. II. Stomatal responses, water use, and hydraulic conductivity in the root/leaf pathway. *Oecologia* **64**: 344–54.

Kuppers, M. (1984c). Carbon relations and competition between woody species in a central European hedgerow. III. Carbon and water balance on the leaf level. *Oecologia* **65**: 94–100.

Kuppers, M. (1985). Carbon relations and competition between woody species in a central European hedgerow. IV. Growth form and partitioning. *Oecologia* **66**: 343–52.

Larcher, W. (1980). *Physiological Plant Ecology*. Springer-Verlag, Berlin.

Leatherman, A. D. (1955). Ecological life-history of *Lonicera japonica* Thunb. Dissertation, University of Tennessee.

Little, S. (1961). Recent results of tests in controlling Japanese honeysuckle. *Hormolog* **3**: 8–10.

Lutz, H. J. (1943). Injuries to trees caused by *Celastrus* and *Vitis*. *Bulletin of the Torrey Botanical Club* **70**: 436–9.

McNab, W. H. & Meeker, M. (1987). Oriental bittersweet: a growing threat to hardwood silviculture in the Appalachians. *Northern Journal of Applied Forestry* **4**: 174–7.

Pallardy, S. G. & Kozlowski, T. T. (1979). Relationship of leaf diffusion resistance of *Populus* clones to leaf water potential and environment. *Oecologia* **40**: 371–80.

Parker, W. C., Pallardy, S. G., Hinckley, T. M. & Teskey, R. O. (1982). Seasonal changes in tissue water relations of three woody species of the *Quercus-Carya* forest type. *Ecology* **63**: 1259–67.

Passioura, J. B. (1982). Water in the soil–plant–atmosphere continuum. In *Physiological Plant Ecology, II. Water Relations and Carbon Assimilation*, ed. O. L. Lange, P. S. Nobel, C. B. Osmond and H. Ziegler, pp. 5–33. Springer-Verlag, Berlin.

Patterson, D. T. (1973). The ecology of oriental bittersweet, *Celastrus orbiculatus*, a weedy introduced ornamental vine. PhD Dissertation, Duke University, Durham, NC.

Patterson, D. T. (1975). Photosynthetic acclimation to irradiance in *Celastrus orbiculatus* Thunb. *Photosynthetica* **9**: 140–4.

Putz, F. E. (1983). Liana biomass and leaf area of a 'tierra firme' forest in Rio Negro basin, Venezuela. *Biotropica* **15**: 185–9.

Putz, F. E. (1984). The natural history of lianas on Barro Colorado Island, Panama. *Ecology* **65**: 1713–24.

Putz, F. E. (1985). Woody vines and forest management in Malaysia. *Commonwealth Forestry Review* **64**: 359–65.
Putz, F. E. (1987). Liana phenology on Barro Colorado Island, Panama. *Biotropica* **19**: 334–41.
Putz, F. E. & Chai, P. (1987). Ecological studies of lianas in Lambir National Park, Sarawak, Malaysia. *Journal of Ecology* **75**: 523–31.
Richards, P. W. (1952). *The Tropical Rain Forest; an Ecological Study*. Cambridge University Press, London.
Roberts, S. W., Knoerr, K. R. & Strain, B. R. (1979). Comparative field water relations of four co-occurring forest tree species. *Canadian Journal of Botany* **57**: 1876–82.
Roberts, S. W., Strain, B. R. & Knoerr, K. R. (1980). Seasonal patterns of leaf water relations in four co-occurring forest tree species: Parameters from pressure-volume curves. *Oecologia* **46**: 330–7.
Ruffner, J. A. (1985). *Climates of the States, Vols I and II*, 3rd edition. Gale Res. Co. Mich.
Sasek, T. W. (1983). Growth and biomass allocation patterns of *Lonicera japonica* Thunb. and *Lonicera sempervirens* L. under carbon dioxide enrichment. Thesis, Duke University, NC.
Sasek, T. W. (1985). Implications of atmospheric carbon dioxide enrichment for the physiological ecology and distribution of two introduced woody vines, *Pueraria lobata* Ohwi (kudzu) and *Lonicera japonica* Thunb. (Japanese honeysuckle). Dissertation, Duke University, Durham NC.
Scholander, P. F. (1957). The rise of sap in lianas. In *The Physiology of Forest Trees*, ed. K. V. Thimann, pp. 3–17. Ronald Press, New York.
Schulze, E.-D. & Kuppers, M. (1979). Short-term and long-term effects of plant water deficits on stomatal response to humidity in *Corylus avellana* L. *Planta* **146**: 319–26.
Schulze, E.-D. & Hall, A. E. (1982). Stomatal responses, water loss and CO_2 assimilation rates of plants in contrasting environments. In *Physiological Plant Ecology II, Encyclopedia of Plant Physiology, n.s., Vol. 12B*, ed. O. L. Lange, P. S. Nobel, C. B. Osmond and H. Ziegler, pp. 181–230. Springer-Verlag, Berlin.
Shackel, K. A., Matthews, M. A. & Morrison, J. C. (1987). Dynamic relation between expansion and cellular turgor in growing grape (*Vitis vinifera* L.) leaves. *Plant Physiology* **84**: 1166–71.
Siccama, T. G., Weir, G. & Wallace, K. (1976). Ice damage in a mixed hardwood forest in Connecticut in relation to *Vitis* infestation. *Bulletin of the Torrey Botanical Club* **103**: 180–3.
Slezak, W. F. (1976). *Lonicera japonica* Thunb., an aggressive introduced species in a mature forest ecosystem. MS thesis, Rutgers University,

New Brunswick, NJ.
Smart, R. E. & Coombe, B. G. (1983). Water relations of grapevines. In *Water Deficits and Plant Growth*, ed. T. T. Kozlowski, pp. 137–96. Academic Press, New York.
Sperry, J. S., Holbrook, N. M., Zimmermann, M. H. & Tyree, M. T. (1987). Spring-filling of xylem vessels in wild grapevine. *Plant Physiology* **83**: 414–17.
Stevens, G. C. (1987). Lianas as structural parasites: The *Bursera simaruba* example. *Ecology* **68**: 77–81.
Tanner, R. D., Hussain, S. S., Hamilton, L. A. & Wolf, F. T. (1979). Kudzu (*Pueraria lobata*): Potential agricultural and industrial resource. *Economic Botany* **33**: 400–12.
Taylor, R. J. & Pearcy, R. W. (1976). Seasonal patterns of CO_2 exchange characteristics of understory plants from a deciduous forest. *Canadian Journal of Botany* **54**: 1094–03.
Thomas, L. K. Jr (1980). The impact of three exotic plant species on a Potomac island. *National Park Service Science Monograph No. 13*. US Department of the Interior, Washington, DC.
Tibbits, T. W. (1979). Humidity and plants. *BioScience* **29**: 358–63.
Trémolières, M., Carbiener, R., Exinger, A. & Turlot, J. C. (1988). Un exemple d'interaction non compétitive entre espèces ligneuses: le cas du lierre arborescent (*Hedera helix* L.) dans la forêt alluviale. *Acta Oecologia/Oecologie Plantarium* **9**: 187–209.
Tsugawa, H. & Kayama, R. (1980). Studies on population structure of kudzu vines (*Pueraria lobata* Ohwi). V. Stem-length proportion of the overwintering stems classified by the number of vascular bundle rings and the proportion of the number of rooted nodes to non-rooted nodes in these stems. *Journal of the Japanese Society of Grassland Science* **27**: 285–9.
Tsugawa, H. & Kayama, R. (1981). Studies on dry matter production and leaf area expansion of kudzu vines (*Pueraria lobata* Ohwi). I. The dry weight and leaf area of the current year's stem produced from the node of the overwintering stem. *Journal of the Japanese Society of Grassland Science* **27**: 267–71.
Tsugawa, H., Tange, M. & Masui, K. (1979). Top and root growth of seedlings of kudzu vines (*Pueraria lobata* Ohwi). *Science Reports of the Faculty of Agriculture, Kobe University* **13**: 203–8.
Tsugawa, H., Tange, M. & Otsuji, J. (1980). Observations on branching and number of leaves in seedlings of kudzu vines (*Pueraria lobata* Ohwi). *Science Reports of the Faculty of Agriculture, Kobe University* **14**: 9–14.
Turner, N. C. (1986). Adaptation to water deficits: A changing

perspective. *Australian Journal of Plant Physiology* **13**: 175–90.
Tyree, M. T., & Sperry, J. S. (1989). Vulnerability of xylem to cavitation and embolism. *Annual Review of Plant Physiology and Molecular Biology* **40**: 19–38.
Vitousek, P. M., Fahey, T., Johnson, D. W. & Swift, M. J. (1988). Element interactions in forest ecosystems: Succession, allometry and input-output budgets. *Biogeochemistry* **5**: 7–34.
Wechsler, N. R. (1977). Growth and physiological characteristics of kudzu, *Pueraria lobata* (Willd.) Ohwi, in relation to its competitive success. Thesis, University of Georgia, Athens, Ga.
Whereat, E. (1990). Seasonal patterns of rooting depth and osmotic adjustment in a sympatric assemblage of *Lonicera japonica* Thunb. and *Smilax rotundifolia* L. Thesis, University of Maryland, College Park, MD.
Whigham, D. (1984). The influence of vines on the growth of *Liquidambar styraciflua* L. (sweetgum). *Canadian Journal of Forest Research* **14**: 37–9.
Wong, S. C., Cowan, I. R. & Farquhar, G. D. (1979). Stomatal conductance correlates with photosynthetic capacity. *Nature* (London) 282: 424–6.
Zimmermann, M. T. (1983). *Xylem Structure and the Ascent of Sap*. Springer-Verlag, New York.
Zimmermann, M. H. & Jeje, A. A. (1981). Vessel-length distribution in stems of some woody American plants. *Canadian Journal of Botany* **59**: 1882–92.

10

Secondary compounds in vines with an emphasis on those with defensive functions

MERVYN P. HEGARTY, ELWYN E. HEGARTY
AND ALWYN H. GENTRY

Introduction

Although the distinction between primary and secondary compounds in plants is imprecise and may on occasions appear arbitrary, the two terms effectively group compounds by both structure and role in plant biochemistry. Primary compounds include sugars, fats, protein amino acids, purines and pyrimidines, which are involved in essential cellular biochemical processes. However, plants also contain many thousand so-called secondary compounds which do not appear to be involved in these processes. The detailed structure and function of many of these is still uncertain, but it is generally accepted that their main roles are the selective deterrence or attraction of various classes of herbivore or pollinator (Bate-Smith, 1972; Rhoades, 1983). Such roles are often suggested, in the absence of other evidence, by a distinctive taste or smell. Other compounds such as the phytoalexins are involved in defenses induced by trauma, still others are precursors for biochemical processes within plants. The development of current theories of secondary metabolism in plants is lucidly reviewed by Haslam (1986). Even when a particular compound is known to be present, there may be large quantitive variations, depending on the part or individual sampled, location, age, season, climate, recent herbivory, and even time of day (McKey, 1974; Swain, 1977; Kingsbury, 1980).

After a brief introduction to the role of secondary compounds in plant defence, we will give examples of the more abundant and well-documented classes of secondary compounds found in vines, particularly those which are known to deter herbivores, comparing where possible their distribution in vines and other life forms. As some families consisting largely or totally of vines are known to be rich in secondary compounds of known toxicity to humans or other animals, we will then discuss whether there is sufficient evidence to suggest that vines, as a life form, should be and/or are especially well protected by secondary compounds.

Classification and defensive roles of secondary compounds

Over evolutionary time, diversification of plants, animals and fungi has been accompanied by the proliferation of protective secondary compounds by plants. So in primeval communities simple phenolic compounds gave protection against fungi, bacteria and viruses, and acted as allelochemicals against other plant organisms. These compounds were joined by various tannins (which extended the range of targets to insects and vertebrates) and an increasing range of other classes of compound including terpenoids, non-protein amino acids, alkaloids, cyanogenic glucosides and glucosinolates (Swain, 1978, Table II), each of which tends to be more effective against some groups of potential herbivores than others. In the invertebrate world, insects confined to certain families eat only plants belonging to given families, each containing a distinctive group of secondary plant substances (Raven, Evert & Eichhorn 1986). Species of a genus tend to share the same expressions of toxicity (Kingsbury, 1980) even though not all congeneric species have the same endowment. Classes of secondary compounds are rarely if ever peculiar to vines.

Protective secondary compounds may also be classified, on the basis of their type of activity, into two groups, (i) **qualitative** (i.e. toxic) compounds such as alkaloids, glucosinolates, cardenolides, cyanogenic substances and non-protein amino acids, and (ii) **quantitative** (i.e. digestibility-reducing) compounds such as polyphenols (syn. tannins), resins, lignin and refractory carbohydrates (Feeny, 1975, 1976; Rhoades & Cates, 1976; Rhoades, 1983). Members of the former group are usually, but not always, present in small quantities, especially in herbaceous species and rapidly flushed leaves, and are more readily synthesized, translocated, and metabolized. Members of the latter, which are less mobile, are thought to be more costly, in carbon and energy, to produce and are characteristically present in large quantities in persistent tissues of woody perennial plants; such plants are very apparent (predictable) resources for herbivores (Feeny, 1976; Rhoades & Cates, 1976; Mooney, Gulmon & Johnson, 1983; Rhoades, 1983). When resources are limited, plants with inherently slow growth, with large investments in herbivore defenses, are favored; then quantitative (e.g. tannins) rather than qualitative (e.g. alkaloids) defenses are employed, and consequently the flexibility of expression in defensive chemicals is reduced (Coley, Bryant & Chapin, 1985).

Although the often bitter-tasting alkaloids are the most prominent compounds in lists of insect feeding deterrents, and seem to inhibit or poison representatives of all major groups of herbivorous insects (Levin, 1976a), sweeping generalizations about the protective function of alkaloids against herbivores are groundless (Robinson, 1979). In Gabon, Hladik & Hladik (1977) noted that in the diet of chimpanzees the percentage of the species which contained alkaloids (15%) was about the same as in the whole flora as sampled.

The widely distributed and often abundant tannins are also characteristically bitter-tasting, and/or astringent. As uniform palatability is an important element in the quantity of a forage consumed by a herbivore, the long-lasting tannins may be at least equal in importance as feeding deterrents to the somewhat more ephemeral alkaloids.

Characteristics and occurrence of secondary compounds, with special reference to families containing vines

The secondary compounds which are of most interest in this context fall into the following main classes:

Alkaloids

Alkaloids are a large group of heterocyclic nitrogen compounds which are derived from α-amino acids. Found in plants and more rarely in animals, some alkaloids have a considerable reputation as toxins with pharmaceutical applications. Alkaloids are widely present in plants, especially in epidermal and actively growing tissues, ovules, seeds and immature fruits (McKey, 1974). They are important deterrents to grazing by insects but sometimes act as attractants. Primitive angiosperms have generally higher incidences of alkaloids than most advanced groups, and insect-pollinated species are richer in alkaloids than wind-pollinated species (Li & Willaman, 1968). Alkaloid concentrations also vary with such factors as time, situation, climate and soil, and the sites of synthesis are not necessarily the sites of accumulation (Robinson, 1979). There is a vast literature on alkaloids, with a series of reviews variously edited by Manske, Holmes, Rodrigo and Brossi (1950–) the major reference source.

Types of alkaloid. The largest group, of about 1200 (in 1980: Gröger 1980), are derived from tryptophan and make up about a quarter of those isolated. They include the **indole alkaloids**, many of which are pharmacologically active. In leguminous vines, they include abrine, isolated from seeds of *Abrus precatorius*, and physostigmine, isolated from Calabar bean

(*Physostigma venenosum*). Other indole alkaloids are used in curare, which in many areas of the Neotropics is prepared chiefly from alkaloids found in liana species of *Strychnos* (Loganiaceae). Alkaloids from the Menispermaceae (which are almost exclusively vines) have been used in other parts of the Neotropics as the main ingredient of curare, or to extend the life and toxicity of New World arrow poisons containing curare (Cheney, 1931). Another indole alkaloid with a different pharmaceutical application – strychine – comes mostly from Old World species of *Strychnos* (Gröger, 1980), including *S. nux-vomica* which is a tree. A ***B*-carboline-derived indole alkaloid**, passiflorine, (Figure 10.1*a*) is very characteristic of the vine family Passifloraceae (Hegnauer, 1962–86; Gibbs, 1974), and these alkaloids have been largely responsible for the famous co-evolutionary radiation of the Muellerian mimic host-specific *Heliconius* butterflies that feed on them (e.g. Brown, Sheppard & Turner, 1974; Gilbert, 1975; Gilbert & Singer, 1975), although the exact relationship between the passionflower chemicals and the distasteful nature of the butterflies that feed on them remains unclear (DeVries, 1987).

Other types of alkaloid include the **pyrrolidine** alkaloids, (e.g. convolvine, dioscorine) which are respectively common in the almost exclusively scandent families Convolvulaceae and Dioscoreaceae. **Pyridine and piperidine** alkaloids are found in the climbing Piperaceae, e.g. *Piper nigrum* from which black and white pepper are prepared (Ayensu 1980). **Pyrrolizidine and quinolizidine** alkaloids are widely distributed in the plant world, being abundant in the Leguminosae, in some genera of Asteraceae (e.g. in the scrambling vine *Senecio mikanioides* and in the largely scandent *Tournefortia* of the Boraginaceae (Gibbs, 1974). Some highly complicated plant–herbivore interactions have been reported for pyrrolizidine alkaloids: for example Edgar & Culvenor (1975) and Edgar (1982) reported on sequestering of such alkaloids (Figure 10.1*b*) from *Parsonsia* vines (Apocynaceae) by larvae and male adults of Danainae butterflies, for use as precursors of pheromones. These alkaloids are also implicated in the distasteful properties of ithomiine butterflies, perhaps the most important models for neotropical mimicry complexes (Boppré, 1984; DeVries, 1987).

Recently a series of polyhydroxy derivatives of pyrrolidine, piperidine, pyrrolizidine and indolizidine alkaloids has been isolated from various leguminous plants. These alkaloids are analogs of sugars, and are potent inhibitors of sugar-metabolizing enzymes (Fellows *et al.*, 1986), and probably contribute to the chemical defenses of the plants in which they occur. For example, 2,5-dihydroxymethyl-3,4-dihydroxypyrrolidine (Figure 10.1*c*), an analog of fructose, occurring in the vine *Derris elliptica*, was as detrimental to the larvae of the bruchid beetle as any alkaloid tested (Evans, Gatehouse & Fellows, 1985).

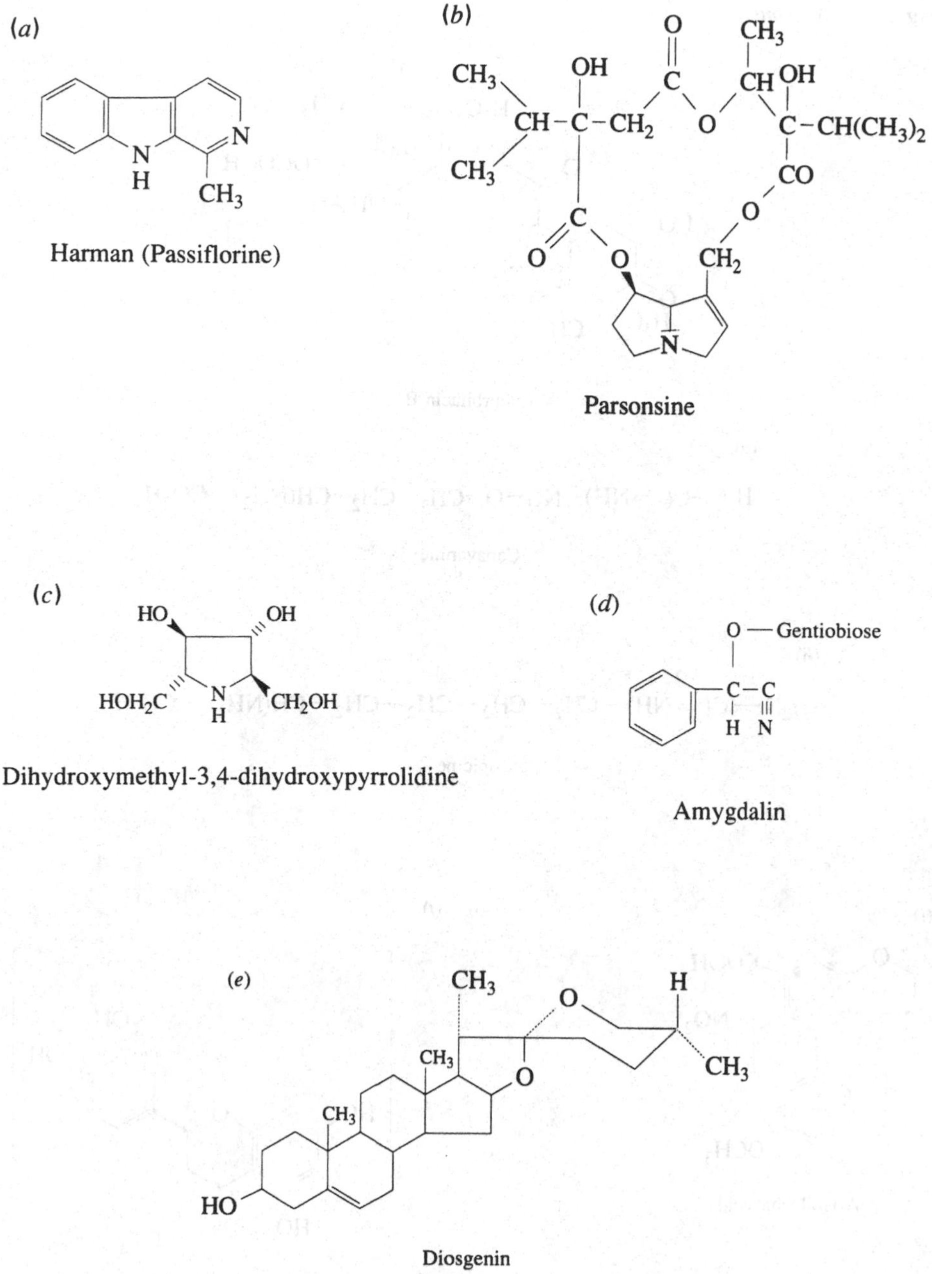

(a)

Harman (Passiflorine)

(b)

Parsonsine

(c)

Dihydroxymethyl-3,4-dihydroxypyrrolidine

(d)

Amygdalin

(e)

Diosgenin

Figure 10.1. Representative examples of the main types of secondary compounds found in vines.

Figure 10.1. (*cont.*)

(*f*)

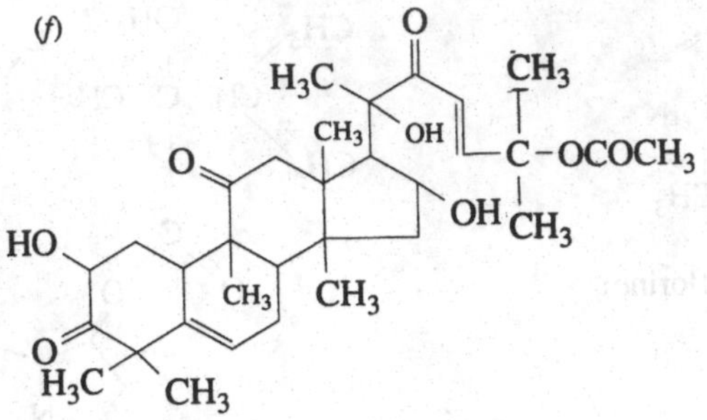

O
H_3C CH_3
CH_3 OH
O C-$OCOCH_3$
OH CH_3
HO
CH_3 CH_3
O
H_3C CH_3

Cucurbitacin B

(*g*)

$H_2N-C(=NH)-NH-O-CH_2-CH_2-CH(NH_2)-CO_2H$

Canavanine

(*h*)

$H_2N-C(=NH)-CH_2-CH_2-CH_2-CH_2-CH(NH_2)-CO_2H$

Indospicine

(*i*)

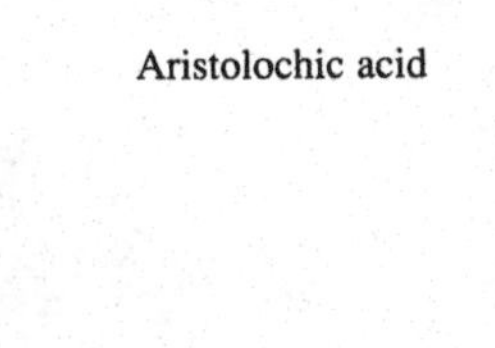

O COOH
O NO_2
OCH_3

Aristolochic acid

(*j*)

OH
OH
HO O
HO
n

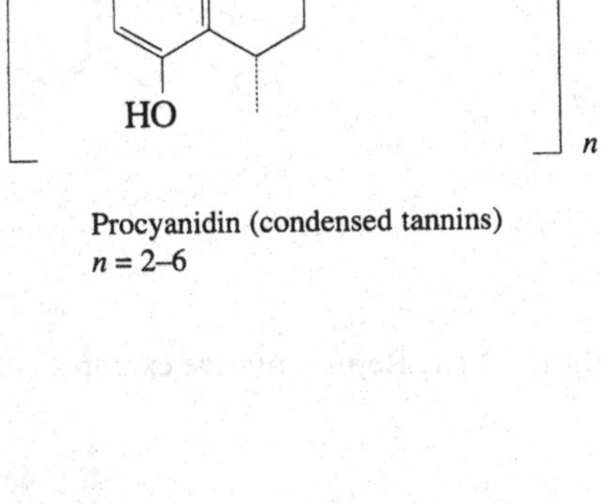

Procyanidin (condensed tannins)
$n = 2–6$

(*k*)

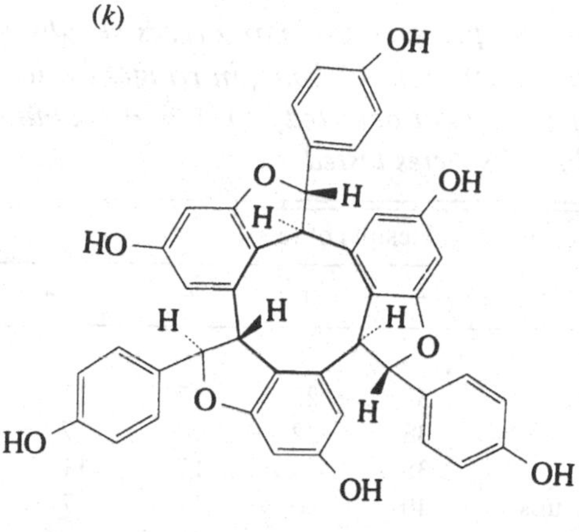

α-Viniferin

The often bitter taste of alkaloids reduces the palatability of foliage. In the case of humans, degradation of potentially toxic alkaloids in yams of *Dioscorea* vines by cooking allows them to be used as a valuable and palatable source of carbohydrate. However, most alkaloid-bearing plants also contain other toxins or feeding deterrents (Levin & York, 1978) which may be heat-stable.

Screening of many species to determine the occurrence and type of alkaloids present in each species has indicated that they are usually found in about 10–20% of species tested in large-scale screenings. They occur in about a third of angiosperm families according to Robinson (1979), who concludes, from a review of studies including those of Levin (1976b) and Levin & York (1978), that the greatest proportions of alkaloid-containing plants are found in the tropics, in rainforest at low altitudes, where pest pressure is strongly developed. Annual plants have a higher incidence of alkaloids than perennials (Levin, 1976b). Hladik & Hladik (1977) reported that alkaloids were more often found in species of disturbed or periodically inundated areas (where populations were dense) than in mature forest, where distributions were patchy, and also remarked upon the very uneven distribution of alkaloids in plant families. There is limited and sometimes inconclusive evidence that shrubs have greater alkaloid presence and toxicity than trees (e.g. Hartley *et al.*, 1973, Levin, 1976b; Hladik & Hladik, 1977; Levin & York, 1978). Results for vines are sometimes included with those for shrubs, complicating comparisons. Records for many species, including some for vines, are listed by Willaman & Schubert (1961), Smolenski, Silinis & Farnsworth (1972, 1973) and Fong *et al* (1972), and systematic summaries of

Table 10.1. *Percentage of species of vines and trees/shrubs containing alkaloids in leaves, from a study in Gabon*[a], *in ranges indicated by precipitated material, on a scale of (−) (not detected) to + + + (abundant or very abundant).* N = *number of species tested*

	Results of tests					
	−	±	+	+ +	+ + +	*N*
Species of liane						
Pluvial or disturbed areas	37	22	17	17	5	40
Mature rainforest	38	32	17	7	7	60
Total for lianes	38	28	17	11	6	100
Total for spp. of trees/shrubs	40	30	17	7	6	256

[a]Adapted from Hladik & Hladik, 1977: p. 545.

these and other data are presented by Gibbs (1974) and Hegnauer (1962–86). More biochemical detail is given by Miller (1973) and in chapters by Leete, Fodor and Gröger in Bell & Charlwood (1980).

Comparison of alkaloid distribution in vines and other life forms. Results from two regional surveys in which life forms were clearly identified, have been examined.

In the first study, the record of alkaloid concentration in leaves of vines (from two separate tests per species in Gabon by Hladik & Hladik, 1977) was compared with results from trees and shrubs. Data from that study have been summarized above (Table 10.1).

These results do not indicate a statistically significant difference between the overall distributions of alkaloids in leaves of vines and supporting life forms.

To find if the vining habit was related to alkaloid presence rather than to the highly variable concentrations of alkaloids per species detected during the above and other studies, we have examined detailed results from New Guinea (Hartley *et al.*, 1973). The presence/absence of alkaloids in leaves and bark was compared for vines and 'other' life forms, which included trees, using only families where one or both sets of life forms contained alkaloids. (Comparisons were not attempted for families where either vines or 'other' forms, but not both, were tested, or where alkaloids were not found in either). In 35 families where both vines and 'other' life-forms contained alkaloids, their presence in species per family was 17% (SE = 32%) for vine species and 16% (SE = 25%) of other life forms, indicating that the presence of alkaloids in vines is not different from that in non-vine species of the same family. Qualifying the original data on which the above calculations have been based, Hartley *et al.* (1973) note frequent variation between the results of field and laboratory tests because of rapid degradation of some alkaloids in harvested samples.

Cyanogens

The presence of compounds capable of releasing hydrogen cyanide (HCN) on chemical or enzymic hydrolysis has been recognized in more than 2000 species of higher plants, including 110 families of ferns, gymnosperms and angiosperms, but the exact source of the HCN has been identified in only about 300 species (Poulton, 1988). Cyanoglycosides are the most common cyanogenic compounds found in plants and about 75 have been characterized, all of which are derived from α-amino acids. While cyanide is now considered to be involved in primary plant metabolism, the cyanogenic glycosides are important defense chemicals against herbivores (Robinson, 1930; Jones, 1972; Conn, 1980). Their poisonous properties in mammals are caused by the HCN released by reaction with plant β-glucosidase usually when the tissues are crushed. While HCN is also considered to be the defensive chemical in sorghum which is active against locusts (Woodhead & Bernays, 1977) there is increasing evidence that the prime deterrents are the volatile carbonyl compounds, e.g. benzaldehyde, released in equimolar amounts with the HCN on cyanogenesis (Figure 10.1*d*) (Jones, 1988). Cyanogenic glycosides are abundant in vines of the Passifloraceae, and are also found in the Leguminosae and Araceae which contain many climbers. Although cyanogens are not abundant in the Sapindaceae, *Cardiospermum* spp., which are vines, contain the cyanogen cardiospermin. Similarly, in the Bignoniaceae, which contain both trees and climbers, cyanogenic compounds (adjudged from the almond vegetative odor) are found only in climbing genera like *Tanaecium* (A. H. Gentry, unpublished data).

Cardiac glycosides (Cardenolides)

About 10 plant families contain these steroidal compounds, traditionally used in Old World arrow poisons (Singh & Rastogi, 1970). Although the mostly herbaceous Scrophulariaceae contain most sources including *Digitalis*, vine-rich families also contain these compounds. Especially noteworthy for their cardiac glycosides are Apocynaceae, the pre-eminent African liana family (Gentry, Chapter 1) and closely related Asclepiadaceae, which are nearly all scandent. An example is the African vine, *Strophanthus kombe* (Apocynaceae), which contains strophanthin, formerly used in arrow poisons but now used as an alternative to digoxin from *Digitalis* in medicine (Ayensu, 1980). A specific group of insect predators, the Danainae, specialize in Apocynaceae and Asclepiadaceae, once again being involved in a complex mimetic system (Brower, Brower & Corvino, 1967; Brower, 1984).

Saponins

Saponins are a class of glycosides that contain a steroidal or triterpenoid sapogenin (the aglycone) linked to one or more sugar residues. They are widely distributed in plants and have been identified in 500 species belonging to more than 80 families (Basu & Rastogi, 1967). They have a bitter taste and readily form foams. Saponins deter or inhibit development of most insect species. Dietary saponins are more toxic to exothermic species (e.g. fish and insects) than to endothermic animals (e.g. mammals and birds) (Applebaum & Birk, 1979). Families containing many vines which are rich in saponins include Apocynaceae, Leguminosae and Dioscoreaceae. Diosgenin (Figure 10.1*e*), extracted from tubers of several American species of vine (*Dioscorea*), has been widely used to manufacture progesterone, birth control pills and other pharmaceuticals.

Cucurbitacins

Cucurbitacins are most abundant in the almost entirely scandent family Cucurbitaceae, but are also found in a few other families in which vines are rare or absent. These compounds, of which 17 have been identified, are tetracyclic triterpenes derived from lanosterol (Figure 10.1*f*) and are responsible for the bitterness and mild toxicity of varieties of zucchini (*Cucurbita pepo*) and melons. Fruits and roots of mature plants appear to contain high concentrations of cucurbitacins, mainly cucurbitacin E. Some cucurbitacins act as attractants to insects – in 15 species of Cucurbitaceae the numbers of leaf-eating beetles attracted were in proportion to the content of cucurbitacins and their glycosides (Sharma & Hall 1973). Similarly, leaf-eating beetles were attracted to *Cucumis sativus* despite its antibiotic effect on spotted mites (da Costa & Jones, 1971). The cucurbitacins have been tentatively identified as the substances responsible for inhibition of phenol oxidase formation in *Botrytis cinerea*, a fungal pathogen (Bar-Nun & Mayer, 1989).

Non-protein amino acids

More than 200 non-protein amino acids have been identified in higher plants. They are so called because they are not usually found in proteins, but may be conveniently considered in the same categories under which protein amino acids are described, e.g. acidic, basic, neutral, aromatic, heterocyclic and sulfur-containing amino acids. The Leguminosae contain the largest number of these compounds, especially in the numerous climbing Faboideae, e.g. canavanine in seeds of *Dioclea megacarpa* and L-3,4-dihydroxyphenylalanine in all species of *Mucuna* tested by Bell & Janzen (1971). Both these compounds are toxic to some seed predators, though canavanine is used as a source of nitrogen by some beetle larvae. Other families containing non-protein amino acids include the Cucurbitaceae (Rosenthal, 1982). Many of these non-protein amino acids have been shown experimentally to inhibit feeding in the grasshoppers *Locusta, Chortoicetes* and *Schistocerca* (Navon & Bernays, 1978).

Although such compounds may be present in high concentrations (0.2–5%) in various tissues, their distribution among species in a genus often varies widely. About 20 non-protein amino acids have been shown to be toxic to humans and livestock. Examples in vines include two analogs of arginine which occur in some creeping legumes, (i) canavanine (Figure 10.1*g*), which occurs in many browsed species of the subfamily Papilionoideae, including *Canavalia ensiformis* (Bell, 1980), and (ii) indospicine (Figure 10.1*h*), which is restricted to some species of *Indigofera* including creeping indigo, *I. spicata*, and is variably toxic to larger herbivores (Hegarty & Pound, 1970). The legume genera *Lathyrus* and *Vicia* contain a large number of herbaceous vine species which, especially in the flowers, fruits and seeds, accumulate a range of unusual amino acids with distinctive biological properties. The consumption by humans of seeds of *Lathyrus* and *Vicia* containing β-cyanoalanine, α,γ-diaminobutyric acid and β-*N*-oxalyl-α,β-diamino propionic acid can give rise to a crippling neurological disease called neurolathyrism. Rosenthal (1982) has summarized much of the information on the chemistry and biological properties of non-protein amino acids.

Aristolochic acids

These are the principal aromatic nitro compounds detected in higher plants. Many species of the genus *Aristolochia*, composed almost entirely of over 300 species of tropical and temperate lianas, contain aristolochic acids, which are nitrophenanthrenic acids (Figure 10.1*i*), together with the closely related aristolactams which occur only in the family Aristolochiaceae. In addition, some species contain the isoquinoline alkaloids (Chen & Zhu, 1986) characteristic of the Magnoliid alliance. The secondary compounds listed above show a wide range of biological properties, and some species of *Aristolochia* have been extensively used in folk medicine (Lewis & Elvin-Lewis, 1977).

It has been suggested that ancestral butterfly larvae feeding on *Aristolochia* species developed the ability to store toxic metabolites and obtain immediate protection from herbivores (Chen & Zhu, 1986). A diversified mimicry complex has arisen, in Papilionoidae (*Battus*, and *Parides*: Rothschild, 1972; DeVries, 1987). When eggs of Australian butterflies which normally oviposit on native species of *Aristolochia* were laid on or placed on the recently naturalized *A. elegans* the larvae succumbed, presumably to toxic metabolites of an alien species (Straatman, 1962).

Polyphenols and lignins

Plants contain several thousands of phenolic compounds of widely diverse structures. These vary from the simple phenols and their derivatives through heterocyclic compounds such as flavonoids and anthocyanins and tannins, the latter so called because of the ability of some of them to react with the proteins in raw animal skins and convert the skins to leather.

Tannins from plants have molecular weights in the range 300–3000 daltons and can be grouped on the basis of properties and structure into proanthocyanidins (condensed tannins) hydrolyzable tannins, oxytannins, and a miscellaneous group (Haslam, 1975). The proanthocyanidins (Figure 10.1*j*) are the most widely distributed in plants, being present in 62% of the dicotyledons and 29% of the monocotyledons and are considered to be of considerable value to the plant as a defense against herbivores. They drastically reduce the availability of nutrients (proteins and polysaccharides) to herbivores; the activity of digestive enzymes in the gut is depressed by tannins (Swain, 1979), which in various forms are present in 80% of woody dicotyledons and 15% of herbaceous dicotyledon species (Rhoades, 1983). The presence of tannins in leaves is of importance in determining the susceptibility of the tissues to attack by pathogens, including viruses (Levin, 1976a; Harborne, 1977). Although many tannins are very bitter-tasting those in grapes (*Vitis vinifera*, Vitaceae) contribute significantly to the flavor sought in wines, and in Greek dishes such as dolmades.

Lignins are phenolic heteropolymers with molecular weights greater than 5000 daltons. They are involved in the structural (fibrous) support of higher plants, especially woody species, and assist plant cell walls to resist bacteria and fungi. When ingested by herbivores they reduce the availability of carbohydrates and protein, in much the same way as tannins. Some breakdown products have been shown to inhibit feeding. These and other aspects of lignin metabolism are reviewed by Swain (1979) and Waterman (1983).

Phytoalexins

These are a chemically heterogeneous group of antimicrobial compounds of low molecular weight, that are synthesized and accumulated in plants after exposure to microorganisms, nematodes and insects. A full discussion of phytoalexins is beyond the scope of this chapter. They include various types of pterocarpans and isoflavonoids, occur in a number of families, particularly Leguminosae, and in general lead to the containment of damage and in some cases to the induction of other protective compounds (Swain, 1977). Phytoalexins have been identified in vines following fungal infections, e.g. (i) dimethylbatatasin IV and a dihydrostilbene from tubers and other parts of *Dioscorea* spp. (Dioscoreaceae) (most recently by Adesanya, Ogundana & Roberts, 1989), and (ii) viniferins, e.g. α-viniferin (Figure 10.1*k*), a trimer of resveratrol (a hydroxystilbene), in the grape *Vitis vinifera* (Pryce & Langcake, 1977). Viniferins are more effective antifungal agents than the parent stilbene (Langcake & Pryce, 1977).

Damage by herbivores also induces reinforcement of defenses by translocation of existing secondary compounds. In turn, some insects have developed means of circumventing these defenses, for example *Epilachne* beetles feeding on cucurbits isolate areas of leaves prior to feeding by trenching, with the result that cucurbitacin B builds up much more outside than inside the feeding area (Tallamy, 1985).

Distribution of secondary compounds in large plant families in relation to proportions of vine species

Although particular toxic compounds tend to be shared by a majority or minority of members of a genus rather than evenly distributed within families, we can also compare the incidence of such compounds at the family level. Particular concentrations of some of the commoner and more easily grouped forms of secondary compounds have been assessed in some large plant families (Table 10.2). For comparative purposes, 30 families were selected, 10 of which are composed mostly or totally of vines, 10 contain a substantial proportion of vines, and 10 in which vines are rare or absent (see Gentry, Chapter 1 for further details of distribution of vines in families).

Alkaloids are characteristic of families of the vine-rich group (Group I of Table 10.2) except for the Vitaceae, in which records of alkaloids are very limited. The Vitaceae appear to be protected most substantially by polyphenols and the induction of phytoalexins following herbivore attack. For *Vitis*, the most widely studied genus of the family, Duke (1977) also lists as phytotoxins a number of organic acids, some of which take part in primary

Table 10.2. *Presence of some classes of secondary compounds in plant families. A comparison of some families which are entirely or mostly composed of vines with those in which they are less abundant, rare or absent*[a]

	Alkaloids	Cyanogens	Cardiac glycosides	Other compounds of note
10 families composed mostly or entirely of vines				
Aristolochiaceae	++	?	?	Aristolochic acid, diterpenes
Asclepiadaceae	++	±	++	Triterpenoid saponins
Convolvulaceae	++	++	?	Polyphenols, tetraterpenoids
Cucurbitaceae	++	?	?	NPAA[b], triterpenoids, cucurbitacins, steroids, saponins, tetraterpenoids
Dioscoreaceae	++	±	?	Steroidal saponins, polyphenols, tetraterpenoids, phytoalexins
Hippocrataceae	+	?	?	
Menispermaceae	++	?	?	
Passifloraceae	++	++	?	
Smilacaceae	+	?	?	Steroidal saponins
Vitaceae	?	±	?	Proanthocyanins, phytoalexins
10 families with many vine species				
Apocynaceae	++	+	++	Steroids, triterpenoid saponins, pentaterpenoids
Araceae	+	++	?	Saponins, pentaterpenoids
Arecaceae	+	±	?	Saponins, tetraterpenoids, polyphenols, flavonoids
Bignoniaceae	+	?	?	Tetraterpenes
Leguminoseae	++	++	?	NPAA, steroidal saponins, tetra- and pentaterpenoids
Malpighiaceae	+	+	?	Saponins
Oleaceae	+	±	?	Triterpenoid saponins, tetraterpenoids
Piperaceae	++	?	?	Mono/sesquiterpenoids
Rubiaceae	++	±	?	Triterpenoid saponins
Sapindaceae	++	+	?	Triterpenoid saponins
10 families with few or no vines				
Brassicaceae	++	?	+	Tetraterpenoids
Cyperaceae	±	+	?	
Dipterocarpaceae	±/?	?	?	Triterpenoid saponins
Fagaceae	±	?	?	NPAA, polyphenols
Lauraceae	++	?	?	Terpenoids
Meliaceae	++	+	?	Saponins, terpenoids
Myrtaceae	±	+	?	Saponins, terpenoids, polyphenols
Primulaceae	±	−	?	Triterpenoid saponins
Proteaceae	±	++	?	Terpenoids
Rutaceae	++	++	?	Polyphenols

[a]++, often present; +, sometimes present; ±, seldom present; ? rare/unreported: few families are fully surveyed and results sometimes conflict.

[b]NPAA, Non-protein amino acids.

Chief sources: Conn (1980); Cronquist (1981); Duke (1977); Everist (1981); Fong *et al.* (1972); Gibbs (1974); Hartley *et al.* (1973); Hegnauer (1962–6); Rosenthal (1982); Singh & Rastogi (1970); Smolenski *et al.* (1972–3); Webb (1949, 1952); Willaman & Schubert (1961).

biochemical processes and so are not considered here. Alkaloids are also quite abundant in Group II, and perhaps somewhat less so in Group III. The latter families, which consist largely of trees, are rich in expensive but long-term protection such as that provided by tannins. As the 10 families of Group III were selected from the approximately half of the families in world flora which contain vines rarely if at all, we cannot presume that they are completely representative of non-scandent families.

Table 10.2 further demonstrates the patchy concentration of various other classes of secondary compounds among plant families. Of the extensively vining families, the Cucurbitaceae and Dioscoreaceae in particular are rich in steroidal saponins, the Convolvulaceae and Passifloraceae have cyanogens, and the Asclepiadaceae cardiac glycosides. These and other families listed are very rich in secondary compounds which are distasteful or toxic to humans. However, the relevance of such compounds to the diet of modern humans is almost insignificant when we consider for how much longer such organisms as fungi, small mobile herbivores such as caterpillars, beetles, slugs and nematodes, and browsing vertebrates, have evolved with and depended on food supplied by the immobile plants.

Discussion

What is unusual about vines that we might expect them to be especially richly or differently endowed from, say, trees, with protective secondary compounds? Vines are a derived life form, having evolved independently in many families (Gentry, Chapter 1), and it is probable that they inherited already established family patterns of secondary metabolism but have selectively developed further useful variations. We might consider that:

1. Defensive chemicals such as alkaloids are concentrated near the surfaces of plants (McKey, 1974, 1979, p. 113). The surface:volume ratio is much higher in vines than in trees. This is because (a) vines have a much greater total leaf biomass and leaf area, in relation to total biomass, than trees (Putz, 1983; Hegarty, 1988; Hegarty & Caballé, Chapter 11), and (b) except in parts of the largest lianes, no part of any stem is more than a couple of centimeters from the surface. So on a quantitative basis the strongly fortified zone should occupy more of the volume of a vine than a tree.
2. Often in vines an extensive canopy and a distant root system are connected by one or two very rambling and twisted stems, containing more vascular and less woody tissue than in trees (Bamber, 1984; Hegarty, 1988, 1989). These connections are occasionally replaced at more favorable locations by cloning, so the need for long-term protection of woody tissues is minimized.

3. Many vines extend and exchange leaves very rapidly (Peñalosa, 1984; Hegarty, 1990), sometimes more continuously than associated trees (Putz & Windsor, 1987) so there is a higher proportion of young stem and leaf to be protected in vines. This involves readily mobilized secondary compounds, rather than lignins and tannins (Gentry & Cook, 1984).
4. There is also the high apparency (Feeny, 1976) of vines, most of which are shade-intolerant and so readily available to passing herbivores, especially during early succession. On the other hand, the diffuse branching of vines makes them less apparent than a tree with an equivalent total surface area of leaf, clustered into a single crown.
5. Many deciduous vines in seasonally dry regions rely on tubers to store much of their resources between growing seasons, unlike deciduous perennials, such as trees, which shed only leaves and twigs. Tubers of such vines need year-round protection but their stems and leaves do not.
6. There is a latitudinal decrease in alkaloid toxicity and presence in plant species (Levin, 1976b, Levin & York 1978). As vines are predominantly tropical, and both vines and alkaloids are especially abundant in the same lowland to medium elevation rainforests (Hartley *et al.*, 1973; Levin, 1976b; Hegarty & Caballé, Chapter 11), we might expect this pattern to be reflected in alkaloid protection in vines at least.
7. Some alkaloid-rich taxa may be 'pre-adapted' to become vines (cf. Gentry, Chapter 1); it is also suggested that some lianas are especially adept at evolving compounds with novel chemical structures (Gentry, 1985).

A combination of all these factors would lead us to expect that vines in general may be protected extremely well from herbivores by qualitative rather than quantitative strategies of anti-herbivory protection. It seems, from Table 10.2 and information presented elsewhere in the text for individual classes of compound, that in general qualitative defenses (alkaloids, cardioglycosides, aristolochic acid, cucurbitacins and phytoalexins), rather than quantitative defences (e.g. polyphenols, i.e. tannins) are important above-ground defenses in most families in which vines predominate.

How is this hypothesis supported by facts? Certainly, in data examined by Levin (1976b), some families in which most or all species climb contained some of the highest percentages of species in which alkaloids were present: these included Menispermaceae (89%) and Aristolochiaceae (74%). On the other hand, values for others such as Bignoniaceae (37%) and Cucurbitaceae (48%) were about the same as in species of tropical and temperate families generally (36–38% in the same study) yet the grape family (Vitaceae) at 5% was very poorly protected by alkaloids. Alkaloids, however, are but one of the chemical defenses available to vines, and in families where alkaloids are not well developed, or absent, other readily mobilized defenses, or alternatively polyphenols, may well be strongly developed.

We have barely mentioned protection of below-ground parts of vines, because in most of the larger surveys only leaves and/or bark have been sampled, from species possibly selected with a bias based on expectations (Hladik & Hladik, 1977). However, there is frequent evidence from folk medicine and individual records that some secondary compounds are concentrated in roots and tubers (Watt & Brandwijk, 1962; Verdcourt & Trump, 1969, Everist, 1981; McKey, 1979; Lopes & Bolzani, 1988).

Although the importance of secondary metabolites is becoming increasingly evident, documentation of unusual concentrations of secondary defensive compounds in vines is largely episodic and certainly very incomplete. This is because (i) many vines are hard to collect, difficult to identify, and of little known economic potential, (ii) results for vines are often assessed with those for other non-tree life forms such as shrubs which do not necessarily occupy similar niches, (iii) concentrations within families are very uneven (although this is characteristic of plants in general).

It is tempting to point to the commercially important, powerful and interesting secondary compounds so far obtained from vines (e.g. curare, strychnine, rotenone), to the circumstantial arguments and limited evidence above that vines are particularly well and diversely defended – and are disproportionately represented in butterfly/plant co-evolutionary interactions – and to infer that vines should have a particularly high potential as sources of insecticides and medicaments. Justification of this opinion still requires much more systematically gathered and collated detail than is presently available. The facts are that novel compounds are still being discovered in species of various habits in which secondary compounds have already been fairly thoroughly investigated, and this is likely to continue indefinitely as methods of isolation and identification increase in sophistication. The evidence thus far available suggests that climbers might well merit special emphasis in the ongoing search for unusual and biologically active secondary compounds.

Summary

Plants contain many thousands of secondary compounds, of which a large number are now considered to have a protective role against potential herbivores. This is achieved by toxicity (e.g. of an extensive range of classes of compound including many of the alkaloids) or reduction of digestibility (e.g. tannins, lignin and resins). Analysis of the results of regional surveys of alkaloid distribution in plants has so far failed to confirm that they are especially abundant in vines, compared with trees. However, some families rich in vines are also rich in alkaloids, and there are reasons to expect that vines as a life-form, with specialised risks due to their form, apparency to herbivores, predominantly tropical distribution, and patterns of leaf production, may have evolved higher levels of some types of protection from herbivores than trees. Such hypotheses may be better supported as numerous further secondary compounds and their probable defensive or other functions continue to be identified.

Note:

Since this review was written the results of a comprehensive chemical and pharmacological survey of plants of the Australian region has been published (D. J. Collins, C. C. J. Culvenor, J. A. Lamberton, J. W. Loder and J. R. Price (1990). *Plants for Medicines: A Chemical and Pharmacological Survey of Plants in the Australian Region*. CSIRO, Australia).

References

Adesanya, S. A., Ogundana, S. K. & Roberts, M. F. (1989). Dihydrostilbene phytoalexins from *Dioscorea bulbifera* and *D. dumentum*. *Phytochemistry* 28: 773–4.

Applebaum, S. W. & Birk, Y. (1979). Saponins. In *Herbivores. Their Interaction with Secondary Plant Metabolites*, ed. G. A. Rosenthal and D. H. Janzen, pp. 539–66. Academic Press, New York.

Ayensu, E. S. (ed.) (1980). *Jungles*. Jonathan Cape, London.

Bamber, R. K. (1984). Wood anatomy of some Australian rainforest vines. In *Proceedings of wood anatomy conference*, ed. S. Sudo, pp. 58–60 Forest and Forest Products Research Institute, Ibaraki, Japan.

Bar-Nun, N. & Mayer, A. M. (1989). Cucurbitacins – repressors of induction of lactase formation. *Phytochemistry* 28: 1369–71.

Basu, N. & Rastogi, R. P. (1967). Terpene saponins and sapinogens. *Phytochemistry* 6: 1249–70.

Bate-Smith, E. C. (1972). Attractants and repellents in higher animals. In *Phytochemical Ecology*, ed. J. B. Harborne, pp. 45–56. Academic Press, New York.

Bell, E. A. (1980). Non-protein amino acids in plants. In *Secondary Plant Products, Encyclopedia of Plant Physiology, n.s. Vol. 8*, ed. E. A. Bell and B. V. Charlwood, pp. 403–32. Springer-Verlag, Berlin.

Bell, E. A. & Charlwood, B. V. (ed.) (1980). *Secondary Plant Products. Encyclopedia of Plant Physiology n.s., Vol. 8*. Springer-Verlag, Berlin.

Bell, E. A. & Janzen, D. H. (1971). Medical and ecological considerations of L-DOPA and 5-HTP in seeds. *Nature (London)* **229**: 136–7.

Boppré, M. (1984). Chemically mediated interactions between butterflies. *Symposia of the Royal Entomological Society of London* **11**: 259–75.

Brower, L. P. (1984). Chemical defences in butterflies. *Symposia of the Royal Entomological Society of London* **11**: 109–33.

Brower, L. P., Brower, J. V. & Corvino, J. M. (1967). Plant poisons in a terrestrial food chain. *Proceedings of the National Academy of Sciences (USA)* 57: 893–8.

Brown, K. S., Sheppard, P. M. & Turner, J. R. G. (1974). Quaternary refugia in tropical America: evidence from race formation in *Heliconius* butterflies. *Proceedings of the Royal Society of London, Ser. B* 187: 369–78.

Chen, Z.-L. & Zhu, D.-Y. (1986). *Aristolochia* alkaloids. In *The Alkaloids. Chemistry and Pharmacology, Vol. 31*, ed. A. Brossi, pp. 29–65. Academic Press, Orlando, FL.

Cheney, R. H. (1931). Geographic and taxonomic distribution of American plant arrow poisons. *Economic Botany* 18: 136–45.

Coley, P. D., Bryant, J. P. & Chapin, F. S. III (1985). Resource availability and plant antiherbivore defense. *Science* 230: 895–9.

Conn, E. E. (1980). Cyanogenetic glucosides. In *Secondary Plant Products, Encyclopedia of Plant Physiology, n.s., Vol. 8*, ed. E. A. Bell and B. V. Charlwood, pp. 461–92. Springer-Verlag, Berlin.

Cronquist, A. (1981). *An Integrated System of Classification of Flowering Plants*. Columbia University Press, New York.

da Costa, C. P. & Jones, C. M. (1971). Cucumber beetle resistance and mite susceptibility controlled by the bitter gene in *Cucumis sativus* L. *Science* 172: 1145–6.

DeVries, P. J. (1987). *The Butterflies of Costa Rica*. Princeton University Press, Princeton NJ.

Duke, J. A. (1977). Phytotoxin tables. In *CRC (Chemical Rubber Company) Critical Reviews in Toxicology*, p. 189–237.

Edgar, J. A. (1982). Pyrrolizidine alkaloids sequestered by Solomon Island Danaine butterflies. The feeding preferences of the Danainae and Ithomiinae. *Journal of Zoology (London)* **196**: 385–99.

Edgar, J. A. & Culvenor, C. C. J. (1975). Pyrrolizidine alkaloids in *Parsonsia* species (family Apocynaceae) which attract danaid butterflies. *Experientia* **31**: 393–4.

Evans, S. V., Gatehouse, A. M. R. & Fellows, L. R. (1985). Detrimental effects of 2,5-dihydroxymethyl-3,4-dihydroxypyrrolidine in some tropical legume seeds on larvae of the bruchid *Callosobruchus maculatus*. *Entomology experimental and applied* 37: 257–61.

Everist, S. L. (1981). *Poisonous Plants of Australia*, 2nd edition. Angus and Robertson, London.

Feeny, P. P. (1975). Biochemical coevolution between plants and their insect herbivores. In *Coevolution of Animals and Plants*, ed. L. E. Gilbert, and P. R. Raven, pp. 3–19. University of Texas Press, Austin.

Feeny, P. P. (1976). Plant apparency and chemical defense. In *Biochemical Interactions between Plants and Insects. Recent Advances in Phytochemistry, Vol. 10*, ed. J. Wallace, and R. Mansell, pp. 1–40. Plenum Press, New York.

Fellows, L. E., Evans, S. V., Nash, R. J. & Bell, E. A. (1986). Polyhydroxy plant alkaloids as glycosidase inhibitors and their possible ecological role. In *Natural Resistance of Plants to Pests: Roles of Allelochemicals, ACS Symposium Series 296*, ed. M. B. Green and P. A. Hedin, pp. 72–8. American Chemical Society, Washington, DC.

Fodor, G. B. (1980). Alkaloids derived from phenylalanine and tyrosine. In *Secondary Plant Products, Encyclopedia of Plant Physiology, n.s., Vol. 8*, ed. E. A. Bell and B. V. Charlwood, pp. 92–127. Springer-Verlag, Berlin.

Fong, H. H. S., Trojánkova, M., Trojánek, J. & Farnsworth, N. R. (1972). Alkaloid screening II. *Lloydia* **35**: 1–34.

Gentry, A. H. (1985). An ecotaxonomic survey of Panamanian lianas. In *Historia Natural de Panama*, ed. W. G. D'Arcy and M. Correa, pp. 29–42. Missouri Botanic Gardens, St Louis, MO.

Gentry, A. H. & Cook, K. (1984). *Martinella* (Bignoniaceae): a widely used eye medicine of South America. *Journal of Ethnopharmacology* **11**: 337–43.

Gibbs, R. D. (1974). *Chemotaxonomy of Flowering Plants*, Vols 1–4. McGill–Queens University, Montreal, Canada.

Gilbert, L. (1975). Ecological consequences of a coevolved mutualism between butterflies and plants. In *Coevolution of Plants and Animals*, ed. L. Gilbert and P. Raven, pp. 210–40. University of Texas Press, Austin.

Gilbert, L. & Singer, M. (1975). Butterfly ecology. *Annual Review of Ecology and Systematics* **6**: 365–97.

Gröger, D. (1980). Alkaloids derived from tryptophan and anthranilic acid. In *Secondary Plant Products, Encyclopedia of Plant Physiology, n.s., Vol. 8*, ed. E. A. Bell and B. V. Charlwood, pp. 128–59. Springer-Verlag, Berlin.

Harborne, J. B. (1977). *Introduction to Ecological Biochemistry*. Academic Press, New York.

Hartley, T. G., Dunstone, E. A., Fitzgerald, J. S., Johns, S. H. & Lamberton, J. A. (1973). A survey of New Guinea plants for alkaloids. *Lloydia* **36**: 217–319.

Haslam, E. (1975). Natural proanthocyanidins. In *The Flavonoids*, ed. J. B. Harborne, T. J. Mabry, and H. Mabry, pp. 505–50. Chapman and Hall, London.

Haslam, E. (1986). Secondary metabolism. Fact or fiction. *Natural Product Reports* 3: 218–49.

Hegarty, E. E. (1988). Canopy dynamics of lianes and trees in subtropical rainforest. PhD thesis, Botany Department, University of Queensland, Australia.

Hegarty, E. E. (1989). The climbers – lianes and vines. In *Tropical Rain Forest Ecosystems: Biogeographical and Ecological Studies. Ecosystems of the World 14B*, ed. H. Lieth and M. J. A. Werger, pp. 339–53. Elsevier, Amsterdam.

Hegarty, E. E. (1990). Leaf life-span and leafing phenology of lianes and associated trees during a rainforest succession. *Journal of Ecology* 78: 300–12.

Hegarty, M. P. & Pound, A. W. (1970). Indospicine, a new hepatotoxic amino acid from *Indigofera spicata*. *Nature (London)* **217**: 354–5.

Hegnauer, R. (1962–86). *Chemotaxonomie der Pflanzen: eine Übersicht über die Verbreitung und die systematische Bedeutung der Pflanzenstoffe*. Berkhauser, Basle.

Hladik, A. & Hladik, C. M. (1977). Signification écologique des teneurs en alcaloides des végétaux de la forêt dense: résultats des testes preliminaires effectués au Gabon. *La Terre et la Vie* **31**: 515–55.

Jones, D. A. (1972). Cyanogenic glucosides and their function. In *Phytochemical Ecology*, ed. J. B. Harborne, pp. 103–24. Academic Press, London.

Jones, D. A. (1988). Cyanogenesis in animal–plant interactions. In *Cyanide Compounds in Biology, Ciba Foundation Symposium 140*, ed. D. Evered and S. Harnett, pp. 151–70. John Wiley and Sons, Chichester.

Kingsbury, J. (1980). Phytotoxicology. In *Toxicology*, 4th edition, ed. L. J. Casarett and J. Doull, pp. 578–90. Macmillan, New York.

Langcake, P. & Pryce, R. J. (1977). A new class of phytoalexins from grapevines. *Experientia* **33**: 151.

Leete, E. (1980). Alkaloids derived from ornithine, lysine and nicotinic acid. In *Secondary Plant Products, Encyclopedia of Plant Physiology, n.s., Vol. 8*, ed. E. A. Bell, and B. V. Charlwood, pp. 65–91. Springer-Verlag, Berlin.

Levin, D. A. (1976a). The chemical defenses of plants to pathogens and herbivores. *Annual Review of Ecology and Systematics* 7: 121–59.

Levin, D. A. (1976b). Alkaloid-bearing plants: an ecogeographic perspective. *American Naturalist* **110**: 261–84.

Levin, D. A. & York, B. M. Jr (1978). The toxicity of plant alkaloids: an ecogeographic perspective. *Biochemical Systematics and Ecology* **6**: 61–76.

Lewis, W. H. & Elvin-Lewis, M. P. (1977). *Medical Botany*. Wiley and Sons, New York.

Li, H. L. & Willaman, J. J. (1968). Distribution of alkaloids in Angiosperm phylogeny. *Economic Botany* **22**: 239–52.

Lopes, L. M. X. & V. da S. Bolzani (1988). Lignins and diterpenes of three *Aristolochia* species. *Phytochemistry* **27**: 2265–8.

McKey, D. (1974). Adaptive patterns in alkaloid physiology. *American Naturalist* **108**: 305–20.

McKey, D. (1979). The distribution of secondary compounds within plants. In *Herbivores. Their Interaction with Secondary Plant Metabolites*, ed. G. A. Rosenthal, and D. H. Janzen, pp. 56–133. Academic Press, New York.

Manske, R. H. F., Holmes, H. L., Rodrigo, R. G. A. & Brossi, A. (variously as editors) (1950–). *The Alkaloids, Vols 1–33*. Academic Press, New York.

Miller, L. P. (ed.) (1973). *Phytochemistry*. Van Nostrand, New York.

Mooney, H. A., Gulmon, S. L. & Johnson, N. D. (1983). Physiological constraints in plant chemical defenses. In *Plant Resistance to Insects, ACS Symposium series 208*, ed. P. A. Hedin, pp. 21–36. American Chemical Society, Washington, DC.

Navon, A. & Bernays, E. A. (1978). Inhibition of feeding in acridids by non-protein amino acids. *Comparative Biochemistry and Physiology*. **59A**: 161–4.

Peñalosa, J. (1984). Basal branching and vegetative spread in two tropical rain forest lianas. *Biotropica* **16**: 1–9.

Poulton, J. E. (1988). Localization and catabolism of cyanogenic glycosides. In *Cyanide Compounds in Biology, Ciba Foundation Symposium 140*, ed. D. Evered and S. Harnett, pp. 67–91. John Wiley and Sons, Chichester.

Pryce, R. J. & P. Langcake, (1977). α-viniferin: an anti-fungal resveratrol trimer from grapevines. *Phytochemistry* **16**: 1452–4.

Putz, F. E. (1983). Liana biomass and leaf area of a 'Tierra firme' forest in the Rio Negro basin, Venezuela. *Biotropica* **15**: 185–9.

Putz, F. E. & Windsor, D. M. (1987). Liana phenology on Barro Colorado Island. Panama. *Biotropica* **16**: 19–23.

Raven, P. H., Evert, R. F. & Eichhorn, S. E. (1986). *Biology of Plants*, 4th edition. Worth, New York.

Rhoades, D. F. (1983). Herbivore population dynamics and plant chemistry. In *Variable Plants and Herbivores in Natural and Managed*

Ecosystems, ed. R. F. Denno and M. S. McClure, pp. 155–220. Academic Press, New York.

Rhoades, D. F. & Cates, R. G. (1976). Toward a general theory of plant antiherbivore chemistry. In *Biochemical Interaction Between Plants and Insects, Recent Advances in Phytochemistry 10*, ed. J. W. Wallace and R. L. Mansell, pp. 168–213. Plenum Press, New York.

Robinson, M. E. (1930). Cyanogenesis in plants. *Biological Reviews* **5**: 126–41.

Robinson, T. (1979). The evolutionary ecology of alkaloids. In *Herbivores. Their Interaction with Secondary Plant Metabolites*, ed. G. A. Rosenthal and D. H. Janzen, pp. 413–48. Academic Press, New York.

Rosenthal, G. A. (1982). *Plant Nonprotein Amino and Imino Acids*. Academic Press, New York.

Rothschild, M. (1972). Secondary plant substances and warning coloration in insects. *Symposia of the Royal Entomological Society of London* **6**: 59–83.

Sharma, G. C. & Hall, C. V. (1973). Relative attractance of spotted cucumber beetle to fruits of fifteen species of Cucurbitaceae. *Environmental Entomology* **2**: 154–6.

Singh, B. & Rastogi, R. P. (1970). Cardenolides – glycosides and genins. *Phytochemistry* **9**: 315–31.

Smolenski, S. J., Silinis, H. & Farnsworth, N. R. (1972). Alkaloid screening I. *Lloydia* **35**: 1–34.

Smolenski, S. J., Silinis, H. & Farnsworth, N. R. (1973). Alkaloid screening III. *Lloydia* **36**: 359–89.

Straatman, R. (1962). Notes on certain Lepidoptera ovipositing on plants which are toxic to their larvae. *Journal of the Lepidopterists' Society* **16**: 99–103.

Swain, T. (1977). Secondary compounds as protective agents. *Annual Reviews of Plant Physiology* **28**: 479–501.

Swain, T. (1978). Plant–animal coevolution: an ecogeographic view of the Paleozoic and Mesozoic. In *Biochemical Aspects of Plant and Animal Evolution, Proceedings of symposium, Phytochemical Society*, ed. J. B. Harborne, pp. 3–19. Academic Press, London.

Swain, T. (1979). Tannins and lignins. In *Herbivores. Their Interaction with Secondary Plant Metabolites*, ed. G. A. Rosenthal and D. H. Janzen. Academic Press, New York.

Tallamy, D. W. (1985). Squash beetle feeding behavior: an adaptation against induced cucurbit defenses. *Ecology* **66**: 1574–9.

Verdcourt, B. & Trump, E. C. (1969). *Common Poisonous Plants of East Africa*. Collins, London.

Waterman, P. G. (1983). Distribution of secondary metabolites in rain forest plants: toward an understanding of cause and effect. In *Tropical Rain Forest: Ecology and Management, British Ecological Society Special Publication* 2, ed. S. L. Sutton, T. C. Whitmore and A. C. Chadwick, pp. 167–79. Blackwell Scientific Publications, Oxford.

Watt, J. M. & Brandwijk, M. G. (1962). *The medicinal and poisonous plants of Southern and Eastern Africa*, 2nd edition. Livingston, Edinburgh.

Webb, L. J. (1949). An Australian phytochemical survey. I. Alkaloids and cytogenetic compounds in Queensland plants. *CSIRO Bulletin 241*.

Webb, L. J. (1952). An Australian phytochemical survey. II. Alkaloids in Queensland flowering plants. *CSIRO Bulletin 268*.

Willaman, J. J. & Schubert, B. G. (1961). Alkaloid-bearing plants and their contained alkaloids. *Technical Bulletin 1234*. USDA Agricultural Research Service, Washington, DC.

Woodhead, S. & Bernays, E. (1977). Changes in release rates of cyanide in relation to palatability of *Sorghum* to insects. *Nature (London)* **270**: 235–6.

PART IV
COMMUNITY ECOLOGY OF VINES

11
Distribution and abundance of vines in forest communities

ELWYN E. HEGARTY AND GUY CABALLÉ

Introduction

Where species diversity of vines is high, usually relative abundance is also high. However, because of their thin stems vines seldom form more than a minor proportion of the biomass of any forest, even in the tropics. Exceptions to this rule are often found during regeneration of very disturbed sites, where one or two species of vine become dominant. This chapter reviews the results of various quantitative studies of vines, mostly from mature tropical and subtropical forests.

As used in this chapter, 'vines' include all non-epiphytic climbers and, unless otherwise noted, those which commence life in the soil but may later become epiphytic, i.e. secondary hemiepiphytes. The vines therefore include the predominantly herbaceous, leafy smaller climbers and the mostly woody, bare-stemmed larger lianas (or lianes), between which there are loosely defined distinctions depending on regional usage.

The contribution of vines to forest structure in various regions

Abundance

Comparisons of the abundance of the thicker-stemmed vines (i.e. those greater than 2.5 cm in diameter) in forests of various regions have been made by Gentry (Chapter 1). However, as shown in studies such as those of Rollet (1969a, b), Proctor *et al.* (1983), Putz (1984) and Hegarty (1988), high proportions of stems of vines are found in the smaller size diameter-classes. Thin vines make an appreciable contribution to species richness (Gentry & Dodson, 1987).

In many vines, a proportion of stems is formed by clonal extension of larger individuals (genets); such vegetative offshoots (ramets) may develop their own root systems, becoming separate individuals, but remain difficult to distinguish from true genets without extensive excavation (as by Caballé 1984, 1986*a*, *b*). Additional sinker roots often form where fallen stems contact the ground, sometimes accompanied by development of robust adventitious climbing shoots, making determination of the location of the principal root system difficult. For this reason, climbing stems which are firmly rooted at the base are often recorded separately, although their complete physiological independence is in doubt (as in Putz, 1984; Putz & Chai, 1987; Hegarty 1988). For most purposes, the diameter of vines is measured at breast height, as for trees, *ad hoc* adjustments are made for stems of irregular cross-section or low branching.

Most direct comparisons of survey records for vines (as in Table 11.1) are also complicated by a lack of uniformity in sample composition, e.g. inclusion/exclusion of the thinner stems, and/or hemiepiphytes. In sampling vines, the areas of sites or plots, and their arrangement within a defined area are also important, because the distribution of vines is strongly clumped (Caballé, 1984, 1986a; Putz, 1984). Rollet (1971) recorded a Poisson-type, clumped distribution for stems of vines in some locations during a study in Venezuela, with vines absent from many sample plots. It has been suggested that there may be few vines associated with buttressed trees or palm trees. Such speculations, briefly reviewed by Hegarty (1989), have been little tested, with some conflicting results.

Very high numbers of stems of vines were recorded (i) in tropical pre-montane rainforest (Holdridge *et al.*, 1971, in a study limited to Costa Rica) with 'woody lianas most conspicuous and abundant in moist semideciduous lowland forests and rare in the wettest ones', (ii) in some regions with a marked dry season, e.g. in Ecuador and Tanzania (from data for lianas >2.5 cm diameter only: Gentry, Chapter 1; see also Gentry, 1982; Gentry & Dodson, 1987), and (iii) in alluvial forest (i.e. subject to occasional inundation) (Proctor *et al.*, 1983; Gentry, chapter 1) (Table 11.1). On the other hand, Ogawa, Yoda & Kira (1961), in a comparison of forests in Thailand, stated that climbing plants were absent from samples of (i) temperate evergreen forest at 1200 m elevation, (ii) seasonally dry mixed savanna forest where most trees are deciduous, and (c) dipterocarp savanna forest.

Some of the few data available for basal areas of vines are also shown in Table 11.1. Vines represented about 2% or less of total basal area in several moist forests and up to 9% in a seasonally dry forest in Mexico. The total basal area of lianas ⩾5 cm in diameter at a site is quite closely related to their abundance (Caballé, 1986a, Figure 17). A direct relationship between basal area of lianas and foliar biomass was found for individuals up to about 12 cm diameter in Venezuelan tierra firme forest (Putz, 1983).

Table 11.1. *Numbers of stems, above-ground biomass* ($t\ ha^{-1}$) *and basal areas* ($m^{-2}\ ha^{-1}$) *of vines/lianas* (v), *compared with total values* (t) *recorded per hectare in various regions. Mean annual precipitation is noted in parentheses after the vegetation type*

	Number of stems			Biomass				Basal area			Sample	
	v	t	%v	v	(a) (b)[a]	t	%v	v	t	%v	Stem diameter (cm)	Total area (ha)
Africa												
Côte d'Ivoire. Rainforest (1800 mm). Bernhard-Reversat, Huttel & Lemée (1979)												
		–		24		562	4.3		–		(large)	0.25
Gabon. Tropical evergreen forest (1700 mm). Caballé (1986a)												
	122	1239	9.8								≥5	10
	149	1125	13.2	–		–			–		≥5	6.16
	80	1422	5.6								≥5	3.84
Ghana. Secondary semi-deciduous forest (1650 mm). Greenland & Kowal (1960)												
		–		14		308	4.5		–		all	0.02
Mean of eleven sites in Africa and Madagascar. Gentry (Chapter 1)												
	1140			–		–			–		≥2.5	11 × 0.1
Asia/Oceania												
Australia. Subtropical evergreen rainforest (1674 mm). Hegarty (1988)												
	5771	16236	35	21	(2, 18.6)	447	4.6	1.6	69.6	2.2	all	1
Malaysia												
Pasoh. High dipterocarp forest (1807 mm). Kato, Tadaki & Ogawa (1978)												
		–		10	(0.5, 9.1)	475	2		–		all	0.2
Sarawak. Lowland rain forests (1) on alluvium (2) on gley soil (3) dipterocarp forest (4) heath forest (5) on limestone (5090–5700 mm). Proctor *et al.* (1983)												
(1)	16652	70600	23								all	1}
(2)	18132	55100	33								all	}
(3)	2600	60600	4	–		–			–		all	1
(4)	2480	63500	4								all	1
(5)	14772	26400	56								all	0.85

Table 11.1 (*cont.*)

	Number of stems			Biomass				Basal area			Sample	
	v	t	%v	v	(a) (b)[a]	t	%v	v	t	%v	Stem diameter (cm)	Total area (ha)
New Guinea. Montane rainforest (*c.* 4000 mm). Edwards & Grubb (1977)												
		–		4	(0.9, 3.3)	505	<1		–		all	0.04
Thailand.												
Monsoon forest (*c.* 1400 mm). Ogawa *et al.* (1965a)												
	93	806	12	–				0.8	36	2	>4.5	0.16
Seasonal evergreen rainforest (2718 mm). Ogawa *et al.* (1965a, b)												
	194	1369	14	20	(2.4, 18)	404	5	1.0	41	2.5	>4.5	0.16
	188	1526	12	–		–		0.5	35	1.6	>4.5	0.16
Six sites in Australia, Borneo, and New Caledonia. Gentry (Chapter 1)												
	400			–		–			–		≥2.5	6 × 0.1
Central and South America												
Brazil. Rainforest (1771 mm). Klinge & Rodrigues (1973); Fittkau & Klinge (1973)												
		–		46		733	6.3 (f/wt)	–			all	0.2
French Guiana.												
Tropical rainforest (3000 mm). Beekman (1981)												
	632	6169	10	13		458	2.8		–		≥1	0.25
	415	5930	7	–		–			–		2	0.15
Mature rainforest (3450 mm). Lescure (1981); Lescure *et al.* (1983)												
		–		3.4		147	2.3	0.5	42.5	1.1	≥1	0.25
Jamaica. Upper montane rainforest. Tanner (1980)												
		–		2.4		338	<1		–		all	0.01
Mexico.												
Deciduous tropical forest (1) upland (2) arroyo (748 mm). Lott *et al.* (1987)												
(1)	410	3940	10	–				0.7	26	2.6	≥2.5	0.1
(1)	550	5060	11	–				1.0	22	4.4	≥2.5	0.1
(2)	1400	4520	31	–				3.5	52	6.7	≥2.5	0.1

(1)	9400	21700	43	–				2.8	30	9.5	all	0.01
(2)	45800	65100	70	–				2.3	33	7	all	0.01
Lowland rainforest (4639 mm). Bongers *et al.* (1988)												
		11208	5.2	–	–				–		(≥0.5 m tall)	1.0
Surinam. Tropical rain forest. Ohler (1980)												
		–		10	(0.5, 10)	406	2.6		–		?	?
Venezuela.												
Bana forest (3600 mm). Bongers *et al.* (1985)												
Tall:		–		1.7	(0.3, 1.4)	182	0.9		–		all	0.015
Low:		–		1.0	(0.2, 0.8)	40	2.5		–		all	0.019
Open:		–		0.4	(0.1, 0.3)	6	7.2		–		all	0.030
Evergreen forest. Rollet (1971)												
	162	939	17	–	–				–		≥1	0.12
Tall caatinga dry forest (3600 mm). Klinge & Herrera (1983)												
		–		2.1		277	0.8		–		all	13 × 0.01
Tierra firme, evergreen tropical rainforest (3500 mm). Jordan & Uhl (1978); Putz (1983)												
		2104	–	15.7		351	4.5		–		all but hemiepiphytes	0.2
Amazon (seven sites in Peru/Venezuela/Brazil). Max. rainfall 3000–4000 mm. Gentry (1988a)												
	20	619	<2–4	–	–				–		≥10	7 × 1
Mean of 56 neotropical sites (Gentry, Chapter 1)												
	700			–		–					≥2.5	56 × 0.1

Costa Rica. Ten types of tropical forest (1450–5800 mm). Holdridge *et al.* (1971). Sample areas ≥0.01 ha, as subplots of sites of up to 0.5 ha.

Number and type of stems	Sample area	Sample size
2800, mostly woody	Dry forest	7 × ≥0.01 ha
1167, woody/herbaceous	Moist forest	7 × ≥0.01 ha
600, mostly herbaceous	Dipterocarp forest	13 × ≥0.01 ha
n.d. mostly woody	Premontane forest	2 × ≥0.01 ha
393, mostly herbaceous	Premontane wet forest	6 × ≥0.01 ha
8980, woody and herbaceous	Premontane rainforest	5 × ≥0.01 ha
n.d. 'soft woody vines'	Lower montane moist forest	1 × ≥0.01 ha
900, woody and herbaceous	Lower montane wet forest	1 × ≥0.01 ha
520, woody and herbaceous	Lower montane rainforest	3 × ≥0.01 ha
100, mostly herbaceous	Montane rainforest	1 × ≥0.01 ha

[a](a) leaf, (b) stem, of vines only.

Table 11.2. *Distribution of stems of vines in size-classes, per hectare*

	Number of stems of diameter (cm)							
	<1	1–<5	5–<10	10–<15	15–<20	≥20	Author	Sample (ha)
Africa								
Gabon. Evergreen rain forest								
Margins	nd	nd	133	10	1	–	Caballé (1984)	2.4
Medium tall	nd	nd	110	15	3	<1	Caballé (1984)	2.16
Mature tall	nd	nd	49	27	4	<1	Caballé (1984)	1.04
Asia–Pacific								
Australia. Evergreen subtropical rainforest								
		........5756		12	2	1	Hegarty (1988)	1.0
Malaysia.								
Pahang, Peninsular Malaysia. Shorea-dipterocarp forest								
	nd	nd	113				Appanah & Putz (1984)	13
Sarawak. Various types of site or forest								
Alluvial	14 400	1960			.............292		Proctor *et al.* (1983)	1 ha total
Alluvial gley soil	13 000	4800			.............332		Proctor *et al.* (1983)	
Dipterocarp	2160	440			...............0		Proctor *et al.* (1983)	1.0
Heath	2040	440			...............0		Proctor *et al.* (1983)	1.0
On limestone	10 700	4040			..............32		Proctor *et al.* (1983)	1.0
Dipterocarp								
Valley	2100	800	50				Putz & Chai (1987)	0.5
Ridge	960	450	20				Putz & Chai (1987)	0.5
Neotropics								
Barro Colorado Island, Panama. Tropical moist forest								
Climbing vines	824	930	40	3	0	0	Putz (1984)	1.0
Free-standing	1568						Putz (1984)	1.0
French Guiana. Evergreen tropical rainforest								
	nd	366	33	10	5	1	Beekman (1981)	1.0
	nd	388	40	12	4	4	Beekman (1981)	0.25
Venezuela. Evergreen forest								
	3257	164	8				Rollet (1971)	0.12

Mexico: seasonally dry forest

	Diameter (cm)					
	0–1.6	>1.6–3.2	>3.2–6.4	>6.4–12.8	Author	Sample (ha)
Valley	42 900	2000	900	0	Bullock (1990)	0.01
Upslope	7200	1200	900	100	Bullock (1990)	0.01

Frequency distributions of stem-diameter classes. These have been compared for vine populations of various areas (Table 11.2). As a further complication of counting vines, unbranched rattans in a rhizomatous clump constitute a single individual in the strict sense of Caballé (1986a), but firmly rooted stems are often counted singly, as by Putz & Chai (1987). With the latter method, rattans may constitute as much as half of all stems at a site (Hegarty, 1988). Because they are of relatively uniform diameter they provide a distinctively modal pattern of distribution of vines in stem-diameter classes. However, stem-diameter classes are not useful to indicate the age-structure of a population of lianas for comparison with that of trees because (i) mean and maximum diameters are smaller (with very few of > 30–40 cm); (ii) patterns of extension growth are complex, with occasional replacement of major shoot axes in many species; (iii) diameters often increase exceedingly slowly (a mean of 1.4 mm a year on Barro Colorado Island: Putz, 1990).

The larger vines tend to inhabit tall forests, and may include tendrillar species of genera such as *Arrabidaea* (Bignoniaceae), *Cissus* (Vitaceae), *Entada* (Leguminosae) and *Paullinia* (Sapindaceae), branch-climbers such as *Monanthotaxis* and *Uvaria* (Annonaceae), *Loeseneriella* and *Salacia* (Hippocrataceae), twiners such as *Dichapetalum* (Dichapetalaceae), *Doliocarpus, Tetracera* (Dilleniaceae), *Landolphia* and *Melodinus* (Apocynaceae) and in the Old World especially, root-climbers of the genus *Piper* (Piperaceae). There are comparatively few large stems (i.e. 10 cm in diameter) in any of the sets of data listed in Table 11.2 except those of Proctor *et al.* (1983) for an alluvial forest. Gentry (1988a) found an average of 20 such lianas per hectare in seven Amazonian sites.

Biomass

In the absence of serious disturbances, lianas seldom contribute more than 5% to the total above-ground biomass of humid tropical or subtropical forests. However, in Venezuela the proportion exceeded 7% in an open bana forest which had a very small total biomass (Bongers, Engelen & Klinge, 1985). Results of a number of studies are compared in Table 11.1. Biomass values based on destructive sampling may underestimate the contribution of vines, especially if the vines are cut so that leaves fall before the trees are felled (Ogawa *et al.*, 1965b).

In general, leaves of vines contribute between 5 and 20% of their total above-ground biomass, (Table 11.1) compared with 1–2% or so usually recorded for supporting trees. Because of this, the percentage contribution of lianas to total foliage biomass is larger but less uniform than their contribution to total above-ground biomass. For example, vines making up less than 5% of the standing biomass of an evergreen rainforest in Thailand were reported to contribute about a third of the total dry weight of foliage (Ogawa *et al.*, 1965*a*, *b*). Foliage of vines made up about 10% (fresh weight) of the canopy in a dipterocarp forest in Cambodia (Hozumi *et al.*, 1969). For individual thin vines (< 1 cm diameter) in French Guiana, leaves made up

25% of the standing biomass, whereas the percentage was only 5% in very thick vines (>21 cm diameter) (Beekman, 1981). Such allometric relationships may also vary by species and season, because many vines are deciduous. A complementary relationship has been shown between vine leaf biomass, and biomass of supporting branches (Kira & Ogawa, 1971) with a predictable maximum total leaf supported by tree branches of a given cross-section.

We may conclude that, because lianas are supported by trees and immobilize little of their resources in woody trunks, they can support a large mass of foliage. Nevertheless, even though stems of lianas are thin, they are long and may often be dislodged from the canopy (Putz, 1990), so that there are replacement costs.

Leaf area

The size of expanded leaves of associated species of vine may vary by a factor of 100 (as in Castellanos *et al.*, 1989). There may also be considerable variation in leaf size, or the number of leaflets per leaf, within individuals selected for sampling. While it has been noted that leaves of lianas are similar in size to those of the tree stratum they reach (Richards, 1964), ratios between leaf area and leaf weight may be higher in lianas than trees (Putz, 1983; Hegarty, 1988). Although mean leaf size of vines is similar to that of associated trees (Castellanos *et al.*, 1989) or in some cases greater (Bullock, 1990), in the former study it was shown that as few as three vine species with very large leaves strongly influenced the result.

The Leaf Area Index (LAI, ha ha^{-1}) of lianas (Table 11.3) is seldom more than about 3 in natural forests and varies locally with canopy maturity. Hegarty (1988) found that the LAI of vines at the sunlit edge of a subtropical rainforest was about twice that of vines of the nearby mature canopy. However, the LAI in mature forest was locally very variable, with one large bole-climbing vine contributing strongly to the overall value for what was otherwise a rather patchy canopy of vines.

Productivity

Lianas account for the greater ratio of litter to wood production in tropical than temperate forests (Jordan & Murphy, 1978; Gentry, 1983). Studies of sorted litterfall have allowed estimation of vine leaf productivity to be made. In the leaf litterfall of two sites in somewhat disturbed rainforest in Gabon, 33–39% was from vines (Hladik, 1974). In an Australian subtropical rainforest, vines contributed 16–40% of monthly litterfall over 2 years, with peaks during autumn, after the rainy season, when some vines were semi-deciduous, but trees in general were not (Hegarty, 1988).

Table 11.3. *Leaf area index of vines in forest communities*

Area/type of forest/author	Leaf Area Index (ha ha^{-1}) Vines	Trees	Vines as % of total
Neotropics			
Barro Colorado Is. Tropical moist forest. Putz (1983)	1.2	5.2	18.7
Old World/Australia			
Japan. Liana communities. Monsi & Ogawa (1977)	2.3–4.1	–	–
Malaysia. High dipterocarp forest. Kato *et al.* (1978)	0.7	7.3	8.7
Thailand. Seasonal evergreen rainforest. Ogawa *et al.* (1965b)	3.3	7.4	30.8
Queensland. Evergreen subtropical rainforest. Hegarty (1988)			
Mature canopy	1.8	8.6	17.3
Site including gaps	2.1	9.0	18.9

In both tropical and temperate zones, a number of vines have short-lived, rapidly replaced leaves (Peñalosa, 1984; Hegarty, 1988). Where very rapidly extending lianas, such as two tropical species studied by Peñalosa (1984) are abundant, local rates of canopy turnover by lianas are undoubtedly very high. However, rapid leaf production may not continue at a high rate through the year in all lianas – for example, in deciduous species. In the Australian study, seasonal leaf production by lianas was bimodal, and a number of species of vine which are found in the mature association had evergreen leaves which persisted for at least as long as those of most associated trees, and much longer than in the semi-deciduous vines of forest gaps. A successional increase in leaf life span was reflected in a decrease in productivity of leaf of vines, in relation to contributing basal areas (Hegarty 1988, 1990). Williams, Field & Mooney (1989) have shown that the light reaching plants of various species of the genus *Piper* is inversely related to leaf longevity, and that net carbon gain and leaf construction costs vary depending on environment. This aspect of the biology of leaf production may contribute to the depletion of shade-intolerant vines in forests where canopy height and density are recovering strongly from some severe damage.

Hegarty (1988) calculated from data for leaf production and retention for the more abundant species in a subtropical rainforest that a biomass equivalent to the standing crop of leaf of lianas was replaced in one year (cf. 1.6 years for associated trees). Because leaves were retained for a longer time in mature-forest vines, slower rates of turnover were recorded in mature forest than at the forest margin.

Sources of variation in vine abundance at the regional scale

Gentry (Chapter 1) has compared some aspects of vine abundance in several regions, and discussed relationships with rainfall and soil conditions. However, most of the studies he reviewed do not include smaller individuals or primary hemiepiphytic climbers (i.e. those germinating above ground as epiphtytes).

It is generally accepted that numbers of vines in general decrease with increasing latitude (with some hemispheric and regional variation) and with altitude (Grubb *et al.*, 1963; Holdridge *et al.*, 1971; Gentry, 1983). Nevertheless, both tropical and subtropical regions have natural communities throughout in which vines are particularly abundant, and which are known locally as liana forests, vine scrubs, *selva de bejucos, bejuquero* or *mata cum cipoal*. Many such communities seem to be of rather low stature and subject to periodic droughts or floods.

On Sumatran river terraces and slopes, where there is at least 50% high-canopy cover in a mature-phase forest, liana abundance decreases with increasing age of the topographical unit (van Schaik & Mirmanto, 1985). Aside from the fact that vines in many areas are virtually confined to moist sites, vine abundance in various regions seems to have a complex relationship with total rainfall, as has been found for vine diversity (Gentry 1988a, b). Also, in seasonally dry regions a high proportion of vines may be both tuberous and deciduous, and so well equipped to survive between wet seasons. It is therefore very difficult to relate the regional abundance of vines to single environmental factors such as latitude, elevation, temperature, soils, geographic isolation, or seasonal rainfall, most of which are not independent. There also appear to be inter-continental differences in densities of vines in forests (Gentry, Chapter 1). We can nevertheless suggest how the height and structure of a forest canopy determines the distribution and abundance of vines in individual communities.

Vines in forest dynamics

Natural communities

Climbing methods in relation to forest structure. Of all the publications on vines, the largest number deal with how they climb. Early authors had three broad objectives: (i) to describe and classify principal climbing mechanisms; (ii) to show how these were used by vines, from observation and field study; (iii) to interpret their ontogeny and morphology. From the end of the 19th century, the essential mechanisms by which vines climb have been recognized (Darwin, 1875, 1882; Treub, 1883a, b; Schenck, 1893; Massart, 1895, 1906; Costantin, 1899; see also Hegarty, Chapter 13). As discussed by Schnell (1970), classifications based on the degree of irritability of climbing structures and/or alterations in direction of growth due to the presence of a support (which are active as distinct from passive responses) are not easily interpreted with reference to morphology or ontogeny. A system shown below simplifies some of these difficulties, in that it recognizes a continuum of climbing mechanisms, while distinguishing active from passive mechanisms.

Passive mechanisms	
scramblers (sarmentose shoots)	Rubiaceae
± *root attachments	Araceae, Marcgraviaceae, Piperaceae, Urticaceae
*thorns	Arecaceae, Combretaceae, Euphorbiaceae, Mimosaceae, Rutaceae
Active mechanisms	
*hooks	Ancistrocladaceae, Annonaceae, Linaceae, Loganiaceae, Olacaceae
*tendrils	Bignoniaceae, Cucurbitaceae, Passifloraceae, Sapindaceae, Vitaceae
branch climbers	Annonaceae, Celastraceae, Leguminosae
twiners	Acanthaceae, Asclepiadaceae, Connaraceae, Convolvulaceae, Dichapetalaceae,
**specialized and differential organs*	Dilleniaceae, Dioscoreaceae, Leguminosae, Malpighiaceae, Menispermaceae

Despite two centuries of research into the structures and mechanisms of attachment, some aspects remain to be studied. The fragmentation and spatial separation of our studies makes it difficult to summarize and compare biogeographic, systematic or ecological data for vines. However, we should no longer ignore the biological and architectural aspects of climbing, a process which is neither continuous nor regular. For example, (i) many forest lianas are capable of living for a long time in the understory as self-supporting plants; (ii) some branches of individual vines have adaptations for climbing, while others do not; (iii) although the chief benefit of climbing seems to be increased access to light, not all climbers thrive best in full sun.

Climbing methods during forest succession. The appropriateness of a climbing method varies with the type of supports available and the height and structure of the forest canopy (Darwin, 1875; Whitmore, 1974; Kelly, 1985). As the tree canopy becomes taller and more continuous, vines must climb higher to reach the light. Twining around saplings and smaller diameter trees is an efficient method of ascent, but requires longer stems than climbing using clinging roots or adhesive tendrils. Hooks also allow vertical ascent, although the support they provide is limited by their number, strength, size, and architecture. Climbers with hooks are found in both young and older, taller forests. Twining branches provide particularly firm attachments, resisting slipping or breaking under the weight of the main stem, and this type of liana can be abundant in tall rainforest (see Table 11.4(a): they made up 16.1% of individuals in a forest in Gabon). Tendrils, on the other hand, are often fragile, particularly when young. The majority attach by coiling around supports to the side of, or below, the extending vine shoot, and so are efficient for traversing canopy foliage, by oblique rather than vertical ascent. In Gabon, only 9% of liana individuals in tall rainforest (with a canopy 15–35 m) were tendrillar, compared with 23.9% in a sector with a 2–25 m canopy (Caballé, 1986a).

In the Australian studies of Hegarty (1988), all 13 species with coiling tendrils were confined to an early-successional, or pioneer (in the sense of Whitmore, 1989) species group. One of these species, *Cayratia eurynema* (Figure 13.3, Chapter 13) develops adhesive pads at the end of tendrils, which prevent slipping, but the tendrils usually make several coils around supports as well. Twiners are the most abundant form of climber in most situations (examples are given in Table 11.4). Herbaceous aroids which use adventitious roots to climb are characteristic of very humid tropical forests, where they are sheltered by supporting trees. A typical assemblage of vines in a mature rainforest uses a range of climbing methods, sometimes two or more per species, and relies heavily on vegetative reproduction. Vines tend to create occasional new direct routes to the canopy via vigorous, initially self-supporting shoots. These shoots typically originate from main stems at or near ground level, and are in many respects unlike the thin stems of juvenile climbers.

Stratification of vines within forest structure is often very marked, with foliage of shade-intolerant lianas tending to inhabit the outer canopy, and that of herbaceous hemiepiphytic vines (e.g. aroids), which are not always included with the vines, remaining below the canopy surface in the more humid rainforests (Rich *et al.*, 1987). In a rainforest in Ecuador, 70–90% of climbers, including many immature individuals, did not climb higher than *c.* 5 m (Grubb *et al.*, 1963). This was the smallest of six equal height intervals in which maximum heights reached by vines were surveyed, and many aroids were included. On Barro Colorado Island, lianas (not including aroids) were distinctly stratified. Those of height classes between 4 and 24 m were underrepresented in closed forest, but were abundant on the edges of

Table 11.4. *Climbing mechanisms of vines in forest communities.*
(a) Percentage of individuals using various methods in an evergreen forest at Makokou and Bélinga, NE Gabon (Caballé, 1986a) (individuals $\geqslant$5 cm in diameter; members of a clonal group with root connections regarded as one individual; N = *number of individuals in samples)*

	Low canopy 2–25 m (Sample area 1.12 ha) %	Medium canopy 6–32 m (Sample area 7.04 ha) %	Tall canopy 15–35 m (Sample area 1.84 ha) %
Tendrils	23.9	16.7	9.0
Twiners	34.8	47.7	53.6
Branch-climbers	9.8	12.6	16.1
With adventitious roots	0	1.4	0.7
Other categories:			
Scrambling (sarmentose)	7.6	2.5	0
With hooks	18.5	10.0	18.7
With thorns	2.2	7.3	1.9
Indeterminate	3.2	1.8	0
Totals:	100.0 (N = 92)	100.0 (N = 964)	100.0 (N = 155)

*(b) Numbers of species (*N*), and the number of climbing, rooted stems employing each climbing method, with numbers of stems shown as a percentage of total stems. From two field studies*

	Sarawak: Putz & Chai (1987). Primary dipterocarp (vines $\geqslant$1 cm diam. in 10 × 0.1 ha plots) of ridge + valley		Queensland: Hegarty (1988). Evergreen rainforest (all vines in 100 × 0.01 ha plots) successional groups				
	All plots		'Early'			'Mature'	
	N	%	*N*	%		*N*	%
Tendrils or hooks	9	15	15	9	+	0	0
Twining main stems	24	48	8	5	+	10	28
Twining/recurved branches	3	14	0	0	+	2	3
Adventitious roots + rattans	1	<1	0	0	+	1	2
	6	23	0	0	+	1	53
Totals:	43	100	23	14	+	14	86

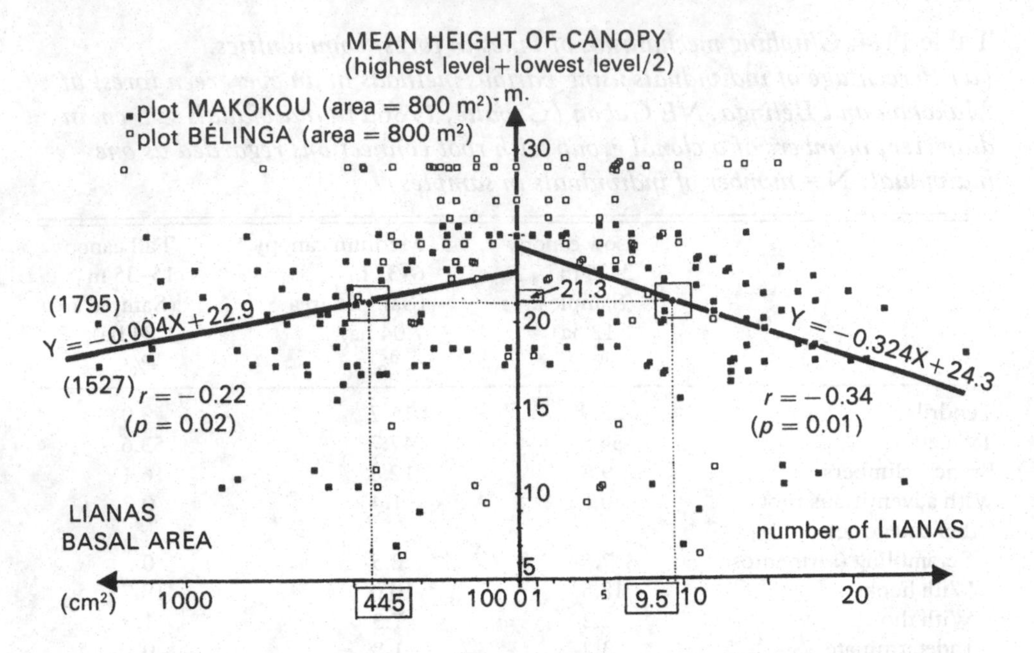

Figure 11.1. Mean canopy height and the distribution of lianas (⩾5 cm in diameter) in two evergreen forests in Gabon. Note: The boundaries of 9 plots were not well defined, therefore values for 116 plots only (9.28 ha) are presented. In calculating Basal Area, the maximum diameter of each liana was used.

treefall gaps (Putz, 1984). Typical distributions of vines in relation to host trees in various types of situation are illustrated by Vareschi (1980) and Hegarty (1989).

In tropical rainforests, the number of lianas has been found to be inversely related to the height of the tree canopy (Caballé (1984, 1986a, b in Gabon: Figure 11.1) and to increasing cover of all tall canopy trees (van Schaik & Mirmanto, 1985, in Sumatra). Hall & Swaine (1981) comment that, in Ghana, small climbers are characteristic of secondary, disturbed forest, only 43% of such species occurring in primary forest as against 87% of species of large climbers.

Severely disturbed communities

While data on vine density in Table 11.1 are mostly from more or less mature communities, regenerating areas of forest are often particularly rich in vines. This is particularly evident after severe disturbances, such as shifting cultivation, which affect soil as well as vegetation. In such circumstances, the forest can be overrun by vines, which sometimes endanger the normal processes of forest regeneration (e.g. Ivory Coast studies by Jaffré & de Namur, 1983; Kahn, 1982).

In temperate forests of Europe, vines are found in predictable locations, such as stream-banks and flood-prone areas, borders of woodlands, hedges and ditches (Walter, 1974, 1979a, b; Beekman, 1984). We must agree with Walter (1974: p. 72) that 'alluvial forests are essentially disturbed forests, and the lianas are part of this unstable forested landscape'. In such regions there is an abundance of vines on old terrace cultivation, now lying fallow or returning to brambles and woodland. These lands may be considered the temperate equivalent of tropical lands recovering from shifting cultivation.

Local variation in vine abundance, basal area and climbing method. Forest vines are often very abundant in proximity to natural gaps. Other situations where vines are aggregated include (i) where the ground is not level and so the forest canopy is of broken contour, e.g. upper slopes, edges of plateaux, or where there are abrupt changes of slope; (ii) flat land where canopies of different heights adjoin; (iii) the edges of tracks and watercourses. Apparent shade-intolerance of most lianes, shown by their great abundance in the situations listed above, leads to their being clumped, where there is direct access to light.

There is some evidence that both density and basal area of lianas of ⩾5cm diameter tend to be least in the largest gaps (Figure 11.2). This may be because fewer trees per unit area are available to give strong support. Beyond a certain threshold, the dynamics of the liana population are no longer determined by access to sufficient light, but by the number of supports available. After all, lianas of the forest survive only by taking advantage of the support of trees.

Although woody vines/lianas are usually less abundant in mature forest than in disturbed areas, liana biomass is often similar in both environments, because in the former a few large individuals may eventually contribute much of the vine biomass (Beekman, 1981). In Gabon, for example, very large individuals (> 10 cm dbh) formed 20% of the liana population, excluding those < 5 cm in diameter, in closed canopy forest, but less than 8% on the forest edge (Caballé, 1984). In an evergreen subtropical rainforest in Australia, the total basal area of all rooted stems of vines (including firmly rooted stems of clones) declined from 5.9% of the total basal area per plot in an area of 80-year regrowth, to 3.2% adjoining a more recently established clearing, to 2.4% in a large treefall gap, and 1.1% in mature forest. Absolute values for vines, however, declined less markedly because some vines persisted to large size despite successional changes to their environment. Big vines (⩾10 cm in diameter) were, however, less diverse in the mature forest than near the forest edge, with one species of root-climber dominating the records (Hegarty,

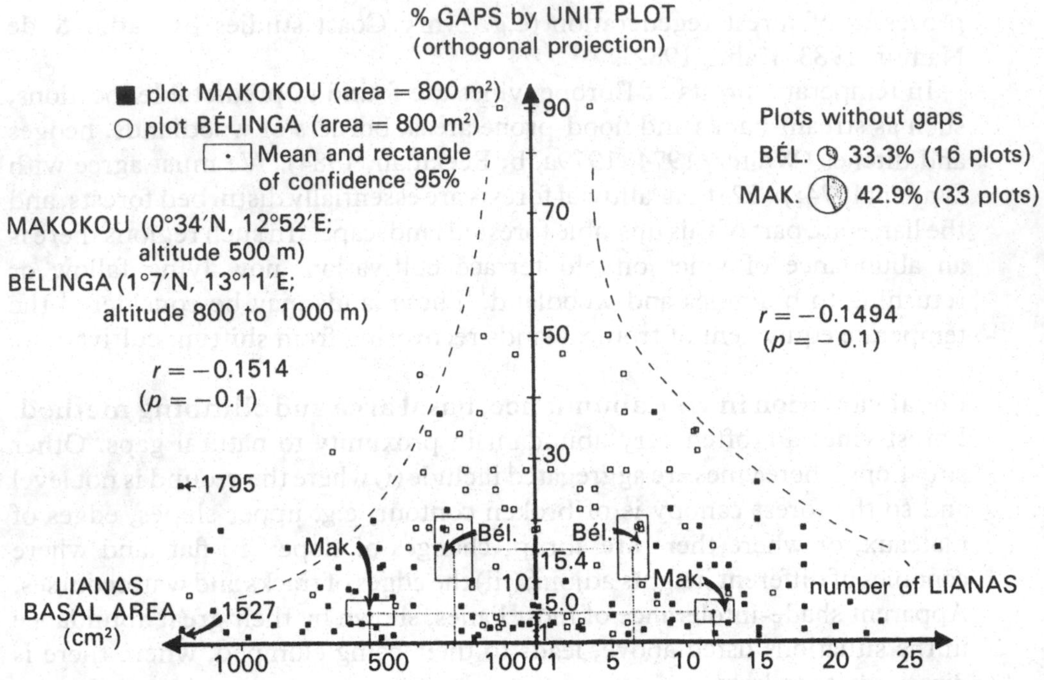

Figure 11.2. Percentage area of gaps in two evergreen forests in Gabon, and the distribution of lianas (⩾5 cm in diameter).

1988). The typical distribution of vines in rainforests containing treefall gaps and tracks is illustrated in Figure 11.3.

Communities where weedy exotic vines are present

An important further source of complication in interpreting the natural succession of vines is the presence of vigorous, naturalized vines. These may be so abundant as to arrest the normal succession more or less indefinitely (Kahn, 1982; Caballé, 1984, 1986a), especially where a large continuous area has been severely damaged. Some small, very isolated areas such as the Hawaiian islands have few native vines, and are readily colonized by exotics (Mueller-Dombois, 1975). Species of *Merremia* and *Ipomoea* (Convolvulaceae), *Cardiospermum* (Sapindaceae), *Anredera* (Basellaceae), *Pueraria* (Leguminosae), *Macfadyena* (Bignoniaceae) and *Passiflora* (Passifloraceae) are typical of disturbed humid forests in various regions remote from their natural distribution, and usual controls such as herbivores. Nearly all such species are very shade-intolerant, fast-growing, fecund and readily dispersed, and so they have the capacity for high productivity.

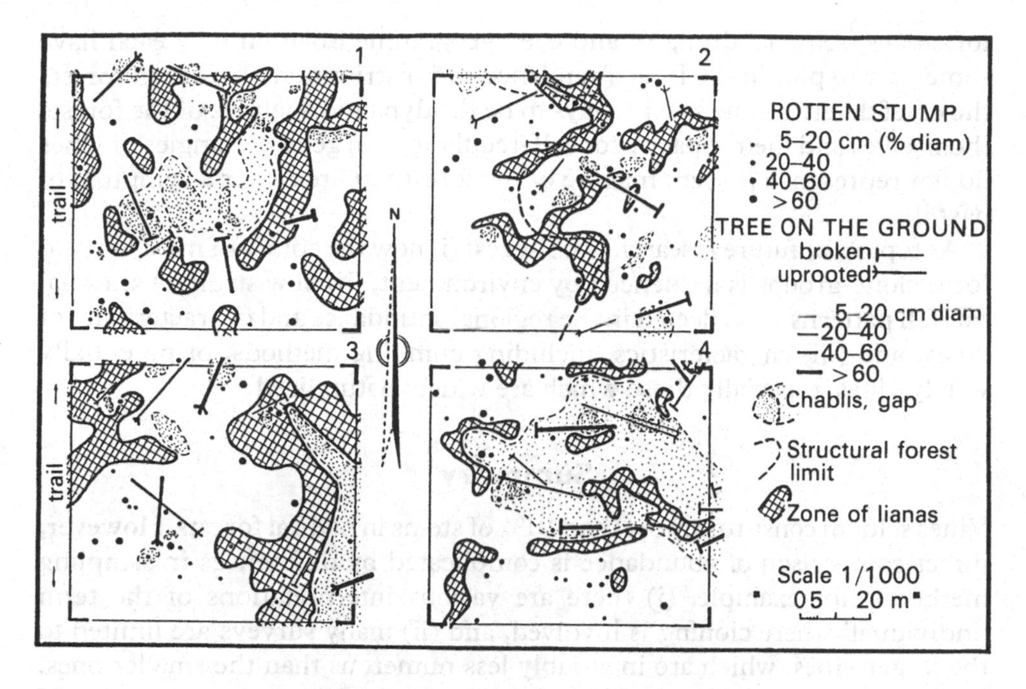

Figure 11.3. Examples of distribution of lianas (⩾5 cm in diameter) in evergreen forest in Gabon (Site: Makokou). Areas of continuous canopy are unshaded.

Conclusions

In woodland and forest, climbing plants combine a diversity of form with opportunistic strategies to place their foliage in its most productive position. The structure of the canopy, which intercepts light and diffuses it to lower strata, chiefly determines the local abundance of vines within individual forests, but the differing abundance of vines in various regions is not so readily explained. For example, light conditions in the canopy of most tropical forests are more or less similar, but it has been observed that vines are not equally abundant in tropical forests. Clearly, access to light, aided by disturbance, is not the only reason for the great abundance of vines in certain tropical forests.

Vines respond to irregularities in the architecture of the forest canopy by forming aggregations (clumps). In general, we can interpret clumps of vines as biological markers of forest disturbance or successive structural crises, signifying some present or former increase in entropy of the system. We conclude that invasion of forest by vines, at the community scale, is determined almost completely by the size and frequency of canopy gaps and by the type and persistence of disturbances. While such factors as individual light requirements of vines, soil characteristics, frequency of storms and tornadoes, seasonal drought, and even geographic isolation may each have some part to play in the local abundance and distribution of vines in forests the chief determinants seem always to be the dynamic qualities of the forests themselves and their capacity for self-regulation. In general, tangles of vines do not represent a present malaise of the forest but are rather a symptom of past ills.

As topics for future research, we suggest (i) how the ability of most vines to form clone-groups is influenced by environment, (ii) how strongly seasonal rainfall patterns are reflected in the regional abundance and characteristics of vines, and (iii) characteristics, including climbing methods, of potentially weedy vines, especially those which are widely naturalized.

Summary

Vines seldom constitute more than 10% of stems in natural forests. However, direct comparison of abundance is complicated by differences in sampling method – for example, (i) there are various interpretations of the term 'individual' where cloning is involved, and (ii) many surveys are limited to the larger vines, which are invariably less numerous than the smaller ones. Despite some reported contributions by rainforest vines of a third or more of total leaf area or litterfall (explained by high leaf:stem ratios and species with short-lived leaves) vines seldom contribute more than 2% of total forest Basal Area, or 5% of standing biomass. These percentages vary depending on the local history of disturbances, with most vines, including the weedy exotic species, tending to clump in and around sites of present and recent disturbances to forest structure. A range of active and passive climbing methods assists vines to reach the canopy, with tendrillar species most abundant in and near sunny gaps, where the tree canopy is low, and root (bole-) climbers most abundant at a later stage when the tree canopy is well developed. The high abundance of vines associated with rainforest tends to decrease with increasing altitude or latitude, or in absence of disturbance, but the probable influences of other factors such as derivation of soils, occasional moisture deficits and geographic isolation are difficult to estimate at present.

References

Appanah, S. & Putz, F. E. (1984). Climber abundance in evergreen dipterocarp forest and the effect of pre-felling climber cutting on logging damage. *Malaysian Forester* 47: 335–42.

Beekman, F. (1981). *Structural and Dynamic Aspects of the Occurrence and Development of Lianes in the Tropical Rain Forest*. Department of Forestry, Agricultural University, Wageningen.

Beekman, F. (1984). La dynamique d'une forêt alluviale rhénane et le rôle des lianes. Pp. 475–501. In *La Végétation des Forêts Alluviales.* Colloques phytosociologiques, Forêts Alluviales Européennes, Strasbourg 1980, édition J. Cramer, Vaduz.

Bernhard-Reversat, F., Huttel, C. & Lemée, G. (1979). Structure et fonctionnement des écosystèmes de la forêt pluvieuse sempervirente de côte d'Ivoire. In *Ecosystèmes Forestiers Tropicaux*, pp. 605–25. UNESCO, Paris.

Bongers, F., Engelen, D. & Klinge, H. (1985). Phytomass structure of natural plant communities on spodosols in southern Venezuela: the Bana woodland. *Vegetatio* **63**: 13–34.

Bongers, F., Popma, J., del Castillo, J. M. & Carabias, J. (1988). Structure and floristic composition of the lowland rain forest of Los Tuxtlas, Mexico. *Vegetatio* 74: 55–80.

Bullock, S. H. (1990). Abundance and allometrics of vines and self-supporting plants in a tropical deciduous forest. *Biotropica* 22: 106–9.

Caballé G. (1984). Essai sur la dynamique des peuplements de lianes ligneuses d'une forêt du Nord-Est du Gabon. *Revue d' Ecologie (Terre et Vie)* **39**; 3–35.

Caballé G. (1986a). Sur la biologie des lianes ligneuses en forêt gabonaise. Thèse Doct. Etat, Université des Sciences et Technique du Languedoc, Montpellier, France.

Caballé G. (1986b). Les peuplements de lianes ligneuses dans un forêt du Nord-Est du Gabon. *Mémoires du Muséum National d'Histoire Naturelle (Paris) Sér. A* **132**: 91–6.

Castellanos, A. E., Mooney, H. A., Bullock, S. H., Jones, C. & Robichaux, R. (1989). Leaf, stem and metamer characteristics of vines in a tropical deciduous forest in Jalisco, Mexico. *Biotropica* 21: 41–9.

Costantin, J. (1899). *La Nature Tropicale.* F. Alcan, Paris.

Darwin, C. (1875). *The Movements and Habits of Climbing Plants.* Murray, London.

Darwin, C. (1882). *La Faculté motrice dans les Plantes* (transl. from English by E. Heckel). Reinwald, Paris.

Edwards, P. J. & Grubb, P. J. (1977). Studies of mineral cycling in a montane rain forest in New Guinea. I. The distribution of organic matter in the vegetation and soil. *Journal of Ecology* **65**: 943–69.

Fittkau, E. J. & Klinge, H. (1973). On biomass and trophic structure of the central Amazonian rain forest ecosystem. *Biotropica* 5: 2–14.

Gentry, A. H. (1982). Patterns of Neotropical plant diversity. *Evolutionary Biology* **15**: 1–84.

Gentry, A. H. (1983). Lianas and the 'paradox' of contrasting latitudinal gradients in wood and litter production. *Tropical Ecology* **24**: 63–7.

Gentry, A. H. (1988a). Tree species richness of upper Amazonian forests. *Proceedings of the National Academy of Sciences (USA)* 85: 156–9.

Gentry, A. H. (1988b). Changes in plant community diversity and floristic composition on environmental and geographical gradients. *Annals of the Missouri Botanical Garden* 75: 1–34.

Gentry, A. H. & Dodson, C. (1987). Contribution of nontrees to species richness of a tropical rain forest. *Biotropica* **19**: 149–56.

Greenland, D. J. & Kowal, J. M. L. (1960). Nutrient content of the moist tropical forest of Ghana. *Plant and Soil* **12**: 154–74.

Grubb, P. J., Lloyd, J. R., Pennington, T. D. & Whitmore, T. C. (1963). A comparison of montane and lowland rain forest in Ecuador. I. The forest structure, physiognomy and floristics. *Journal of Ecology* **51**: 567–601.

Hall, J. B. & Swaine, M. D. (1981). Distribution and ecology of vascular plants in a tropical rainforest: Forest vegetation in Ghana. *Geobotany 1*. Dr W. Junk, The Hague.

Hegarty, E. E. (1988). Canopy dynamics of lianes and trees in subtropical rainforest. PhD thesis, Botany Department, Univerity of Queensland.

Hegarty, E. E. (1989). The climbers – lianes and vines. In *Tropical Rain Forest Ecosystems: Biogeographical and Ecological Studies. Ecosystems of the World 14B*, ed. H. Lieth and M. J. Werger, pp. 339–53. Elsevier, Amsterdam.

Hegarty, E. E. (1990) Leaf life-span and leafing phenology of lianes and associated trees during a rainforest succession. *Journal of Ecology* **78**: 300–12.

Hladik, A. (1974). Importance des lianes dans la production foliaire de la forêt équatoriale du Nord-Est du Gabon. *Comptes Rendus, Académie des Sciences (Paris), Sér. D* **278**: 2527–30.

Holdridge, L. R., Grenke, W. C., Hatheway, W. H., Liang, T. & Tosi, J. A. Jr (1971). *Forest Environments in Tropical Life Zones*. Pergamon Press, New York.

Hozumi, K., Yoda, K., Kokawa, S. & Kira, T. (1969). Production ecology of tropical rain forests in south-west Cambodia. 1. Plant biomass. *Nature and Life in South-East Asia* **6**: 1–51.

Jaffré, T. & de Namur, C. (1983). Evolution de la biomasse épigée au cours de la succession secondaire dans le Sud-Ouest de la Côte d'Ivoire. *Acta Oecologica: Oecologia Plantarum* **4**: 259–72.

Jordan, C. F. & Murphy, P. G. (1978). A latitudinal gradient of wood and litter production, and implication regarding competition and species diversity in trees. *American Midland Naturalist* **99**: 431–4.

Jordan, C. F. & Uhl, C. (1978). Biomass of a 'tierra firme' forest of the Amazon basin. *Oecologia Plantarum* **13**: 387–400.

Kahn, F. (1982). La reconstitution de la fôret tropicale humide, Sud-Ouest de la Côte d'Ivoire. *Office Recherche Scientifique Technique Outre Mer, Mémoire 97.*

Kato, R., Tadaki, Y. & Ogawa, H. (1978). Plant biomass and growth increment studies in Pasoh forest. *Malayan Nature Journal* **30**: 211–24.

Kelly, D. L. (1985). Epiphytes and climbers of a Jamaican rainforest: vertical distribution, life forms and life histories. *Journal of Biogeography* **12**: 223–41.

Kira, T. & Ogawa, H. (1971). Assessment of primary production in tropical and equatorial forests. In *Productivity of Forest Ecosystems, Proceedings of Brussels Symposium 1969*, ed. P. Duvigneaud, pp. 309–21. UNESCO, Paris.

Klinge, H. & Herrera, R. (1983). Phytomass structure of natural plant communities on spodosols in southern Venezuela: the tall Amazon caatinga forest. *Vegetatio* **53**: 65–84.

Klinge, H. & Rodrigues, W. W. (1973). Biomass estimation in a central Amazonian rain forest. *Acta Cientifica Venezolana* **24**: 225–37.

Lescure, J. P. (1981). La Végétation et la flore dans la région de la Piste de St Elie. Bulletin de liaison du groupe de travail Ecosystème forestier Guyanais. *Office Recherche Scientifique Technique Outre Mer Cayenne, 3*: pp. 21–24.

Lescure, J. P., Puig, H., Riera, B., Leclerc, D., Beekman, A. & Beneteau, A. (1983). La phytomasse épigée d'une forêt dense en Guyane française, *Acta Oecologica: Oecologia Generalis* **4**: 237–51.

Lott, E. J., Bullock, S. H., & Magallanes, A. J. Arturo S. (1987). Floristic diversity and structure of upland and aroyo forests of coastal Jalisco. *Biotropica* **19**: 228–35.

Massart, J. (1895). Sur la morphologie du bourgeon. I. La différentiation raméale chez les lianes. *Annales du Jardin Botanique de Buitenzorg* **13**: 121–36.

Massart, J. (1906). *Les lianes, leurs moeurs, leurs structures*. Bulletin de la Société Centrale Forestière, Brussels.

Monsi, M. & Ogawa, K. (1977). Ecological considerations on some characteristics of liana communities. In *Proceedings of international symposium on protection of the environment, Tokyo 1974*, pp. 325–36. Maruzen, Tokyo.

Mueller-Dombois, D. (1975). Some aspects of island ecosystem analysis. In *Tropical Ecological Systems, Ecological Studies 11*, ed. F. B. Golley and E. Medina, pp. 353–66. Springer, Berlin.

Ogawa, H., Yoda, K. & Kira, T. (1961). A preliminary survey on the vegetation of Thailand. *Nature and Life in South-East Asia* **1**: 21–157.

Ogawa, H., Yoda, K., Kira, T., Ogino, K., Shidei, T., Ratanawongse, D. & Apasatuya, C. (1965a). Comparative ecological study on three main types of forest vegetation in Thailand. I. Structure and floristic composition. *Nature and Life in South-East Asia* 4: 13–48.
Ogawa, H., Yoda, K., Ogino, K. & Kira, T. (1965b). Comparative ecological study on three main types of forest vegetation in Thailand. II. Plant Biomass. *Nature and Life in South-East Asia* 4: 49–80.
Ohler, F. M. J. (1980). *Phytomass and Mineral Content in Untouched Forest*. CELOS Report 132, University of Surinam.
Peñalosa, J. (1984). Basal branching and vegetative spread in two tropical rain forest lianas. *Biotropica* **16**: 1–9.
Proctor, J., Anderson, J. M., Fogden, S. C. L. & Vallack, H. W. (1983). Ecological studies in four contrasting lowland rain forests in Gunung Mulu National Park, Sarawak. II. Forest environment, structure and floristics. *Journal of Ecology* 71: 237–60.
Putz, F. E. (1983). Liana biomass and leaf area of a 'Tierra Firme' forest in the Rio Negro basin, Venezuela. *Biotropica* 15: 185–9.
Putz, F. E. (1984). The natural history of lianas on Barro Colorado Island, Panama. *Ecology* **65**: 1713–24.
Putz, F. E. (1990). Liana stem growth and mortality rates on Barro Colorado Island, Panama. *Biotropica* 22: 103–5.
Putz, F. E. & Chai, P. (1987). Ecological studies of lianas in Lambir National Park, Sarawak, Malaysia. *Journal of Ecology* 75: 523–31.
Rich, P. M., Lum, S., Muñoz, E. L. & Quesada, A. M. (1987). Shedding of vines by the palms *Welfia georgii* and *Iriartea gigantea*. *Principes* **31**: 31–40.
Richards, P. W. (1964). *The Tropical Rain Forest*. Cambridge University Press, Cambridge.
Rollet, B. (1969a). Etudes quantitatives d'une forêt dense humide sempervirente de plaine de la Guyane Vénezuelienne. Thèse Doct. Etat., University of Toulouse, 4 volumes.
Rollet, B. (1969b). La régeneration naturelle en forêt dense humide sempervirente de plaine de Guyane Vénézuelienne. *Bois et Forêts des Tropiques* **114**: 19–38.
Rollet, B. (1971). La regeneración naturel en bosque denso siempreverde de llanura de al Guayana Venezolana. *Instituto Forestal Latino-Americano de Investigacion y Capacitacion, Merida Venez* **35**: 39–73.
Schenck, H. (1892). Beiträge zur Biologie und Anatomie der Lianen im Besonderen der in Brasilien einheimischen Arten. 1. Beiträge zur Biologie der Lianen. In *Botanische Mittheilungen aus den Tropen, 4*, ed. A. F. W. Schimper, pp. 1–253. G. Fischer, Jena.
Schenck, H. (1893). Beiträge zur Biologie und Anatomie der Lianen im

Besonderen der in Brasilien einheimischen Arten. 2. Beiträge zur Anatomie der Lianen. In *Botanische Mittheilungen aus der Tropen, 5*, ed. A. F. W. Schimper, pp. 1–271. G. Fisher, Jena.

Schnell, R. (1970). *Introduction à la phytogéographie des pays tropicaux. 1. Les flores – les structures.* Gauthier-Villars, Paris.

Tanner, E. V. J. (1980). Studies on the biomass and productivity in a series of montane rain forests in Jamaica. *Journal of Ecology* **68**: 573–88.

Treub, M. (1883a). Sur une nouvelle catégorie de plantes grimpantes. *Annales du Jardin Botanique de Buitenzorg* **3**: 44–75.

Treub, M. (1883b). Observations sur les plantes grimpantes du jardin botanique de Buitenzorg. *Annales du Jardin Botanique de Buitenzorg* **3**: 160–82.

van Schaik, C. P. & Mirmanto, E. (1985). Spatial variation in the structure and litterfall of a Sumatran rain forest. *Biotropica* **17**: 196–205.

Vareschi, V. (1980). *Vegetations-ökologie der Tropen.* Eugen Ulmer, Stuttgart.

Walter, J. M. (1974). Arbres et forêts alluviales du Rhin. *Bulletin Société Histoire Naturelle, Colmar* **55**: 37–88.

Walter, J. M. (1979a). Etude des structures spatiales en forêt alluviale rhénane. I. Problèmes structuraux et données expérimentales. *Oecologia Plantarum* **14**: 345–59.

Walter, J. M. (1979b). Etude des structures spatiales en forêt alluviale rhénane. V. L'architecture forestière observée. *Oecologia Plantarum* **14**: 401–10.

Whitmore, T. C. (1974). *Change with time and the role of cyclones in tropical rain forest on Kolombangara, Solomon Islands.* Paper 46, Commonwealth Forestry Institute, University of Oxford.

Whitmore, T. C. (1989). Canopy gaps and the two major groups of forest trees. *Ecology* **70**: 536–8.

Williams, K., Field, C. B. & Mooney, H. A. (1989). Relationships among leaf construction cost, leaf longevity and light environment in rain-forest plants of the genus *Piper*. *American Naturalist* **133**: 198–211.

12

Vines in arid and semi-arid ecosystems

PHILIP W. RUNDEL AND TAMARA FRANKLIN

While vines and woody lianas are important growth forms in tropical forest ecosystems, and a noticeable component in many temperate forest ecosystems, they are largely ignored in discussions of the floras of arid and semi-arid ecosystems of the world. Nevertheless, vines represent a small but fascinating component of these floras, and provide insights into modes of adaptive strategies in the tolerance of drought and heat stress. In this chapter we first review the variation in growth forms of arid zone vines, then describe the floristic diversity of vines in mediterranean-type and desert ecosystems, and finally discuss the adaptations of arid zone vines to environmental stress.

Growth forms

Vine species in arid and semi-arid areas of the world occur in a remarkably broad range of growth forms although none of these is unique to such areas. At one extreme are woody lianas, which may occur as both evergreen and deciduous species. Species with this growth form may grow up over shrubs to form their own canopies in full sun, as do tropical lianas, or may grow high into woodland trees while remaining within the shaded canopy of their host species. *Ephedra pedunculata*, unusual as a liana in the Gymnospermae (Cutler, 1939), has been observed to use telephone poles as a substrate in the Chihuahuan Desert region of south Texas (S. Carlquist, personal communication).

At the other extreme are annual vine species that commonly trail along the ground or grow up into the lower branches of shrubs or suffrutescent species. These annual vine species, most notable in the Fabaceae, Convolvulaceae and Cucurbitaceae, are important in the mediterranean-climate regions of Europe and California, and in the Australian and Sonoran Deserts (see below). Annual vines lack the carbohydrate storage tissues present in stems of lianas and in succulent roots of herbaceous perennial vines.

The great majority of arid zone vines are herbaceous perennials with perennial storage tissues below ground and ephemeral stems and leaves above ground. With few exceptions, these herbaceous perennial vines are characterized by succulent storage tissues, usually as below-ground tubers. These below-ground storage organs have been classified into a number of morphological forms by Pate & Dixon (1982). For vines, these include perennial or ephemeral below-ground stem tubers, root tubers on true roots, and root tubers on adventitious roots. To these forms can be added a group of African and North American vine species, largely in the Cucurbitaceae, with above-ground succulent trunks and ephemeral stems. The succulent root tubers of many arid zone Cucurbitaceae may be massive (Stocking, 1955; Dittmer & Rosen, 1963; Dittmer & Talley, 1964), weighing as much as 100 kg. The phylogenetic distribution of root and stem succulent species, many of which are vines, are described by Jacobsen (1960) and Rowley (1987).

The habitats of herbaceous perennial vines are highly variable. In desert ecosystems, many of these vines are trailing species which spread out over rocks on the ground surface, sprawling without rooting at nodes. In shrublands, however, many species with a viney growth form are slender ephemerals which grow up into the shade canopy of low shrubs.

A final growth form of arid vines includes parasitic plants which climb over host shrubs and even herbaceous perennials. Only two genera are represented in this growth form, *Cassytha* in the Lauraceae and *Cuscuta* in the Convolvulaceae.

Floristic diversity

Vines are not treated as an unambiguous growth form in most floristic treatments, and thus data on the floristic diversity of vines in arid and semi-arid ecosystems are difficult to obtain. While lianas and climbing perennial vines are readily evident in the species descriptions of most floras, annual vines and trailing herbaceous vines are not always designated by their growth form. Scandent shrub and suffrutescent species also present problems of interpretation. We have chosen to classify obligately scandent species as vines. In this section we describe general floristic patterns of occurrence for vine growth forms in representative areas within mediterranean-type and desert ecosystems of the world (Table 12.1), and highlight unusual biogeographic and phylogenetic occurrences of vine growth forms.

Mediterranean-climate vine floras

The flora of the Mediterranean Basin of Europe and North Africa has a moderately large number of vine species, with 94 vines among the approxi-

Table 12.1. *Floristic distribution of vine species in mediterranean-climate and desert ecosystems. Data are given for representative areas within each region on the relative frequency of vine growth forms in the native vascular plant flora, the relative frequency of woody vines among all vines, and the relative frequency of parasitic vines among all vines. See text for sources of data*

	Total angiosperms	Total vine frequency (%)	Woody vine frequency (%)	Parasitic vine frequency (%)
Mediterranean-type ecosystems				
Mediterranean basin	2400	4.0	10.6	3.2
California	1390	2.4	41.2	20.6
South Africa	2622	1.4	50.0	11.1
Western Australia	1510	2.0	61.3	13.1
Chile	654	8.5	19.2	3.8
Desert ecosystems				
Mojave Desert	1093	1.3	7.1	7.1
Central Australia	1992	2.0	17.9	28.2
Arabian Desert	334	2.7	11.1	22.2
Namib Desert	127	1.6	0.0	0.0
Sonoran Desert	2441	7.0	24.1	11.2

mately 2400 species of vascular plants from the Iberian Peninsula (Polunin & Smythies, 1988). Woody climbers are rare, however, and limited to species of *Clematis* (Ranunculaceae), *Vitis* (Vitaceae), and *Lonicera* (Caprifoliaceae), the latter the only evergreen liana. Nearly 70% of the vine flora of south-western Europe comes from two herbaceous genera of legumes, *Vicia* and *Lathyrus*. Two important genera of ornamental vines, *Hedera* (Araliaceae) and *Vinca* (Apocynaceae) come from the Mediterranean Basin, and have become invasive weeds in other mediterranean-climate regions.

The mediterranean-climate region of California is relatively poor in vine species. The flora of the Santa Barbara region, for example, includes only 34 species of vines among 1390 native vascular plants (Hoover, 1970). The most widespread and abundant woody vine in mediterranean-climate regions of California is *Toxicodendron diversilobum* (poison oak) in the Anacardiaceae, a family not notable for vines. This morphologically plastic species may grow as an herbaceous perennial, shrub or woody vine twining to 30 m or more in forest habitats. Other woody genera include *Clematis* (Ranunculaceae). *Lonicera* (Caprifoliaceae) and *Rubus* (Rosaceae). *Lonicera* species are again the only evergreens among these. Herbaceous species of the genus *Marah*

(Cucurbitaceae) with large fleshy tubers are important elements in many chaparral and woodland habitats, particularly following disturbance.

Despite its very high overall species diversity, the mediterranean-climate region of South Africa is not notably rich in vines. Only 36 of the 2622 species in the flora of the Cape Peninsula are vines (Levyns, 1966). Woody vines include species of *Asparagus* (Liliaceae), *Clematis branchiata* (Ranunculaceae), *Antizoma capensis* (Menispermaceae), *Rhoicissus tomentosa* (Vitaceae), and *Secamone alpinii* (Asclepiadaceae). These are most common in forested areas rather than in shrubby fynbos communities. Only two species of Cucurbitaceae and six species of viney legumes are present, relatively small numbers for these families.

The mediterranean-climate region of Western Australia is likewise not floristically rich in vines. Only 30 vine species are present among the 1510 species included in the flora of the Perth region (Marchant *et al.*, 1987). More than 60% of these are woody or semi-woody species, a higher percentage than in any other mediterranean region (Table 12.1). Surveys of coastal dune communities near Perth have reported 4.0% of vascular plant species as vines, while heathland communities ranged from 1.6 to 2.3% vines in their floras (Beard, 1980). While this diversity of vines is not high, Western Australia is notable for the evolution of viney growth forms in families not otherwise well known for climbing. One such family is the Pittosporaceae with twining shrubs present in the genera *Sollya*, *Pronaya*, *Cheiranthera* and *Billardiena*. Another unusual twining shrub is *Comesperma volubile* in the Polygalaceae. The carnivorous sundews comprising the genus *Drosera* are normally thought of as small herbs, but a number of trailing species have evolved in Western Australia where they grow for a meter or more into shrubs. Two species of *Stylidium* (Stylidiaceae) in the Franklin River area of southwestern Australia twine upward over shrubs and restiods (Carlquist, 1974). At least two species of *Thysanotus*, a genus of typically tufted monocotyledons in the Liliaceae, have twining stems which extend up and over adjacent shrubs. The vine habit is present even in aquatic plants in Western Australia, with *Utricularia volubilis* forming twining flowering shoots which may climb for nearly a meter up the stems of aquatic monocotyledons (Erickson, 1968). Even in a family such as the Fabaceae where woody vines are not unusual, as in *Hardenbergia* and *Kennedia*, the presence of two species of vines in the genus *Acacia* in southwestern Australia is unexpected. Climbing species of *Acacia* are otherwise limited to the tropics.

On a relative basis, the richest mediterranean-climate region for vine species is that of central Chile. Vines comprise nearly 8% of the vascular plant flora of the Santiago basin (Navas, 1973–9). This is four times the frequency of vine species in many other mediteranean-climate regions (Table 12.1). The diversity of vines in Chile is notable not only for the number of species but also for the interesting patterns of phytogeographic and phylogenetic distribution of both herbaceous and woody vines. One unusual element in this flora is the presence of viney Asteraceae, a family not noted for this growth form. Woody vines occur in two genera of the tribe Mutisiae. These include a number of species of *Mutisia*, and *Proustia pyrifolia*, the latter unusual as an evergreen liana in a genus otherwise composed entirely of woody, xeromorphic shrubs. The presence of the genera *Lardizabala* and *Boquila* (Lardizabalaceae) in central and austral Chile represent the only New World occurrence of this Asian family of woody vines (Carlquist, 1984). *Muhlenbeckia hastulata*, a common woody shrub in central Chile with twining branches and evergreen leaves, is an unusual (but not unique) example of a vine growth form in the Polygalaceae.

Herbaceous vines are no less interesting in central Chile. The Bignoniaceae, an important tropical family of woody vines, is represented in central Chile by the suffrutescent *Eccremocarpus scaber*, the only member of this family to occur in a mediterranean climate. An interesting component of the Chilean vine flora is the genus *Tropaeolum* (Tropaeolaceae), a diverse assemblage of tuberous, herbaceous vines from central and northern Chile. The Loasaceae, a temperate family of herbs and small shrubs, is represented by three genera with herbaceous climbers in central Chile: *Loasa*, *Cajophora* and *Scyphanthus*. The Cucurbitaceae, commonly an important vine family, has only a single species of *Sicyos* in central Chile with this growth form.

Desert vine floras

Warm desert floras commonly include only 1–3% vines, a figure comparable to the range for floras in mediterranean-type ecosystems. One notable difference, however, lies in the relative paucity of woody vines in desert regions and the relative importance of annual herbaceous vines.

The Mojave and adjacent Great Basin Desert in South–central Nevada include only 14 species of vines among the 1093 species present (Beatley, 1976). The only woody vine, a species of *Clematis*, is restricted to the highest elevations and pinyon–juniper margins of this region.

The arid flora of central Australia has 40 species of vines present within a vascular plant flora of 1992 species (Jessop, 1981), with woody vines largely restricted to stream-banks and some acacia stands. These woody species include *Tinospora smilacina* (Menispermaceae), *Jasminum* species (Oleaceae), and *Porana sericea* and *Ipomoea costata* (Convolvulaceae). Members of the Convolvulaceae are particularly important in the vine flora of central Australia, comprising almost half of the species distributed among eight genera with growth forms ranging from annuals to herbaceous perennials to woody vines. Unlike most arid zones where the great majority of vines are perennials, nearly one-third of the vine flora from central Australia are annuals.

Qatar, an area of the Saharan Desert on the Arabian Peninsula, has nine species of native vines in a flora of 334 species. Only a single species of woody vine, *Cocculus pendulus*, is present, and it is quite rare (Batanouny, 1981).

The Namib Desert in southwestern Africa shows little more diversity of vines. Only two of the 127 species reported by Giess (1962–7) have vine growth forms. Both of these are trailing perennials with underground rootstocks growing in washes or dry watercourses. There may be evidence in the Namib for selection against the vine growth form in the presence of woody shrubs in families where this growth form is rare. *Acanthosicyos horrida* (Cucurbitaceae) is an ecologically important woody shrub with leafless spiney branches in dune areas of the Namib. *Ipomaea arenoides* (Convolvulaceae) is an upright shrub with few branches.

Desert regions, however, are not inherently poor in vine species. An exception to the general pattern of low vine diversity and few vines present can be seen in the rich vine flora of the Sonoran Desert. While vines are not abundant in the winter rainfall and hyper-arid portions on the temperate margin of this desert in the USA, the less xeric southern margin of the Sonoran Desert represents the northern extension of many subtropical vine groups including a large diversity of woody lianas. The Sonoran Desert flora (Shreve & Wiggins, 1964) includes 62 genera and 170 species of vines, 7% of the flora. More than a quarter of these species are woody or suffrutescent. The three most important vine families in this flora are the Cucurbitaceae (15 genera, 29 species), the Convolvulaceae (9 genera, 35 species), and the Asclepiadaceae (7 genera, 26 species). Legumes contribute only 5 genera and 18 species. Examples of tropical families represented in this desert vine flora include the Malpighiaceae (*Mascagnia*, *Gaudichaudia*, *Heteropteris* and *Janusia*), Euphorbiaceae (*Tragia* and *Dalechampia*), Sapindaceae (*Serjania*, *Cardiospermum*, *Paullinia* and *Gousania*), Passifloraceae (*Passiflora*), and Bignoniaceae (*Bignonia*). Two cactus genera with fleshy underground storage tissues (*Wilcoxia*, *Peniocereus*) become vine-like in growth habit. Annual vines are relatively numerous in the Sonoran Desert flora. They occur in many families including the Fabaceae (*Vicia*), Convolvulaceae (*Jacquemontia*, *Ipomoea*, *Quamoclit*), and the Cucurbitaceae (*Luffa*, *Schizocarpum*, *Echinopepon*, *Sicyosperma* and *Sicyos*).

Adaptations to environmental stress

Our knowledge of the form and functional modes by which herbaceous and woody vines adapt themselves to the environmental stresses of drought and heat in arid environments is very limited. Much of the small amount of existing data on the physiological ecology of desert vines comes from studies of perennial Cucurbitaceae. This group is characterized by large leaves, ephemeral summer or winter deciduous stem and leaf tissues, and large succulent roots. These specialized characteristics mean that their adaptive strategies to tolerate environmental stress may be quite different from those present in herbaceous vines without perennial storage tissues and in woody lianas.

Water relations

Almost nothing is known about the water relations of arid and semi-arid vines. The few data that exist suggest that many of these species demonstrate remarkable buffering against physiological drought stress. Comparative studies of the water relations of wash plants in the Sonoran Desert of California found that midday summer water potentials of *Cucurbita digitata* were only −1.7 MPa, an unusually high level for such an environment in summer (Nilsen, Sharifi & Rundel, 1984). In contrast, drought-deciduous shrubs and winter-deciduous woody legumes at the same site had mean midday potentials of about −4.0 MPa, while evergreen shrubs were below −6.0 MPa (Figure 12.1).

One component of physiological buffering against water stress may lie in the large storage root which characterizes most species of arid zone perennial vines with ephemeral above-ground tissues. The massive taproot of *Cucurbita foetidissima*, for example, may weigh as much as 72 kg, while that of *Apodanthera undulata* reaches 57 kg (Dittmer & Rosen, 1963). Despite its buffered water potentials, *C. digitata* has been found to have smaller taproots weighing less than 3 kg (Dittmer & Rosen, 1963). Tubers up to 90 kg have been reported in *Marah* (Stocking, 1955). It would appear, therefore, that water storage in these fleshy taproots may have an important function in moderating diurnal shifts in water stress, as has been shown for stem-succulent species of *Espeletia* in the paramo of Venezuela (Goldstein, Meinzer & Monasterio, 1984), but not sufficient water storage to buffer seasonal drought stress in arid and semi-arid climates. Deep root systems must also be present in species which maintain active photosynthetic tissues through the summer drought period. Deep roots have been reported for *Citrullus colocynthis* (Cucurbitaceae) in Indian desert areas (Sen & Bhandari, 1974).

An interesting comparison can be seen in the water relations of *Marah macrocarpus* and *Cucurbita foetidissima* growing together in the coastal sage scrub of southern California. While both species have fleshy storage roots, their phenology is different. *Marah* initiates new shoots in December, but dies back in May. In contrast, *C. foetidissima* is dormant over the winter and

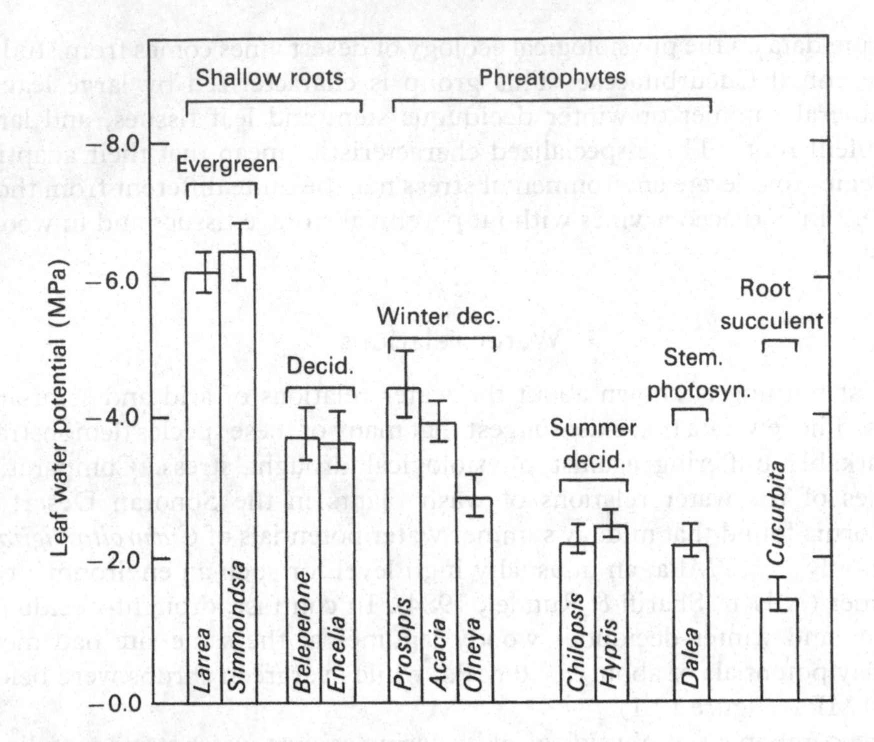

Figure 12.1. Comparative midday water potentials in 11 wash woodland species, including *Cucurbita digitata* in August 1981. Error bars equal ±2 standard errors of the means of six measurements for each species. From Nilsen *et al.* (1984).

initiates new growth with higher temperatures in spring. It maintains physiological activity and growth all through the summer. Both species showed relatively high midday water potentials (low stress) in April: −0.9 MPa for *Marah* and −0.35 MPa for *C. foetidissima* (Figure 12.2). Aboveground senescence, nevertheless, occurred in *Marah* by mid-May. Midday water potentials in *Cucurbita* remained between −0.3 and −0.5 MPa through July, at which time they increased to −0.8 MPa. This moderately high level of water potential (low stress) was maintained until sensecence began in late August.

Pressure–volume curve analyses of leaves of these two species have indicated that midday water potentials at the onset of senescence were close to calculated values of the osmotic potential at zero turgor. *Marah macrocarpus* had an osmotic potential at zero turgor of −1.05 MPa, compared with −1.00 at full turgor. For *C. foetidissima* these values were −0.77 and −0.67 MPa, respectively. This small difference between osmotic potentials at zero and full turgor suggests that these plants are very sensitive to small changes in

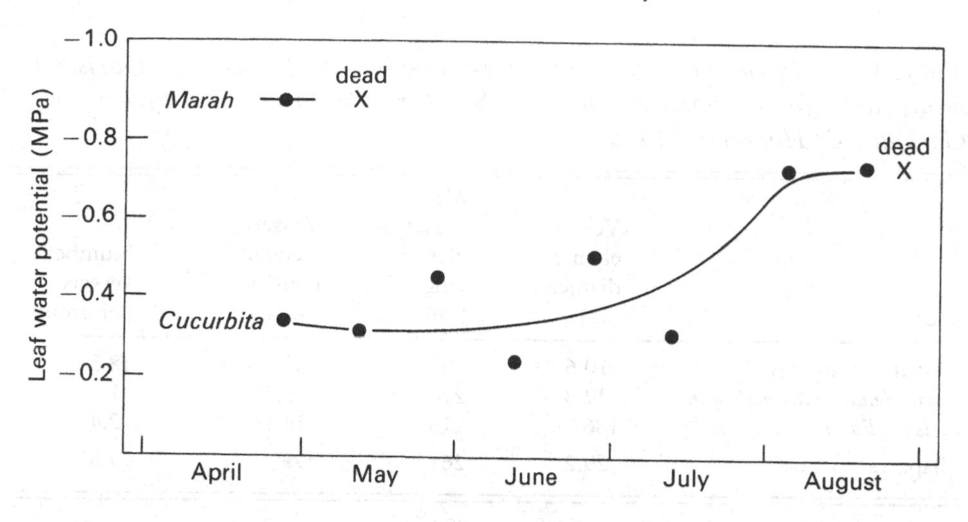

Figure 12.2. Seasonal change in midday leaf water potential in *Marah macrocarpus* and *Cucurbita foetidissima* growing in the Ecological Reserve at the University of California, Irvine during spring and summer 1981.

tissue water potential. It is not clear, however, if water stress is a causal factor in the senescence of ephemeral tissues.

Classic studies on the water relations of succulent plants carried out by the Carnegie Institute of Washington in the early part of this century clearly demonstrated the tremendous buffering capacity of tubers in arid zone vines (Macdougal & Spaulding, 1910; Macdougal, 1912). Tuberous stems of *Ibervillea sonorae* (Cucurbitaceae) were removed from soil and placed on a dry laboratory shelf where they continued to produce green shoots each spring for eight years. At the end of this period these tuberous stems still retained 50% of the original fresh weight. Macdougal & Spaulding (1910) speculated that this species might be expected to maintain some growth capacity without water for a quarter of a century. Although *Ibervillea* appeared to demonstrate greater moisture retention and carbohydrate reserves after prolonged desiccation than the succulent stems of cacti, the root tubers of *Marah fabaceus* dried out more rapidly when removed from the soil, losing 75% of their water content in the first year.

The storage of carbohydrates in the fleshy tissues of perennial vines is arguably the primary function of these tissues (Mooney & Gartner, Chapter 6). Many vines, interestingly, appear to have large carbohydrate reserves well in excess of amounts needed for an annual cycle of new growth. It has been suggested that an additional function of these carbohydrates may be to provide osmoticum to generate sufficient root pressure to overcome xylem embolisms (Sperry *et al.*, 1987).

Table 12.2. *Xylem characteristics of three woody vine species in comparison to mean values for 41 chaparral shrubs in Southern California. Data from Carlquist & Hoekman (1985)*

Species	Vessel element diameter (μm)	Mean vessel element length (μm)	Vessel element density (no. mm⁻²)	Number of vessels per group
Clematis lasianthus	10.6	198	425	18.3
Toxicodendron diversilobum	37.3	216	65.7	1.9
Vitis girdiana	186	113	24.1	2.4
Chaparral shrubs	29.2	261	299	4.8

The hydraulic architecture of vines in arid and semi-arid environments has not been widely investigated in the past. The studies of several groups, now in progress, should add greatly to our knowledge in this area. Three species of vines were included in a survey of xylem structure in the native woody flora of southern California (Carlquist & Hoekman, 1985). Mean values for vessel element diameter, vessel element length and density of vessels per mm^2 of these three woody vines are compared in Table 12.2 with mean values for 41 chaparral shrubs. The vessel diameter of *Vitis girdiana* was the largest of any of the 207 species surveyed, with a size more typical of vessel diameters in mesic site lianas. Large vessel diameters have been shown to be more vulnerable to cavitation (Ewers, 1985), and may explain the importance of the relatively mesic microsites where *Vitis* grows. Vessel density was low in *Vitis*, and vessel element length was relatively short. Mean vessel element diameters and lengths in *Toxicodendron diversilobum* and *Clematis lasianthus* were well within the range of typical chaparral shrubs. *Clematis* was distinctive, however, in its relatively high density of vessels per mm^2 and number of vessels per group. Both of these characteristics, as well as the presence of vasicentric tracheids in *Clematis*, can be considered as adaptations to a xeric environment (Carlquist & Hoekman, 1985).

The significance of vasicentric tracheids as a drought survival mechanism has been stressed by Carlquist (1985). Three-quarters of the chaparral shrubs surveyed by Carlquist & Hoekman (1985) had this characteristic. Baas & Schweingruber (1987) included 13 species of woody vines from the Mediterranean Basin in a survey of ecological trends in the wood anatomy of more than 500 European species. None of these vines, including species of *Clematis* (see Carlquist, 1985), were reported to have vascicentric tracheids,

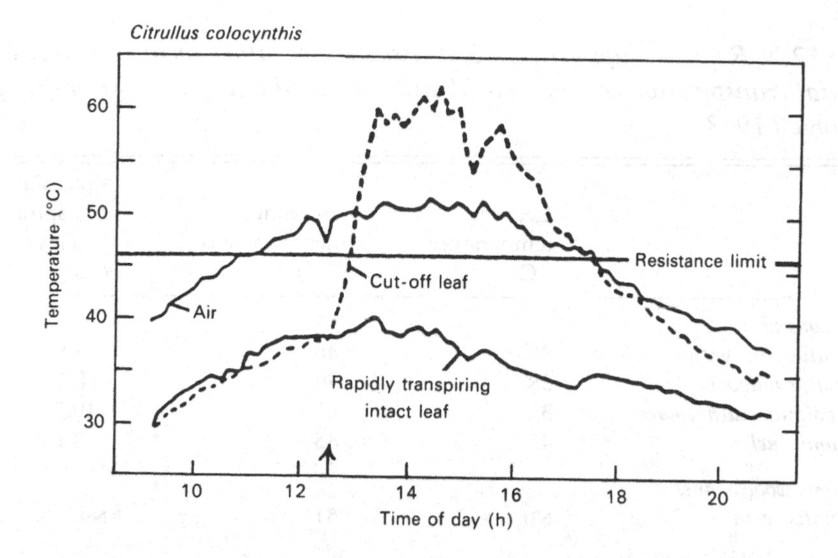

Figure 12.3. Effect of transpiration cooling on leaves of watered *Citrullus colocynthis* in the Saharan Desert. The effect of cutting a petiole on leaf temperature is shown by the dashed line. The heat resistance for leaf temperature is 46 °C. After Lange (1959).

suggesting that woody vines may be less drought-resistant in this aspect of their xylem anatomy than woody shrubs in the same habitats.

Energy balance

The broad, ephemeral leaves of many arid zone Cucurbitaceae may at first seem poorly adapted to the high irradiance levels present in these environments. However, they are able to utilize their large leaves for photosynthetic production without problems of heat damage by maintaining high rates of transpirational cooling. Studies of *Citrullus colocynthis*, a mesomorphic xerophyte common across the Sahara Desert to the Indian arid zone, have illustrated the significance of this transpirational cooling. The maximum heat tolerance of *C. colocynthis* is 46 °C, more than 6 °C lower than maximum ambient air temperatures during the summer in Mauritania (Lange, 1959). High rates of transpiration allow leaves to maintain midday temperatures 10–12 °C below ambient and thus remain well below lethal limits for heat. Classic experiments by Lange (1959) demonstrated that a leaf whose petiole was severed at midday increased rapidly in temperature to more than 60 °C in less than 60 min (Figure 12.3). Althawadi & Grace (1986) have described water use by this same species in Saudi Arabia where transpirational cooling is less extreme.

Table 12.3. *Relationships between leaf tissue temperature, heat tolerance and potential transpiration cooling in arid and semi-arid vines. Data from Lange & Lange (1963)*

	Leaf temperature (°C)	Maximum heat tolerance (°C)	Potential transpirational cooling (°C)
Herbaceous vines			
Citrullus colocynthis	36	46	8.8
Citrullus vulgaris	38	46	8.1
Convolvulus althaeoides	37	47	10.7
Cucumis melo	42	45	8.1
Deciduous woody vines			
Clematis recta	ND	51	ND
Evergreen woody vines			
Asparagus acutifolius	ND	54	ND
Lonicera implexa	48	52	1.6
Smilax aspera	46	55	−1.8

The heat tolerances of a number of vines and trailing species in an arid mediterranean-climate region of Spain were studied by Lange & Lange (1963). Four species of herbaceous vines, all native or cultivated Cucurbitaceae, had relatively modest heat tolerances of 45–47 °C compared to maximum leaf temperatures of 36–42 °C (Table 12.3). These species were able to cool their leaves 8–11 °C below ambient through transpiration. Leaves of evergreen woody vines, in contrast, had only a very small potential for transpirational cooling, and behaved quite differently from the herbaceous Cucurbitaceae. These species had maximum leaf temperatures of 46–48 °C, and lethal temperature limits at 52–55 °C (Table 12.2). In these temperatures, they behaved very similarly to leaves of evergreen maquis sclerophylls (Lange & Lange, 1963).

Not all ephemeral leaves of Cucurbitaceae maintain temperatures well below ambient. Schulze, Lange & Koch (1972) found that leaves of cultivated *Vitis vinifera* reached 44 °C during midday in July, three degrees over ambient air temperature. We have looked at seasonal patterns of leaf temperature in two species of Cucurbitaceae in the semi-arid coastal zone of southern California. *Marah macrocarpus*, a mesophytic vine which initiates growth as early as December, has large leaves 3–4 °C over midday ambient air temperature in the spring (Figure 12.4). The thick, pubescent leaves of *Cucurbita foetidissima* were generally close to ambient temperature during

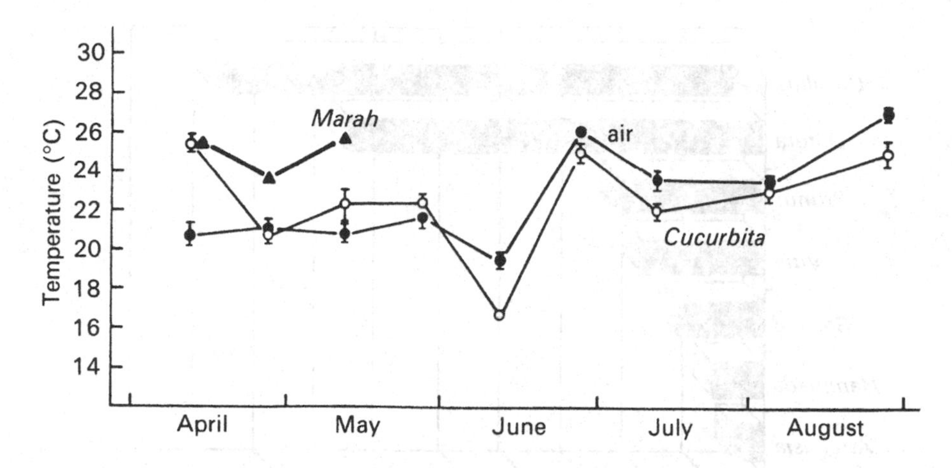

Figure 12.4. Seasonal change in midday leaf temperatures for *Marah macrocarpus* and *Cucurbita foetidissima* growing in the Ecological Reserve at the University of California, Irvine during spring and summer 1981.

the spring and dropped 2–3 °C below ambient during the summer months. During April and May 1982, when these data were collected, midday transpiration rates were more than 12 times higher in *Curcurbita* than in *Marah*.

Photosynthetic capacity

Many arid zone vines have notably rapid rates of shoot growth when soil moisture is available, suggesting that they may also have great photosynthetic capacity. High rates of transpirational cooling present in a number of arid Curcurbitaceae also suggests high levels of photosynthetic capacity, since leaf conductance and assimilation are closely linked (Schulze & Hall, 1982).

Citrullus colocynthis, previously described for its high degree of transpirational cooling, fits this model in maintaining a high maximum assimilation rate at the end of the summer dry season in the Negev Desert of Israel (Schulze *et al.*, 1972). It reached a peak assimilation rate of 45 mg g^{-1} h^{-1}, considerably higher than any other species studied (Figure 12.5), with little or no midday stomatal closure. This assimilation rate is comparable to that of agricultural crop plants. Midday leaf temperatures at this time were maintained 2–3 °C below ambient air temperatures. High photosynthetic capacity, however, is not necessary for the success of vines. A cultivated vine studied at this location, *Vitis vinifera*, reached a peak photosynthetic rate of only 10 mg g^{-1} h^{-1} in mid-morning before dropping sharply to rates of 1 mg g^{-1} h^{-1} at midday (Schulze *et al.*, 1972). Leaf temperatures of *Vitis* during the middle of the day were 2–3 °C above ambient air temperature.

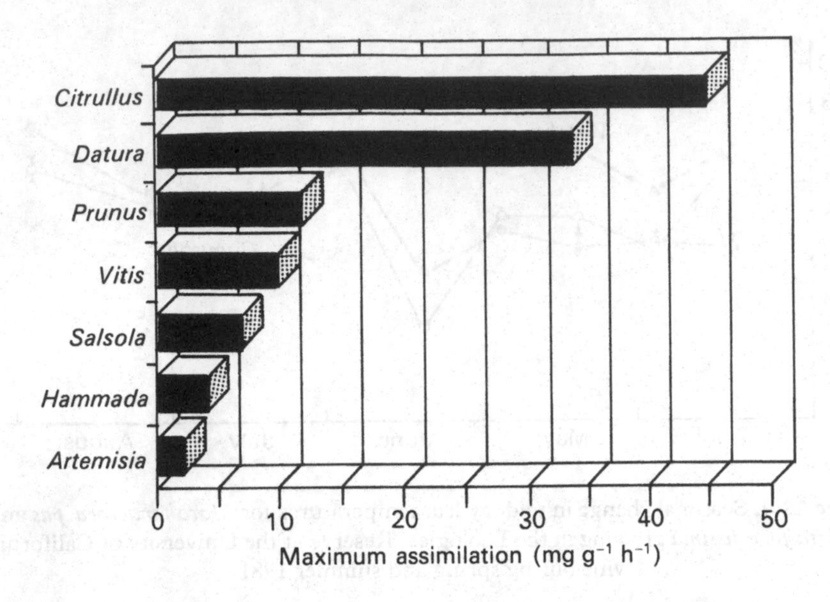

Figure 12.5. Comparative rates of maximum photosynthesis for seven Negev Desert species at the end of the dry season in September 1967. Data from Schulze et al. (1972). *Citrullus colocynthis* is a native vine, while *Vitis vinifera* is a cultivated vine of this site. *Datura stramonium* is a cultivated tree. *Hammada scoparia* and *Artemisia herba-alba* are native shrubs.

High levels of photosynthetic capacity are present in summer-active species of Cucurbitaceae in North American warm deserts as well. We have measured net assimilation rates in the field for *Cucurbita digitata* as high as 44 $\mu m\ m^{-2}\ s^{-1}$ (M. R. Sharifi *et al.*, unpublished data). Photosynthetic rates of 56 mol $m^{-2}\ s^{-1}$ have been reported for *Cucurbita palmae* from Death Valley, California (Seemann, Downton & Berry, 1974). High levels of foliar nitrogen in the leaves of *Cucurbita foetidissima* (Shields & Mangum, 1954) and the previously described transpirational cooling in this species suggest that *C. foetidissima* also maintains high assimilation rates.

While it is difficult to draw definitive conclusions about photosynthetic capacity in arid zone vines with the minimal data base that exists, some preliminary suggestions can be made. Large-leaved Cucurbitaceae and other taxa that exhibit large amounts of transpirational cooling can be expected to maintain high rates of net assimilation. However, vines with evergreen or drought-deciduous leaves should be expected to have ranges of photosynthetic capacity comparable with those plants of other growth forms with similar leaf morphologies. Carter, Teramura & Forseth (1989) came to similar conclusions about photosynthetic capacity in temperate forest vines.

CAM metabolism

Crassulacean acid metabolism (CAM) is well known as an important mechanism of physiological adaptation to drought in many succulent plants (e.g. Kluge & Ting 1978). In CAM metabolism, stomata open at night and close during the day, increasing the efficiency of water use in the fixation of carbon. Succulent storage tissues and typically thick cuticles in CAM plants also function to minimize water loss and maintain positive carbon balances under drought conditions.

Two families which are important as arid and semi-arid zone vines, the Vitaceae and Cucurbitaceae, have species that utilize CAM metabolism. In the Vitaceae, CAM metabolism has been found in stem-succulent species of *Cyphostemma*. which are not vines in growth form, and in some viney species of *Cissus* (Schutte, Steyn & Van der Westhuizen, 1967; Milburn, Pearson & Ndeqwe, 1968; Mooney, Troughton & Berry, 1977, Virzo de Santo *et al.*, 1980; Olivares *et al.*, 1984). The occurrence of CAM metabolism in *Cissus trifoliata* has been studied in some detail in a semi-arid region near Coro in Venezuela (Olivares *et al.*, 1984). At this site, vines of *C. trifoliata* climb to 5 m or more in height on arborescent cacti. Under well irrigated conditions, *C. trifoliata* fixes all of its carbon during the day using normal C_3 metabolism. As water stress increases, dark CO_2 fixation associated with CAM metabolism gradually increases in importance until it totally replaces fixation during the day. Under extreme water deficits, the plants shed their leaves and the stems remain dormant until the next rains. Although the drought-deciduous leaves of *C. trifoliata* are not notably succulent, there is little differentiation between palisade and spongy parenchyma, with both exhibiting large vacuoles and chloroplasts.

Drought-deciduous CAM plants are known in other families of stem succulents, but in these cases only the stem performs CAM metabolism and the mesophytic leaves are limited to C_3 photosynthesis (Lange & Zuber, 1977). Such a pattern of physiological specialization in photosynthetic tissues is also present in *Seyrigia*, a stem-succulent Cucurbitaceae from arid savanna regions in southwestern Madagascar (DeLuca, Alfani & Virzo de Santo, 1977). CAM metabolism is present in the succulent stems of this climber, while the mesophytic leaves are so short-lived that the genus was originally reported as aphyllous (Keraudren, 1960). The stems in *Seyrigia* have a specialized anatomical structure consisting of compact and relatively uniform chlorenchyma without differentiation of cortex (or vascular cylinder). The inner chlorenchyma is composed of large water-storage cells with few chloroplasts (DeLuca *et al.*, 1977). CAM metabolism is also present in another genus of Cucurbitaceae, *Xerosicyos*, in southwestern Madagascar (DeLuca *et al.*, 1977; Winter, 1979). In *Xerosicyos*, however, the CAM process takes place in large succulent leaves with large water-storage cells in the inner mesophyll (DeLuca *et al.*, 1977). While many other genera of arid Cucurbitaceae exhibit succulent above-ground stems (e.g. *Gerardanthus*, *Ibervillea*, *Momordica*, and *Tumamaca*), there are no reports of CAM metabolism from these species. None of these taxa have succulent leaf tissues.

Herbivory defense

Little is known about herbivore defense in arid zone vines where fleshy tissues would seem to present an attractive resource for vertebrate and invertebrate herbivores. Existing data are available only for the Cucurbitaceae. Arid zone Cucurbitaceae commonly maintain succulent leaf and stem tissue through the summer period when resource availability for phytophagous insects and herbivores is highly limited. One mode of chemical defense against herbivory in the Cucurbitaceae is the presence of oxygenated tetracyclic triterpenes, referred to as cucurbitacins. These extremely bitter compounds effectively deter non-adapted invertebrate and vertebrate herbivores (Guha & Sen, 1975).

Studies of the distribution and concentration of cucurbitacins within a diverse range of species of *Cucurbita* have found that perennial xerophytic species, despite their water- and potentially nutrient-limited habitats, had no significantly different concentrations of these compounds in either leaves or fruits than those in annual mesophytic species (Whitaker & Bemis, 1975; Metcalf *et al.*, 1982; Tallamy & Krischik, 1989). Root tissues of xerophytic species, however, were higher in cucurbitacins than those of mesophytic species, possibly suggesting the importance of carbohydrate storage in below-ground tissues to the survival of these arid zone plants.

Conclusions

Vines represent a small, but morphologically and ecologically diverse component of the mediterranean-climate and desert floras of the world. This grouping under the designation of 'vines' covers a range of growth forms, from annuals, to herbaceous perennials with succulent storage organs (geophytes and hemicryptophytes), to woody vines, and even to herbaceous parasites. This range of growth forms, and varied phenological development patterns within any single growth form, provides evidence of the range of adaptive strategies involved. Some arid zone vines are woody lianas which grow up and over host shrubs or woodland trees (passively or actively climbing using tendrils) into a full sun position. Other vines are slender herbaceous perennials that preferentially grow up above the warm soil surface into the shaded inner canopy of shrubs. In open arid land environments, many vines are trailing species that spread widely over the ground surface or, at most, clamber over rocks.

While vines commonly represent no more than 1–2% of the floras of mediterranean-climate and desert ecosystems, they are notably diverse in the mediterranean region of central Chile and in the Sonoran Desert of Mexico. Vines represent more than 8% of the floras of these regions. In central Chile, the diversity of vines is largely a function of high rates of endemism in herbaceous perennials. The high Sonoran Desert diversity, however, is very different, with relatively low endemism among vine species. Vine diversity in this region is largely due to northern extensions of subtropical floristic elements from the south into the most favorable desert habitats.

Our knowledge of the adaptations of arid zone vines to environmental stress is largely limited to studies of Cucurbitaceae with fleshy below-ground storage tissues. Many of these species are characterized by remarkably buffered seasonal patterns of tissue water potential, by high rates of transpirational cooling, and by high levels of photosynthetic capacity. Other ecologically successful vines in arid zone ecosystems appear to utilize entirely different adaptive strategies, but these have not been well investigated. Future research should greatly increase our knowledge of this fascinating group of species by allowing us to understand the nature of plant form and functional response under the multiple environmental stresses presented by arid ecosystems.

REFERENCES

Althawadi, A. M. & Grace, J. M. (1986). Water use by the desert cucurbit *Citrullus colocynthis* (L.) Schrad. *Oecologia* 70: 475–80.

Baas, P. & Schweingruber, F. H. (1987). Ecological trends in the wood anatomy of trees, shrubs and climbers from Europe. *International Association of Wood Anatomists Bulletin* 8: 245–74.

Batonouny, K. H. (1981). *Ecology and Flora of Qatar*. University of Qatar, Doha, Qatar.

Beard, J. S. (1980). Biogeography of the kwongan. In *Kwongan: Plant Life of the Sandplain*, ed. J. S. Pate and J. S. Beard, pp. 1–26. University of Western Australian Press, Nedlands.

Beatley, J. (1976). *Vascular Plants of the Nevada Test Site and Central–Southern Nevada: Ecologic and Geographic Distributions*. US Energy Research and Development Administration, Washington, DC.

Carlquist, S. (1974). *Island Biology*. Columbia University Press, New York.

Carlquist, S. (1984). Wood and stem anatomy of Lardizabalaceae, with comments on the vining habit, ecology and systematics. *Botanical*

Journal of the Linnean Society 88: 257–77.
Carlquist, S. (1985). Vasicentric tracheids as a drought survival mechanism in the woody flora of southern California and similar regions: review of vasicentric tracheids. *Aliso* 11: 37–68.
Carlquist, S. & Hoekman, D. A. (1985). Ecological wood anatomy of the woody southern California flora. *International Association of Wood Anatomists Bulletin* 6: 319–47.
Carter, G. A., Teramura, A. H. & Forseth, I. N. (1989). Photosynthesis in an open field for exotic versus native vines of the southeastern United States. *Canadian Journal of Botany* 67: 443–6.
Cutler, H. C. (1939). Monograph of the North American species of *Ephedra*. *Annals of the Missouri Botanical Garden* 26: 373–424.
DeLuca, P., Alfani, A. & Virzo de Santo, A. (1977). CAM, transpiration, and adaptive mechanisms to xeric environments in the succulent Cucurbitaceae. *Botanical Gazette* 138: 474–8.
Dittmer, H. J. & Rosen, M. L. (1963). The periderm of certain members of the Cucurbitaceae. *Southwestern Naturalist* 8: 1–9.
Dittmer, H. J. & Talley, B. P. (1964). Gross morphology of tap roots of desert cucurbits. *Botanical Gazette* 125: 121–6.
Erickson, R. (1968). *Plants of Prey in Australia*. Lamb Publ., Osburn Park, Western Australia.
Ewers, F. W. (1985). Xylem structure and water conduction in conifer trees, dicot trees, and lianas. *International Association of Wood Anatomists Bulletin* 6: 309–17.
Geiss, W. (1962). Some notes on the vegetation of the Namib Desert. *Cimbebasia 1962(2)*: 1–35.
Ghua, J. & Sen, S. P. (1975). The cucurbitacins – a review. *Plant Biochemical Journal* 2: 12–28.
Goldstein, G., Meinzer, F. C. & Monasterio, M. (1984). The role of capacitance in the water balance of Andean giant rosette species. *Plant, Cell and Environment* 7: 179–86.
Hoover, R. F. (1970). *The Vascular Plants of San Luis Obispo County, California*. University of California Press, Berkeley, Calif.
Jacobsen, H. (1960). *A Handbook of Succulent Plants*. Blandford, London.
Jessop, J. (ed.). (1981). *Flora of Central Australia*. Reed, Sydney.
Keraudren, M. (1960). Une cucurbitacée aphylle de Madagascar: *Seyrigea* gen. nov. *Bulletin de la Société Botanique de France* 107: 298–9.
Kluge, M. & Ting, I. P. (1978). *Crassulacean Acid Metabolism: An Analysis of an Ecological Adaptation*. Heidelberg: Springer-Verlag.
Lange, O. L. (1959). Untersuchungen über Warmehaushalt und Hitzeresistenz mauritanischer Wusten- und Savannenpflanzen. *Flora* 147: 595–651.

Lange, O. L. & Lange, R. (1963). Untersuchungen über Blattentemperaturen, Transpiration und Hitzeresistenz an Pflanzen mediterraner Standorte (Costa Brava, Spanien). *Flora* **153**: 387–425.

Lange, O. L. & Zuber, M. (1977). *Frerea indica*, a stem succulent CAM plant with deciduous C_3 leaves. *Oecologia* **31**: 67–72.

Levyns, M. R. (1966). *A Guide to the Flora of the Cape Peninsula*. Juta, Cape Town.

Macdougal, D. T. (1912). The water balance of succulent plants. *Annals of Botany* **26**: 71–93.

Macdougal, D. T. & Spaulding, E. S. (1910). The water balance of succulent plants. *Carnegie Institute of Washington Publication* 141.

Marchant, N. G., Wheeler, J. R., Rye, B. L., Bennett, E. M., Lander, N. S. & Macfarlane, T. D. (1987). *Flora of the Perth Region*. Western Australian Herbarium, Perth.

Metcalf, R. L., Rhodes, A. M., Metcalf, R. A., Ferguson, J., Metcalf, E. R. & Lu, P.-Y. (1982). Cucurbitacin contents and diabroticite (Coleoptera: Chrysomelidae) feeding upon *Cucurbita* spp. *Environmental Entomology* **11**: 931–7.

Milburn, T. R., Pearson, D. J. & Ndeqwe, N. A. (1968). Crassulacean acid metabolism under natural tropical conditions. *New Phytologist* **67**: 883–97.

Mooney, H. A., Troughton, J. H. Berry, J. A. (1977). Carbon isotope ratio measurements of succulent plants in southern Africa. *Oecologia* **30**: 295–305.

Navas, L. E. (1973–79). *Flora de la Cuenca de Santiago de Chile*, Vols 1–3. Universidad de Chile, Santiago.

Nilsen, E. T., Sharifi, M. R. & Rundel, P. W. (1984). Comparative water relations of phreatophytes in the Sonoran Desert of California. *Ecology* **65**: 767–78.

Olivares, G., Urich, R., Montes, G., Coronel, I. & Herrera, A. (1984). Occurrence of Crassulacean acid metabolism in *Cissus trifoliata* L. (Vitaceae). *Oecologia* **61**: 358–62.

Pate, J. & Dixon, K. (1982). *Tuberous, Cormous and Bulbous Plants*. University of Western Australia Press, Nedlands.

Polunin, O. & Smythies, B. E. (1988). *Flowers of South-West Europe*. Oxford University Press, Oxford.

Rowley, G. D. (1987). *Caudiciform and Pachycaul Succulents*. Strawberry Press, Mill Valley, Calif.

Schulze, E.-D. & Hall, A. E. (1982). Stomatal responses, water loss and CO_2 assimilation rates of plants in contrasting environments. In *Physiological Plant Ecology II, Encyclopaedia of Plant Physiology, n.s., Vol. 12B*, ed. O. L. Lange, P. S. Nobel, C. B. Osmond and

H. Ziegler, pp. 181–230. Springer-Verlag, Berlin.
Schulze, E.-D., Lange, O. L. & Koch, W. (1972). Ökophysiologische Untersuchungen an Wild- und Kulturpflanzen der Negev-Wuste. III. Tageslaufe von Nettophotosynthese und Transpiration am Ende der Trockenzeit. *Oecologia* 9: 317–40.
Schutte, K. H., Steyn, R. & Van der Westhuizen, M. (1967). Crassulacean acid metabolism in South African succulents: a preliminary investigation into its occurrence in various families. *Journal of South African Botany* 33: 107–10.
Seemann, J. R., Downton, W. J. S. & Berry, J. A. (1974). Field studies of acclimation to high temperature: winter ephemerals in Death Valley. *Carnegie Institute of Washington Yearbook* 78: 157–62.
Sen, D. N. & Bhandari, M. C. (1974). On the ecology of a perennial cucurbit in the Indian arid zone – *Citrullus colocynthis* (Linn.) Schrad. *Int. J. Biometeor.* 18: 113–20.
Shields, L. M. & Mangum, W. K. (1954). Leaf nitrogen in relation to structure of leaves of plants growing in gypsum sand. *Phytomorphology* 4: 27–38.
Shreve, F. & Wiggins, I. L. (1964). *Vegetation and Flora of the Sonoran Desert*, Vols. 1 and 2. Stanford University Press, Stanford, Ca.
Sperry, J. S., Holbrook, N. M., Zimmermann, M. H. & Tyree, M. T. (1987). Spring filling of xylem vessels in wild grapevine. *Plant Physiology* 83: 414–17.
Stocking, K. M. (1955). Some taxonomic and ecological considerations of the genus *Marah* (Cucurbitaceae). *Madroño* 13: 113–44.
Tallamy, D. W. & Krischik, V. A. (1989). Variation and function of cucurbitacins in *Cucurbita*: an examination of current hypotheses. *American Naturalist* 133: 766–86.
Virzo de Santo, A., Alfani, A., Greco, L. & Fioretto, A. (1980). Environmental influences on CAM activity of *Cissus quadrangularis*. *Journal of Experimental Botany* 31: 75–82.
Whitaker, T. W. & Bemis, W. P. (1975). Origin and evolution of the cultivated *Cucurbita*. *Bulletin of the Torrey Botanical Club* 102: 362–8.
Winter, K. (1979). ^{13}C values of some succulent plants from Madagascar. *Oecologia* 40: 104–12.

13
Vine–host interactions

ELWYN E. HEGARTY

Introduction

Some vines scramble and climb and some creep or trail. The latter do not necessarily interact with living hosts, and in fact some may not climb even when opportunities are offered (Jones & Gray, 1977). Creeping stems of vines seen in forests are most often stolons of species which have already reached the canopy, and which subsequently give rise to climbing stems which allow new territory to be colonized (e.g. Peñalosa, 1982), or young vines which have not yet located a host. The action of climbing by such stems usually involves interactions between stems and foliage of both vine and host. Once in the canopy, some vines may extend to use adjacent hosts.

This chapter considers how climbing methods differ in suitability to ascend hosts of particular sizes and shapes, and the effectiveness of anti-climber defenses.

Classification of vines

Although vines differ so much in size and form, it has been traditional to classify climbing methods into only four or five major categories. Climbers were classified by authors from Darwin (1875) and Schenck (1892, 1893) to Richards (1952) on the basis of the most obvious climbing technique. These usually include:

1. twining using stems, branches or petioles/petiolules
2. use of tendrils, including leaf tendrils
3. scrambling, often assisted by hooks to avoid slipping
4. use of adventitious roots

Each has a wide range of architectural expression, although closely related species tend to develop similar climbing attributes. Many species combine several methods, and remain difficult to classify exactly. While in general tendrillar and root-climbers do not conspicuously twine, vines are such a large, heterogeneous group that there are exceptions to almost every rule that can be formulated. More recent variations from parts or the whole of the above classification include those of:

Menninger (1970), who used a set of more colorful, but mostly less commonly applied terms, including leaners, weavers, rooters, thorn clingers, and graspers – including twiners, stickers and clingers;

Putz (1984a), who grouped root-climbers with those with adhesive tendrils. The latter offer greater opportunities for vertical ascent than non-adhesive (twining) tendrils;

Hegarty & Clifford (1990), who separated twiners, using stem apices, and so forming close relationships with the stems of hosts, from those using lateral parts of the plant, e.g. branches, petioles and petiolules which support from below, rather than guide from the top, the direction of growth of the climbing stem;

Putz & Chai (1987), who separated twiners into two groups, depending on whether or not the leading shoot was determinate, i.e. was one of a series of modules, replaced in sequence. Each such determinate module coiled round a host at decreasing angles and, having produced leaves, ceased growth. Its supported coils provided a firm base for the next shoot segment. This behavior was more widespread in Malaysia than on Barro Colorado Island. Indeterminate twiners, in which leading shoots were not replaced, only ascended a lower range of diameters of supports.

In a more complex approach, Cremers (1973, 1974) found that existing architectural models for trees were applicable to many young vines, and developed three further models for others. The models are based on branch orientation, arrangement and rhythms of growth, as well as climbing method. These models form the basis of further discussions of vine architecture by Hallé, Oldeman & Tomlinson (1978).

Climbing methods found in some vine-rich families are compared in Table 13.1. In most families, including the Araceae, Cucurbitaceae and Passifloraceae, only one or two methods have been developed, but in others, for example the Bignoniaceae and Leguminosae, variations on three or four basic methods are found.

Further bases of classification of vines have been developed, reflecting recent increasing interest in their ecology. These do not indicate the method of climbing. Richards (1952) considered that, from an ecological point of view, the size and maximum height of a climber were more important than the method, and distinguished large, mainly woody from small, mainly

Table 13.1. *Principal methods of climbing adopted by vines in some more speciose families*

	Method of climbing						
	Twining			Use of tendrils			
Family	Apices	Branches	Petiol(ul)es	Coiling	adhesive or hooked	Scrambling (± hooks/thorns)	Use of roots
Apocynaceae	+	–	–	+	–	+	–
Araceae	–	–	–	–	–	–	+
Arecaceae	–	–	–	–	–	+	–
Aristolochiaceae	+	–	–	–	–	–	–
Asclepiadaceae	+	–	–	–	–	+	+
Bignoniaceae	+	–	–	+	+	+	+
Combretaceae	+	–	–	–	–	+	–
Convolvulaceae	+	–	–	–	–	–	–
Cucurbitaceae	±	–	–	+	+	+	–
Leguminosae	+	+	–	+	+	+	–
Menispermaceae	+	–	?	–	–	–	–
Passifloraceae	–	–	–	+	–	–	–
Piperaceae	–	–	–	–	–	–	+
Ranunculaceae	±	–	+	–	–	–	–
Sapindaceae	+	–	–	+	–	–	+
Vitaceae	–	–	–	+	+	–	–

herbaceous vines. All four of the traditionally classified methods of climbing were found in both groups. However, Grubb *et al.* (1963) considered that 'photophytic' and 'skiophytic' climbers, i.e. those with preferences for exposure or shade respectively, were more appropriate to describe the behaviour of vines than size. Box (1981) classified vines/lianas by seasonal leafing behaviour, relating evergreen-ness or deciduousness to climatic limitations.

In general, the types of climber vary little from region to region, with twiners being the most abundant and those using adventitious roots the least, except in some very humid tropical forests (Rich *et al.*, 1987). The distribution of various climbing methods during rainforest succession is discussed by Hegarty & Caballé (Chapter 11).

The application of methods of climbing in forest conditions

When a young vine, with limited resources, begins to climb for the first time, the presence of suitable supports and absence of tip herbivory are important for its rate of growth (Putz, 1984a) and survival (Janzen, 1971). Some vines, however, delay climbing until they have established a shrubby base and supportive root structure, e.g. *Embelia australiana* (Myrsinaceae) and the rosette-forming *Calamus* spp. (Arecaceae), the first ascent by which may be by a robust self-supporting stem which begins to twine around or lean on hosts only at some distance from its origin. Hegarty (1988) observed that vines of the mature phase of a subtropical evergreen rainforest were often shrubby for a time, whereas the seedlings of vines in disturbed areas usually began to climb before branching.

The methods of climbing summarized below are characteristic of those used by vines for their first ascent. This avoids the complication that, in many species, the basal portions of reiterative ('water' or coppice) shoots, and/or potentially fertile, short-internode canopy shoots may exhibit no particular adaptation for obtaining the support of a host (Figure 13.1).

Twiners

Twining is achieved by movements of (a) shoot tips or (b) lateral branches or parts.

(a) The tips of extending shoots of twiners rotate ('nutate') in irregular, but roughly circular or elliptical and unidirectional patterns (circumnutation), part of the motion being autonomous and part an exogenous, geotropic reaction. Growth takes place in pulses or waves which travel round the stem while a more or less constant direction of extension growth is maintained (Johnsson, 1979). The rate of circumnutation also varies, even within species, with temperature, shoot orientation, and restriction to length of free-moving stem, and is greater in light than in darkness (literature reviewed by Johnsson, 1979). However, mechanical stimulation of plant stems, such as by rubbing, is rapidly followed by a retardation of growth in almost all vascular plants ('thigmomorphogenesis': Jaffe, 1976), complicating the design of experiments involving the effects of contact.

Baillaud (1957) states that a tendency to twine is almost always associated with common important characteristics of stem morphology, plus indefinite apical extension, constancy of twining direction and absence of sensitivity to touch. The latter aspect, noted also by Darwin (1875) is distinct from the response by a twiner to interruption of circumnutation by contact with a potential support, which results in shortening of the length of free-moving stem, a subsequent reduction in apical twining circle and coiling around the support, with the stem of the twiner usually also twisting around its own axis at the same time (Baillaud, 1962a). The maximum twining radius of vines, which limits

Figure 13.1. Canopy branch of *Malaisia scandens* (Moraceae) showing short internodes and virtual absence of twining by branches.

the size of host stem that can be encircled, varies by species (Darwin, 1875). Most twiners can only climb supports of a maximum diameter less than 10 cm (Putz, 1984a), with much smaller minimum diameters, but a proportion, including especially the determinate twiners, can climb quite large diameter tree trunks (Putz & Chai, 1987). Most vines readily climb other vines, including those which provide direct access to the tree canopy (Putz, 1984a).

About 95% of twiners invariably coil to the right, and the direction is usually maintained throughout genera and families. *Dioscorea* is a major exception (Baillaud, 1957, 1962a) in that some of its many species twine to the left and some to the right. At least 20 species have been reported to twine unreliably in either direction. Those discussed by Darwin (1875) Baillaud (1962a) and Cribb (1985) are characteristically weakly twining species of initially shrubby or semi-prostrate form, so some observed reversals of direction in climbing (as distinct from temporary reversals during what is always essentially unidirectional nutation) may be an accident of position of supports.

It has been suggested that left-handed twining is indicative of a considerable measure of taxonomic isolation from other twining species, and that such isolation would be expected if left-handed twining were less efficient, due to some inherent spirality of structure, so not promoting diversification into numerous species (Dormer, 1972). Although left-hand twiners readily intertwine with free-hanging right-hand twiners to reach the canopy, paths to the canopy created by right-hand vines spiraling round saplings create a very uneven track to the canopy for a left-hand vine, while offering some support to subsequent right-hand twining stems.

(b) Other twiners. In a minority of twiners, the axillary portions of lateral branches (e.g. in Annonaceae, Loganiaceae), or petioles/petiolules (e.g. in Ranunculaceae, Tropaeolaceae), are the means of securing vines to hosts. Rather like tendrils in that their direction of movement responds to touch, such portions of vines can also twine only around fairly slender supports. The attachments then thicken and harden, to support the adjacent main axis of the vine near to a support.

Twining leaf apices, as in *Gloriosa* (Liliaceae) and *Flagellaria* (Flagellariaceae) function as, and are usually included with, tendrils.

Tendrillar plants

Tendrils are developed from stems, stipules, leaves and flower pedicels. Tendrils usually appear in seedlings only after 5–10 leaves are expanded, and the stem becomes unsteady. However, in vigorous seedlings from the large-seeded *Entada phaseoloides* (Leguminosae), paired tendrils expand at the end of the first leaf rachis. Expansion of associated leaflets is delayed until a series of tendrils is available for support (Figure 13.2).

Some tendrils attach by twining alone (e.g. most Cucurbitaceae and Vitaceae) others by terminal adhesive pads (e.g. a minority of the same families) or by terminal hooks (e.g. some Bignoniaceae and Leguminosae). Jaffe & Galston (1968) list typical movements by tendrils as follows:

circumnutation (some exceptions listed: Jaffe & Galston, 1968, pp. 420–1);
bending and terminal coiling following contact with a support (an 'irritable' reaction: Darwin, 1875);
free coiling along the axis of the tendril, not necessarily as the result of stimulation.

The maximum circumnutation of a tendril usually coincides with its period of maximum irritability (Sachs, 1888). Aspects of a proposed physiological mechanism of contact coiling by tendrils are summarized by Jaffe & Galston (1968: p. 432). Post-attachment spiral contraction of thickening tendrils in most species draws the stem of the vine closer to the support,

Figure 13.2. Extending shoot of *Entada phaseoloides* (Leguminosae) showing position of shoot tip and delayed expansion of leaflets associated with tendrils.

usually with one or more reversals of coiling, involving similar numbers of coils (Darwin, 1875). Thus damaging twisting is avoided, and some independence is retained during swaying movements. The ends of many tendrils which fail to attach to any support also coil, but in one direction only. Such tendrils are usually then shed. There have been many studies of the mechanics and physiology of these processes (e.g. Darwin, 1875; Sachs, 1888; Baillaud, 1957, 1962b; Jaffe & Galston, 1966, 1968).

The youngest tendrils normally extend beyond the tip of the extending vegetative shoot that carries them. Where non-adhesive tendrils and associated leaves expand together, unsupported shoots carrying them tend to grow obliquely under the influence of gravity. As such tendrils elongate, they also droop, so that their free ends are below, or obliquely to the side of, the axis of the stem, while the tip of the shoot extends. The shoots are thus effectively directed in a close-to-horizontal plane which is ideal for anchoring to and traversing foliage, or spanning short gaps, and can take advantage of such tenuous supports as leaves, petioles and twigs (Figure 13.3). Because in such cases the shoot apex of a vine is often extending at a distance from the support afforded by tendrils, efficient upward growth is not promoted.

Most tendrillar species characteristically establish in disturbed, sunny

Figure 13.3. Extending shoot of *Cayratia eurynema* (Vitaceae) showing shoot tip, expanding leaves and much-branched descending tendrils.

areas (Bews, 1925; Kelly, 1985) and their foliage expands in full sun. Exceptions include the tendrillar climbers which can climb vertically up tree boles – e.g. species with adhesive pads or fine terminal hooks which can explore crevices in bark, and expand within them to form strong supports – and perhaps those in which tendrils are infrequently present (*Passiflora pittieri*: Killip, 1938; Longino, 1986).

Scramblers and scandent shrubs

These are imprecise descriptions for the rambling climbers which lean rather than climb, either with no particular climbing facility other than long stems and gravity, or with hooks or thorns which catch in supports. They include the rattans, which are long-stemmed and extensively armed with grappling hooks and spines. Scramblers with long rambling stems are sometimes difficult to determine as scramblers or twiners, or even shrubs, and arbitrary, conflicting classifications are found in surveys and floras.

In some loosely twining species, extensive woody thickening of the recurving bases of lateral shoots or retained petioles takes place, e.g. in *Palmeria scandens* (Monimiaceae), known as 'anchor vine', *Quisqualis* and *Combretum* spp. (Combretaceae). Like thorns or hooks, such developments are effective for stopping vines from slipping as they climb and scramble over branches of hosts.

Opposite branches and leaves are common among the vine flora of Barro Colorado Island, and it has been suggested that this has resulted because opposed branches assist anchorage of vines to forks of trees (Croat, 1978). Elsewhere, however (e.g. Australia), the proportion of opposite-branching vines is smaller, perhaps because some families, such as Bignoniaceae, in which they are abundant are poorly represented. So many factors are involved that the theory is difficult to test.

Root climbers

These use adventitious roots to cling to hosts, and like those discussed above which have adhesive or hooked tendrils, can ascend supports of almost any diameter or texture, e.g. cliffs, palm trees, buttresses and boles of large trees (= 'bole-climbers': Whitmore, 1975). Not all aerial roots of vines are developed in direct contact with a potential host, for climbing; in a number of species free-hanging contractile roots develop at nodes (e.g. *Sicyos sicyoides*: van den Honert, 1941). Root-climbers may be woody or herbaceous. The latter type may lack a connection with the soil early or late in life, or may have intermittent access (Ayensu, 1980; Kelly, 1985; Putz & Holbrook, 1986). The Araceae and Cyclanthaceae, many of which began life as epiphytes –

'primary' as opposed to 'secondary' hemiepiphytes: Putz & Holbrook, 1988 – are often routinely excluded, particularly when surveying Neotropical vines (e.g. Putz, 1984b; Gentry & Dodson, 1987). However, the distinction between primary and secondary epiphytism is sometimes difficult to make in field surveys, and a number of other authors have included some or all climbing hemiepiphytes with the vines (Grubb *et al.*, 1963; Givnish & Vermeij, 1976; Hegarty, 1988).

Seedlings of at least some root-climbers have been shown to grow towards dark surfaces, such as tree buttresses ('skototropism': Strong & Ray, 1975; Gentry, 1983). Effectively, stems grow away from sources of strong light, although juvenile foliage at each stage continues to face the strongest source of incident light (Givnish & Vermeij, 1976; Hegarty, 1988). The chance of survival of one such species (*Piper novae-hollandiae*, Piperaceae) is enhanced if it climbs up the side of the tree which is least exposed to severe seasonal desiccation (Hegarty, 1988), in the same way that mosses are concentrated on the sheltered side of trees. Some of the larger dicotyledonous root-climbers (e.g. Bignoniaceae, and Piperaceae) shed their original climbing roots and become free-hanging once they reach the support of the lower tree crown. It is unusual for root-climbers, which inhabit tree boles and inner branches, to extend to additional host trees.

Distances between supports

The tips of root-climbers climb vertically and are supported against the host while adventitious clinging roots extend from each lower internode. Other forms of climber, however, have leader shoots that are partially unsupported; the distance such a stem can bridge is determined by the length of stem it can hold erect. Scramblers and hook-climbers tend to have thicker leader shoots than other climber types, the maximum unsupported span exceeding 3m in *Combretum decandrum* (Combretaceae) (Putz, 1984a). Leaf expansion on leader shoots is frequently delayed (French, 1977), which in twiners at least can result in greater attachment success than for leafy stems (as found by Penãlosa, 1982).

Anti-climber defences of trees and shrubs

A variety of 'anti-climbing' defenses of trees has been proposed (Table 13.2). However, the efficiency of individual defenses is inherently difficult to test. Complications include not only the lack of independence of single factors (see Putz, 1984b) but also how to sample interactions between non-standard pairs or groups of patchily distributed species. Of the various hypotheses for defense of trees from vines that have been offered and tested, only large

Table 13.2. *Possible defenses of trees and shrubs against invasion by vines*

Attribute of host	Discussed ●/Discounted ○	Test result + / ○[a]
Canopy		
Large long compound leaves, e.g., palms, which are shed as units	Putz, Lee & Goh (1984) ●	Maier (1982) + ; Putz (1984b) + Rich *et al.* (1987) +
Drooping leaves, branches remote from central shoot tip	Janzen (1975) ●	
Retained dead leaves		Page & Brownsey (1986) +
Crown depth		Putz (1984b) ○
Unstratified crowns	Smith (1973) ●	
Stems		
Rapid stem thickening	Putz (1980) ●	Putz (1984b) ○; Putz, Lee & Goh (1984) + / ○
Rapid increase in height	Putz (1980) ●; Putz, Lee & Goh (1984) ●	
Spiny stems		
Sawing		Putz (1984b) ○
As fragile barriers	Putz (1984b) ●	Maier (1982) +
Smooth bark	Putz (1980) ●	Boom & Mori (1982) ○
Long branch-free boles	Putz (1984b) ●	
Buttresses	Black & Harper (1979) ●	Boom & Mori (1982) ○
	Putz (1980) ○	
Asynchronous swaying	Putz (1980) ●	
Flexible main stems	Rich *et al.* (1987) ●; Putz Lee & Goh (1984) ●	Putz (1984b) +
Harbouring protective insects	Janzen (1969) ● Putz (1984b) ●	Janzen (1973) +
	Putz & Holbrook (1988) ●	
	Putz, Lee & Goh (1984) ●	

[a]The results are often qualified by circumstances and inter-correlations with other factors; + / ○ indicates significance/ non-significance.

compound leaves, shed as units, such as those of palms or ferns, have been fully supported by data from more than one observer. Flexible trunks, rapid stem thickening, the presence of fragile spines or protection by aggressive ants have each been found effective in separate studies, but a need for confirmation by further testing is indicated in some cases because other studies were less conclusive (Table 13.2).

Another complication of such studies is that many hosts cannot be climbed directly by all vines in their vicinity because some employ climbing methods that are inappropriate to trunk diameter. For example, root-climbers can climb thicker tree trunks than twiners, or the tendrillar vines without positive attachments to bark. Vines confined to the trunk of a host are not as readily dislodged when large compound leaves (e.g. of palms) are shed (Rich *et al.*, 1987). However, palms in general appear to support rather few vines, apart from root climbers (from data given in Maier, 1982; Emmons & Gentry, 1983; Gentry & Dodson, 1987; Rich *et al.*, 1987) (Figure 13.4). The dearth in palm groves of large woody vines may reflect a general lack of diversity in both trees and vines, which is probably determined as much by other factors as by efficient vine-shedding. These might include waterlogging of soils, which creates anaerobic conditions for deep roots of vines as well as of trees.

It is also difficult to deal with the probabilities that if one vine can climb a tree, it may assist others to do so (Putz, 1984a), or that vines may invade hosts via inter-crown connections. Nevertheless, only about 30–50% of the canopy trees on Barro Colorado Island carry vines (Putz, 1984a) and a similar proportion may be assumed, from values for vine-leaf loads, for larger trees in tropical rainforest in French Guiana (Beekman, 1981).

As vines evolved from trees or shrubs (Hutchinson, 1959; Savinykh, 1986) yet continue to share the same set of resources, it may be expected that those trees and shrubs which are difficult for vines to climb, or which deter their growth, should have some competitive advantage. Yet there is so far no clear evidence that species which have some ability in some situations to repel vines are locally abundant as a direct result.

It appears that in more humid, dense rainforests where vines remain a minority group (Hegarty & Caballé, Chapter 11, Table 11.1), even if temporarily abundant in disturbed areas, the continued dominance of trees is only partly due to anti-climber defense strategies of individual host trees. The development of mature community structure – typified by increasing canopy height and continuity – allows a number of potential host trees to combine to exercise other forms of control on the abundance and persistence of vines. These may include:

1. Poor regeneration of most species of vine once the floor of the forest is deeply shaded (UNESCO, 1978; Peñalosa, 1984). Apart from the difficulties of seedlings from small seeds growing in shade, successional

Figure 13.4. Palm trees (*Archontophoenix cunninghamiana*) with aroids and ferns climbing boles below crown level. Shrubs and other trees (lower half of photograph) are heavily infested with tendrillar and twining vines in their canopies.

changes in soil pH, and in nitrogen cycling in soils and plants (Lamb, 1980; Stewart, Hegarty & Specht, 1988), and even allelopathy, might limit survival of young vines.

2. High costs of clonal renewal in tall forest because many vines lack a permanent trunk, and it is expensive to replace and maintain stems of ever-increasing length. The problem is exacerbated by a reduction in the productivity of vines, related to an increasingly shaded environment during succession (shown in relation to population basal areas for species of a subtropical rainforest: Hegarty, 1991).

Detrimental effects of vines on hosts

Where vines are abundant and trees have failed to deter them from invading the canopy, the leaf area and fecundity of trees can be seriously reduced (Kira & Ogawa, 1971; Stevens, 1987; Hegarty & Caballé, Chapter 11; Putz, Chapter 18). Twiners may also interfere with downward translocation in trees by constricting boles. Apart from these circumstances, which take place during normal cyclic regeneration of forests, there are numerous instances of invasion of disturbed communities by fast growing, naturalized vines such as *Cardiospermum*, *Ipomoea*, *Macfadyena*, *Passiflora* and *Pueraria* ('kudzu') spp. when such species are released from natural controls by herbivores or some environmental factors, or even by absence of competition for a specialized niche (Mueller-Dombois, 1975). Such invasions may limit a community's chance of recovery from disturbance.

It has rarely been suggested that sharing a pool of resources with colonizing vines is advantageous for trees, except where vines help rapid closure of forest canopies following damage, thus discouraging regeneration of weedy tree species and retaining humidity (Kochummen & Ng, 1977).

Conclusion

Vine–host relationships result from the evolution of a number of climbing methods, each applicable to different forms of host. Trees, the most usual hosts, are often disadvantaged by vines. In some species of tree, the anti-climber defenses that have been proposed may, separately or more probably together, reduce competition from those vines with the ability to invade them. However, this should be considered as part of a broader frame of interaction, in which trees acting in conjunction develop a canopy which may hinder the proliferation and persistence of vines below, yet at the same time foster upper-level pathways so that vines which have already reached the canopy of one tree can extend their territory to others.

Summary

To obtain the support of hosts, vines use tendrils, hooks, adhesive attachments, twining shoots, leaves or petioles, scrambling, or combinations of these and other methods. Twining is the most usual method of climbing. Most twiners everywhere coil from left to right, and are most efficient when climbing thin stems, such as those of saplings. Most vines with coiling tendrils thrive at sites of disturbance, and may extend across a number of hosts, whereas vines which employ adhesive roots or tendrils can climb vertically up the boles of large trees, but are usually confined to their original host.

Vines can have detrimental effects on trees and shrubs – for example, by creating excessive shade for hosts, so that their normal shoot growth is reduced, or by constriction of growing trunks so that downward translocation is hindered. Despite the many means of climbing that have evolved, usually a proportion of potential hosts remains free of vines. Possible explanations, for some host species, have included protection by a habit of growth or form that discourages efficient climbing, branch shedding, creation of an understory environment unfavorable to vine regeneration, or even resident, herbivorous insects. As several factors may be involved, such theories require careful testing. However the vines secure positions in various parts of forest structure, their host trees remain the dominant life-form except when disturbance favors the rapid proliferation of weedy and/or alien vines.

References

Ayensu, E. S. (1980). *Jungles*. Stranglers and climbers, pp. 142–3. Jonathan Cape, London.

Baillaud, L. (1957). Récherches sur les mouvements spontanes des plantes grimpantes. Thesis, Faculté des Sciences de Besançon. Carrère, Rodez.

Baillaud, L. (1962a). Mouvements d'exploration et d'enroulement des tiges volubiles. *Handbuch der Pflanzenphysiologie, Vol. 17*, ed. W. Ruhland, pp. 635–715. Springer, Berlin.

Baillaud, L. (1962b). Mouvements autonomes des tiges, vrilles et autres organes à l'exception des organes volubiles et des feuilles. *Handbuch der Pflanzenphysiologie, Vol. 17*, ed. W. Ruhland, pp. 562–634. Springer, Berlin.

Beekman, F. (1981). *Structural and Dynamic Aspects of the Occurrence and Development of Lianes in the Tropical Rain Forest*. Department of Forestry. Agricultural University, Washington.

Bews, J. W. (1925). *Plant Forms and their Evolution in South Africa.* Longmans Green and Co., London.
Black, H. L. & Harper, K. T. (1979). The adaptive value of buttresses to tropical trees: additional hypotheses. *Biotropica* **11**: 240.
Boom, B. M. & Mori, S. A. (1982). Falsification of two hypotheses on liana exclusion from tropical trees possessing buttresses and smooth bark. *Bulletin of the Torrey Botanical Club* **109**: 447–50.
Box, E. O. (1981). Macroclimate and Plant Form. An introduction to predictive modelling in phytogeography. *Tasks for Vegetation Science. 1.* Dr W. Junk, The Hague.
Cremers, G. (1973). Architecture de quelques lianes d'Afrique tropicale. 1. *Candollea* **28**: 249–80.
Cremers, G. (1974). Architecture de quelques lianes d'Afrique tropicale. 2. *Candollea* **29**: 57–110.
Cribb, A. B. (1985). The direction of twining in some Queensland plants. *Queensland Naturalist* **25**: 122–4.
Croat, T. B. (1978). *Flora of Barro Colorado Island.* Stanford University Press, Stanford, Calif.
Darwin, C. (1875). *The Movements and Habits of Climbing Plants.* Murray, London.
Dormer, K. J. (1972). *Shoot Organization in Vascular Plants.* University Press, Syracuse, NY.
Emmons, L. H. & Gentry, A. H. (1983). Tropical forest structure and the distribution of gliding and prehensile-tailed vertebrates. *American Naturalist* **121**: 513–24.
French, J. C. (1977). Growth relationships of leaves and internodes in viny angiosperms with different modes of attachment. *American Journal of Botany* **64**: 292–304.
Gentry, A. H. (1983). Macfadyena. In *Costa Rican Natural History*, ed. D. H. Janzen, pp. 272–3. University of Chicago Press, Chicago, Ill.
Gentry, A. H. & Dodson, C. (1987). Contribution of nontrees to species richness of a tropical rain forest. *Biotropica* **19**: 149–56.
Givnish, T. J. & Vermeij, G. J. (1976). Sizes and shapes of liane leaves. *American Naturalist* **110**: 743–78.
Grubb, P. J., Lloyd, J. R. Pennington, T. D. & Whitmore, T. C. (1963). A comparison of montane and lowland rain forest in Ecuador. I. The forest structure, physiognomy and floristics. *Journal of Ecology* **51**: 567–601.
Hallé, F., Oldeman, R. A. A. & Tomlinson, P. B. (1978). *Tropical Trees and Forests.* Springer-Verlag, Berlin.
Hegarty, E. E. (1988). Canopy dynamics of lianes and trees in subtropical rainforest. PhD thesis, Department of Botany, University of

Queensland, Brisbane, Australia.

Hegarty, E. E. (1991). Leaf litter production by lianes and trees in a sub-tropical Australian rain forest. *Journal of Tropical Ecology* 7 (in press).

Hegarty, E. E. & Clifford, H. T. (1990). Climbing angiosperms in the Australian Flora. In *The Rainforest Legacy, Vol. 2*, ed. A. P. Kershaw and G. L. Werren. Australian Government Publishing Service, Canberra (in press).

Hutchinson, J. (1959). *The Families of Flowering Plants. I. Dicotyledons.* Clarendon Press, Oxford.

Jaffe, M. J. (1976). Thigmomorphogenesis: a detailed characterization of the response of beans (*Phaseolus vulgaris* L.) to mechanical stimulation. *Zeitschrift für Pflanzenphysiologie* 77: 437–53.

Jaffe, M. J. & Galston, A. W. (1966). Physiological studies on pea tendrils. 1. Growth stimulation following mechanical stimulation. *Plant Physiology* 41: 1014–25.

Jaffe, M. J. & Galston, A. W. (1968). The physiology of tendrils. *Annual Review of Plant Physiology* 19: 417–34.

Janzen, D. H. (1969). Allelopathy by myrmecophytes: The ant *Azteca* as an allelopathic agent of *Cecropia*. *Ecology* 50: 147–53.

Janzen, D. H. (1971). Escape of juvenile *Dioclea megacarpa* (Leguminosae) vines from predators in a deciduous tropical forest. *American Naturalist* 105: 97–112.

Janzen, D. H. (1973). Dissolution of mutualism between *Cecropia* and its *Azteca* ants. *Biotropica* 5: 15–28.

Janzen, D. H. (1975). *Ecology of Plants in the Tropics, Studies in Biology 58*. Edward Arnold, London.

Johnsson, A. (1979). Circumnutation. In *Encyclopedia of Plant Physiology, n.s., Vol. 7*, ed. W. Haupt and M. E. Feinleib, pp. 627–46. Springer-Verlag, Berlin.

Jones, D. L. & Gray, B. (1977). *Australian Climbing Plants*. Reed, Sydney.

Kelly, D. L. (1985). Epiphytes and climbers of a Jamaican rainforest: vertical distribution, life forms and life histories. *Journal of Biogeography* 12: 223–41.

Killip, E. P. (1938). The American species of Passifloraceae. *Field Museum of Natural History, Botanical series* 19: 1–613.

Kira, T. & Ogawa, H. (1971). Assessment of primary production in tropical and equatorial forests. In *Productivity of Forest Ecosystems, Proceedings of Brussels Symposium 1969*, ed. P. Duvigneaud, pp. 309–21. UNESCO, Paris.

Kochummen, K. M. & Ng, F. S. P. (1977). Natural plant succession after farming in Kepong. *Malaysian Forester* 40: 61–78.

Lamb, D. (1980). Soil nitrogen mineralisation in a secondary rainforest

succession. *Oecologia (Berlin)* 47: 257–63.
Longino, J. T. (1986). A negative correlation between growth and rainfall in a tropical liana. *Biotropica* **18**: 195–200.
Maier, F. E. (1982). Effects of physical defenses on vines and epiphyte growth in palms. *Tropical Ecology* **23**: 212–17.
Menninger, E. A. (1970). *Flowering Vines of the World*. Hearthside, New York.
Mueller-Dombois, D. (1975). Some aspects of island ecosystem analysis. In *Tropical Ecological Systems*, ed. F. B. Golley and E. Medina, pp. 353–66. Ecological Studies 11. Springer, Berlin.
Page, C. N. & Brownsey, P. J. (1986). Tree fern skirts: a defence against climbers and large epiphytes. *Journal of Ecology* **74**: 787–96.
Peñalosa, J. (1982). Morphological specialization and attachment success in two twining lianas. *American Journal of Botany* **69**: 1043–5.
Peñalosa, J. (1984). Basal branching and vegetative spread in two tropical forest lianas. *Biotropica* **16**: 1–9.
Putz, F. E. (1980). Lianas vs trees. *Biotropica* **12**: 224–5.
Putz, F. E. (1984a). The natural history of lianas at Barro Colorado Island, Panama. *Ecology* **65**: 1713–24.
Putz, F. E. (1984b). How trees avoid and shed lianas. *Biotropica* **16**: 19–23.
Putz, F. E. & Chai, P. (1987). Ecological studies of lianas in Lambir National Park, Sarawak, Malaysia. *Journal of Ecology* **75**: 523–31.
Putz, F. E., Lee H. S. & Goh, R. (1984). Effects of post-felling silvicultural treatments on woody vines in Sarawak. *Malaysian Forester* 47: 214–26.
Putz, F. E. & Holbrook, N. M. (1986). Notes on the natural history of hemiepiphytes. *Selbyana* **9**: 61–9.
Rich, P. M., Lum, Shawn, Leda Muñuz, E. & Mauricio Quesada, A. (1987). Shedding of vines by the palms *Welfia georgii* and *Iriartea gigantea*. *Principes* **31**: 31–40.
Richards, P. W. (1952). *The Tropical Rain Forest*. Cambridge University Press, Cambridge.
Sachs, J. (1888). *Lectures on the Physiology of Plants* (transl. from German by H. M. Ward). Clarendon Press, Oxford.
Savinykh, N. P. (1986). The origin of climbing herbaceous plants. *Byull. Mosk. O-V. Ispyt. Prir. Otd. Biol.* **91**: 64–71 (from abstract: Biological Abstracts 82 (7) AB–104: 60491).
Schenck, H. (1892). Beiträge zur Biologie und Anatomie der Lianen in Besonderen der in Brasilien einheimischen Arten. 1. Beiträge zur Biologie der Lianen. In *Botanische Mitteilungen aus den Tropen*, ed. A. F. W. Schimper, pp. 1–253. G. Fischer, Jena.
Schenck, H. (1893). Beiträge zur Biologie und Anatomie der Lianen in

Besonderen der in Brasilien einheimischen Arten. 2. Beiträge zur Anatomie der Lianen. In *Botanische Mitteilungen aus den Tropen 5*, ed. A. F. W. Schimper, pp. 1–271. G. Fischer, Jena.

Smith, A. P. (1973). Stratification of temperate and tropical forests. *American Naturalist* **107**: 671–83.

Stevens, G. C. (1987). Lianas as structural parasites: the *Bursera simaruba* example. *Ecology* **68**: 77–81.

Stewart, G. R., Hegarty, E. E. & Specht, R. L. (1988). Inorganic nitrogen assimilation in plants of Australian rainforest communities. *Physiologia Plantarum* **74**: 26–33.

Strong, D. R. & Ray, T. S. (1975). Host tree location and behaviour of a tropical vine *Monstera gigantea* by phototropism. *Science* **190**: 804–6.

United Nations Educational, Scientific and Cultural Organisation (1978). *Tropical Forest Ecostystems*. Natural resources research 14. UNESCO, Paris.

van den Honert, T. H. (1941). Experiments on the water household of tropical plants. I. Water balance in *Cissus sicyoides* L. *Annales du Jardin Botaniques de Buitenzorg* **51**: 58–82.

Whitmore, T. C. (1975). *Tropical Rain Forests of the Far East*. Clarendon Press, Oxford.

14

Seasonality of climbers: a review and example from Costa Rican dry forest

PAUL A. OPLER, HERBERT G. BAKER
AND GORDON W. FRANKIE

Introduction

Conducting a study of the seasonality of flowering, fruiting, and foliation, often termed a phenological study, of species within a plant community is often not an end in itself. It may serve as a way to discover the component species of the community and when each species may be expected to be in a reproductive or vegetative stage suitable for study.

Knowledge of the seasonality of reproductive and vegetative activities by the species in a community is fundamental to other questions. How do the plants respond to environmental variables, notably the climate? Are the phenological patterns of some species affected by competition between plant species for pollinating or dispersal agents? Does foliation occur in such a way as to minimize herbivory? Are the flowering and fruiting of plants with particular pollinatory or dispersal strategies limited to particular seasons? Alternatively, the zoologist may wish to know the seasonal availability of nectar, fruit, or leaf resources for particular animals.

Our approach in this chapter is to emphasize the phenological patterns of reproduction and foliation by many locally sympatric climbers, with a review of the literature and from our own fieldwork in northwestern Costa Rica. We also consider phenological studies of plants (climbers and others), that are adapted for certain groups of pollinators or frugivores.

Community studies

In addition to our work there have been only three other studies on the phenology of climber communities: those by Croat (240 species: 1975) and Putz & Windsor (43 species: 1987), both on Barro Colorado Island, Panama; and that of Hegarty (1988) on the climbers of a Queensland rainforest.

Lieberman (1982) included about 33% (11 of 33 species) of the lianas from a dry tropical forest in Ghana. Nevling (1971) reported on the flowering phenology of an elfin forest in Puerto Rico, but the flora there was depauperate and included only four or five climber species. In a phenological study of woody plants in a northern Japanese forest, Hirabuki (1984) included five species of lianas, but the number of lianas in this community was very small. There have been no previous phenological studies of entire herbaceous vine communities.

Panamanian studies

In a study of extensive herbarium material collected on Barro Colorado Island, Croat (1975) tallied fruiting and flowering of 809 species, including 909 herbaceous vines and about 150 species of lianas. Croat's study included some cultivated species and was subject to the vagaries of collection statistics. It is possible that more collections may have been made during the dry season, when conditions for biologists are most pleasant and visits by botanists are more frequent.

Croat's results showed flowering peaks for both herbaceous vines and lianas from December to March during the early part of the 4–5 month dry season. Fruiting by both herbaceous vines and lianas peaked from February to April, showing that maturation of fruits occurred during or immediately after the onset of flowering.

For all herbaceous plants of open areas – the primary habitat for herbaceous vines – Croat's results demonstrated year-round flowering with a dip in November, while fruiting levels were fairly equitable with a dip in April. Thus, the seasonal patterns for vines were dissimilar to those for herbaceous plants as a group.

Flowering seasonality of trees was unlike that of lianas and showed peaks at the end of the dry season extending into early wet season (February–June), while fruiting by canopy trees was bimodal with peaks in late dry season (March–May) and mid-wet season (July–August).

In contrast to Croat's (1975) study of herbarium material, Putz & Windsor (1987) followed the phenology, including leaf production, of marked individuals of 43 liana species and contrasted the results with those obtained for 26 species of trees. Their results differed significantly from those of Croat. Flowering by lianas was bimodal with peaks in mid-dry season (March) and again in mid-wet season (August). Fruiting by lianas was concentrated in the dry season (February–March) with a lower peak in late wet season (November). The dry season fruiting peak falls within that reported by Croat (1975), but the late wet season peak did not appear in Croat's results. All lianas were evergreen, although individuals of two normally evergreen species were deciduous for short periods. Foliation by lianas was greatest during the dry season and somewhat lower during the wet season.

In a study of fruit fall on Barro Colorado Island, Foster (1982) found that most liana seeds appeared in his basket traps from March to June and that average seed weights for the vast majority of lianas ranged between 0.01 and 1 g. Foster's results indicate a fruiting peak for lianas that agrees in part with that reported by Croat (1975) and Putz & Windsor (1987), except that it extends further into the early wet season. Foster divided the dispersal types between fleshy, presumably animal-dispersed fruits – 33 species (42%), and winged wind-dispersed fruits – 45 species (58%).

Other tropical studies

In an African study of woody plant phenology, Lieberman (1982) emphasized trees, but included 11 liana species, of 33 listed for the local flora. Her dry forest study site in Ghana had two dry seasons and two wet seasons each year. Her data presentation gave phenological information only on the flowering of lianas. Almost equal numbers of lianas flowered either exclusively in the dry seasons (3) or wet seasons (4), or flowered more or less continuously encompassing both wet and dry seasons (4).

Nevling (1971) displayed results for a study of flowering seasonality by all spermatophyte plants in an elfin forest community in Puerto Rico. His results for the climber component of the flora showed only minor oscillations in the number in flower. Since only four or five climber species were involved the peaks or valleys shown in the annual pattern may not be significant. For the community as a whole, flowering peaked in July and October and seemed to be inversely correlated with rainfall.

In Queensland, Australia, Hegarty (1988) studied the phenology of 45 climber species for almost 3 years. These species occurred either in or adjacent to a rainforest and included both lianas and herbaceous vines. Most species were involved in a long fall–winter–spring peak (most were flowering in September–October), with a relative lull in the cool season (June–July). Fruiting peaked in April–July, and had a low point in September. Wind-dispersed propagules matured in dry weather. Most were evergreen, but a few were deciduous during the Southern Hemisphere winter.

Temperate studies

Most temperate studies do not include climbers, due to lack of adequate representation. A study in northern Japan included five deciduous lianas (Hirabuki, 1984). As one might expect, no overt activity was observed in the winter. The seasonal leafing, flowering, and fruiting schedules were alike.

Flowering was recorded from May to mid-July; fruit maturation and dissemination occurred in late summer and fall. Leaf production and expansion occurred in spring before flowering and leaf shedding was in the fall. These patterns are probably typical of climbers in most winter-cold temperature ecosystems.

The Costa Rican study

In two lowland tropical dry forest locations in Guanacaste Province, northwestern Costa Rica, a series of plant community phenological studies was undertaken as a part of a comparative ecosystems study by the Organization for Tropical Studies. Results of studies of tree phenology (Frankie, Baker & Opler, 1974), and treelets and shrubs (Opler, Frankie & Baker, 1980b) have been previously reported, while the results from our studies of climber phenology are reported here for the first time. A study of herbaceous plant phenology remains unreported (P. Opler, H. Baker and G. Frankie, unpublished data). Studies of climber reproductive and leafing phenology were conducted during 1972 and 1973.

General descriptions of the study sites and climate were given by Frankie *et al.* (1974) for COMELCO – now largely part of the Lomas Barbudal Biological Reserve, and Daubenmire (1972) and Turner (1975) for Hacienda La Pacifica – now Centro Ecologico La Pacifica. The methods for monitoring lianas are generally those described by Frankie *et al.* (1974). We monitored lianas along approximately linear transects, each through riparian forest, or hill forest, or derived savanna; 113 individuals of 33 species were observed every 4–6 weeks. Calculation of phenological activity levels follows the methods of Opler *et al.* (1980b). The seasonality of herbaceous vines was determined by monthly observations in secondary succession plots at Hacienda La Pacifica (Opler, Baker & Frankie, 1977, 1980a).

The climate of the study sites is characterized by a marked dry season usually extending from mid-November to mid-May. Normally, rainfall is sparse during the dry season. Mean annual precipitation is approximately 1.5 m, with somewhat more rainfall at La Pacifica. The rainy season usually begins with heavy convectional storms in May and early June, but a relatively dry period often follows in July and early August. Mean monthly temperatures vary little during the year (mean = 28 °C); photoperiod differs only by 30 min between the longest and shortest days.

For each observation of a climber, the reproductive and foliation statuses were recorded. Flower buds, open flowers, immature fruits, and mature fruits were subjectively recorded as none, few, or many. Foliation status was recorded as bare, full leaves (full complement of mature leaves) or new leaves (few or many).

Analysis of the data was carried out by assigning a number for the status of each individual monitored in a given month. For plants having many of a category (buds, open flowers, immature fruits, mature fruits, or new leaves) the value 2 was assigned; for plants having few of a category the value 1 was assigned; and for no activity the value 0 was assigned. For each species the sum of the values for a given month was divided by the number of individuals monitored; this gave the activity value for the month.

To obtain the community values, the mean species activity levels for all species monitored that month were summed and divided by the number of species under observation. A mean community value of 2 would indicate that all individuals of all climber species observed had many flowers, fruits, or leaves. A value of 1 would indicate that, on average, all climbers had few of the category under scrutiny. Because so few individuals were monitored, the variance in activity levels would not be meaningful.

The lianas and herbaceous vines found in the study areas are listed on Table 14.1. Where possible, the names follow those given by Janzen & Leisner (1980). Asterisked species are those that conform best with the term herbaceous vine, while the others are more or less woody lianas that have perennially woody above-ground stems. Note that among woody species (63) only Asclepiadaceae (6 spp.), Bignoniaceae (17 spp.), and Sapindaceae (7 spp.) were represented by five or more species, while among herbaceous vines (40 spp.) Convolvulaceae (13 spp.) and Leguminosae (15 spp.) dominate. Most of the species in the above families have animal-dispersed fruits and all rely on gravity, or explosive dehiscence, or wind for dispersal. Of course, other families with poorer representation have fleshy animal-dispersed fruits (e.g. Cucurbitaceae, Passifloraceae, Vitaceae). The preponderance of physical dispersal is typical of the dry forest flora in general.

The community patterns for flowering by lianas and herbaceous vines were quite different (Figure 14.1). Herbaceous vine flowering was highly seasonal with a peak in late wet season and early dry season. This peak also agrees with that for terrestrial herbs in the same community. The pattern for lianas is relatively aseasonal with slight peaks in May, July, and January; these peaks may not be statistically significant.

The fruiting patterns are more similar for the two classes of climbers (Figure 14.2), although that for herbaceous vines is more seasonally variable than that observed in lianas. Both groups have peaks in the dry season, with that for herbaceous vines in early dry season and that for lianas in late dry season. Fruiting by herbaceous vines immediately follows flowering, but lianas that flower at other seasons delay fruit maturation and dispersal until late dry season. This pattern is shared by many dry forest canopy trees that have gravity- or wind-dispersed diaspores (Frankie *et al.*, 1974). Guanacaste climbers with animal-dispersed fruits mature their fruits in the wet season. Again, this is a pattern shared by trees and shrubs in the same community.

Both herbaceous vines and lianas produced new leaves predominantly in

Table 14.1. *Systematic listing of climber species from the Guanacaste study sites. Herbaceous vines are preceded by an asterisk*

Amaranthaceae
- **Pfaffia paniculata* (Mart.) O. Ktze.

Apocynaceae
- *Echites tuxtlensis* Standl.
- *Fernaldia pandurata* (A. DC.) Woodson
- *Forsteronia spicata* (Jacq.) G. Mey.
- *Prestonia acutifolia* (Benth.) K. Schum.

Asclepiadaceae
- **Blepharodon mucronatum* (Schlecht.) Dcne.
- *Cynachum rensonii* (Pitt.) Woodson
- *Marsdenia trivirgulata* Bartl.
- *Matelea guirosii* (Standl.) Woodson
- *M. trianae* (Don.) Spellm.
- *Sarcostemma bilobum* Hook. & Arn.
- **S. glaucum* HBK.
- *S. odoratum* (Hemsl.) Holm.

Asteraceae
- **Calea prunifolia* HBK.

Bignoniaceae
- *Adenocalymma inundatum* Mart. ex DC.
- *Amphilophium paniculatum* (L.) HBK.
- *Anemopaegma chrysoleucum* (HBK.) Sandw.
- *Arrabidaea conjugata* (Vell.) Mart.
- *A. corallina* (Jacq.) Sandw.
- *A. mollissima* (HBK.) Btlr. & K. Schum.
- *A. patellifera* (Schlecht.) Sandw.
- *Callichlamys latifolia* (L. Rich.) K. Schum.
- *Clytostoma binatum* (Thunb.) Sandw.
- *Cydista aequinoctialis* (L.) Miers. var. *aequinoctialis*
- *C. aequinoctialis* var. *hirtella* (Benth.) A. Gentry
- *C. diversifolia* (HBK.) Miers
- *C. heterophylla* Seibert
- *Macfadyena unguis-cati* (L.) A. Gentry
- *Mansoa hymenaea* (DC.) AG.
- *Pithecoctenium crucigerum* (L.) A. Gentry
- *Xylophragma seemannianum* (O. Ktze.) Sandw.

Boraginaceae
- *Tournefortia volubilis* L.

Caesalpiniaceae
Bauhinia glabra Jacq.

Combretaceae
Combretum decandrum Jacq.
C. farinosum HBK.

Connaraceae
Rourea glabra HBK.

Convolvulaceae
Bonamia sulphurea (Brandeg.) T. Myint.
Convolvulus nodiflorus Desv.
Ipomoea carnea Jacq.
I. hederifolia (L.) G. Don.
I. meyeri (Spreng.) G. Don.
I. minutiflora (Mart. & Gal.) House
I. nil (L.) Roth
I. trichocarpa × lacunosa
I. trifida (HBK.) Don.
I. umbracticola House
Jacquemontia pentantha (Jacq.) G. Don.
Merremia aegyptica (L.) Urban
M. cissoides (Lam.) Hall.
M. umbellata (L.) Hall.

Cucurbitaceae
Cayaponia attenuata (Hook. & Arn.) Cogn.
Cucumis anguria L.
Rytidostylis carthaginensis (Jacq.) Ktze.
Sicydium tamnifolium (HBK.) Cogn.

Dilleniaceae
Davilla kunthii St. Hil.
Tetracera volubilis L.

Dioscoreaceae
Dioscorea convolvulacea C. & S.

Euphorbiaceae
Dalechampia scandens L.

Fabaceae
Calopogonium caeruleum (Benth.) Sauvelle
C. mucunoides Desv.
Canavalia brasiliensis Mart. ex Benth.
Centrosema plumieri (Turp.) Benth.
C. pubescens Benth.
C. sagittatum (Humb. & Bonpl.) Brandeg.
Dalbergia glabra (Millsp.) Standl.
Desmodium barclayi Benth.
Dioclea megacarpa Rolfe
Mucuna pruriens (L.) DC.
Pachyrrhizus erosus (L.) Urb.

Table 14.1 (*cont.*)

**Phaseolus atropurpureus* DC.
**P. lunatus* L.
**Rhynchosia calycosa* Hemsl.
**R. edulis* Griseb.
**R. minima* (L.) DC.
**Teramnus uncinatus* (L.) Sw.
Hippocrateaceae
Hippocratea volubilis L.
Liliaceae
Smilax spinosa Mill.
Loasaceae
**Gronovia scandens* L.
Malpighiaceae
Banisteriopsis muricata (Cav.) Cuatr.
Heteropteris beechyana Adr. Juss.
Hiraea reclinata Jacq.
Stigmaphyllon ellipticum (HBK.) Adr. Juss.
Mimosaceae
Entada polystachya (L.) DC.
Mimosa albida Humb. & Bonpl.
M. quadrivalis L.
Passifloraceae
**Passiflora foetida* L.
P. platyloba Killip.
P. pulchella HBK.
Polygalaceae
Securidaca sylvestris Schlecht.
Rhamnaceae
Gouania polygama (Jacq.) Urban
Sapindaceae
Cardiospermum grandiflorum Sw.
Paullinia costaricensis Radlk.
P. cururu L.
Serjania atrolineata Sauv. & Wr.
S. mexicana (L.) Willd.
S. schiedeana Schlecht.
Urvillea ulmacea HBK.
Trigoniaceae
Trigonia rugosa Benth.
Verbenaceae
Petrea volubilis L.
Vitaceae
Cissus biformifolia Standl.
C. rhombifolia Vahl.
C. sicyoides L.
Vitis tiliifolia H. & B. ex. Roem. & Schult.

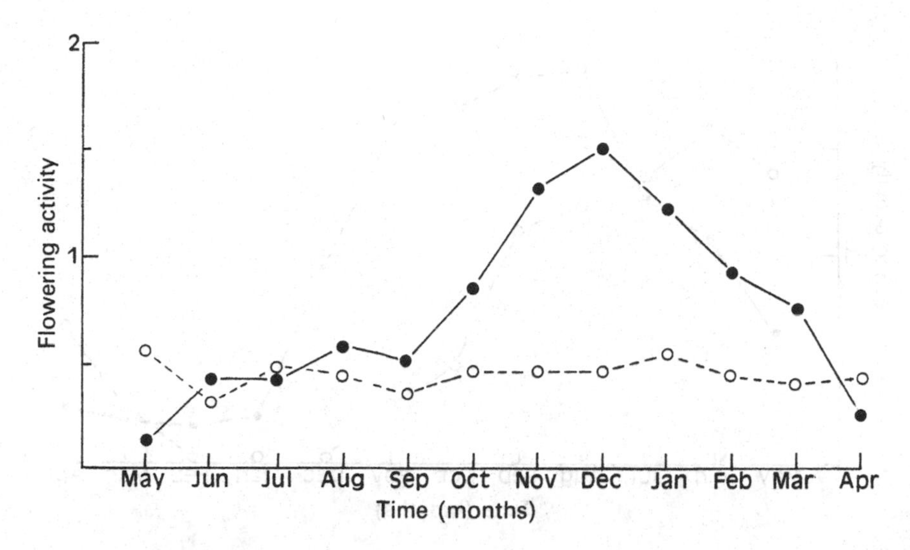

Figure 14.1. Flowering phenology for climbers from Costa Rican tropical dry forest study sites. Solid circles and solid lines indicate herbaceous vines; open circles and dashed lines indicate lianas. See text for description and calculation of activity levels.

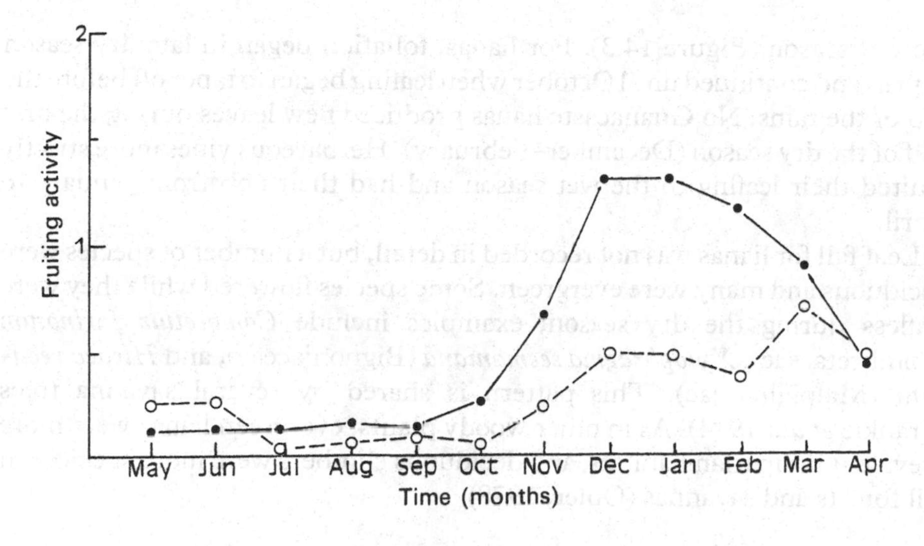

Figure 14.2. Fruiting phenology for climbers from Costa Rican tropical dry forest study sites. Solid circles and solid lines indicate herbaceous vines; open circles and dashed lines indicate lianas. See text for description and calculation of activity levels.

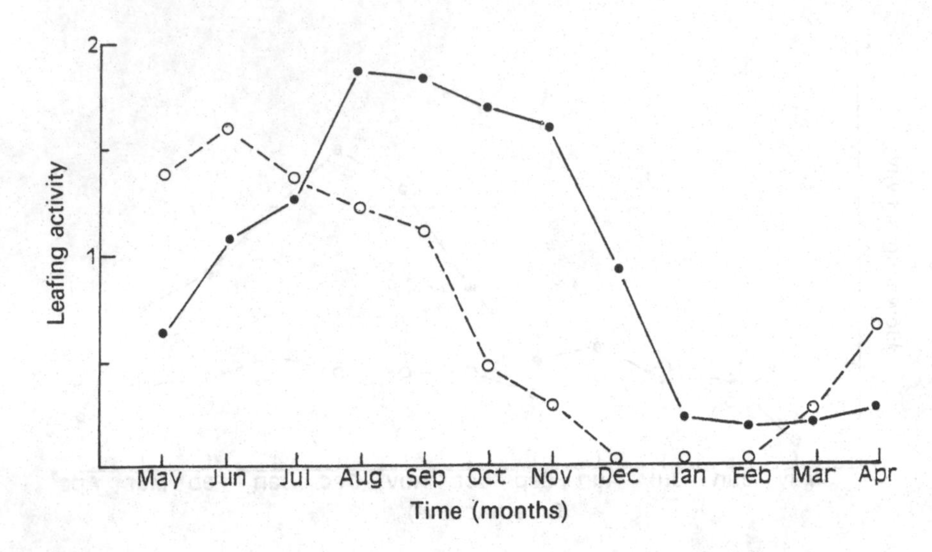

Figure 14.3. Leafing phenology for climbers from Costa Rican tropical dry forest study sites. Solid circles and solid lines indicate herbaceous vines; open circles and dashed lines indicate lianas. See text for description and calculation of activity levels.

the wet season (Figure 14.3). For lianas, foliation began in late dry season (April) and continued until October when leafing began to taper off before the end of the rains. No Guanacaste lianas produced new leaves during the first half of the dry season (December–February). Herbaceous vines more strictly limited their leafing to the wet season and had their ebb from January to April.

Leaf fall for lianas was not recorded in detail, but a number of species were deciduous and many were evergreen. Some species flowered while they were leafless during the dry season; examples include *Combretum farinosum* (Combretaceae), *Xylophragma seemaniana* (Bignoniaceae), and *Hiraea reclinata* (Malpighiaceae). This pattern is shared by several savanna trees (Frankie *et al.*, 1974). As in other woody plants, evergreen lianas were more prevalent in riparian habitats, and deciduous climbers were more speciose in hill forests and savannas (Opler, 1978).

Comparison with Panamanian studies

The climate at Barro Colorado Island is characterized by a briefer dry season than that of the Guanacaste study areas, and mean annual precipitation there, at 3 m, is almost double that of our Costa Rica study sites. Our results for herbaceous vines agree well with those presented by Croat (1975). Flowering by lianas in Panama is more seasonally limited than that we found for dry forest lianas in Guanacaste. This is puzzling since in Guanacaste there is a longer dry season and much lower annual precipitation. The relative aseasonality of our results for liana flowering is strongly influenced by the pattern for Bignoniaceae, the dominant family (Gentry, 1974). Most of these species are adapted to pollination by large bees (Frankie *et al.*, 1983).

The fruiting peaks for lianas reported by Putz & Windsor (1987) were March–May and again in November. In Guanacaste, our data show the same dry season peak, but we did not find a separate late wet season peak. Perhaps the specific taxa studied may have contributed to this difference, or it is possible that a wet season peak may be typical of wetter habitats. Future studies of liana phenology in a wet forest habitat could help confirm this postulate.

The leafing pattern reported for Barro Colorado Island lianas by Putz & Windsor (1987) differs strikingly from that we found for lianas in Guanacaste. We found a highly seasonal pattern of leafing in our dry forest environment, and they found leaf production relatively aseasonal. We feel that this difference is likely to be attributable to the greater similarity of Barro Colorado Island to rainforest environments. For example, Frankie *et al.* (1974) report that foliation by rainforest trees in Costa Rica is also relatively aseasonal.

Pollinator and frugivore studies

Flower-visiting animals (insects, hummingbirds, bats) that utilize nectar or pollen often depend on groups of plants that include several life forms. Zoologists have chosen to study the phenology of the plants that their subjects depend on in order to determine the seasonality of resource use, and to determine whether the timing of flowering or fruiting seasons might have been determined by interspecific competition.

Several such studies have included climbers among the plants monitored, and this work provides additional insights into the selective forces behind the phenological patterns we can observe. Gentry (1974) described five basic flowering patterns for the Bignoniaceae of Costa Rica and Panama. One of these was for the bat-pollinated *Crescentia* trees, while the other four were for woody climbers. His 'cornucopia' pattern (Type 3) was most frequent, and involves a flowering episode of several weeks. Type 2, the 'steady state' pattern, involves species that produce a few flowers each day over several months. The 'big bang' pattern (Type 4) is shown by species that produce a massive crop of flowers whose anthesis is limited to just a few days, while the 'multiple bang' pattern (Type 5) consists of several discrete brief massive flowerings. All four of these patterns were displayed by one or more species in our study area.

Frankie *et al.* (1983) report on the phenology of plants adapted for pollination by large bees in the same Guanacaste dry forest habitats that we reported on above. Together with trees, climbers, especially lianas, provided nectar and pollen resources for large bees – primarily large Anthophoridae – throughout the year. Trees and lianas adapted for these bees that flowered in the dry season had brief periods (3–6 days) of mass flowering, while those that flowered in the wet season had extended flowering periods (greater than 7 days) with few flowers produced each day. A few Bignoniaceae had more than one episode of mass flowering in a season. Frankie *et al.* (1983) did not find evidence to suggest that the spacing of flowering periods had evolved in response to interspecific competition for pollinators.

Studies of the flowering phenology of plants adapted for hummingbirds in lowland and pre-montane wet forests in Costa Rica included several climbers (Stiles, 1978, 1985). Stiles found several groups of plants that cut across life forms, each adapted to groups of hummingbirds with different bill morphologies and foraging strategies. Stiles felt that the hummingbirds and the morphology of flowers they utilized were co-evolved and that the spacing of flowering periods by the plants had been in part determined by interspecific competition for pollinators. Periods when most floral resources were available coincided with the courtship and reproduction of the hummingbirds. Based upon statistical tests for randomness and regularity, Poole & Rathcke (1979) disputed Stiles' assertion about the co-evolved relation between hummingbird reproductive cycles and the flowering seasonality of their food plants. Stiles (1979) rebutted Poole & Rathcke's results by asserting that their assumptions were not biologically realistic.

Further north in southeastern Mexico, Toledo (1975) found the flowering of hummingbird-adapted plants (with 2 climbers among 26 species) peaked during periods of courtship and reproduction by the birds, but that few hummingbird plants were in flower from September to December.

Similarly, if flowering patterns of plants adapted to specific pollinators can cut across life forms, so can the fruiting behavior of plants adapted to frugivorous animals. In a study of fruiting by *Smilax aspera*, a woody climber of Mediterranean scrub, Herrera (1981) generalized that competition for frugivores that can disperse its seeds is an important selective factor determining fruit size and seed number. We feel that breeding system, canopy position, dispersal type, and habitat are much more important determinants of fruit size, as well as of seed size and number (Baker, 1972; Bawa & Opler, 1975; Frankie *et al.*, 1974; Opler *et al.* 1980a).

Conclusion

We have presented available information on flowering, fruiting, and leafing pattern of climber communities, including the results of our study of a tropical dry forest climber community from northwestern Costa Rica. We have also discussed other studies of plants adapted for specific flower visitors.

Relatively few studies of climber phenology have been reported and there is still not sufficient information to theorize on the relative importance of climate, local environments, or biotic factors on the seasonal patterns of flowering, fruiting, and leafing that we see in different plant communities. We need more species-by-species analysis, with the relating of events within individual plants. Most authors make only empirical hypotheses based upon apparent correlations between the phenological patterns observed and some recorded biotic or abiotic environmental factors known about each study area. In-depth ecophysiological studies such as that by Longino (1986) should reveal more information about the proximate factors that are responsible for the patterns we see.

Acknowledgements

The Costa Rican research was supported by the National Science Foundation (Grant number GB–7805, GB–25592, GB–25592A 2, and GB–40747X). We thank the Organization for Tropical Studies for facilities and co-operation. Landowners at La Pacifica and COMELCO cooperated in several aspects of the study; Ing. Werner Hagnauer of La Pacifica provided land for the successional studies that provided most information on herbaceous vines. Although many botanists provided determinations of the plants, those at the Field Museum (Chicago) and the Missouri Botanical Garden identified most climbers. Voucher specimens are deposited at the Museo National de Costa Rica and the University of California at Berkeley, in addition to the two institutions named above.

References

Baker, H. G. (1972). Seed weight in relation to environmental conditions in California. *Ecology* 53: 997–1010.

Bawa, K. S. & Opler, P. A. (1975). Dioecism in tropical forest trees. *Evolution* 29: 167–79.

Croat, T. B. (1975). Phenological behavior of habit and habitat classes on Barro Colorado Island (Panama Canal Zone). *Biotropica* 7: 270–7.

Daubenmire, R. (1972). Phenology and other characteristics of tropical semi-deciduous forest in northwestern Costa Rica. *Journal of Ecology* 60: 147–70.

Foster, R. B. (1982). The seasonal rhythm of fruitfall on Barro Colorado Island. In *The Ecology of a Tropical Forest: seasonal rhythms and long-term changes*, ed. E. G. Leigh, Jr, A. S. Rand and D. M. Windsor, pp. 151–72. Smithsonian Institution Press, Washington, DC.

Frankie, G. W., Baker, H. G. & Opler, P. A. (1974). Comparative phenological studies of trees in tropical wet and dry forests in the lowlands of Costa Rica. *Journal of Ecology* **62**: 881–919.

Frankie, G. W., Haber, W. A., Opler, P. A. & Bawa, K. S. (1983). Characteristics and organization of the large bee pollination system in the Costa Rica dry forest. In *Handbook of Experimental Pollination Biology*, ed. C. E. Jones and R. J. Little, pp. 411–47. Van Nostrand Reinhold Co., New York.

Gentry, A. H. (1974). Flowering phenology and diversity in tropical Bignoniaceae. *Biotropica* **6**: 64–8.

Hegarty, E. E. (1988). Canopy dynamics of lianas and trees in subtropical rainforest. PhD thesis, University of Queensland, St Lucia, Australia.

Herrera, C. M. (1981). Fruit variation and competition for dispersers in natural populations of *Smilax aspera*. *Oikos* **36**: 51–8.

Hirabuki, Y. (1984). Phenology of woody plants in a fir forest in Sendai, northeast Japan. *Japanese Journal of Ecology* **34**: 235–8.

Janzen, D. H. & Leisner, R. (1980). Annotated check-list of plants of lowland Guanacaste Province, Costa Rica, exclusive of grasses and non-vascular cryptograms. *Brenesia* **18**: 15–19.

Lieberman, D. (1982). Seasonality and phenology in a dry tropical forest in Ghana. *Journal of Ecology* **70**: 791–806.

Longino, J. T. (1986). A negative correlation between growth and rainfall in a tropical liana. *Biotropica* **18**: 195–200.

Nevling, L. I. Jr (1971). The ecology of an elfin forest in Puerto Rico, 16. The flowering cycle and an interpretation of its seasonality. *Journal of the Arnold Arboretum* **52**: 586–613.

Opler, P. A. (1978). Interaction of plant life history components as related to arboreal herbivory. In *Ecology of Arboreal Folivores*, ed. G. G. Montgomery, pp. 23–31. Smithsonian Institution Press, Washington, DC.

Opler, P. A., Baker, H. G. & Frankie, G. W. (1977). Recovery of tropical lowland ecosystems. In *Recovery and Restoration of Damaged Ecosystems*, ed. J. Cairns, Dickson and Herricks, pp. 379–419. University Press of Virginia, Charlottesville.

Opler, P. A., Baker, H. G. & Frankie, G. W. (1980a). Plant reproductive characteristics during secondary succession in neotropical lowland forest ecosystems. *Biotropica* (supplement), pp. 40–6.

Opler, P. A., Frankie, G. W. & Baker, H. G. (1980b). Comparative phenological studies of treelet and shrub species in tropical wet and dry forests in the lowlands of Costa Rica. *Journal of Ecology* **68**: 167–88.

Poole, R. W. & Rathcke, B. J. (1979). Regularity, randomness, and

aggregation in flowering phenologies. *Science* **203**: 470–1.
Putz, F. E. & Windsor, D. M. (1987). Liana phenology on Barro Colorado Island, Panama. *Biotropica* **19**: 334–41.
Stiles, F. G. (1978). Temporal organization of flowering among the hummingbird foodplants of a tropical wet forest. *Biotropica* **10**: 194–210.
Stiles, F. G. (1979). Regularity, randomness, and aggregation in flowering phenologies. *Science* **203**: 471.
Stiles, F. G. (1985). Seasonal patterns and coevolution in the hummingbird-flower community of a Costa Rican subtropical forest. *Neotropical Ornithology*, ed. P. A. Buckley, M. S. Foster, E. S. Morton, R. S. Ridgely and F. G. Buckley, pp. 757–85. Ornithological Monograph 36.
Toledo, V. M. (1975). La estacionalidad de las flores utilizadas por los colibries de una selva tropical humeda en Mexico. *Biotropica* 7: 63–70.
Turner, D. C. (1975). *The Vampire Bat – a study in behavior and ecology*. Johns Hopkins University Press, Baltimore, Md.

15
Breeding and dispersal systems of lianas

ALWYN H. GENTRY

The organization and floristic composition of tropical plant communities is strongly and intricately influenced by interactions with pollinators and dispersers (Ashton, 1969; Frankie, 1975; Stiles, 1977, 1978; Gentry, 1982a, 1983; Frankie *et al.*, 1983; Feinsinger, 1983; Bawa, Perry & Beach, 1985b). Data on tropical pollination and dispersal systems have begun to accumulate rapidly only in the last few years, but many of the available data have specifically excluded climbers. In this chapter I summarize what is known of climber reproductive biology from the literature, supplemented by my own observations and those of colleagues.

This chapter is a summary largely based on the liana subset of the data of Gentry (1982a, 1988, Chapter 1). These data are for plants > 2.5 cm dbh (2.5 cm greatest diameter for lianas) in 0.1 ha. The first half of the chapter summarizes pollination systems for the lianas and liana communities represented in these samples and compares them with pollination systems for other habit groups. The second half focuses on the dispersal strategies of these lianas and liana communities.

Tropical pollination systems are known to be highly diverse, with hummingbirds, perching birds, bees, hawkmoths, settling moths, bats, butterflies, wasps, flies, and beetles all known to play a role (Faegri & van der Pijl, 1966; Baker, 1973; Bawa *et al.*, 1985a). Knowledge of the diversity of tropical pollination systems continues to increase, most recently with non-flying mammals (Lumer, 1980; Janson, Terborgh & Emmons, 1981; Steiner, 1981) and thrips (Ashton, 1979; Thien, 1980; Appanah & Chan, 1981) discovered to play significant roles. Even more recently, apomixis has been suggested to be important in tropical forest breeding systems (Ashton, 1977; Kaur *et al.*, 1978; Renner, 1989; see also Gentry, 1989a).

However, much of the diversity of pollination systems in the tropics occurs in understory or second growth species. Most authors have focused on a single pollination guild rather than on the plant community as a whole (e.g. large bees: Frankie *et al.*, 1983; Hawkmoths: Haber & Frankie, 1989; hummingbirds: Feinsinger, 1978; Stiles, 1978; butterflies and hummingbirds: Heithaus, 1974; bees and wasps: Heithaus, 1979a, b; bats: Heithaus, Fleming & Opler, 1975). The closest to a complete survey of the pollination systems of a tropical forest is the study of La Selva, Costa Rica, trees by Bawa *et al.* (1985a; see also Kress & Beach, 1990). An earlier comparison of La Selva and Guanacaste trees, shrubs, and treelets (Frankie, Baker & Opler, 1974; Opler, Frankie & Baker, 1980) focused on phenology. Hubbell (1979) also discussed how pollination biology relates to the structure and composition of a dry forest community.

According to Bawa *et al.* (1985a), canopy pollination mechanisms in La Selva canopy trees are 'monotonous', with the great majority of canopy tree species pollinated by large to medium-sized bees (44%) or by a group of small diverse 'generalist' insects (23%). Among the 52 canopy tree species at La Selva only 1 is hummingbird-pollinated, 1 sphingid-pollinated, 1 butterfly-pollinated, 2 bat-pollinated, 2 wasp-pollinated, and 4 small-bee-pollinated; thrip pollination and wind pollination are apparently non-existent. Subcanopy tree species at La Selva showed a much greater diversity of pollination systems, with only 20% of the species pollinated by large to medium-sized bees, 17% by small bees, 11% by sphingids, 3% by bats, 5% by hummingbirds, 6% by butterflies, 1% by thrips, and perhaps as many as 3.6% by wind (but see Gentry, 1989a).

How do lianas fit into this picture? Do their breeding systems show any distinct tendencies? Unfortunately, direct data similar to those for La Selva trees are unavailable for liana communities. Indeed, as late as 1976 Jacobs could summarize liana ecology as 'virtually a blank'.

It is possible, however, to extract some data on liana pollination and breeding systems from the literature, which is mostly focused on single taxonomic groups or guilds. In this brief review of climber pollination, I will try to follow the classification of pollination systems of Bawa *et al.* (1985a) except for combining small bees (such as halictids and trigonids) with small diverse insects. I will first look at data for different pollination guilds (which is highly skewed toward Costa Rica), then survey the most important families of climbers (strongly biased toward the Neotropics), and finally look at pollination in whole liana communities.

Overview of liana pollination by guild

Some lianas are bat-pollinated, e.g. *Cobaea*, *Trianaea*, *Calycophysum*, *Marcgravia*, some *Mucuna*, *Freycinetia*, and *Cayaponia* spp.); indeed, five of the 12 bat-pollinated neotropical genera discussed by Vogel (1958) are lianas, mostly of cloud forests. Dobat (1985) lists 18 scandent genera with bat-pollinated species; some of these records are erroneous but at least 14 climbing genera have at least one bona fide bat-pollinated species. However, no climbers are included among the bat-pollinated species studied by Heithaus *et al.* (1975) and they make up only a tiny fraction (7%) of the 270 reputedly bat-pollinated genera collated by Dobat (1985). On the whole, bat pollination seems rather uncommon among neotropical lianas, except possibly in cloud forests. It may be even rarer in paleotropical lianas where only four scandent genera have been reported to have bat-pollinated species (Dobat, 1985), although one of these is the important climbing genus *Freycinetia* (Vogel, 1981b; Dobat, 1985).

Only one climber (*Ipomoea alba*) is included in the 31-species, sphingid-pollinated dry forest community described by Haber & Frankie (1989), and hawkmoth pollination is also rare in La Selva canopy trees (only 1 species). Although two climbing Cucurbitaceae genera, *Psiguria* and *Gurania*, are famous for their pollination interactions with *Heliconius* butterflies (Gilbert, 1972, 1980), only one vine is included among 18 butterfly-pollinated dry forest species studied by Heithaus (1974). Overall, pollination by Lepidoptera does not seem especially prevalent among lianas.

Similarly, hummingbird pollination is rare in lowland forest climbers (5 climbers (8%) among the 59 La Selva hummingbird flowers (Stiles, 1978), just as it is in La Selva canopy trees (only 1 species, an *Erythrina*). Heithaus (1974) lists several hummingbird-pollinated climbers (4 of 18 hummingbird-pollinated species) in a dry forest community but only one of these (*Combretum farinosum*) seems morphologically adapted for bird pollination. At middle and upper elevations, hummingbird-pollinated climbers seem more prevalent, especially in Passifloraceae and Ericaceae (Snow & Snow, 1980; Feinsinger, 1983). For example, two of the 10 hummingbird-pollinated species studied by Feinsinger (1978) at Monteverde, Costa Rica, were vines as were 16.7% of that site's 600 hummingbird-pollinated species (Feinsinger, 1983).

In contrast, beetle pollination is more prevalent among climbers than in any other habit group, especially in the subcanopy, with 53% of the beetle-pollinated La Selva species being scandent (aroids, cyclanths, *Desmoncus*) along with 37.5% of those on Barro Colorado Island (Schatz, 1989).

Pollination by large and middle-sized bees may be unusually important in climbers. For example, large and middle-sized bees pollinate 38 of 130 (30%) of the climbers in a Guanacaste dry forest, compared with only 35 of 160 (22%) of the trees (and only 15 shrubs, herbs, and epiphytes combined) (Frankie *et al.*, 1983). Frankie *et al.* (1983) suggest that large–medium bee-pollinated climbers have relatively low visitation rates compared to trees with the same pollinators, and consequently often have buildup of non-removed nectar. A significant segment of the large–medium bee-pollinated climbers (e.g. Malpighiaceae, *Dalechampia*, some cucurbits) have oils or resins as rewards rather than nectar (see summary in Simpson & Neff, 1981) and a majority of the plants with these otherwise rather uncommon rewards are scandent. Perhaps there is something about scandent habit that puts these unusual attractants at an adaptive premium. On the other hand, there are few climbers, but many shrubs, herbs, and trees, among the Guanacaste species pollinated by small bees (Frankie *et al.*, 1983).

We may conclude that climbers are highly diversified in pollination systems, including some members of all the major pollination guilds. There are hints that some guilds like large–medium bees and beetles may be disproportionately well represented among climbers. It is not clear from the available evidence whether such apparent peculiarities in frequency of pollination of climbers by particular pollinator guilds are significant.

Taxonomic overview of liana pollination

In a survey of the plant taxon-focused pollination literature the pre-eminence of large and medium-sized bees as liana pollinators again stands out.

Bignoniaceae, the most important neotropical liana family (Gentry, Chapter 1) is typical. It is overwhelmingly pollinated by large to middle-sized bees (Gentry, 1989b). There is some evolutionary diversification in pollination among different climbing bignon genera, with eight small genera exclusively hummingbird-pollinated and another mostly so, two small genera (*Tanaecium* and *Leucocalanthe*) hawkmoth-pollinated (along with miscellaneous species of a few other genera), and two genera mostly butterfly-/small-bee-pollinated (along with miscellaneous species of a few other genera) (Gentry, 1989b). However, 26 of the climbing genera are exclusively pollinated by large and middle-sized bees and several more predominantly so. Altogether 85% of the 400 scandent neotropical bignon species are pollinated by medium-sized to large bees.

Leguminosae, the next most important liana family in the Neotropics, is also overwhelmingly pollinated by large to middle-sized bees. Although the family has diversified with respect to pollination mode, most of the climbers are papilionoids (see Gentry, Chapter 1) whose strongly zygomorphic, bilaterally symmetric flowers are the epitome of the large-bee syndrome (Frankie *et al.*, 1983). An important exception is the large pantropical genus *Mucuna*, which is largely bat-pollinated (van der Pijl, 1941; Vogel, 1958; Dobat, 1985), although there are also large-bee-pollinated species like *M. rostrata* (personal observation). The relatively few scandent caesalpinioids (e.g. *Bauhinia*) are also mostly pollinated by medium–large bees.

Mimosoid climbers are mostly small-flowered and pollinated by small diverse insects and/or butterflies.

Perhaps the next most important neotropical climbing family, at least at the community level, is Malpighiaceae, pollinated by large and middle-sized bees that gather oil from the calyx glands rather than nectar (Vogel, 1974, 1989; Simpson & Neff, 1981).

Several families of climbers with more diversity in pollination systems have a preponderance of pollination by large and medium-sized bees. For example, the great majority of Passifloraceae species are large- to medium-bee-pollinated, although there have been frequent shifts to hummingbird pollination (at least 12 different times according to MacDougal, 1983; see also Janzen, 1968; Snow & Snow, 1980; Gentry, 1981; Escobar, 1985). There is also at least one bat-pollinated species (Sazima & Sazima, 1978), a clade which has specialized in wasp pollination (MacDougal, 1983), and some of the smallest-flowered species may be pollinated by small diverse generalist insects.

Apocynaceae, like Passifloraceae, are mostly bee-pollinated (at least in the Neotropics) but also encompass numerous other pollination systems. Most African species have floral morphologies and fragrances suggesting pollination by settling moths. Genera like *Forsteronia* with tiny flowers are probably pollinated by small diverse insects (*sensu* Bawa *et al.*, 1985a). *Mandevilla* and *Odontadenia*, mostly large-bee-pollinated, have a few species with red or deep purple flowers that appear to be hummingbird-pollinated (e.g. Feinsinger, 1978; Gentry, 1984).

Convolvulaceae, the second most diverse family of climbers (i.e. including slender vines) in the Neotropics (Gentry, Chapter 1) includes many lianas (e.g. *Maripa*) pollinated by large and middle-sized bees, but also some taxa presumably pollinated by small diverse insects (e.g. *Dicranostyles*). The floral diversification within the large genus *Ipomoea* is especially interesting. Most species of the genus are bee-pollinated but some are sphingid-pollinated (*I. alba*), and others hummingbird-pollinated (e.g. *I. quamoclit*, *I. coccinea* and relatives and very distantly related *I. mirandina*). A few scandent *Ipomoea* species (*I. peduncularis*, southern African *I. albivenia*) are bat-pollinated (Dobat, 1985), but even in this overwhelmingly scandent genus bat-pollinated species are mostly trees (e.g. *I. arborescens*, *I. murucoides*); Southeast Asian *Erycibe ramiflora* may also be bat-pollinated (Kock, 1972 *fide* Dobat, 1985). In Convolvulaceae there is a noticeable trend for liana taxa to be pollinated by large to middle-sized bees (or small diverse insects: *Dicranostyles*) whereas taxa of the forest edge or open areas have more diverse pollination systems.

A taxonomic survey of pollination syndromes of climbing families also indicates numerous taxa with small inconspicuous flowers pollinated more or less consistently by small diverse insects (*sensu* Bawa *et al*., 1985a). These include Hippocrateaceae, Menispermaceae, and Sapindaceae, which, at the community level, tie with Malpighiaceae as the most important neotropical lianas after Bignoniaceae and Leguminosae (Gentry, Chapter 1). The other two liana taxa most important at the community level, Connaraceae and Dilleniaceae, have similar flowers. Other less important liana families presumably pollinated by small diverse insects include Dioscoreaceae, Dichapetalaceae (some also pollinated by larger bees?), Smilacaceae, Polygonaceae, Gnetaceae, and Vitaceae. Although the pollination of these groups is essentially unstudied, their small inconspicuous flowers strongly suggest that small diverse insects, perhaps including small bees, are the principal pollinators. Why many of these tiny flowers have such elaborate and species-specific floral structures, a situation reminiscent of trees like Lauraceae or Sapotacae, remains an outstanding open question of tropical pollination biology.

Cucurbitaceae might be assigned to the families mostly pollinated by small diverse insects, but shows more variation. For example, *Psiguria* and *Gurania* are mostly butterfly-pollinated (Gilbert, 1972, 1980), and *Calycophysum* and some *Cayaponia* spp. bat-pollinated (Vogel, 1958; Dobat, 1985). Many of the small-flowered species like *Sicydium* and *Fevillea* are certainly pollinated by the small diverse generalist pollination guild. In Africa *Momordica* and *Thladiantha* have oil glands and are pollinated by large to medium-sized oil-collecting bees (Vogel 1981a). *Cucurbita* itself is famous for pollination by oligolectic bees of the genera *Peponapis* and *Xenoglossa* (Hurd & Linsley, 1964; Michener, 1979).

Strychnos, the main Loganiaceae climbing genus, has many species with small flowers which may also be pollinated by small diverse insects. However, some species (e.g. *S. panamensis*) have longer tubular flowers and are likely to be pollinated by settling moths or possibly butterflies. Moreover, I have observed butterflies and small bees avidly visiting small-flowered *Strychnos*. In Table 15.1 unidentified *Strychnos* spp. have been rather arbitrarily classified as butterfly-pollinated on the theory that this represents a kind of compromise between small diverse insects and settling moths. Verbenaceae, where butterfly pollination is especially prevalent (e.g. *Clerodendron*, *Lantana*, *Sphenodesme*) is similar with most of the climbing species probably butterfly-pollinated, but a few like the smaller-flowered *Aegiphila* species probably pollinated by small diverse insects. The previously unreported pollinator of the important verbenaceous climber *Petrea*, suggested as large–medium bees on account of the conspicuous, slightly zygomorphic flowers (Croat, 1978), are euglossine bees (S. Renner, personal communication).

Several scandent families have a single characteristic type of pollen vector. One important climbing family is uniformly fly-pollinated. Aristolochiaceae, nearly all scandent, have a unique and complex flower that resembles rotting meat in color and odor and traps small flies which serve as its pollinators (Vogel, 1963). Likewise, climbing figs, like other *Ficus* species, have their own highly specialized and species-specific pollination system, involving chalcidoid wasps of the family Agaonidae (Janzen, 1979; Ramírez, 1970). *Cobaea*, the only truly scandent genus of Polemoniaceae is bat-pollinated (Vogel, 1958; Dobat, 1985). Three scandent families are beetle-pollinated: Araceae (the climbers pollinated by dynastine scarabs), Cyclanthaceae (the climbers pollinated by derelomine weevils and Nitidulidae) and Annonaceae, mostly scandent only in the Old World (Schatz, 1989). The only climbing New World palm genus, *Desmoncus*, is also beetle-pollinated (Schatz, 1989).

The climbers of only one family, Cactaceae, may be mostly sphingid pollinated, although the largest (more or less) scandent genus *Rhipsalis* has small, perhaps autogamous flowers and some taxa (e.g. *Schlumbergera*, *Rhipsalidopsis*) are hummingbird-pollinated (Barthlott, 1983). At any rate, climbing Cactaceae are only marginally lianescent. Perhaps their association with dry areas is related to the prevalence of sphingid pollination, otherwise relatively rare in climbers, in Cactaceae. Two of the main liana genera of Polygalaceae – *Diclidanthera* and *Moutabea* – are probably pollinated by settling moths, but several other important climbing genera of the family are pollinated by large and medium-sized bees.

Several important climbing families are mostly hummingbird-pollinated. These include Tropaeolaceae, Gesneriaceae (especially *Columnea sensu lato*), Ericaceae, Acanthaceae (though small-flowered *Mendoncia* species are not hummingbird-pollinated), and Amaryllidaceae (*Bomarea*) (Stiles, 1978, 1980, 1981). Miscellaneous genera or species of numerous other climbing families are also pollinated by hummingbirds. It is noteworthy that hummingbird-pollinated climbers are concentrated in middle-elevation cloud forests where non-hermit Trochilinae hummingbirds are the main pollinators (Stiles, 1981). Although many climbers do have long tubular hermit-type flowers, perhaps their floral displays (too clumped, either through numerous flowers per plant or resulting from vegetative reproduction?) are not compatible with the traplining behavior of hermits.

We are left with several major climbing families which can only be classified as having highly diversified pollination systems lacking a dominant theme. For example, in Marcgraviaceae each genus has a different pollination syndrome: *Marcgravia*, bat-pollinated (Vogel, 1958, 1969; Sazima & Sazima, 1980; *Souroubea*, butterfly-pollinated (personal observation), *Norantea sensu stricto*, hummingbird-pollinated. It is unclear what pollinates the *Norantea* segregates, although *Schwarzia* and *Macrgraviastrum* are likely to be bat-pollinated, to judge from their floral morphology. However, many Marcgraviaceae are probably autogamous despite their elaborate flowers and inflorescences (Sazima & Sazima, 1980; H. Bedell, personal communication).

Table 15.1. *Pollination strategies of some trans-Andean forests*

Site	M-L Bees		Small diverse insects		Butterfly		Hummingbird		Moth		Beetle		Other[a]		Indet.[b]	Ferns 'wind'		Total[c]
	No.	%	No.	%	No.	%	No.	%	No.	%	No.	%	No.	%	No.	No.	%	
Trans-Andean moist and wet forest (from Gentry, 1982a)																		
Curundu, Panama	10	43	10	43	1	4					1	4	F	4				23
Madden Forest, Panama	18	62	10	34	1	3												29
Rio Palenque, Ecuador	9	35	10	38			1	4			3	11	W	4	1	2	8	27 (26)
Pipeline Road, Panama	18	49	16	43	1	3	1	3										37
Moist/wet average	14	47	11.5	39.5	1	3	0.5	2			1	3	0.5	2	0.25			29 (29)
Choco pluvial forest (from Gentry, 1982a, 1986)																		
Tutunendo, Colombia	13	29	17	39	3	7	3	7			4	9	2B	5	2			46 (44)
Bajo Calima, Colombia	14	30	15	32	5	11	2	4	1	2	6	13	2B 2W	4 4	4			51 (47)
Pluvial average	13.5	29	16	35	4	9	3	7	1	1	5	11	3	7	3			48.5 (45.5)

Dry forest (from Gentry, 1982a, 1986)															
Galerazamba, Colombia	7	44	6	38	1	6			2	13			2		18 (16)
Llanos, Venezuela (500 m²)	7	100													7
Boca de Uchire, Venezuela	13	81	3	19											16
Blohm Ranch, Venezuela	8	50	7	44							P	6			16
Capeira, Ecuador	9	45	9	45	1	5	1	5							20
Guanacaste (upland), Costa Rica (700 m²)	5	83	1	17											6
Guanacaste (Gallery), Costa Rica (800 m²)	5	63	3	38											8
Tayrona, Colombia	11	65	3	35									1		18 (17)
Dry forest average (incl. incomplete sites)	8	62	4	31	0.25	2	0.13	1	0.25	2	0.13	1			13.6 (13)

[a]B, bat; F, fly; M, non-flying mammal; P, perching bird; W, wasp.
[b]Includes species identified to genus when that genus has more than one pollination syndrome. Identified taxa included here are *Norantea s.l.*, *Manihot*, *Sarcopera*, most *Anthurium*, *Drymonia*, Asclepiadaceae, *Blakea/Topobaea*.
[c]Number in parentheses = number of species analyzed (excluding indets).

Combretaceae has two main neotropical climbing genera. *Thiloa* and some *Combretum* species are part of the small diverse insect syndrome. A few *Combretum* species seem to be unequivocably hummingbird-pollinated. Most *Combretum* species, however, have an unusual pollination strategy employing perching birds and/or non-flying mammals (Prance, 1980; Janson *et al.*, 1981). Paleotropical *Quisqualis* is presumably sphingid-pollinated.

Rubiaceae have independently evolved the scandent habit in many largely unrelated genera with an equivalent diversification in pollination systems. The largest climbing genus, paleotropical *Mussaenda*, often has conspicuously enlarged calycine flags and probably is pollinated by both sunbirds and butterflies. *Schradera* is presumably moth-pollinated as are miscellaneous scandent species of such genera as *Chomelia* and *Randia*. At least one scandent genus of Rubiaceae, *Manettia*, is largely hummingbird-pollinated (Feinsinger, 1978). *Chiococca* is probably pollinated by butterflies and small bees; *Uncaria* is probably mostly butterfly-pollinated since it has flowers similar to *Cephalanthus* which is visited by butterflies (personal observation).

Euphorbiaceae have only miscellaneous climbing genera but some of these have very unusual and interesting pollination systems. Most climbing euphorbs have small nondescript flowers and are presumably pollinated by small diverse insects. Climbing *Mabea* species, however, are probably pollinated in large part by small arboreal marsupials like their non-climbing relatives (Steiner, 1981). *Dalechampia* usually has a pair of large colored bracts subtending the flower clusters and is mostly pollinated by resin-collecting large and medium-sized bees (Armbruster & Webster, 1979).

The largest climbing composite genus, *Mikania*, has small white flowers and is probably pollinated by small diverse insects, including small bees and probably also butterflies. The second largest climbing genus, *Mutisia* is hummingbird-pollinated and *Pseudogynoxys* perhaps predominantly butterfly-pollinated (personal observation).

Solanaceae climbers, most important in middle-elevation cloud forests, have genera pollinated by bats (*Trianaea* (Vogel, 1958), some *Markea* spp. (Vogel, 1958, 1969; Humphrey & Bonaccorso, 1979 *fide* Dobat 1985), probably *Solandra* (van der Pijl, 1961; Vogel, 1969)), hummingbirds (*Juanulloa*), and large–medium bees collecting nectar (*Markea*) or pollen (*Solanum, Lycianthes*).

Melastomataceae (Renner, 1984, 1986, 1989) may not fit the normal large–middle-sized-bee vs. small-diverse-insect dichotomy. Pollen rather than nectar is usually the reward and is taken by females of a wide variety of bees, both large and small, the unusual anther and appendage features so important in the family's taxonomy serving as a kind of landing platform. Indeed, Renner (1984) found that a third of the Central Amazonian apifauna visited melastomes, generally quite non-specifically. Some taxa, including largely hemiepiphytic and climbing *Blakea*, produce nectar, with various nectar-producing species pollinated by hummingbirds, bats, and rodents, as well as bees (Lumer, 1980; Renner, 1986, 1989); these pollination systems mostly occur in middle-elevation cloud forest taxa.

I have saved the best for last. Pollinators of scandent Asclepiadaceae, by far the largest family of climbers, have been little studied (in contrast to the temperate herbs of that family); among the few tropical asclepiad climbers with pollinations records are *Matelea* (flies), and *Sarcostemma* (a wide variety of insects including bees, butterflies, and flies) (S. Liede, personal communication). Asclepiadaceae are in many ways the vine equivalent of orchids among epiphytes in that both have pollinia, unusually complicated floral morphology, and are exceedingly speciose. It is highly likely that the unusual Asclepiadaceae pollination system, wherein the pollinia ensure that the seeds of a given fruit share fathers as well as mothers, preadapts them for what I have called explosive speciation (Gentry, 1982b, 1989a; Gentry & Dodson, 1987). If genetic transilience via founder effect in small colonizing populations (Templeton, 1980) plays a major role in tropical plant speciation as has been suggested (Gentry, 1982b, 1989a), asclepiads, like orchids, should be especially prone to speciation via rapid genetic drift in closely related founding populations.

Janzen (1977) has already asked what orchids and asclepiads have in common that has generated their convergence in a uniquely low ratio of fathers to seeds. The evolutionary results are evident in the spectacular speciation of both families and also in the typical sparse distributions and rarity of individuals of a given species. The convergence may result from both families being an unusual combination of generally *K*-selected plants with strongly *r*-selected seed dispersal. In this light, it is noteworthy that Asclepiadaceae, at least in neotropical forests, are almost exclusively plants of ephemeral habitats like forest edges and light gaps. In my experience, fruit set in Asclepiadaceae is a rarer occurrence than in almost any other family. Once set, a fruit is large and produces numerous small wind-dispersed seeds. Their many small seeds are optimal for a weed-like high risk dispersal strategy appropriate for an adaptive milieu in which quick colonization of newly available habitats is at an adaptive premium (see discussion in Gentry & Dodson, 1987). It is no doubt significant that Old World Periplocoideae, which lack pollinia and are presumably primitive in the family, are forest lianas as opposed to the herbaceous vine-strategy that characterizes most pollinium-bearing taxa.

A safe conclusion from the preceding survey is that vine pollination systems are highly diversified, both within the climbing guild and within many families of climbers. Nevertheless, an unusually high number of climbing families seems characterized by evolutionary diversification based on a single unusual pollination mode, e.g. fly-pollinated Aristolochiaceae. Whether this tendency is significantly higher in climbers than in other habit groups remains unclear as does whether certain specific pollination strategies, like beetle pollination or large-bee pollination, are unusually prevalent in climbers.

Liana community patterns

Bawa *et al.* (1985a) predicted that canopy vines will prove to have similar flowering strategies to canopy trees with a preponderance of pollination by large–medium bees and small generalist insects, while subcanopy vines and epiphytes may have a greater diversity of pollination modes. In the next section I will test this prediction with transect data from a series of neotropical forests.

The conspicuous/inconspicuous pollination classes of Gentry (1982a) are here more finely subdivided according to the categories of Bawa *et al.* (1985a). My procedure has been almost entirely armchair speculation, based on floral morphology. Although it seems reasonable to suppose that floral morphology accurately reflects pollination strategies as Bawa *et al.* (1985a) have shown for La Selva, all conclusions must be taken as somewhat tentative. I am unable to hazard guesses from floral morphology as to 'small-bee' vs. 'small-diverse-insect' pollination guilds and so have lumped these two categories together. Several other genera – including *Sarcopera*, small-flowered *Mendoncia*, *Salacia alwynii*, and several curcubits) – are excluded from these tabulations because I am unable to suggest probable pollinators, even after consultation with K. Bawa, R. Foster, and F. Putz. Some species at each site are excluded because of incomplete identifications; for some families, identification to genus or even family is adequate to assign a probable pollinator, but in other cases specific identification is necessary.

The four moist and wet lowland forest samples of Gentry (1982a) average 49 conspicuous-flowered species and 62 inconspicuous-flowered species >2.5 cm diameter; thus the majority (56%) of species have inconspicuous, presumably generalist-pollinated, flowers. However, this is exactly reversed for climbers, which average 16 species with conspicuous flowers and 12 with inconspicuous flowers, well over half (57%) of the climbing species having conspicuous flowers. Taken separately, the trees and shrubs at these four sites average only 40% conspicuous-flowered species. These very preliminary figures imply that climbers might generally tend to have a higher percentage of conspicuous flowers (and be more specialist-pollinated) than trees.

Tables 15.1 and 15.2 provide data for a more detailed examination of climber pollination. When the presumed pollinators of the climbing species of the four trans-Andean moist and wet forests of Gentry (1982a) are assigned to the categories of Bawa *et al.* (1985a) (except for combination of small bees and small generalist insects), the great majority (73–96%, $\bar{X}=87\%$) of the species are pollinated by middle-sized to large (= M–L) bees (9–18 species per sample, constituting 35–62% of the climbers) or small generalist insects (10–16 species, constituting 34–43% of the climbers). All of the climbers pollinated by other than M–L bees, small diverse insects, or butterflies are either hemiepiphytic (Araceae, Cyclanthaceae, *Ficus schippii*) or cauliflorous (the only two hummingbird-pollinated climbers in the samples plus a fly-pollinated *Aristolochia*). Thus Bawa *et al.*'s (1985a) predictions are supported. Understory climbers (like understory trees) have relatively diversified pollination systems. Canopy lianas have exactly the same higher proportion (44%) of large-bee pollination as canopy trees.

A similar analysis for a series of Amazonian lowland sites supports the generality of this pattern but there are also interesting differences. For the 15 Amazonian sites an average of 83% of the climbers are pollinated by M–L bees or small generalist insects (Table 15.2). There is a reversal of the roles of M–L bees and small generalists, however, with the Amazonian sites generally having a higher proportion of small diverse insect pollinators than M–L bees. The reason for this difference is readily apparent. There are exactly the same number of M–L bee-pollinated species in the Amazon samples (average = 14) as in the trans-Andean moist and wet forests (average = 14), but the Amazonian sites are richer in species and nearly all of the 'additional' species belong to families pollinated by small generalist insects like Hippocrateaceae, Connaraceae, and Menispermaceae which are underrepresented outside Amazonia (Gentry, Chapter 1).

There is also another difference from the trans-Andean forests. Restriction of the other, less common pollination systems to the subcanopy and understory is less pronounced in the Amazonian forests. Except for one cauliflorous hummingbird-pollinated *Passiflora*, the beetle-pollinated aroids and cyclanths, and perhaps some *Marcgravia* species, all of the moth, hummingbird, bat, wasp, perching bird, and non-flying mammal-pollinated species of climbers represented in these 15 samples, bloom in the canopy.

Another way that pollination systems vary is with forest type (Frankie *et al.*, 1974). In an earlier study (Gentry, 1982a), I categorized the species of plants > 2.5 cm diameter in 0.1 ha. samples into two very broad classes: those with conspicuous (mostly pollinated by specialist pollinators) and inconspicuous (mostly pollinated by small generalist pollinators) flowers. In that paper I concluded that the increasing species richness in progressively wetter neotropical forests is mostly accounted for by additional species with inconspicuous flowers and presumably small generalist pollinators. If the lianas from that data set (Gentry 1982a: Appendix) are analyzed separately,

Table 15.2. *Pollination strategies of some Amazonian forests (see Gentry, 1989, for sites)*

Site	M-L bees		Small diverse insects		Butterfly		Hummingbird		Moth		Beetle		Other[a]		Indet.[b]	Ferns 'wind'		Total[c]
	No.	%	No.	%	No.	%	No.	%	No.	%	No.	%	No.	%	No.	No.	%	
Indiana, Peru	18	35	25	49	2	4					1	1	B F 2W	8	6	1	1	57 (51)
Sucursari, Peru	13	34	21	55	4	11									1			39 (37)
Jenaro Herrera, Peru	12	34	19	54	1	3			2	6			W	3	6			41 (35)
Von Humboldt, Peru	15	48	15	48	1	3									5			36 (31)
Tambopata (lateritic 1), Peru	18	45	16	40	2	5			1	3	1	3	2P	5	2			42 (40)
Tambopata (lateritic 2), Peru	12	32	16	42	3	8					3	8	2P W	8	2	1	3	40 (38)
Tambopata (sandy) Peru	14	47	13	43	2	7					1	3	P	3	1			32 (30)
Cabeza de Mono, Peru	11	41	11	41	2	7	1	4	1	4			B	4	4			31 (27)
Mishana (sand), Peru	11	48	8	35	2	9					1	4	M	4	3			26 (23)
Mishana (sandy lowland), Peru	16	30	28	52	4	8	1	2	2	4	1	2	M B	4				54
Yanamono No. 1, Peru	14	34	18	44	4	10					1	2	M P B	7	2	1	2	43 (41)

Yanamono No. 2, Peru	10	25	19	48	7	17					1	3	W P	5	1	1	3	41 (40)
Cocha Cashu. Peru	15	35	23	53	4	9							M	2	1			44 (43)
Jatun Sacha, Ecuador	17	30	25	44	4	7	3	5			3	5	B P W		2	2	4	59 (57)
Shiringamazu. Peru	18	40	20	44	2	4	2	4			1	2	M	2	1			46 (45)
Average	14	37	19	46	3	7	0.5	1	0.4	1	0.9	2	1.6	4	–	0.4	1	39 (39)

[a]B, bat; F, fly; M, non-flying mammal; P, perching bird; W, wasp.
[b]Includes species identified to genus when that genus has more than one pollination syndrome. Identified taxa included here are small-flowered *Mendencia*, most *Anthurium*, *Salacia alwynni*, *Sarcopera*, *Cayaponia*, *Selysia* (see text).
[c]Number in parentheses = number of species analyzed (excluding indets).

they show a similar pattern but with a generally higher percentage of conspicuous-flowered species than the trees. Moist and wet forest samples average 16 species of conspicuous-flowered climbers vs. 12 species with inconspicuous flowers. Conspicuous-flowered species thus account for less than 57% of the climbing species. Dry forests average 6.25 conspicuous-flowered liana species vs. 2.75 inconspicuous-flowered species; over 69% of the dry forest lianas are conspicuous-flowered. In contrast, a pluvial forest site had 25 (54%) inconspicuous-flowered climbers but only 21 conspicuous-flowered climbers. There thus seems to be a clear trend among climbers, just as in trees, for increasingly wetter forests to have a greater proportion of inconspicuous-flowered, presumably generalist-pollinated, species.

The 'conspicuous-flowered' climbers of Gentry (1982a) are nearly all pollinated by large and medium-sized bees (Table 15.1). Thus the driest sites have up to 100% climber pollination by large and middle-sized bees. In Table 15.1 the dry forest samples from Gentry (1982a) are re-tabulated according to the pollinator categories of Bawa *et al.* (1985a) along with data from several additional dry forest sites. The dry forest lianas are overwhelmingly pollinated by M–L bees and small generalist insects, which together account for an average of 93% of the dry forest climbers. The eight dry forests of Table 15.1 all together have a total of only six sampled taxa with other pollination syndromes. Perhaps the most interesting of these are the two sphingid-pollinated climbers (a *Capparis* and a cactus) at Galerazamba, Colombia, the driest sample site. Perhaps this constitutes very tentative support for the generalization that sphingid pollination is more prevalent in drier areas.

At the opposite extreme, pluvial forest lianas have a much more diversified spectrum of pollination systems than other lowland forests. Pollination by M–L bees and small diverse insects together occurs in only 64% of the pluvial forest climbers (vs. averages of 83–93% in other forest types). Although pluvial forests have almost exactly the same number of sampled M–L bee-pollinated ($\bar{X} = 13.5$ spp.) and small diverse insect-pollinated ($\bar{X} = 16$ spp.) climbers, they have more species, and the additional climbing species represent a broad miscellany of families and pollination syndromes.

I have not attempted a similar analysis of pollinators of Andean forest communities since I do not know their flora as well as the lowland one and am far less confident about armchair assignment of probable pollinators.

Finally, I will hazard a few remarks on intercontinental differences in climber pollination biology. In Africa, moth pollination, especially by settling moths, is surely more prevalent than in the Neotropics due to the greater prevalence of Apocynaceae and Rubiaceae lianas. Beetle pollination perhaps holds constant, with many beetle-pollinated Annonaceae climbers taking the place of neotropical cyclanths and climbing aroids. Most noteworthy is the paucity of M–L bee-pollinated climbers, perhaps not surprising on a continent notably depauperate in the large and middle-sized solitary bees that are such important neotropical pollinators (Michener, 1979). It is tempting to suppose that there is a causal relationship between the lack of Bignoniaceae climbers in African forests and the impoverished apifauna of solitary bees.

In Asian forests where Annonaceae climbers are especially abundant, beetle pollination is presumably more important. Since diversity of climbing Apocynaceae and Rubiaceae in Asia is intermediate between the Neotropics and Africa one might predict that moth pollination will be also. Pollination by both M–L bees and small diverse insects seems remarkably low by neotropical standards (together accounting for *c.* a third (perhaps 35–38%) of my Bornean liana samples; this is hardly surprising in view of the very small populations of small insects that are thought to be typical of Asian dipterocarp forests (Janzen, 1977).

Another aspect of liana breeding systems, phenology, has been addressed in several recent papers (Putz & Windsor, 1987; Opler, Chapter 14) and will not be discussed here. It should be noted, however, that liana phenology is exceptionally interesting, with such taxa as Bignoniaceae showing extremely intricate and complex phenological patterns (Gentry, 1974). There may be a general tendency for lianas to have shorter flowering periods than other habit groups (cf. Stiles, 1978, for hummingbird-pollinated species) and to be prone to forming floral mimicry groups (Gentry, 1974; Frankie *et al.*, 1983).

Sexual systems in lianas

One feature of tropical forest reproductive biology that has received a large share of recent attention is the frequency of different sexual systems (Bawa & Opler, 1975; Ashton, 1976, 1977; Bawa, 1980, 1981; Bawa & Beach, 1981; Sobrevila & Arroyo, 1982). Sexual systems that enforce outcrossing may be important in sparse populations like those of many rainforest trees. Protogyny, self-incompatibility, monoecy, and ultimately dioecy, might thus be important in lianas as they are in trees (e.g. Ashton, 1969; Bawa, Perry, & Beach, 1985b). The extent of dioecy in tropical forests has been especially emphasized (Bawa & Opler, 1975; Bawa, 1980; Bawa *et al.*, 1985b), and it is now well known that tropical forest trees are characterized by high levels of dioecy, generally about a quarter (21–26%) of the tree species of a given forest (Bawa *et al.*, 1985b), and sometimes up to 31% (Sobrevila & Arroyo, 1982).

The only complete data sets on sexuality of a tropical flora are for Barro Colorado Island (Croat, 1978) and Chamela, Mexico (Bullock, 1985). Based on the fact that 21% of the BCI trees are dioecious but only 9% of the total flora (Croat, 1978; see also Bawa, 1980), Bawa *et al.*, (1985b) suggest that dioecy may be twice as prevalent in trees as in the rest of the flora. They note that large perennial plants, especially trees, are generally more prone to dioecy (Bawa, 1980), which could suggest that large perennial climbers would also have high levels of dioecy. Indeed, the Flora of the Carolinas data tabulated by Conn, Wentworth & Blum (1980; see also Bawa, 1980) suggest that vines have a higher frequency of dioecy (16%) than any other group, even trees. However, as tabulated by Croat (1978), only 11% of the BCI climbers (= herbaceous vines + lianas, but excluding hemiepiphytic *Clusia* and *Coussapoa*) are dioecious. Nevertheless, since the 30 dioecious BCI climbers constitute 26% of the island's dioecious species, but climbers themselves only 21% of the flora, we might conclude that climbers are somewhat more dioecious, even on BCI.

However, dioecy may have been overestimated in these data. It now seems probable that the eight cucurbit species considered dioecious by Croat are in fact sequentially monoecious (Condon & Gilbert, 1988) (as well as his 20 'polygamous' Sapindaceae climbers (cf. Ruiz & Arroyo, 1978; S. Bullock, personal communication)), while the three species of 'andromonoecious' *Tetracera* are in fact functionally dioecious (Kubitzki & Baretti-Kuipers, 1969; Simpson & Neff, 1981). Thus only 25 BCI climbers (9.4% of the island's scandent species) should be considered dioecious, these constituting 21.7% of the dioecious flora, exactly the same as the 21% of the overall flora made up of climbers.

On the other hand, climbers are disproportionately represented in the monoecious flora of BCI, with 31 scandent species listed by Croat plus the scandent Sapindaceae and Cucurbitaceae, constituting over twice as many monoecious (59 spp.) as dioecious (25 spp.) scandent plants on BCI. In fact, a full 35% of the monoecious BCI plants are scandent. The BCI figures for climber sexual systems are strikingly similar to those found by Bullock (1985 and personal communication) at Chamela where 10% of the vines are dioecious but 18% monoecious. While the dioecy rate among Chamela climbers is lower than in shrubs or trees, the monoecy rate among climbers is the highest for any habit group. Except for unusually frequent monoecy, the sexual systems of climbers at both BCI and Chamela are not very distinctive and closely parallel those of the flora as a whole.

There may also be a correlation of hemiepiphytic and strangling habits with monoecy in *Ficus*. The free-growing monoecious subgenus *Pharmacosycea* has given rise to nearly all the climbers and stranglers (subgenus *Urostigma*), likewise monoecious, whereas the large dioecious subgenus *Ficus* has produced only a few climbers and stranglers (Corner, 1958; Ramírez, 1977).

Perhaps the surprisingly low rate of dioecy among (neo?)tropical climbers is related to their evolutionary emphasis on medium-to-large-bee pollination and wind dispersal (see above and below), as opposed to the usual correlation of dioecy with small generalist-pollinated flowers and fleshy animal-dispersed fruits (Givnish, 1980; Bawa, 1980). In fact the 9–10% dioecy rate among neotropical climbers is so low that temperate zone climbers are apparently more predominantly dioecious (16% of the Carolina flora vines) than (neo)tropical ones! On the other hand, the lower frequency of large-bee pollination and wind dispersal in paleotropical climbers suggests that they may prove to have a greater frequency of dioecy.

It is noteworthy that *Smilax* and *Dioscorea*, the two largest monocotyledon vine genera, are both dioecious. Indeed, it is their prevalence in temperate zone floras that accounts for the relatively high frequency of dioecy among temperate climbers. Perhaps dioecy is somehow related to their evolutionary success with a growth form that has been notably unsuccessful for most other monocotyledons. It is also interesting that the supposedly dioecious taxa which Condon & Gilbert (1988) found to be actually sequentially monoecious are vines (*Adenia*, *Psiguria*, *Gurania*); perhaps there is something about scandent habit that in some way favors the heretofore confusing sexual system of sequential monoecy.

Liana dispersal ecology

Lianas tend to have much higher levels of wind dispersal than do trees or shrubs, at least in the Neotropics (Gentry, 1982a, 1983). There are also inter-community changes in liana dispersal, with dry forest liana communities more predominantly wind-dispersed than wetter forests and wetter forests more bird- and mammal-dispersed (Gentry, 1982a, 1983). Even though wind dispersal is less prevalent in montane forests, Sobrevila (1978, *fide* Armesto, 1986) found that lianas in a Venezuelan forest at 1700 m were 37% anemochorous, a higher percentage than for any other habit group.

It has been suggested that there are also more subtle differences in dispersal ecology between different neotropical forests. For example, forests on white sand have been suggested as having larger, more toxic, and better protected seeds than on other substrates (Janzen, 1974) and the relatively small seeds of wind-dispersed lianas would seem inappropriate to this habitat. On the other hand, Macedo (1977) and Prance (1978) suggested that higher than normal levels of wind and other long-distance dispersal mechanisms in Amazonian campinas were evidence that these function as ecological islands colonized by long distance dispersal. But what constitutes the normal dispersal spectrum of terra firme forest against which other communities might be evaluated? Here I will try to establish these trends for the liana component of lowland tropical forest communities.

Figure 15.1 summarizes the data on liana dispersal ecology for a wide array

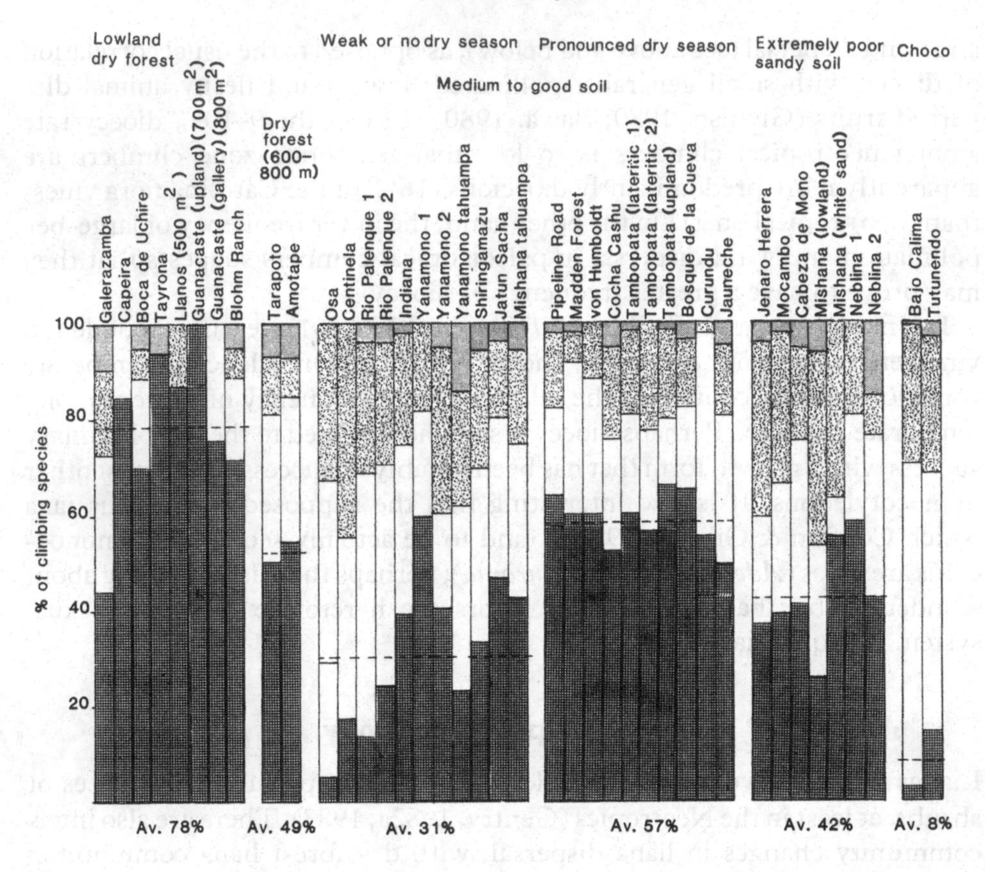

Figure 15.1. Dispersal strategies of neotropical lowland lianas. For samples of climbers $\geqslant 2.5$ cm diameter in 0.1 ha. Site data given in Gentry, 1988, 1989. Cross-hatched, percentage wind-dispersed; white, percentage bird-dispersed; light stipple, percentage mammal-dispersed; dark stipple, other (water, autochorous, etc.) Dashed line, wind-dispersal $\bar{X}$ for group of ecologically similar sites (repeated below bars).

of neotropical lowland forests (from Gentry, 1982a, 1988, Chapter 1). Included in this data set are 10 dry forests, 28 moist to wet forests, and two pluvial forests. Twenty-one of the moist and wet forests are on poor to relatively rich lateritic to alluvial soils and seven on poor to very poor sandy soils. Half of the poor to good soil sites are in areas with a relatively pronounced dry season (= moist forest) and half in areas with relatively little dry season stress (= wet forest). As suggested from the much smaller data set of Gentry (1982a), lowland dry forest lianas have higher levels of wind dispersal ($\bar{X} = 78\%$ of the species) than do moist and wet forests, which average 57 and 31% wind dispersal, respectively (Mann-Whitney $U = 106$, $P < 0.005$, $U = 114$, $P < 0.001$). Moist forest on good soil has significantly higher levels of wind dispersal than does wet forest (Mann–Whitney $U=148$, $P<0.001$). Forests on very nutrient-poor sandy soils are intermediate between moist and wet forest ($\bar{X}=42\%$) but significantly different only from the former ($U=129$; $P<0.001$). Choco area pluvial forests have much lower levels of wind dispersal (8%) than do other lowland forests.

Gentry (1983) suggested that a forest on ultra-poor white sand near Iquitos, Peru had a somewhat lower percentage of wind dispersal than nearby forests on other substrates. For the forest as a whole, Iquitos area samples ranged from 10 to 15% wind dispersal, while lianas averaged 45% wind dispersal (Figure 15.1), a rate 3–4 times that for the communities as a whole. Amazonian liana communities are just as predominantly wind-dispersed as trans-Andean and Central American ones, and prevalence of wind dispersal is a normal feature of the community rather than a dispersal disharmony (contrary to Prance, 1978). Moreover, the sample from upland white sand has a higher percentage of wind-dispersed lianas than an adjacent lowland site, contrary to my earlier suggestion or the scenario postulated by Janzen (1974). That lianas are much more strongly wind-dispersed than trees is strongly supported by these data, but between-community differences in neotropical liana dispersal, except for those associated with major differences in rainfall regime, are extremely subtle, if they exist at all.

Although different neotropical lowland moist, wet, or dry forests have similar dispersal spectra, equivalent paleotropical forests do not. Wind dispersal is much less prevalent among Old World lianas (Table 15.3). The two Borneo samples average only 17% wind-dispersed lianas. Even relatively dry forests in continental Africa have only 24–30% wind-dispersed lianas compared with well over 50% in comparable neotropical sites. Even some liana families like Hippocrateaceae and Apocynaceae that are mostly wind-dispersed in the Neotropics switch to mostly mammal dispersal in the Old World. Madagascar may be closer to the neotropical dispersal spectrum, with the 42% wind-dispersed lianas in a dry forest sample only marginally less than in the neotropical dry forest extreme.

The only exception to the paucity of wind dispersal among paleotropical lianas is a sample from Omo Forest, Nigeria, which is also anomalous in such other aspects as floristic composition and species richness and, along with most of west tropical Africa, may represent old secondary forest (see Okali & Ola-Adams, 1987; Gentry, 1988). Even the Omo site has only 52% wind-dispersed lianas, which would be low for a neotropical forest with an equivalent 1800 mm of annual rainfall.

The low representation of wind dispersal among Old World lianas may be related to a fundamentally different forest structure wherein large treefall gaps are rarer and in which the forests may be generally less dynamic than in the Neotropics (e.g. Putz & Appanah, 1987).

Table 15.3. *Liana wind-dispersal in Old World tropics*

Site	Number of presumed wind-dispersed spp.	Percentage of wind-dispersed spp.
Asia		
Semengoh, Sarawak	2	10
Bako, Sarawak	6	23
Average	4	17
Australia		
Davies River, Queensland	4	24
Madagascar		
Ankarafantsika (dry forest)	12	42
Perinet (wet forest)	8	24
Africa		
Makokou No. 1 (moist)	14	30
Makokou No. 2 (moist)	11	30
Pugu (dry)	8	24
Omo (moist; old 2°?)	14	52
Average	12	34 (28 without Omo)

In temperate zone forests anemochory is much less pronounced among vines. All of the north temperate zone lianas included in my samples are bird-dispersed and only a third (12 spp.) of the southeastern USA lianas listed by Duncan (1975) are wind-dispersed, although many north temperate trees are. In the Valdivian area of Chile only a tiny fraction of lianas are wind-dispersed (only *Campsidium* in my samples) as emphasized by Armesto (1986) who points out that wind dispersal is commonest in large-canopy trees in southern Chile and even in shrubs is more prevalent than in climbers.

Conclusion

We may summarize that while liana breeding systems do not appear to be very different from those of canopy trees, they apparently do have some characteristic tendencies. Although most lianas are pollinated by medium-sized to large bees or small diverse insects, just as are canopy trees, there are hints that medium-sized to large bees may be somewhat more important in liana than in tree pollination. Most liana families are characterized by 1–few characteristic pollination strategies. While lianas have many different pollination strategies, none are known to be thrip- or wind-pollinated, even in the temperate zone. Liana pollination systems are less diversified in dry forest and more diversified in the subcanopy and in pluvial forest. These features might be anticipated from liana population structure, since liana individuals are almost certainly more widely scattered than are trees, at least when genetically identical, vegetatively reproduced liana clumps are counted as single individuals. A population of widely scattered individuals would put pollination by specialized pollinators, especially trapliners, at an adaptive premium; conversely, reproductive strategies like wind (and perhaps thrip) pollination, that are efficient only in gregarious populations, would be at a selective disadvantage in lianas.

The prevalence of hemiepiphytic rather than free-climbing lianas in cloud forests and pluvial forest (Gentry, 1988) would automatically make the individual plants more isolated and accentuate the importance of traplining pollinators, since hemiepiphytes are prevented by their growth strategy from forming clonal clumps that climb over various trees. Similarly, the conspicuous olfactory cues associated with beetle pollination may be especially favorable in a population of widely scattered hemiepiphytic individuals. Lianas that flower in the subcanopy may have more diversified pollination strategies, in part because conspicuous olfactory cues associated with beetle pollination in taxa like Cyclanthaceae, aroids, and Annonaceae are especially effective compared with visual cues in the dark forest understory. The same arguments can be made for climbers in cloud forests, which by definition have low light levels much of the time, and lowland pluvial forests are known to share many typically cloud forest features (Gentry, 1986).

Neotropical lianas are more prone to wind dispersal than neotropical trees. Paleotropical lianas have much lower levels of wind dispersal than in the Neotropics, perhaps relating to generally smaller treefall gaps and generally less dynamic forests.

Despite being popularly conceived of as typical of second growth areas, lianas are extremely prone to reproduction by vegetative means rather than from seeds (Peñalosa, 1984; Putz, 1984), and often are much longer-lived than trees. Although I know of no data on this point, the liana growth form might be expected to be prone to form genotypic mosaics via the accumulation of somatic mutations in each independent apical bud. If liana clones accumulate somatic mutations, as some long-lived temperate plants apparently do, they might even be expected (cf. the 'Red Queen hypothesis' of Van Valen, 1973) to evolve intra-generationally in response to specific parasites or predators. There is some evidence that pest and parasite pressures are higher in tropical forests (Levin, 1975) and that lianas are especially prone to development of host-specific predators (Hegarty, Hegarty & Gentry, Chapter 10; Gentry, 1991). Thus it is tempting to suppose that intraplant selection and even evolution of the kind suggested by Whitham & Slobodchikoff (1981) might occur unusually frequently in lianas. Perhaps individual lianas are, in a sense, able to 'span evolutionary time' (Whitham & Slobodchikoff, 1981). Thus their pollination and dispersal systems may be almost redundant in ecological time.

One might expect that such long-lived organisms, which need to replace themselves only once over many hundreds of years, might be an extreme case of *K*-selection, even more than canopy trees, with breeding systems emphasizing long-term flexibility over immediate fitness. If so, lianas should have especially high levels of heterozygosity enforced by obligate outcrossing, incompatibility mechanisms, dichogamy, and/or sexual differentiation. It is gratifying that these are, in fact, the features that characterize liana breeding systems (see also Gentry, 1989a), although whether they do so to an even greater extent than they do tropical trees is as yet unknown.

In contrast to their pollination biology, liana dispersal, at least in the Neotropics, seems more *r*-selected than that of trees. This is especially evident in the relatively numerous small and less protected seeds that are characteristic of wind dispersal. The apparent paradox is readily explained when the incredible longevity that characterizes individual lianas is considered. To an even greater degree than trees, lianas can afford to be relatively profligate in the release of genetic variability, since they can achieve reproductive success even when replacing themselves only once over many hundreds of years. The individual long-lived plants must be adapted to tolerate short-term environmental change (in the case of lianas often by vegetative regeneration after a host tree falls), while their sexual reproduction serves to adapt them to relatively far distant future changes, at a time-scale exceeding that generally associated with ecological time.

Acknowledgements

Supported by a series of grants from the National Geographic Society and National Science Foundation. Amazonian field work also supported by the Mellon Foundation and MacArthur Foundation. This general study of lianas was in part an outgrowth of studies of systematics and ecology of Bignoniaceae funded by NSF BSR-8607113. I thank K. Bawa, S. Renner, F. Putz, and B. Simpson for review comments, S. Renner for pointing out several additional references, and R. Clinebell for assistance with data analysis.

References

Appanah, S. & Chan, H. T. (1981). Thrips: the pollinators of some dipterocarps. *Malaysian Forester* 44: 234–52.

Armbruster, W. S. & Webster, G. (1979). Pollination of two species of *Dalechampia* (Euphorbiaceae) in Mexico by Euglossine bees.

Biotropica **11**: 278–83.
Armesto, J. J. (1986). Mecanismos de diseminación de semillas en el bosque de Chiloe: una comparación con otros bosques templadas y tropicales. *An. IV Congr. Latinoam. Bot.* **2**: 7–24.
Ashton, P. (1969). Speciation among tropical forest trees: some deductions in the light of recent evidence. *Biological Journal of the Linnean Society* **1**: 155–96.
Ashton, P. (1976). An approach to the study of breeding systems, population structure and taxonomy of tropical trees. In *Tropical Trees: Variation, Breeding, and Conservation*, ed. J. Burley and B. Styles, pp. 35–42. Academic Press, New York.
Ashton, P. (1977). A contribution of rainforest research to evolutionary theory. *Annals of the Missouri Botanical Garden* **64**: 694–705.
Ashton, P. (1979). Some geographic trends in morphological variation in the Asian tropics and their possible significance. In *Tropical Botany*, ed. K. Larsen and L. Holm-Nielsen, pp. 35–48. Academic Press, London.
Baker, H. G. (1973). Evolutionary relationship between flowering plants and animals in American and African tropical forests. In *Tropical Forest Ecosystems in Africa and South America: A comparative review*, ed. B. Meggers, E. Ayensu and W. Duckworth, pp. 145–59. Smithsonian Institution Press, Washington, DC.
Barthlott, W. (1983). Biogeography and evolution in Neo- and Paleotropical Rhipsalinae (Cactaceae). *Sonderbd. Naturwiss. Ver. Hamburg* **7**: 241–8.
Bawa, K. S. (1980). Evolution of dioecy in flowering plants. *Annual Review of Ecology and Systematics* **11**: 15–39.
Bawa, K. S. & Beach, J. H. (1981). Evolution of sexual systems in flowering plants. *Annals of the Missouri Botanical Garden* **68**: 254–74.
Bawa, K. S. & Opler, P. A. (1975). Dioecism in tropical forest trees. *Evolution* **29**: 167–79.
Bawa, K. S., Bullock, S. H., Perry, D. R., Coville, R. E. & Grayum, M. H. (1985a). Reproductive biology of tropical lowland rain forest trees II. Pollination systems. *American Journal of Botany* **72**: 346–56.
Bawa, K. S., Perry, D. R. & Beach, J. (1985b). Reproductive biology of tropical lowland rain forest trees. I. Sexual systems and incompatibility mechanisms. *American Journal of Botany* **72**: 331–45.
Bullock, S. (1985). Breeding systems in the flora of a tropical deciduous forest in Mexico. *Biotropica* **17**: 287–301.
Condon, M. A. & Gilbert, L. E. (1988). Sex expression of *Gurania* and *Psiguria* (Cucurbitaceae): neotropical vines that change sex. *American Journal of Botany* **76**: 875–84.

Conn, J. S., Wentworth, T. R. & Blum, U (1980). Patterns of dioecism in the flora of the Carolinas. *American Midland Naturalist* **103**: 310–15.

Corner, E. J. H. (1958). An introduction to the distribution of *Ficus*. *Reinwardtia* **4**: 15–45.

Croat, T. (1978). *Flora of Barro Colorado Island*. Stanford University Press, Stanford, Ca.

Dobat, K. (1985). *Bluten und Fledermäuse*. Waldemar Kramer, Frankfurt am Main.

Duncan, W. H. (1975). *Woody Vines of the Southeastern United States*. University of Georgia Press, Athens, Ga.

Escobar, L. (1985). Biologica reproductivo de *Passiflora manicata* e hibridación con la curuba, *Passiflora mollissima*. *Actual. Biol. Univ. Antioquia* **14**: 111–21.

Faegri, K. & van der Pijl, L. (1966). *The Principles of Pollination Ecology*. Pergamon, Oxford.

Feinsinger, P. (1978). Ecological interactions between plants and hummingbirds in a successional tropical community. *Ecological Monographs* **48**: 269–87.

Feinsinger, P. (1983). Coevolution and pollination. In *Coevolution*, ed. D. Futuyma and M. Slatkin, pp. 282–310. Sinauer, Sunderland, Mass.

Frankie, G. W. (1975). Tropical forest phenology and pollinator plant coevolution. In *Coevolution of Animals and Plants*, ed. L. Gilbert and P. Raven, p. 192–209. University of Texas Press, Austin.

Frankie, G. W., Baker, H. & Opler, P. (1974). Comparative phenological studies of trees in tropical wet and dry forests in the lowland of Costa Rica. *Journal of Ecology* **62**: 881–919.

Frankie, G. W., Haber, W., Opler, P. & Bawa, K. (1983). Characteristics and organization of the large bee pollination system in the Costa Rican dry forest. In *Handbook of Experimental Pollination Ecology*, ed. C. E. Jones and R. J. Little, pp. 411–47. Van Nostrand and Reinhold, New York.

Gentry, A. H. (1974). Flowering phenology and diversity in tropical Bignoniaceae. *Biotropica* **6**: 64–8.

Gentry, A. H. (1981). Distributional patterns and an additional species of the *Passiflora vitifolia* complex: Amazonian species diversity due to edaphically differentiated communities. *Plant Systematics and Evolution* **137**: 95–105.

Gentry, A. H. (1982a). Patterns of neotropical plant species diversity. *Evolutionary Biology* **15**: 1–84.

Gentry, A. H. (1982b). Neotropical floristic diversity: phytogeographical connections between Central and South America, Pleistocene climatic fluctuations, or an accident of the Andean orogeny? *Annals of the*

Missouri Botanical Garden **69**: 557–93.

Gentry, A. H. (1983). Dispersal ecology and diversity in neotropical forest communities. *Sonderbd. Naturwissenschaftlichen Vereins in Hamburg* **7**: 303–14.

Gentry, A. H. (1984). New species and combinations in Apocynaceae from Peru and adjacent Amazonia. *Annals of the Missouri Botanical Garden* **71**: 1075–81.

Gentry, A. H. (1986). Species richness and floristic composition of Choco region plant communities. *Caldasia* **15**: 71–91.

Gentry, A. H. (1988). Changes in plant community diversity and floristic composition on environmental and geographical gradients. *Annals of the Missouri Botanical Garden* **75**: 1–34.

Gentry, A. H. (1989a). *Speciation in Tropical Plants*. Academic Press, New York.

Gentry, A. H. (1989b). Evolutionary patterns in neotropical Bignoniaceae. *Memoirs of the New York Botanical Garden* **55**: 118–29.

Gentry, A. H. (1991). A symposium of Bignoniaceae ethnobotany and economic botany. *Annals of the Missouri Botanical Garden* (in press).

Gentry, A. H. & Dodson, C. (1987). Diversity and biogeography of neotropical vascular epiphytes. *Annals of the Missouri Botanical Garden* **74**: 205–33.

Gilbert, L. E. (1972). Pollen feeding and reproductive biology of *Heliconius* butterflies. *Proceedings of National Academy of Sciences (USA)* **69**: 1403–7.

Gilbert, L. E. (1980). Food web organization and conservation of neotropical diversity. In *Conservation Biology*, ed. M. Soulé & Wilcox, pp. 11–34. Sinauer, Sunderland, Mass.

Givnish, T. J. (1980). Ecological constraints on the evolution of breeding systems in seed plants: dioecy and dispersal in gymnosperms. *Evolution* **34**: 959–72.

Haber, W. A. & Frankie, G. W. (1989). A tropical hawkmoth community: Costa Rican dry forest Sphingidae. *Biotropica* **21**: 155–72.

Heithaus, E. R. (1974). The role of plant–pollinator interactions in determining community structure. *Annals of the Missouri Botanical Garden* **61**: 675–91.

Heithaus, E. R. (1979a). Flower visitation records and resource overlap of bees and wasps in northwest Costa Rica. *Brenesia* **16**: 9–52.

Heithaus, E. R. (1979b). Community structure of neotropical flower visiting bees and wasps: diversity and phenology. *Ecology* **60**: 190–202.

Heithaus, E. R., Fleming, T. H. & Opler, P. A. (1975). Foraging patterns and resource utilization in seven species of bats in a seasonal tropical

forest. *Ecology* **56**: 841–54.
Hubbell, S. P. (1979). Tree dispersion, abundance, and diversity in a tropical dry forest. *Science* **203**: 1299–309.
Hurd, P. D. & Linsley, E. G. (1964). The squash and gourd bees – genera *Peponapis* Robertson and *Xenoglossa* Smith – inhabiting America north of Mexico. *Hilgardia* **35**: 373–477.
Janson, C. H., Terborgh, J. & Emmons, L. H. (1981). Non-flying mammals as pollinating agents in the Amazonian forest. *Biotropica* **13** (suppl.) 1–6.
Janzen, D. H. (1968). Reproductive behavior in the Passifloraceae and some of its pollinators in Central America. *Behavior* **32**: 33–48.
Janzen, D. H. (1974). Tropical blackwater rivers, animals, and mast fruiting by the Dipterocarpaceae. *Biotropica* **6**: 69–103.
Janzen, D. H. (1977). Promising directions of study in tropical plant-animal interactions. *Annals of the Missouri Botanical Garden* **64**: 706–36.
Janzen, D. H. (1979). How to be a fig. *Annual Review of Ecology and Systematics* **10**: 13–51.
Kaur, A., Ha, C., Jong, K., Sands, V., Chan, H., Soepadmo, E. & Ashton, P. (1978). Apomixis may be widespread among trees of the climax rain forest. *Nature (London)* **271**: 440–2.
Kock, D. (1972). Fruit bats and bat-flowers. *Bull. East African Nat. Hist. Soc.* 7: 123–6.
Kress, W. J. & Beach, J. H. (1990). Flowering plant reproductive systems at La Selva Biological Station. In *La Selva: Ecology and Natural History of a Neotropical Rainforest*, ed. L. McDade, K. Bawa, H. Hespenheide and G. Hartshorn. University of Chicago Press, Chicago (in press).
Kubitzki, K. & Baretta-Kuipers, T. (1969). Pollendimorphie und Androdiozie bei *Tetracera* (Dilleniaceae). *Naturwissenschaften* **56**: 219–20.
Levin, D. A. (1975). Pest pressure and recombination systems in plants. *American Naturalist* **109**: 437–51.
Lumer, C. (1980). Rodent pollination of *Blakea* (Melastomataceae) in a Costa Rican cloud forest. *Brittonia* **32**: 512–17.
MacDougal, J. M. (1983). Revision of *Passiflora* L. section *Pseudodysosmia* (Harms) Killip emend MacDougal, the Hooked Trichome group (Passifloraceae). PhD dissertation, Duke University, Durham, NC.
Macedo, M. (1977). Dispersão de plantas lenhosas de uma campina Amazonica. *Acta Amazonica* 7(Suppl.): 40–6.
Michener, C. D. (1979). Biogeography of the bees. *Annals of the Missouri Botanical Garden* **66**: 277–347.

Okali, D. U. & Ola-Adams, B. A. (1987). Tree population changes in treated rain forest at Omo Forest Reserve, south-western Nigeria. *Journal of Tropical Ecology* 3: 291–313.

Opler, P. A., Frankie, G. W. & Baker, H. G. (1980). Comparative phenological studies of treelet and shrub species in tropical wet and dry forests in the lowlands of Costa Rica. *Journal of Ecology* **68**: 167–88.

Peñalosa, J. (1984). Basal branching and vegetative spread in two tropical rain forest lianas. *Biotropica* **16**: 1–9.

Prance, G. T. (1978). The origin and evolution of the Amazon flora. *Interciencia* 3: 207–22.

Prance, G. T. (1980). A note on the probable pollination of *Combretum* by Cebus monkeys. *Biotropica* **12**: 239.

Putz, F. E. (1984). The natural history of lianas on Barro Colorado Island, Panama. *Ecology* **65**: 1713–24.

Putz, F. E. & Appanah, S. (1987). Buried seeds, newly dispersed seeds, and the dynamics of a lowland forest in Malaysia. *Biotropica* **19**: 326–33.

Putz, F. E. & Windsor, D. M. (1987). Liana phenology on Barro Colorado Island. *Biotropica* **19**: 334–41.

Ramírez B., W. (1970). Host specificity of fig wasps (Agaonidae). *Evolution* 24: 680–91.

Ramírez B., W. (1977). Evolution of the strangling habit in *Ficus* L., Subgenus *Urostigma* (Moraceae). *Brenesia* **12/13**: 11–19.

Renner, S. (1984). Pollination and breeding systems in some central Amazonian Melastomataceae. *Proc. 5th Intern. Symp. Pollination*, pp. 175–280. INRA publ.

Renner, S. (1986). The neotropical epiphytic Melastomataceae: phytogeographic patterns, fruit types, and floral biology. *Selbyana* **9**: 104–11.

Renner, S. (1989). A survey of reproductive biology in neotropical Melastomataceae and Memecylaceae. *Annals of the Missouri Botanical Garden* **76**: 496–518.

Ruiz Z., T. & Arroyo, M. T. K. (1978). Plant reproductive ecology of a secondary deciduous forest in Venezuela. *Biotropica* **10**: 221–30.

Sazima, M. & Sazima, I. (1978). Bat pollination of the passion flower, *Passiflora mucronata*, in southeastern Brazil. *Biotropica* **10**: 100–9.

Sazima, M. & Sazima, I. (1980). Bat visits to *Marcgravia myriostigma* Tr. et Planch. (Marcgraviaceae) in southeastern Brazil. *Flora* **169**: 84–8.

Schatz, G. E. (1990). Some aspects of pollination biology in Central American forests. In *Reproductive Ecology of Tropical Forest Plants*, ed. K. Bawa, pp. 69–84. UNESCO, Paris.

Simpson, B. B. & Neff, J. (1981). Floral rewards: alternatives to pollen and nectar. *Annals of the Missouri Botanical Garden* **68**: 301–22.
Snow, D. & Snow, B. (1980). Relationships between hummingbirds and flowers in the Andes of Colombia. *Bulletin of the British Museum of Natural History (Zoology)* 38(2): 105–39.
Sobrevila, C. & Arroyo, M. T. K. (1982). Breeding systems in a montane tropical cloud forest in Venezuela. *Plant Systematics and Evolution* **140**: 19–37.
Steiner, K. E. (1981). Nectarivory and potential pollination by a neotropical marsupial. *Annals of the Missouri Botanical Garden* **68**: 505–13.
Stiles, F. G. (1977). Coadapted competitors: the flowering seasons of hummingbird-pollinated plants in a tropical forest. *Science* **198**: 1177–8.
Stiles, F. G. (1978). Temporal organization of flowering among the hummingbird food plants of a tropical wet forest. *Biotropica* **10**: 194–210.
Stiles, F. G. (1980). The annual cycle in a tropical wet forest hummingbird community. *Ibis* **122**: 322–43.
Stiles, F. G. (1981). Geographical aspects of bird-plant coevolution, with particular reference to Central America. *Annals of the Missouri Botanical Garden* **68**: 323–51.
Templeton, A. (1980). The theory of speciation via the founder principle. *Genetics* **94**: 1011–38.
Thien, L. B. (1980). Patterns of pollination in the primitive angiosperms. *Biotropica* **12**: 1–13.
van der Pijl, L. (1941). Flagelliflory and cauliflory as adaptations to bats in *Mucuna* and other plants. *Annales du Jardin Botanique de Buitenzorg* **51**: 83–93.
van der Pijl, L. (1961). Ecological aspects of flower evolution II. Zoophilous flower classes. *Evolution* **15**: 44–59.
Van Valen, L. (1973). A new evolutionary law. *Evolutionary Theory* **1**: 1–30.
Vogel, S. (1958). Fledermausblumen in Sudamerika. *Oesterreichische Botanische Zeitung* **104**: 491–530.
Vogel, S. (1963). Duftdrusen im Dienste der Bestaubung. Uberbau und Funktion der Dimophoren. *Akademie der Wissenschaften (Mainz), Abhandlung der Mathematisch-Naturwissenschaftlichen Klasse* 1962: 600–736.
Vogel, S. (1969). Chiropterophilie in der neotropischen Flora. Neue Mitteil II. *Flora, Abt. B*, **158**: 185–222.
Vogel, S. (1974). Olblumen und olsammelnde Biennen. *Trop. Subtrop.*

Pflanzenwelt 7: 283–457.

Vogel, S. (1981a). Abdominal oil-mopping – a new type of foraging in bees. *Naturwissenschaften* **68**: 627–8.

Vogel, S. (1981b). Bestaubungskonzepte der Monokotylen und ihr Ausdruck im System. *Berichte der Deutschen Botanischen Gesellschaft* **94**: 663–75.

Vogel, S. (1990). History of the Malpighiaceae in the light of pollination ecology. *Memoirs of the New York Botanic Garden* **55**: 130–42.

Whitham, T. G. & Slobodchikoff, C. N. (1981). Evolution by individuals, plant-herbivore interactions, and mosaics of genetic variability: the adaptive significance of somatic mutations in plants. *Oecologia* **49**: 287–92.

PART V
ECONOMIC IMPORTANCE OF VINES

16
The ethnobotany and economic botany of tropical vines

OLIVER PHILLIPS

Introduction

What makes a plant useful? Two distinct components are needed: firstly, the evolutionary processes that shape the plant's biochemistry, physiology and physiognomy, and secondly, the human quality of taking a curious interest in these plants and so discovering the edible and the poisonous, the remedial and the toxic. Both requisites are most fully satisfied in the tropics, and in particular the tropical rainforest, and this is one reason why tropical forests are such an invaluable resource. Here the intensity of biotic interactions has resulted in a myriad of secondary plant products (see Levin, 1976), and this process may have been especially prevalent in the woody climbers, a life form essentially restricted to the tropical forest. And it is here, even today, that the greatest diversity of indigenous cultures persists, maintaining a treasure-trove of ethnobotanical information. Still, indigenous peoples have been worse than decimated (the Indian population of Brazil is only 0.1% of its 1492 level and current tribal extinction rates are more than one per year (Posey, 1983)), representing an irreversible loss in cultural diversity.

This indigenous knowledge is also the foundation for the use of plants as medicine and food by the wider rural population of the South. The World Health Organization estimates that 80% of the population of developing countries relies on traditional medicine for primary health care, and that 85% of traditional medicine involves the use of plant extracts. This means that 3.5–4 billion people in the world rely on plants as a major source of drugs (Farnsworth *et al.*, 1985), including the 1.2 billion people in industrialized nations where 40% of prescribed medicines contain chemicals originally isolated from plants. A number of biochemicals from tropical vines, such as curare and diosgenin, now have major roles in Western clinical medicine.

It has been suggested (Gentry 1984; Gentry & Cook, 1984) that lianas are especially prone to evolving protective compounds with novel chemical structures. The theoretical reasons to expect tropical climbing plants in particular to have a high level of biodynamic compounds, and hence to be prominent features of native pharmacopeias, stem from their fast growing and light-demanding nature (Gentry, 1984; F. Putz, personal communication). Because of these features, the leaves of climbers are likely to be shorter-lived than those of many other forest plants. This will tend to make them less 'apparent' to herbivores (*sensu* Feeny (1976), apparency of a plant resource to herbivores is a composite measure of, *inter alia*, (i) how easy it is to locate, (ii) how predictable it is in time and space, and (iii) how long it is available for). (Note that the unapparent nature of climbers is enhanced by the fact that their leaves tend to be in smaller concentrations than those of most trees, making them potentially more difficult to locate.) Less apparent plants or parts of plants therefore tend to have to resist a smaller variety of herbivores. Thus we might expect them to invest more in the highly active 'qualitative' defenses as opposed to the more energetically expensive, broad-spectrum 'quantitative' defenses. Furthermore, since qualitative defenses are effective at much lower concentrations, the plant can more quickly build them up to a concentration that deters herbivores. Presumably there is a premium on the rapid synthesis of effective herbivore deterrents, since there may simply not be enough time to manufacture defenses such as polyphenolics, lignins, or sclerophyllous tissue, to protect the continous production of tender new growth.

Now, qualitative defenses, for example alkaloids and cyanogenic glucosides, tend to exhibit much more specific biological activity than quantitative defenses such as tannins, resins and latex. For this reason they are not only more toxic, but are also much more likely to have medicinal applications.

A further reason why we might expect climbers to have more qualitative defenses, and hence to be useful medicinally, is that their stem weight is limited by the constraints of supporting plants (Putz, 1983). Therefore it makes more evolutionary sense to invest in defenses that can be effective in low concentrations, so as not to add excessively to the weight that has to be supported. In this context it is interesting to note that, at least in the Neotropics, most large woody families conspicuous for their copious latex or resin production have few or no liana representatives (e.g. Euphorbiaceae, Moraceae, Guttiferae, Myristicaceae, Sapotaceae).

Hegarty, Hegarty & Gentry (Chapter 10) discuss in greater depth the factors affecting the nature and concentration of secondary chemicals in vines, as contrasted with other life forms.

Of course there are practical factors, as well as ecological ones, which shape patterns of climber use in indigenous medicine. For example, the fact that most reported uses for lianas involve just the stem is surely related to the inaccessibility of the leaves and reproductive structures of many large lianas. In contrast, reports exist for the traditional use of every party of herbaceous climbers. In a sense, then, lianas are the ethnobotanical equivalents of trees, just as herbaceous vines are of shrubs and herbs.

Another important category of use for tropical climbers is for all kinds of fiber, whether it be for general cordage, lashing together the walls and roof of houses, or for weaving baskets, hammocks, fishing nets, etc. The best known fibrous lianas are the rattans (climbing palms, especially from Southeast Asia), but numerous liana taxa are used in this way. A prime attribute of lianas is their flexibility, for surviving the swaying or breakage of their living supports, so it is hardly surprising that their use for fiber should cut across taxonomic boundaries.

Yet another important attribute of vines that has resulted in widespread use is the possession by some herbaceous tropical climbers of starchy tubers as survival structures. Many of these are edible, particularly the various species of *Dioscorea* and *Ipomoea* which make a major contribution to the diet of many people, especially in the tropics.

It is clear that just as the vast spectrum of indigenous uses of tropical vines has been an integral part of those cultures, so it is proving with our own industrialized society. We live in an increasingly interdependent world, exemplified by the wide variety of products from tropical climbers that many of us are able to enjoy this century for the first time. Whilst trade in some products, such as pepper, rattans and vanilla, dates from the last century (much earlier in the case of pepper), for others it is much more recent. For example, curare has been used medically for barely 50 years, and diosgenin-base oral contraceptives are a phenomenon of the last 30 years. As the 21st century rapidly approaches, the potential contributions of tropical vines to fields as diverse as family planning, new foods, and biodegradable pesticides, will be crucial as we try to develop more sustainable lifestyles.

In this chapter I have reviewed the literature for each family with useful, climbing, tropical species, to compose concise summaries. Sixteen families with many reported uses are also summarized genus by genus in the accompanying tables to help readers interested in the ethnobotanical literature on individual taxa. As well as considering many lianas and herbaceous climbers I include climbing hemiepiphytes, but exclude strangling *Ficus* and *Clusia* species. I have consulted primary and secondary sources for uses of climbers as foods, in crafts and construction, in physiologically active preparations, and as ornamentals; especially helpful were compilations of useful plants by Uphof (1959), von Reis Altschul (1975), Perry (1980), von Reis and Lipp (1982), Kunkel (1984), Oliver-Bever (1986), and Mabberley (1987), most of whom cite primary references.

The family-by-family approach I have used is an appropriate way to deal with the diffuseness of the ethnobotanical literature. Given the current database, it is practically impossible to test theories about geographical, ecological, or anthropological patterns of plant use. There are almost as many techniques adopted as there are research papers, and only very recently have ethnobotanists attempted to test well defined hypotheses. Moreover, a brief appraisal of the family tables suggests huge gaps in the literature: for example, every report I could discover on the medicinal use of *Mandevilla* is attributed to the pioneering work in Colombia of one man, Richard Schultes (Schultes, 1979), yet this is a widespread neotropical genus of apocynaceous lianas, for which there are surely many undocumented uses elsewhere. On a continental scale, most references are to research in Latin America (with the least in Africa). Whether this is mostly a function of (i) relative intensity of ethnobotanical effort, (ii) greater vine species richness, (iii) richer ethnobotany, or (iv) a relative preference for using vines among Latin Americans, is unclear.

Families with useful climbing plants in the tropics

Amaryllidaceae

Bomarea has about 150 Neotropical species, mostly climbers. The perennating starchy tubers of at least four species are eaten: *B. acutifolius* (Link and Otto) Herb. in Mexico, *B. ovata* Mirb. and *B. edulis* Herb. in Santo Domingo, and *B. glaucescens* Baker in Ecuador (Kunkel, 1984).

Annonaceae

Although Annonaceous lianas are very important ecologically in the Old World, they provide few useful products (see Table below). *Artabotrys* spp., *Desmos* spp., and *Popowia congensis* Engl. and Diels all provide edible fruits, with those of *Uvaria rufa* especially esteemed in tropical Asia (Uphof, 1959; Kunkel, 1984; Mabberley, 1987).

Artabotrys	stimulant tea from flowers	Old World: Lewis & Elvin-Lewis (1977)
	edible fruits	Old World: Mabberley (1987)
Cyathostemma	tonic, aphrodisiac	Indo-China: Perry (1980)
	protective post partum	Malaysia: Perry (1980)
	antispasmodic	Indonesia: Perry (1980)
Popowia	edible fruits	Indo-China, Africa: Kunkel (1984)

Desmos	edible fruits	SE Asia: Kunkel (1984)
	increases lactation	SE Asia: von Reis Altschul (1975)

Apocynaceae

Schultes (1979) considers this family, possibly the richest of all in alkaloids, to be underrepresented in native Amazonian pharmacopeias, possibly because many species are just too toxic for medical use. Still, there is a sizeable ethnobotanical literature on the Apocynaceae, and phytochemical studies on the little-known neotropical climbing genera and species would be especially worthwhile.

For example, Schultes (1979) records numerous uses in Colombian Amazonia for *Mandevilla*, in spite of the absence of reported alkaloids in this large genus. Medical uses include supposed aphrodisiac flowers (*M. steyermarkii* Woodson), antifungal latex (*M. vanheurckii* (Muell. Arg.) Margraf, and *M. neriodes* Woodson), depilatory latex (*M. scabra* (R. et S.) Schumann), antisyphilictic latex (*M. stephanotidifolia* Woodson ex Schultes), and leaves that soothe scorpion stings and ant-bites (*M. stephanotidifolia*). *Mandevilla steyermarkii* in particular has multiple uses in the Vaupes region of Colombian Amazonia (Schultes, 1979). Apart from Schultes's extensive report, apparently few other ethnobotanical data exist for this genus.

In contrast to *Mandevilla*, *Strophanthus*, which includes well-known ornamentals, is used widely in the Paleotropics, and contributes significantly to our own pharmacopeia (Uphof, 1959, von Reis Altschul, 1975, Morton, 1977). In Nigeria, *S. gratus* Franch is used for gonorrhea, and poultices are applied to wounds, ulcers and Guinea worm (Morton, 1977). Ouabain, a cardiac glycoside extracted from the seeds, is used as an intravenous cardiac stimulant in Europe and the USA for emergency treatment of heart failure and pulmonary edema. K-strophanthoside is extracted from the seeds of *S. kombe* Oliver, and has pharmaceutical applications similar to ouabain (Morton, 1977); the species is used for arrow poison in Tanzania and Malawi (Uphof, 1959). In Nigeria, *S. sarmentosus* DC. provides an arrow poison, and is a source of commercial cortisone, whilst two more *Strophanthus* species in the Philippines are also used for arrow poisons (von Reis Altschul, 1975).

Aganosma	coffee/tea substitute	SE Asia: Kunkel (1984)
Allamanda	febrifuge	French Guiana: Grenand (1980)
Alyxia	headache	Sumatra: Elliott & Brimacombe (1985)

Anodendron	fibre (esp. fishing nets)	SE Asia: Uphof (1959)
Anechites	improves memory	Colombia: Schultes (1979)
Chonemorpha	binding	Philippines: Bodner & Gereau (1988)
	fibre (esp. fishing nets)	India, Malaysia: Uphof (1959)
Condylocarpon	febrifuge	French Guiana: Grenand (1980)
Landolphia	edible fruits	Tanzania: FAO (1983)
	rubber	Africa, Madagascar: Uphof (1959)
	cicatrasing latex; magical uses	French Guiana: Grenand (1980)
Mandevilla	*many uses – see text*	Colombia: Schultes (1979)
Mesichites	treat 'susto' (= soul loss)	Colomba: Schultes (1979)
Odontadenia	flea, lice, mosquito repellant, gum bleeding, toothache, drink flavour	Colombia: Schultes (1979)
Parameira	fibre for rope	Philippines: Uphof (1959)
Parsonsia	applied to leg swellings	Solomon Is: Perry (1980)
	disinfectant, tuberculosis, vulnerary febrifuge, rheumatism, kidneys	Philippines: Perry (1980) Indo-China: Perry (1980)
Prestonia	anticoagulant	Colombia: Forero Pinto (1980)
	hallucinogen	Peru: Lopez Guillen & Kiyan de Cornelio (1974)
Strophanthus	*many uses – see text*	Tropics: Morton (1977) Tropics: von Reis Altschul (1975) Tropical Africa; Uphof (1959)
Tabernaemontana	stimulant	Colombia: Schultes (1979)

Urceola	rubber	SE Asia: Uphof (1959)
Willughbeia	edible fruit	SE Asia: Uphof (1959), Kunkel (1984)
	rubber	SE Asia: Uphof (1959)

Araceae

Philodendron provides the widest range of uses among aroid epiphytic lianas. In Colombia *P. craspedodromum* Schultes is used by the Desana Indians for a fish poison (Schultes, quoted in Davis & Yost, 1983). The Waorani in Ecuador use a *Philodendron* to treat the snake-bite of *Bothrops castelnaudi* (Davis & Yost, 1983). *P. hederaceum* removes warts in Mexico (Martínez, 1984) and an unidentified species is used by the Guaymi in Panama to treat skin rashes (Joly *et al.*, 1987), Finally, Schultes (1962) and Garcia-Barriga (1974) record *P. dysicarpum* Schultes as a female oral contraceptive in Colombia.

Anthurium	cordage	Costa Rica: Williams (1981)
Epiprenum	dart poison	SE Asia: Uphof (1959)
Heteropsis	strong cordage, edible fruit	Ecuador: Davis & Yost (1983)
Monstera	edible fruit	Neotropics: Uphof (1959)
	stomach pains	Colombia: Forero Pinto (1980)
	ornamental ('cheeseplant')	World: Mabberley (1987)
Philodendron	*many uses and reports – see text*	
Pothos	worms, smallpox, asthma	Malaysia: Perry (1980)
	postpartum	Indo-China: Perry (1980)
Syngonium	*Paraponera* ant-bite, witchcraft	Ecuador: Vickers & Plowman (1984)

Arecaceae

Palms provide a uniquely diverse range of services to humans. As well as their specialist uses – as spices, waxes, gums, poisons and medicines – they also provide the basic necessities of food, shelter, fuel and fiber (Plotkin & Balick, 1984). Climbing palms contribute their fair share to both indigenous and industrial cultures. They are especially important in construction; indeed, the value of the end-products of the rattan industry has been estimated as high as US $4 billion. They are Southeast Asia's most valuable export after timber (Myers, 1984).

Desmoncus, the only neotropical climbing palm genus, provides both edible fruit (*D. longifolius*, *D. prunifer* Poepp., *D. macroanthos* Mart. and *D. prostratus*) (Bodley 1978; von Reis & Lipp, 1982; Kunkel, 1984), and fiber for weaving by Indians in both South America (A. H. Gentry, personal communication) and Central America (Uphof, 1959). A rattan wicker work cottage industry based on a *Desmoncus* species is prospering in Iquitos, Peru; the local artisans claim that it has better qualities than the imported Asian rattans (Gentry, 1986). In Africa, the fiber from local rattans such as *Calamus* spp. and *Ancistrophyllum secundiflorum* Mann et Wendl. is also used locally (Uphof, 1959).

The harvesting of rattan for fiber has the potential to develop into a sustainable industry in heavily logged areas, providing tropical Asian countries with foreign exchange without further depleting their natural resources (Myers, 1984). However, it remains mostly a wild-harvested product and intensive exploitation has led to heavy pressure on well-known canes such as *Calamus manan*, a large diameter cane that dominates the furniture trade (Dransfield 1985, Manokaran, 1985). Basu (1985) reports that in India the populations of at least six *Calamus* species should be regarded as 'vulnerable' to extinction, owing to overcollection, and all 13 of the endemic commercial rattan species should be regarded as 'threatened'.

One way to lessen the pressure on rarer species might be to promote trade and cultivation of promising lesser known species, and Dransfield (1985) singles out *Calamus* sections *Podocephalus* and *Phyllanthectus* as having special potential. However, this would be unlikely to have a permanent beneficial effect if traditional concepts of communal landownership continue to erode, so that there is no incentive for individual harvesters to conserve populations of valuable species. This implies, for example, that scarce conservation money will be better spent in giving communities a financial stake in the future of their local forests, for example by encouraging village cooperatives that process rattan *in situ*, than in researching ways to grow rattans in plantations.

Asian rattans have several important non-construction uses. *Calamus ornatus* Blume ex Schult. var *philippinensis* Becc. and *C. merrillii* Becc. have been singled out as having special potential for cropping in Southeast Asia for their nutritious fruit (Food and Agriculture Organization, 1984). At least 10 more *Calamus* species have edible fruits, and the stems and/or young leaves of still other *Calamus* species and five *Daemonorops* species are used as vegetables in the region (Kunkel, 1984). Furthermore, *Calamus* species have numerous medicinal uses, ranging from malaria treatments to reducing childbirth pain. China imports 'dragon's blood', an astringent product of various Southeast Asian *Daemonorops* species, for use as a sedative and a tonic (Perry, 1980); the same product is used locally as a varnish and a red dye (Jacobs, 1982).

Ancistrophyllum	fiber used locally	W Africa: (Uphof, 1959)
Calamus	*numerous uses and references – see text*	
Daemonorops	stems and young leaves eaten	SE Asia: Kunkel (1984)
	sedative, tonic	SE Asia: Perry (1980)
	varnish, red dye	SE Asia: Jacobs (1982)
Desmoncus	edible fruit	S America: Bodley (1978), von Reis Altschul & Lipp (1982), Kunkel (1984)
	weaving baskets	S America: Gentry (1986)
	weaving	Central America: Uphof (1959)

Aristolochiaceae

Aristolochia, the principal genus in the family, consists largely of neotropical climbers with a range of medicinal uses. In this region *Aristolochia* is used in snake-bite remedies (at least 14 tropical species are used thus (Uphof, 1959; Williams, 1981; Vickers & Plowman, 1984; Joly *et al.*, 1987)), possibly a case of the 'doctrine of signatures' based on twining habits and unusual flowers. Other widespread uses include as an emmenagogue (i.e. to stimulate the menstrual flow), for which Uphof, 1959, lists six tropical species, as a cold cure and febrifuge (i.e. to relieve fever) (4 tropical species in Uphof), and as an abortifacient (3 tropical species in Uphof).

Asclepiadaceae

The family that owes its name to Asclepius, the classical god of healing, features several medicinally important vines as well as a number of toxic plants.

Hoya, for example, has a varied range of medicinal applications in Southeast Asia, as well as several important ornamental species (see Mabberley, 1987). The latex of *H. rumphii* Bl. and other species are used to treat poisonous fish stings in Indonesia (Uphof, 1959; Perry, 1980). In Indonesia *H. rumphii* is also used to soothe gonorrhea, and in the Philippines *H. imbricata* Decne helps to mature boils and to encourage healing of wounds and varicose ulcers (Perry, 1980). In Peninsular Malaysia *H. coriacea* Bl.

treats coughs and asthma, and *H. diversifolia* Bl. is used for rheumatoid patients (Perry, 1980).

Marsdenia species are used to stimulate the appetites of people in both Ecuador and Colombia (*M. reichenbachii* Triana, Uphof (1959)), and in Indonesia (*M. tinctoria* (Roxb.) R. Br., Perry (1980)). Perry (1980) also reports a claim that in Indonesia *M. tinctoria* is used to promote children's hair growth. Other uses include as an indigo dye in Southeast Asia, for which *M. tinctoria* was cultivated until very recently (Uphof, 1959; Mabberley, 1987), the use of *M. zimapanica* Hemsl. Tequamp. as a poison for coyotes in Mexico (Uphof, 1959) and *M. hamiltonii* Wight for its edible fruits in India (Kunkel, 1984).

Cryptostegia	rubber, ornamental	Central America: Williams (1981)
Cynastrum	liver complaints	Costa Rica: Nuñez (1975)
Dischidia	gonorrhea, fish stings	Indonesia: Perry (1980)
	eczema, herpes, boils, goiter	Philippines: Perry (1980)
	diuretic, blennorrhea	Indo-China: Perry (1980)
Funastrum	latex used as fly-killer	Costa Rica: Nuñez (1975)
Gymnema	emetic	Philippines: Perry (1980)
Hoya	*many uses – see text*	SE Asia: Uphof (1959), Perry (1980)
		Tropics: Mabberley (1987)
Marsdenia	*many uses – see text*	SE Asia: Uphof (1959), Perry (1980)
		Ecuador, Colombia: Uphof (1959)
		India, Mexico: Kunkel (1984)
Parquetina	stuns fish	W Africa: Oliver-Bever (1986)
	rickets, diarrhea, skin lesions	Ghana, Liberia: Oliver-Bever (1986)
Periploca	rubber	Ivory Coast, Zaïre: Uphof (1959)
Sarcostemma	kills cow skin parasites, twine	Central America: Williams (1981)
Telosma	whole plant edible	ES Asia: Uphof (1959)

	kidney and bladder diseases	Indo-China: Perry (1980)
Vincetoxicum	nutritious fruit	Costa Rica: Nuñez (1975)

Asteraceae

Mikania, a large genus of tropical climbers, provides a number of indigenous remedies. In Brazil both *M. sessiflora* and *M. officinalis* are used to make soothing teas for stomach trouble, whilst *M. smilacina* is used to treat rheumatism (von Reis Altschul, 1975; von Reis & Lipp, 1982). In Ecuador *M. guaco* and *M. vitifolia* are regarded as snake-bite cures, in Africa *M. cordata* provides an antimalarial, and in Fiji squashed *M. micrantha* leaves are put on sores and insect bites (von Reis Altschul, 1975; von Reis & Lipp, 1982).

Basellaceae

The tubers and leaves of *Anredera cordifolia* (Ten.) Steenis are eaten in South America, where it is also grown as an ornamental (Uphof, 1959; Mabberley, 1987). Similarly, *A. diffusa* (Moq.) Soukup tubers are eaten in Peru. In Indonesia, the leaves of *Basella alba* are also eaten, and *B. rubra* L. is cultivated in the tropics as a pot-herb and for the fruit sap which is used as a food coloring and as edible agar (Uphof, 1959; von Reis Altschul, 1975).

Begoniaceae

In Ecuador the Waorani chew the succulent stems of a scandent *Begonia* species as a condiment; the plant may also be good for colds (Davis & Yost, 1983).

Bignoniaceae

The most speciose neotropical liana family has an impressive variety of ethnobotanical uses. Indeed, nearly every Bignoniaceous liana genus seems to have a specific use reported for it (A. Gentry, personal communication). This suggests that this family, although scarcely looked at by chemists, is characterized by high levels of biodynamic compounds (Gentry & Cook, 1984; Lewis, Elvin-Lewis & Gnerre, 1987). As examples of liana genera with especially widespread and varied uses I will consider *Arrabidaea*, *Mansoa*, and *Martinella*.

The Jívaro in northern Peru use an *Arrabidaea* in a cure for thrush, a fungal disease of the throat and mouth (Lewis *et al.*, 1987). In Belem, Brazil, *A. chica*

(HBK) Verlot is sold as an anti-inflammatory agent and as a tonic (van den Berg, 1984). The genus is also an important source of dyes such as carajurin, a flavan derivative (Gottlieb, 1984), which are used for various colors depending on the species and the mode of preparation. The Siona in eastern Ecuador cultivate *A. chica* for a brown dye from the leaves for face paint and to decorate the traditional robes known as 'cushmas' (Vickers & Plowman, 1984); in Central America, leaves of the same species are used to dye palm fiber dark brown before making it into hats, and to dye rushes red before weaving them into mats (Williams, 1981). Similarly, the Jívaro color their teeth with an *Arrabidaea* (Lewis *et al.*, 1987). Finally, in Central America the stems of Bignoniaceous lianas like *A. pubescens* P. DC. are used as rope substitutes, especially to construct the framework of houses (Williams, 1981).

Both species of *Martinella* (*M. obovata* (HBK) Bur. et Schum., and *M. iquitoensis* Sampaio) are widely used as eye medicines in South America (Gentry & Cook, 1984; Lewis *et al.*, 1987). Juice extracted from the root is used in drops to treat eye inflammation or irritation. The fact that the same part is used in the same way by many disparate indigenous groups strongly suggests that *Martinella* root is an effective eye medication (Gentry & Cook, 1984). The same genus is also an ingredient in an arrow poison in Colombia (Schultes, 1970a), and is recorded as a febrifuge in both Colombia (Schultes, 1970a) and Peru (Ayala Flores, 1984).

A decoction of garlic-smelling *Mansoa standleyi* (Steyermark) Gentry leaves is drunk by the Waorani of eastern Ecuador to treat fever (Davis & Yost, 1983). *M. standleyi* and *M. alliacea* (Lam.) Gentry are the bignons most widely used by the Jívaro, for treating aching and swollen parts of the body, arthritis and/or rheumatism, as well as for colds, influenza, coughs and sore throats (Lewis *et al.*, 1987). Ayala Flores (1984) also reports the same use of *M. alliacea* as an antirheumatic by the Achual Jívaro subtribe.

Adenocalymma	incense and ritual baths	Brazil: van den Berg (1984)
Anemopaegma	aphrodisiac	Brazil: van den Berg (1984) Lewis & Elvin-Lewis (1977)
Arrabidaea	*many uses – see text*	Peru: Lewis *et al.* 1987) Brazil: van den Berg (1984) Ecuador: Vickers & Plowman (1984) Central America: Williams (1981)

Clytostoma	euphoric and coca additive	Bolivia: Davis (1983)
Distictella	arrow poison ingredient	Colombia: Schultes (1970b)
Macfadyena	conjunctivitis	Peru: Lewis *et al.* (1987)
	heavy menstrual bleeding	Peru: Lewis *et al.* (1987)
	ornamental	Guatemala: Williams (1981)
Mansoa	febrifuge, eliminates fatigue	French Guiana: Grenand (1980)
	many uses – see text	Peru: Lewis *et al.* (1987)
		Peru: Ayala Flores (1984)
		Ecuador: Davis & Yost (1983)
Martinella	conjunctivitis	Peru: Lewis *et al.* (1987)
	eye medicine	Amazonia: Gentry & Cook (1984)
	arrow poison ingredient	Colombia: Schultes (1970a)
	febrifuge	Peru: Ayala Flores (1984)
Memora	skin ailments	Peru: Lewis *et al.* (1987)
Mussatia	euphoric and coca additive	Bolivia, Peru: Davis (1983)
	eliminates fatigue	Bolivia: Oblitas Poblete (1969)
		Bolivia, Peru: Plowman (1980)
Pithecoctenium	skin disease	Guatemala: Williams (1981)
	rope substitute	Central America: Williams (1981)
	toys and kitchen tools	Central America: Williams (1981)
Pleonotoma	conjunctivitis	Colombia: Schultes (1984)
Tanaecium	hallucinogenic snuff	Brazil: Prance (1972)
	skin problems	French Guiana: Grenand (1980)
	aphrodisiac	Colombia: Schultes (1984)

Boraginaceae

Scandent *Tournefortia* species have a variety of uses (see Table below). Interestingly, in Ecuador *Tournefortia* infusions are seen as providing physical, and perhaps spiritual, cleansing treatments. The Siona take *T. angustifolia* R. and P. as a purgative to prepare themselves before drinking the hallucinogen ayahuasca (see Malpighiaceae) (Vickers & Plowman, 1984); also in Ecuador, *T. breviloba* tea is drunk for 1–2 days after childbirth to 'clean everything out' (von Reis Althschul, 1975). Unfortunately, this second report does not record the ethnic group involved.

Tournefortia	purgative	Ecuador: Vickers & Plowman (1984)
	administer postpartum	Ecuador: von Reis Altschul (1975)
	snake-bites	Mexico: von Reis & Lipp (1982)
	influenza, insect stings, snake-bites	Mexico: von Reis Altschul (1975)
	diving goggles, poisonous fish	Marshall Is: von Reis & Lipp (1982)
	cough	Tanzania: von Reis Altschul (1975)
	sores and boils, 'female trouble'	W Indies: von Reis Altschul (1975)
	edible fruit	Colombia: von Reis Altschul (1975)

Cactaceae

In Mexico *Pereskia aculeata* Mill is cultivated for its fruit (Uphof, 1959), and at least eight neotropical climbing *Hylocereus* species have edible fruit (Kunkel, 1984). The fruits of three *Selenicereus* species are eaten in the West Indies and Central America (Kunkel, 1984); *S. grandiflorus* (L.) Britton and Rose is cultivated in Mexico for an antirheumatic drug, and used in Costa Rica as a heart stimulant (Nuñez, 1975).

Celastraceae (including Hippocrateaceae)

There are several cultivated ornamental *Celastrus* lianas (Mabberley, 1987). *Celastrus paniculatus* Willd. is used in both Burma and the Philippines as an opium antidote and as a stimulant (Perry, 1980). *Hippocratea* scandent shrubs are sometimes used as 'rope-bridges' in Africa (Mabberley, 1987); in Malaysia a root decoction of *H. indica* Willd. is a tonic, an emmenagogue, and is given post partum; in Indonesia the sap is drunk for fever and the leaves are used to treat rheumatism (Perry, 1980). *Hippocratea comosa* Sw. seeds are eaten from the West Indies to Guyana (Kunkel, 1984). Several scandent *Salacia* species have edible fruits (Kunkel, 1984).

Combretaceae

In Madagascar, *Combretum coccineum* Lam. bark is used for fiber (Uphof, 1959). *Quisqualis indica* L. is a widespread ornamental ('Rangoon Creeper': Mabberley, 1987), its stems are used for cordage in the Philippines (Bodner & Gereau, 1988), and in Southeast Asia the dried fruits and seeds are important anthelmintics (Perry, 1980). Quisqualic acid from *Q. indica* is now used in clinical medicine as an anthelmintic (Farnsworth, 1988).

Connaraceae

In Sumatra, Borneo, and the Nicobar Islands, a root decoction of *Rourea mimosoides* Planch. is used for colic (Uphof, 1959). In Southeast Asia, *Agelaea trinervis* (Llanos) Merr. is used as an aphrodisiac and for rheumatism (Perry, 1980); Ayensu (1978) reports that in West Africa an infusion of *A. obliqua* (P. Beauv.) Baill. leaves is administered postpartum, and the fruit is used for teeth cleaning.

Convolvulaceae

This mostly tropical climbing family provides a variety of products. Most important is *Ipomoea*, of which Kunkel (1984) lists at least 16 climbing tropical and subtropical species with edible tubers; there are probably many more, and the leaves of yet others are eaten as pot-herbs. The genus also gives us medicinals, hallucinogens, and ornamentals. *Ipomoea batatas* (L.) Lam. (sweet potato) is the best known of the edible tubers, probably originating in Central America and being found in both Hemispheres in pre-Columbian times. It is now grown throughout the tropics as a subsistence and a cash crop, and in warm temperate regions as a cash crop (Williams, 1981). The annual harvest from *Ipomoea batatas*, 110 million tonnes, represents more than one fifth of the total global root crop; many cultivars exist and wild genes from *I. trifida* have been successfully introduced to provide nematode resistance in Japan (Prescott-Allen & Prescott-Allen, 1988). The Káyapo, who like many indigenous subsistence farmers are important conservers of food plant diversity, maintain at least 16 cultigens of *I. batatas* (Posey, 1983).

Another important food plant is *I. aquatica* ('Chinese cabbage'), cultivated for its edible shoots in Southeast Asia (Uphof, 1959). Medicinal applications include the following. In West Africa *I. pes-caprae* is used for edema and rheumatism, whilst in Indonesia the same species is used for sores, ulcers, and boils, and *I. batatas* for dysentery and diabetes (Perry, 1980; Oliver-Bever, 1986). In Indochina *I. nil* is used as an insecticide and an anthelmintic (Perry, 1980). *Ipomoea batatas* is also used to treat epilepsy and hysteria in the western Ghats of India (Pushpangadan & Atal, 1984). The seeds of *I. purga* are well known in the tropics as purgatives (e.g. Nuñez, 1975; Mabberley, 1987). *Ipomoea tricolor* Cav. (cultivated as several varieties of 'Morning Glory', such as 'Heavenly Blue') from tropical America has ergoline alkaloids in its seeds. This was a sacred plant of the Aztecs, and while the seeds are still used around Oaxaca they have been adopted elsewhere since the 1960s for their LSD-type effects (Emboden, 1972; Mabberley, 1987). Another genus with hallucinogenic seeds is *Rivea* (Albert-Puleo, 1979); Emboden (1972) records *Rivea corymbosa* as another sacred Aztec plant. Conquistadors and priests were unable to stamp out its use, and today in Mexico female shamen ('curanderas') use it as a teacher plant in much the same way as Amazonian shamen use *Banisteriopsis* (see Malphigiaceae).

Argyreia	edible fruits and leaves	SE Asia: Kunkel (1984)
Calonyction	edible tubers	Tropics: Kunkel (1984)
Convolvulus	edible succulent roots	Mexico: Uphof (1959)
Cuscuta	shampoo, purgative	India: Pushpangadan & Atal (1984)
Erycibe	skin rash, speeds labor delivery	SE Asia: Perry (1980)
Ipomoea	*numerous uses and reports – see text*	
Jacquemontia	swellings, syphilis	Tanzania: von Reis & Lipp (1982)
Maripa	edible fruits	French Guiana: Grenand (1980)
Merremia	edible tubers, leaves	SE Asia, India: Kunkel (1984)
	purgative, snake-bite, thrush, diuretic, anthelmintic, *inter alia*	SE Asia: Perry (1980)
Piptostegia	purgative	Brazil: Uphof (1959)
Rivea	hallucinogenic seeds	Mexico: Albert-Puleo (1979)
	hallucinogenic seeds since Aztecs	Mexico: Emboden (1972)

Cucurbitaceae

This predominantly tropical family includes several climbing genera of huge economic significance as food plants. Cucurbits produce cucumbers and gherkins (these two apparently very different vegetables come from the fruits of the same species, *Cucumis sativus* L.), gourds (*Benincasa*, *Cucurbita*, *Lagenaria*, *Trichosanthes*, plus the Bignoniaceous tree *Crescentia*), marrows (*Cucurbita pepo* L.), squashes and pumpkins (*Cucurbita* spp.), and watermelons (*Citrullus lanatus* (Thunb.) Mansf.), as well as a number of less cosmopolitan foods, oils and medicines. The ethnobotanical literature is extensive and I have selected one New World and one Old World genus to illustrate some of the exciting properties of less well known cucurbits.

Fevillea is a small neotropical liana genus widely reported to have purgative and emetic seeds and is used as an antidote for poisoning (Gentry & Wettach, 1986, and references therein, e.g. Ayensu (1981)). Other known applications include the use by the Kuna Indians in northern Colombia of the sap of a *Fevillea* species as a stimulant (Forero Pinto, 1980). However, the most potentially significant property of the genus is the very high oil content of the large seeds, for which Pio Correa (1931) and Gentry & Wettach (1986) report values in excess of 50% of seed weight. Reports of indigenous use of the seeds as candles come from the Campa in central Peru (*F. pedatifolia* (Cogn.) Jeffrey, Gentry & Wettach, 1986) and the Siona in eastern Ecuador (*F. cordatifolia* L., Vickers & Plowman, 1984). Gentry & Wettach (1986) suggest that different *Fevillea* species' seeds could be used commercially to produce fuel oil, edible oils and drying oils, and to treat waxes and rubber; they calculate that if unimproved *Fevillea* was grown in a structurally intact rainforest at the density of all the naturally occuring lianas, the annual yield would be 800 kg oil per hectare – without felling a single tree. This would be equivalent to most figures for seed oil species in plantations.

Momordica, a genus of scramblers from the Old World tropics, has a wide variety of uses. Fruits from cultivated plants are cooked and eaten in both Africa (Uphof, 1959), Southeast Asia (Bodner & Gereau, 1988), and in the New World (Uphof, 1959; Forero Pinto, 1980; Vickers & Plowman, 1984). Medicinal uses abound: in Colombia it is used to treat measles (Forero Pinto, 1980), in West Africa, diabetes (Oliver-Bever, 1986), whilst in Southeast Asia Perry (1980) catalogs an astonishing range of uses. *Momordica charantia* L. is used, for example, for sprue, asthma and skin afflictions in Malaysia, as a laxative and vermifuge in Indonesia, and for dysentery, colitis and children's cough in the Philippines. In Indochina *M. cochinchinensis* (Lour.) Spreng is used as a resolvent of abscesses and mumps, and to treat edema and rheumatism. In Nigeria *M. charantia* leaves are used to treat breast cancer, (Ayensu, 1978, who lists numerous other uses). *Momordica* is also used as a soap substitute and a parasiticide in Southeast Asia (Perry, 1980), and for insecticide in West Africa (Uphof, 1959).

Taking cucurbit vines as a whole, we can conclude that it would be hard to find a community in the tropics for which they aren't a valuable source of food or medicines, or both.

Anisosperma	emetic, anthelmintic	Brazil: Uphof (1959)
Apodanthera	antisyphilictic	Brazil: Uphof (1959)
Benincasa	edible fruits	Tropical Africa, Asia: Uphof (1959)
	bruises, wounds, gonorrhea, hystero-epilepsy	Malaysia: Perry (1980)
	tonic, styptic wounds	Indonesia: Perry (1980)
	respiratory organs, inflamed eyes	Philippines: Perry (1980)
Cephalandra	cures addiction to narcotics	India: Pushpangadan & Atal (1984)
Citrullus	edible fruit (water-melons)	Tropics: Uphof (1959)
	post-abortion bleeding	Malaysia: Perry (1980)
	rheumatism	Indonesia: Perry (1980)
	beri-beri, cystitis	Palau: Perry (1980)
Coccinia	scabies, smallpox, febrifuge	SE Asia: Perry (1980)
	edible fruits	Sudan, SE Asia: Uphof (1959)
	delayed childbirth	Tanzania: von Reis & Lipp (1982)
Cucumis	edible fruit (melons, gherkins)	Pantropical: Uphof (1959)
	edible fruit	Peru: Vasquez & Gentry (1989)
	dysentery, sprue, gallstones	SE Asia: Perry (1980)
Cucurbita	edible fruit (squashes, pumpkins)	Pantropical: Uphof (1959)
	anthelmintic	SE Asia, Palau: Perry (1980)
	febrifuge, gastritis	SE Asia: Perry (1980)
	domestic utensils, burns, neuralgia	W Africa: Oliver-Bever (1986)
Cyclanthera	edible fruit ('pepino')	Peru: Vasquez & Gentry (1989)
	edible fruit	Peru, Bolivia: Uphof (1959)

Elaterium	edible fruit	Central America: Uphof (1959)
Fevillea	*numerous uses and reports – see text*	S America
Hodgsonia	seeds edible and yield cooking oil	SE Asia: Kunkel (1984)
Lagenaria	children's toy	Ecuador: Vickers & Plowman (1984)
	kitchen and musical instruments	Tropical Africa: Uphof (1959)
Luffa	edible fruit, kitchen utensil	Philippines: Bodner & Gereau (1988)
	edible fruit, febrifuge	Sumatra: Elliott & Brimacombe (1985)
	purgative, emetic	SE Asia: Perry (1980)
	edible fruit, kitchen utensils	Tropics: Uphof (1959)
Melothria	edible fruit	Ecuador: Vickers & Plowman (1984)
	edible fruits	Caribbean, South Asia: Kunkel (1984)
	purgative, diuretic, diarrhea, edema, stomachic, itching, edible fruit	SE Asia: Perry (1980)
Momordica	*numerous uses and reports – see text*	Pantropical
Ourania	edible fruit	French Guiana: Grenand (1980)
Polakowskia	edible fruit	Costa Rica: Uphof (1959)
Psiguria	catarrh	Colombia: Forero Pinto (1980)
Sicana	edible fruit	Peru: Vasquez & Gentry (1989)
Telfairia	seed oil for soap and candles	Tropical Africa: Uphof (1959)
Trichosanthes	astringent, tonic, vermifuge	Malaysia: Perry (1980)
	edible fruit	Tropical Asia: Uphof (1959)
	colic, kitchen utensils	India: Pushpangadan & Atal (1984)

Cyclanthaceae

The pendulous aerial roots of *Evodianthus funifer* (Poit.) Lindm. ssp. *funifer* are used by the Ecuadorian Secoya to weave baskets (Vickers & Plowman, 1984).

Dilleniaceae

In Central America the leaves of *Davila kunthii* St., *Tetracera sessiliflora* Tr. and Pl., and *T. volubilis* L. are all used as sandpaper (Williams, 1981). Silica bodies in their leaves make them rough, and probably helps these non-twining lianas to climb. In Africa the leaves of three *Tetracera* species are eaten (Kunkel, 1984), and *T. alnifolia* Willd. is used in Ghana to treat gonorrhea, and children's rickets (Ayensu, 1978). In Borneo the Iban use leaves from *Tetracera* lianas to file their teeth (F. Putz, personal communication). Several *Tetracera* species are also used in herbal medicine in Southeast Asia. For example, *T. scandens* (L.) Merr. is used for boils, snake-bites and dysentery, and is given both as an emmenagogue and postpartum (Perry, 1980).

Dioscoreaceae

Dioscorea is a pantropical genus, producing annual twining stems from tubers that provide us with important foods and steroids.

Most widely grown for their edible tubers are *D. alata* L. (tubers up to 50 kg), *D. esculenta* (Lour) Burkill, *D. trifida* L., *D. batatas* Decne, and *D. cayenensis* Lam. (Mabberley, 1987). Yams have been independently domesticated around the world, and are cultivated, for example, by the Bontoc in the Philippines (Bodner & Gereau, 1988), and the Siona and Waorani in Ecuador (Davis & Yost, 1983; Vickers & Plowman, 1984). Kunkel (1984), Uphof (1959), von Reis Altschul (1975), and von Reis & Lipp (1982) list a total of 92 mostly tropical and 18 mostly subtropical/warm temperate species with edible tubers, many of which need to be treated first to rid them of toxic alkaloids. One Amazonian tribe, the Káyapo, maintain 17 distinct varieties of yams (Posey, 1983). Even a group like the Bomagu-Angoiang in New Guinea, for whom yams are not an important food source, grow five different species (Clarke, 1971).

Yams are the pre-eminent source of steroidal sapogenins – used as starting points in the synthesis of cortisone, hydrocortisone, androgens, estrogens, progestins and oral contraceptives (Myers, 1983). One hundred and eighty tons of diosgenin a year was harvested from Mexico in the mid-1970s (much more now), mostly from wild *D. composita* Hemsl and *D. floribunda* Mar. and

Gal., and retail sales of the final products totalled over US $1 billion by the early 1980s (Morton, 1977; Myers, 1983). Many other *Dioscorea* species are candidates for novel sources of diosgenin (Morton, 1977; Myers, 1983), and not surprisingly the genus features strongly in native pharmacopeias (see Perry, 1980, for a complete listing for East Asia). For example, *D. alata* is used for digestive problems and as a febrifuge in Costa Rica, and in Indochina to speed recovery from kidney and spleen disorders (Nuñez, 1975; Perry, 1980). *Dioscorea esculenta* and *D. persimilis* are also used in Southeast Asia to treat kidney troubles, while *D. hispida* Dennst. is used throughout its tropical Asia range for syphilictic sores, and for arthritic and rheumatic pain (Perry, 1980).

Euphorbiaceae

This is a family that is rich in secondary compounds, but poor in lianas. In tropical Africa *Plukenetia conophora* Muell. Arg. is cultivated for a drying oil (Uphof, 1959). In Peru, *P. volubilis* ('wild peanuts') are sold in markets to be roasted and eaten (Vasquez & Gentry, 1989). In Zaïre *Manniophyton africanus* Muell. Arg. provides fiber for ropes and fishing nets (Uphof, 1959).

Glagellariaceae

Flagellaria indica L. stems are used in Thailand and Peninsular Malaysia for basket making and fish traps, and the young leaves as shampoo (Uphof, 1959; Mabberley, 1987).

Gnetaceae

Climbing *Gnetum* species are distributed pantropically. The seed and/or fruit of at least 13 species is eaten in all three equatorial continents. (Uphof, 1959; von Reis & Lipp, 1982; Kunkel, 1984; FAO, 1984, 1986; Vasquez & Gentry, 1989). Frequently the seeds are roasted or ground into flour before eating. In the Philippines *G. indicum* (Lour.) Merr. seeds are eaten as sweets by cooking them with brown sugar (FAO, 1984).

Icacinaceae

The tubers of scandent *Humiranthera* species (*H. amplu* (Miers) Baehni, *H. duckei* Huber, and *H. rupestris* Ducke) are eaten in tropical South America after treatment (Kunkel, 1984). In Africa, the tubers, fruits and seeds of *Icacina senegalensis* Juss and *I. trichantha* Oliv. are edible (Kunkel, 1984).

Leguminosae

Legumes provide humankind with a huge variety of important products, ranging from vital sources of vegetable protein, through fibers, medicines and insecticides, to the poisons used by traditional cultures to capture animal protein. Useful climbing taxa are most prevalent in the Papilionoideae, from which we get various beans (in *Phaseolus* and *Vigna*). Other uses of legumes include as arrow poisons, insecticides, fiber, and in many traditional medicines. I will concentrate on the use of *Derris* and *Lonchocarpus* as fish poisons and insecticides.

Derris consists of 80 tropical vine species. Three are widely cultivated for rotenones (*D. elliptica* (Wallich) Benth., *D. malaccensis* Prain, and *D. ulignosa* Benth.). Rotenone is exported as a natural insecticide; because it is relatively harmless to mammals, biodegradable, and effective (toxic to fish at only 0.05 ppm: Oliver-Bever, 1986), its future seems assured in a world increasingly concerned about synthetic pesticides. *Lonchocarpus* from South America is also harvested for rotenones, and it has been a commercially valuable source since 1932 (Krukoff & Smith, 1937). Both genera are used as indigenous fish poisons. *Derris* use is reported from tropical Asia, the Philippines, Micronesia, and Brazil (Krukoff & Smith, 1937: Uphof, 1959; Prance, Campbell & Nelson, 1977; von Reis & Lipp, 1982; Rickard & Cox, 1986). *Lonchocarpus* is used by South American Indians, who dam streams before soaking the leaves and stems of the 'barbasco' in the trapped water, temporarily paralyzing the fish, which float to the surface (Davis & Yost, 1983; Balick, 1984). Fish which are not collected usually recover, so ensuring that the river ecosystem is less seriously perturbed, in contrast to the commercial method of mass-dynamiting, which has already wiped out many local Amazonian fisheries (Balick, 1984; O. Phillips, personal observation). However, in Southeast Asia excessive use of *Derris* has seriously damaged many rivers (F. Putz, personal communication). Krukoff & Smith (1937) also reported that root extract of the neotropical *Lonchocarpus rariflorus*. Mart. is used by tribal people to exterminate 'Sauba' (*Atta* spp.) – leaf-cutter ants – which are a major agricultural pest in the region, so pointing to the potential for safe use of barbasco on Amazonian land as well as in Amazonian water.

Caesalpinioideae

Bauhinia	febrifuge	Colombia: Forero Pinto (1980)
	dysentery, venereal diseases, astringent, vermifuge, expectorant	Colombia: Garcia-Barriga (1974)

	fish poison	Guyana: von Reis & Lipp (1982)
	diarrhea	French Guiana: Grenand (1980)
Caesalpinia	tonic, antiperiodic, colic	Tropics: Uphof (1959)
	red dye	Africa: Uphof (1959)
Mimosoideae		
Entada	astringent, snake-bites	Costa Rica: Nuñez (1975)
	hair treatment	Central America: Williams (1981)
	hair treatment	S Asia: Uphof (1959)
	soap	French Guiana: Grenand (1980)
Mimosa	insomnia	Colombia: Forero Pinto (1980)
Papilionoideae		
Abrus	conjunctivitis	Costa Rica: Nuñez (1975)
Canavalia	human and livestock food	Pantropical: Uphof (1959)
Dalbergia	joss-sticks	Malaysia: Mabberley (1987)
Derris	fish poison (*many reports – see text*)	SE Asia, Melanesia, Brazil
Dolichos	edible pods	Central America: Williams (1981)
	human and livestock food	Tropical Asia: Uphof (1959)
	insecticide	Africa: Uphof (1959)
Lonchocarpus	fish poison (*many reports – see text*)	S America
	jaundice, tonic, skin afflictions, venereal disease	W Africa: Oliver-Bever (1986)
	indigo dye	W Africa: Uphof (1959)
Milletia	stuns fish	Malaysia: Uphof (1959)
Mucuna	twine, edible seeds, fruits as toys	Philippines: Bodner & Gereau (1988)

	hemorrhoids, intestinal parasites	Costa Rica: Nuñez (1975)
	black dye, edible seeds	Central America: Williams (1981)
	L-dopa extracted (Parkinson's Disease)	Farnsworth (1988)
Pachyrhizus	edible tubers since pre-Columbian times	New World: Mabberley (1987)
	root eaten	Philippines: Bodner & Gereau (1988)
Phaseolus	edible beans (many varieties)	Pantropical: Uphof (1959)
	edible beans	Ecuador: Vickers & Plowman (1984)
Physostigma	physostigmine = poison antidote, glaucoma, tetanus *inter alia*	Europe and USA: Morton (1977)
	edema, rheumatism, skin diseases	West Africa: Morton (1977)
Pueraria	(tropical kudzu) fodder and cover crop	Tropics: Ocana *et al.* (1988)
Sphaenostylis	edible seeds and tubers	West Africa: Uphof (1959)
Vigna	edible seeds, pods (many varieties)	Pantropical: Uphof (1959)
	edible seeds, pods, stems, leaves, fiber for weaving	Philippines: Bodner & Gereau (1988)

Liliaceae

Gloriosa superba L. is an ornamental ('Flame-Lily': Mabberley, 1987). In the western Ghats of India its tubers are crushed into a paste, heated slightly, and applied to the abdomen and vulva as an aid to childbirth; they are also used to induce abortion (Pushpangadan & Atal, 1984).

Loganiaceae

Strychnos predominates in ethnobotanical reports and is represented in the Neotropics by about 70 climbing species. Many of these are used as ingredients in native arrow poisons (see Menispermaceae for a discussion of curare). The genus has an extensive range of toxic alkaloids (see Krukoff,

1965; Ohiri, Verpoorte & Baerheim Svendsen, 1983; Oliver-Bever, 1986: pp. 68–70). Krukoff (1965) reviews the 18 species known to be used as ingredients in curare. Since then at least one species can be added to the list (*S. diaboli*, used by the Mayongong Indians in Brazil (von Reis & Lipp, 1982), and Prance *et al.* (1977) found that *S. solimoesana* was used by the Paumarí near the Rio Purus, far from the previous Brazilian report of its use in the Rio Tocantins basin (Krukoff, 1965). The sole reported medical use from the Neotropics is for bee bites, by the Guaymi in western Panama (Joly *et al.*, 1987).

Loranthaceae

In the Philippines the Bontoc eat *Helixanthera parasitica* Lour fruit (Bodner & Gereau, 1988).

Malpighiaceae

This large family, centred in the Neotropics, includes a number of liana genera. Ethnobotanical interest centers on *Banisteriopsis caapi* (Spruce ex Griseb) Morton, which is used in a culturally important hallucinogenic drink by Indians and mestizos throughout western Amazonia.

'Ayahuasca' (Peru), 'Caapi' (Brazil), and 'Yajé' (Colombia) are the commonest names given to the hallucinogenic drink prepared from cultivated and wild *Banisteriopsis caapi*, the bark of which contains three beta-carboline alkaloids (Schultes, 1978). Schultes (1978) lists the following confamilial lianas which are also used to prepare ayahuasca: *B. cabrerana* Cuatr., *B. martiana* (Juss.) Cuatr. var. *laevis* Cuatr., *B. quitensis* Ndz., *Mascagnia glandulifera* Cuatr., *Tetrapteris methystica* Schultes, *T. mucronata* Cav. One or more additives are needed for maximum effect; most widely used are two malpighs (*Banisteriopsis rusbyana* and *Diplopterys cabrerana* (Cuatr.) Gates), and a rubiaceous shrub (*Psychotria viridis* R. et P.). All three contain dimethyltrytamines (DMT), which probably react synergistically with β-carbolines (Lewis & Elvin-Lewis, 1977; Ayala Flores & Lewis, 1978; Schultes & Hoffmann, 1980). It remains a major ethnobotanical mystery as to how indigenous people selected these essential additives from the enormous Amazonian flora. (For a detailed chemical investigation of ayahuasca see Rivier & Lindgren, 1972.)

The use of ayahuasca is a fundamental feature of many Amazonian cultures. It is used by shamen to communicate with the spirit world in order to diagnose and treat illness. These beliefs are very powerful, often surviving in urbanized cultures which have some access to western medicine (Dobkin de Rios, 1972; Luna, 1984). Naranjo (1979) has published a comprehensive review of the belief systems connected to ayahuasca use among tribal groups in Ecuadorean Amazonia. The drug is used by a shaman when his patient's illness cannot be treated by herbal and/or clinical preparation, and when the origin of the illness may be magical. Some conditions treated by shamen would be classified as psychological by Western medicine, and indeed, several authors have pointed out the potentially valuable role hallucinogens such as ayahuasca could play in Western psychotherapy (e.g. Myers, 1984).

Ayahuasca use is not restricted to therapy. For example, Sharanahua and Culina men take the drink socially in order to have visions (Rivier & Lindgren, 1972). Well-practised shamen claim to attain an exceptional level of perception – telepathy, allowing them to communicate with absent families and friends, whether living or dead (Ayala Flores & Lewis, 1978). In times of individual or group uncertainty Indians may be likely to use the drug more frequently in a social context, to affirm shared traditional values and thus perhaps resist acculturation. Emboden (1972) also reports that the drug may also be taken for its aphrodisiac effects. It should be emphasized that not all traditional uses of ayahuasca are 'benign'. Many tribes see most illnesses as a consequence of witchcraft. For example, among the Shuar (Jívaro) Indians of Ecuador, apprentice shamen are taught by the ayahuasca drink how to cast spells, as well as how to rid patients of evil spirits (Naranjo, 1979). According to Davis & Yost (1983) the neighboring Waorani, who use *Banisteriopsis muricata* to prepare their hallucinogen, consider the use of this drug to be an aggressive act.

Further evidence for the fundamental role of ayahuasca is the large number of varieties of *B. caapi* that Indians are able to distinguish. As Schultes (1986) has discussed, this ability is somewhat of a mystery, since varieties consistently distinguished by Indians are often indistinguishable to botanists in the field. This suggests that taxonomically we may have much to learn from native people like the Tukanoa in Colombian Amazonia, under whose guidance Langdon collected 18 named varieties of *B. caapi* (Schultes, 1986).

Banisteriopsis	hallucinogenic drink: *numerous reports – see text*	
	headache, fevers,	Panama: Joly *et al.* (1987)
Diplopterys	additive to hallucinogen	Ecuador: Vickers & Plowman (1984)
Heteropteris	gonorrhea, diarrhea	Colombia: Schultes (1975)
Hiraea	conjunctivitis	Colombia: Schultes (1975)

Mascagnia	boils and infections	Colombia: Schultes (1975)
	hallucinogenic drink	NW Amazonia: Schultes (1978)
Mexia	diuretic, laxative, emetic	Colombia: Schultes (1975)
Stigmaphyllon	suppresses vomiting	French Guiana: Grenand (1980)
Tetrapteris	weak curare by boiling bark with *Strychnos* bark	Colombia: Schultes (1975)
	hallucinogenic drink	Colombia: Schultes (1975)
	hallucinogenic drink	Brazil: Schultes (1975)
	skin infections	Colombia: Schultes (1975)

Marcgraviaceae

This family is of minor ethnobotanical significance. Joly *et al.* (1987) recorded the use of *Marcgravia nepenthoides* Seemann by the Guaymi in Panama for treatment of headaches after bad dreams; the same group use a climbing *Souroubea* species as an antidiarrhoeic. In Colombia the Waunana use the climbing *Marcgravia myriostigma* Tr. and Pl. to treat headaches (Forero Pinto, 1980). The Waorani eat the fruit of an unidentified climbing *Marcgravia* in Ecuador (Davis & Yost, 1983).

Melastomataceae

In Indonesia, stem sap from *Macrolenes mucosa* (Bl.) Backh is applied to infected eyes; also, the young shoots are eaten with rice and the berries are made into jellies, preserves and drinks (Uphof, 1959; Kunkel, 1984). Young leaves of *Medinilla hasseltii* Bl. are eaten in Sumatra with rice, fruits, and fish (Uphof, 1959). A *Medinilla* species is used in New Guinea for lashing fences together (Clarke, 1971).

Menispermaceae

This large pantropical family is rich in medicinal and toxic lianas; the best known uses are for the South American arrow poison 'curare' and its extracted chemical components, a story which illustrates the heavy dependence of Western medicine on the traditional lore of indigenous rainforest peoples (see McIntyre, 1947).

Chondrodendron tomentosum Ruiz et Pavon is the sole source of D-tubocurarine, or curare, a muscle relaxant that is essential in surgery and not yet (1982) synthesized artificially (Morton, 1977; Mabberley, 1987); the lianas are still all wild-harvested (A. Gentry, personal communication). In Amazonia this species is one of many menispermaceous and loganiaceous (*Strychnos*) taxa that various indigenous groups use to prepare curare, but the Jívaro in western Amazonia are credited with giving us the particular benefits of *C. tomentosum* (Lewis *et al.*, 1987). The deadly arrow poison was noted as long ago as 1541 when a Spanish explorer Francisco de Orellana, lost a companion who was shot with a curare-tipped arrow. He later wrote, 'The arrow did not penetrate half a finger, but as it has poison on it, he gave up his soul to our Lord' (quoted in Kreig, 1965). Exactly how the poison works remained a mystery until 1842, when Claude Bernard discovered that it has a blocking effect on the neurotransmitters to the muscles, which then relax completely. When the respiratory muscles relax death soon occurs by asphyxiation. Yet it was not until 1935 that curare could be used medically, when Harold King isolated the chief paralyzing ingredient from curare, D-tubocurarine.

Other menisperms reported as curare ingredients include: in Peru *Curarea toxicofera* and *Abuta rufescens* (Ayala Flores, 1984), in Colombia three *Abuta* species (Schultes, 1979), in Ecuador *Curarea tecunarum* (Davis & Yost, 1983), and in Brazil *Curarea toxicofera* (Krukoff & Barneby, 1970; Prance, 1972) and *Telitoxicum minutiflorum* (Uphof, 1959). Krukoff & Barneby (1970) list a total of 11 menisperm liana and treelet species from four genera that are used to prepare curare. Curare is often prepared from more than one species. The Waorani, neighbors of the Jívaro, extract the ingredients by percolating water through a leaf compress of liana bark, and boiling the dark liquid until it thickens. After cooling and reheating, the viscous surface scum is removed and applied to the tip of the dart or arrow (Davis & Yost, 1983). In Brazil, the Jamamadi treat in a similar fashion the combined barks from *Curarea*, *Strychnos* (Loganiaceae), *Guatteria* (Annonaceae), and *Zanthoxylon* (Rutaceae), to get a deadly sticky residue (Prance, 1972). Also in Brazil, both the men and women of the Dení Indians drink a decoction from *Curarea tecunarum* Krukoff and Barneby as a long-term contraceptive (Prance, 1972).

An altogether different use exists for the West African *Dioscoreophyllum cumminsii* (Stapf) Diels (Oliver-Bever, 1986). The most potent sweetening agent known is the protein 'monellin', found in the berries of this high forest climber; monellin is 9000 times as sweet as sucrose on a molecular weight basis. Being so sweet, it can be used in minute concentrations, and preliminary trials have confirmed its potential in low calorie drinks and food (Oliver-Bever, 1986).

Abuta	edible fruits	Brazil: von Reis & Lipp (1982)
	edible fruits	Colombia: von Reis Altschul (1975)
	malaria remedy	Colombia: von Reis Altschul (1975)
	malaria remedy	Peru: Phillips & Wilkin (unpublished)
	anaesthetic for toothache	French Guiana: Grenand (1980)
	cold remedy	Venezuela, Colombia, Ecuador: von Reis & Lipp (1982)
	arrow poison: *Curare – see text*	Brazil: Krukoff & Moldenke (1938)
		Peru: Schultes (1977)
		Colombia: von Reis & Lipp (1982)
Anamirta	fish poison	Philippines: von Reis Altschul (1975)
	fish poison, parasiticide	SE Asia: Uphof (1959)
	skin parasites, relieves malaria	India: Morton (1977)
	picrotoxin extracted – various clinical uses, e.g. epilepsy	Morton (1977)
Chondrodendron	snake-bite	Brazil: Grieve (1967)
	curare arrow poison: *many reports – see text*	
	D-tubocurarine extracted: *see text*	Lewis & Elvin-Lewis (1977)
Cissampelos	fish bait	Brazil: Posey (1984)
	eye diseases	Peru: von Reis & Lipp (1982)
	snake-bite	Colombia: von Reis & Lipp (1982)
	snake-bite	Panama: Joly *et al.* (1987)
	venereal diseases	Brazil: von Reis & Lipp (1982)
	febrifuge, diarrhea, colic	Tanzania: von Reis & Lipp (1982)

	astringent, tonic, febrifuge	Brazil: Uphof (1959)
	relieves menorrhagia	Tropics: Uphof (1959)
	prevents threatened abortion	Tropics: Uphof (1959)
	urinalysis	Panama: von Reis & Lipp (1982)
	febrifuge, emmenagogue, abortifacient, diuretic	W Africa: Oliver-Bever (1986)
	venereal diseases, diuretic	Costa Rica: Nuñez (1975)
Cocculus	snake- and insect-bites, skin allergies	India: Pushpangadan & Atal (1984)
	snake-bites	Brazil Uphof (1959)
	febrifuge	Senegal: Uphof (1959)
	promotes menstruation	Brazil: Lewis & Elvin-Lewis (1977)
Coscinum	yellow dye, tonic	India: Lewis & Elvin-Lewis (1977)
Curarea	contraceptive	Brazil: Prance (1972)
	arrow poison: *Curare – see text – many reports*	
Dioscoreophyllum	potent sweetener: *see text*	W Africa: Oliver-Bever (1986)
Jateorhiza	tonic, diarrhea, dysentery, colic	Tropical Africa: Uphof (1959)
	burns, snake-bites	Tropical Africa: Oliver-Bever (1986)
	depressant and stimulant alkaloids	clinical & veterinary medicine: Oliver-Bever (1986)
Kolobopetalum	insomnia	W Africa: Oliver-Bever (1986)
Pycnarrhena	tonic, cicatrizant, snake-bites	Philippines: Oliver-Bever (1986)
Sciadotenia	toothache	Brazil: von Reis & Lipp (1982)
	arrow poison: *Curare – see text*	Peru, Ecuador: von Reis & Lipp (1982)
	hunting charm	French Guiana: Grenand (1980)
Sphaenocentrum	aphrodisiac, cough, wounds	W Africa: Oliver-Bever (1986)

Stephania	anthelmintic, sedative	W Africa: Oliver-Bever (1986)
Telitoxicum	arrow poison: *Curare – see text*	Brazil: Uphof (1959)
Tinospora	febrifuge	India, Senegal: Oliver-Bever (1986)
	febrifuge, tetanus and jaundice	Indo-China: Perry (1980)
Triclisia	anemia, leg edema	W Africa: Oliver-Bever (1986)

Myrsinaceae

Embelia philippensis A.DC. provides cordage, and edible fruit and leaves in the Philippines (FAO, 1984). In India and Malaysia, *E. ribes* Burm. dried fruits are used to adulterate black pepper, and as a stomachic, tonic, astringent and anthelmintic (Uphof, 1959). The leaves of several African and Asian species of *Embelia* are eaten (Kunkel, 1984).

Nepenthaceae

Several climbing *Nepenthes* species are used to make ropes, baskets, etc., for example *N. distillatoria* L. (Sri Lanka) and *N. reinwardtiana* Miq. (Southeast Asia) (Uphof, 1959).

Oleaceae

Jasminum includes some cultivated ornamental scandent shrubs, for example *J. sambac* Ait., 'Arabian Jasmine', the flowers of which are used to scent tea (Uphof, 1959).

Orchidaceae

Although highly speciose, Orchids are mostly of minor economic importance. The principal exceptions are the use of vanilla (from the liana genus *Vanilla*) as a flavoring, and the several tropical epiphytic genera that are cultivated for the lucrative horticultural trade. Vanillin was synthesized in 1874, but the pods of the Central American *Vanilla planifolia* Jackson are still used as flavorings and this originally neotropical species is widely cultivated (Mabberley, 1987). Vanilla extract is obtained by macerating the cured 'beans' (immature fruits) in alcohol (Purseglove *et al.*, 1981). Vanilla is used to flavor a wide variety of confectionery, especially ice-cream. Most cultivation today is in the Old World, where the plants grow well, although the initial trials in Java, Réunion and Mauritius were unsuccessful because of the absence of natural pollinators. Commercial production in the eastern hemisphere became possible in 1841 when Edmond Albius, a former Réunion slave, developed a practical method of artificial pollination that is still used today (Purseglove *et al.*, 1981). Former French possessions in the Indian Ocean, especially the Malagasy Republic, and in the West Indies, are now the major vanilla producers. Mexico exports some of the spice (Purseglove *et al.*, 1981). Other New World species with mostly local importance in flavoring are *V. gardneri* Rolf., *V. guianensis* Reich, *V. pitheri* Schitr. and *V. pompona* Schiede (Williams, 1981; Mabberley, 1987).

Pandanaceae

The young inflorescences of *Freycinetia funicularis* Merr are boiled and eaten in the Moluccas (Uphof, 1959).

Passifloraceae

The family is dominated numerically and ethnobotanically by *Passiflora*. Of the 350 New World and 20 Old World species in this genus of lianas and herbaceous vines I found references to the fruit of at least 56 being edible – whether raw, or as fruit juice, or in ice-cream (Uphof, 1959; Nuñez, 1975; Forero Pinto, 1980; Williams, 1981; von Reis & Lipp, 1982; Kunkel, 1984; Vickers & Plowman, 1984; Bodner & Gereau, 1988; Vasquez & Gentry, 1989). Most of the edible species are found in Amazonia and Central America. At present only a very few are grown on a subsistence or commercial basis elsewhere (*P. edulis* Sims, *P. foetida* L., *P. laurifolia* L., and *P. quadrangularis* L. are now the most widespread, although many less well known species such as *P. ligularis* Juss, and *P. mollissima* (Kunth) L. Bailey are reported to have superior flavors (Williams, 1981; Mabberley, 1987)). Still more species are prized for their spectacular flowers. Their preference for secondary habitats should make more *Passiflora* species suitable for widespread cultivation, and possibly for reclamation of deforested areas.

The complex phytochemistry of *Passiflora* is thought to be the result of a co-evolutionary arms race with heliconiid butterflies, whose larvae eat the leaves of these vines (Gilbert, 1975). This has presumably helped to make *Passiflora* an important medicinal genus. Particularly noteworthy is the use for snake-bites of three different species in Panama (Joly *et al.*, 1987), reports of anti-inflammatory activity from Colombia (Glenboski, 1975) and from Panama (Joly *et al.*, 1987), and three reports of wound healing: in the Bahamas (Eldredge, 1975), in Southeast Asia (Perry, 1980), and in El Salvador (von Reis Altschul, 1975).

Adenia	fish poison, lumbago	Nigeria, Ghana: Oliver-Bever (1986)
	fish poison	Guinea: Uphof (1959)
	to produce amnesia	Liberia: Oliver-Bever (1986)
	headache, ringworm, conjunctivitis	SE Asia: Perry (1980)
	headache, snake-bite	Tanzania: von Reis & Lipp (1982)
	stomach trouble	Philippines: von Reis Altschul (1975)
Passiflora	edible fruits: *numerous reports – see text*	
	snake-bite, hypertension	Panama: Joly *et al.* (1987)
	anti-inflammatory, diabetes	Panama: Joly *et al.* (1987)
	aphrodisiac, emmenagogue, hysteria	Costa Rica: Nuñez (1975)
	bone breakages	Colombia: Forero Pinto (1980)
	cuts	Bahamas: Eldredge (1975)
	eye inflammation	Colombia: Glenboski (1975)
	itching, wounds	SE Asia: Perry (1980)
	wounds, swellings, gonorrhea, purge	El Salvador: von Reis Altschul (1975)

Piperaceae

Climbing Asian *Piper* species have two principal uses. *Piper betle* L. is cultivated widely in Southeast Asia, and used as a masticatory. *P. nigrum* L. provides the black and white peppers of commerce.

Piper betle leaves are used by covering them with lime and 'cutch' (wood from *Acacia catechu* Willd.), and then placing slices of the seeds of the palm *Areca catechu* L. on the leaf and adding flavorings (e.g. gambir, see Rubiaceae: *Uncaria*). This mixture, or quid, is chewed for its mild stimulant and euphoric effect. An incidental effect is that the dye from the gambir darkens the teeth (Uphof, 1959). The leaves are also used in traditional medicine for cleaning wounds, and as a stomachic and astringent. In Sumatra crushed leaves are used in a pessary (vaginal suppository) by mothers after childbirth; possibly this serves as a protection against infection, since the leaves' essential oil contains numerous bacteriostatic phenols (Perry, 1980; Elliott & Brimacombe, 1985).

Piper nigrum is cultivated in Madagascar, Sri Lanka, and parts of Southeast Asia which together produce more than 20 000 tonnes per annum. Within the last 20 years there has been a spectacular increase in the Brazilian industry. The ground dried fruits produce black pepper; with the fleshy fruit coating removed they yield white pepper (Uphof, 1959). Pepper is today the most consumed spice in the world (Mabberley, 1987), and excavations in the Indus Valley – hundreds of miles from its place of origin in the western Ghats – have shown that its use as a spice dates back to before 1000 BC (Purseglove *et al.*, 1981). Three thousand pounds of pepper was part of a ransom demanded of Rome in AD 408 by the King of the Goths to prevent the sacking of the city. Unfortunately for the Romans, the ransom only bought the city two more years before it fell to the Goths in AD 410, precipitating the collapse of the western Roman Empire (Purseglove *et al.*, 1981). The pepper trade assumed an even greater importance in the Middle Ages and the Renaissance, helping make the fortunes of the merchants who brought back the rare spice from Southeast Asia to the European markets, where it was in demand as a flavoring for salted or ageing meat. 'Oil of pepper' from the seeds is also widely used as a food flavoring, and medicinally in Southeast Asia as a stimulant, febrifuge, tonic, irritant and abortifacient (Uphof, 1959, Perry, 1980). Another spice, 'long pepper', comes from cultivated *P. chaba* Hunter, *P. longum* L. and *P. retrofractum* Vahl.; furthermore, *P. cubeba* L. is cultivated in Java and Singapore for 'cubeb berries', used to flavor cigarettes, food, bitters and throat lozenges (Uphof, 1959).

Polypodiaceae

In Southeast Asia, leaves from the climbing fern *Stenochlaena palustris* (L.) Mett are used for braiding, anchor ropes and fish traps, and in Java the young leaves are a delicacy (Uphof, 1959). The Secoya, in Ecuador, use the rough leaves of *Lomariopsis japurensis* (Mart.) J. Sm. as a cloth for hand-scrubbing (Vickers & Plowman, 1984); in southern Peru an infusion from the climbing rhizome of the same species is drunk as a diarrhea cure (O. Phillips and P. Wilkin, unpublished data).

Ranunculaceae

Clematis lianas have various medicinal uses in Southeast Asia (Perry, 1980). *Clematis gouriana* Roxb. and *C. chinensis* DC. are both used as diuretics in Indochina, as is *C. dioica* L. in Costa Rica (Nuñez, 1975; Perry, 1980). Another important use for *Clematis* is for cordage in Central America

(*C. dioica* and *C. grossa* L.: Williams, 1981) and in the Philippines (*C. javana* DC.: Bodner & Gereau, 1988). *Clematis* spp. are also important in the horticultural trade.

Rhamnaceae

On both sides of the Pacific *Gouania* roots and shoots are crushed to yield a soapy liquid: in Malaysia *G. leptostachya* DC. is used to wash hair, and in the Philippines roots of *G. tilaefolia* yield soap (Uphof, 1959); in tropical America chewing sticks of *G. lupuloides* (L.) Urban give a soapy mouthwash (Mabberley, 1987). In tropical Asia *Ventilago maderaspatana* Gaertner is harvested for its reddish root dye ('ventilagin'), which is used to color wool, cotton and silk (Uphof, 1959), and in India *V. calyculata* Tul. seeds are used for a cooking oil (Kunkel, 1984). Finally, the Bontoc in the Philippines use the stems of *Berchemia philippinensis* S. Vidal for jar casings and the rims of winnowing baskets (Bodner & Gereau, 1988).

Rosaceae

Some scandent *Rubus* species have edible fruits, for example *R. elmeri* Focke in the Philippines (Uphof, 1959). In New Guinea, *R. moluccanus* L. is used as rope (Clarke, 1971).

Rubiaceae

The alkaloid-rich family that has given us coffee, quinine and ipecac has few useful climbers. An important exception is *Uncaria gambir* Roxb., which is cultivated in its native Southeast Asia. Here, a solidified leaf extract called 'gambir' is used as an ingredient in betel quids (see Piperaceae: *Piper betle*). Gambir is also used in a wide range of local medicinal preparations (Perry, 1980; Elliott & Brimacombe, 1985). In Sumatra the most common use of gambir is in remedies for various disorders of the alimentary canal, as well as for urinary tract problems (Elliott & Brimacombe, 1985). In Indonesia gambir is an important astringent (van Steenis-Kruseman, 1953); similarly, in India an infusion of fresh leaves is given to treat diarrhea and dysentery (Perry, 1980). Gambir is also used industrially in southern Asia as a tanning agent and for applications in dyeing, printing, and clearing beer (Uphof, 1959). Another useful vine in this genus is *U. africana* G. Don., which is used to treat coughs in West Africa, and syphilis and stomach pains in Zaïre (Oliver-Bever, 1986). In Peru, the Jívaro and other indigenous groups use two climbing *Manettia* species for blackening their teeth by chewing the leaves, a habit which is likely to account for the excellent dental health of those native people who have not abandoned the habit under the pressure of disapproval by the dominant culture (Lewis & Elvin-Lewis, 1984).

Lecontea	black dye	Madagascar: Uphof (1959)
Manettia	febrifuge	French Guiana: Grenand (1980)
Paederia	prodtitis, gout, dysentery, flatulence, eye infections, toothache, emetic, herpes, ringworm, malaria	Philippines: Perry (1980)
	antirheumatic, digestive, stomachic	SE Asia: Perry (1980)
	tonic, anti-inflammatory, febrifuge	Indo-China: Perry (1980)
Sabicea	dysentery	French Guiana: Grenand (1980)
Schradera	blacken teeth	Colombia: Uphof (1959)
Uncaria	(*gambir extracted – see text*)	SE Asia: many reports

Sapindaceae

Two large neotropical climbing genera in this family are widely used as fish poisons ('barbasco', see also Leguminosae: *Derris*, *Lonchocarpus*). References to this use exist for seven identified *Serjania* species – from Mexico through Peru and Brazil the stems are beaten into the water and release the serjanosides which are toxic to fish (Gottlieb, 1984) – and for four *Paullinia* species from Central America to Peru (von Reis Altschul, 1975; Williams, 1981; von Reis & Lipp, 1982).

Two *Paullinia* species are also used to prepare important stimulating beverages. *Paulinia yoco* Schultes and Killip provides a caffeine-rich drink, 'Yoco', drunk predawn by Indians in northwest Amazonia (Vickers & Plowman, 1984). According to Schultes (1942) this is the most important non-alimentary plant in the economy of the natives of the Putomayo river. Seeds from *P. cupana* HBK and *P. cupana* var. *sorbitus* (Mart.) Ducke have a caffeine content 3–5 times that of coffee, and make an Amazonian beverage called 'guaraná' which is consumed either hot or cold (Balick, 1984). Carbonated and sweetened, guaraná is a popular soft drink throughout Brazil and has caught on in other Latin American countries and in the United States, so that it is becoming an important crop for Brazil. Traditional uses of guaraná in Amazonia include as a tonic, analgesic, and aphrodisiac (Balick, 1984).

Cardiospermum	diuretic, emetic, purgative, buboes, sore eyes, aperient, rheumatism	SE Asia: Perry (1980)
	ornamental	Neotropics: Mabberley (1987)
Paullinia	fish poison in Neotropics: *see text*	von Reis Altschul, 1975, Williams, 1981
	stimulating beverage: *see text* – numerous reports	
	edible aril	Neotropics: von Reis Altschul (1975) Central America: Williams (1981) Ecuador: Vickers & Plowman (1984) French Guiana: Grenand (1980)
	emetic	Colombia: Schultes (1977)
	haemostatic, dysentery, febrifuge	W Africa: Oliver-Bever (1986)
Serjania	fish poison in Neotropics: *see text*	von Reis Altschul, 1975, Williams, 1981, von Reis & Lipp, 1982, Gottlieb, 1984
	malaria	Mexico: von Reis Altschul (1975)
	vaginal wash	Peru: von Reis & Lipp (1982)
	weave coffee baskets	Costa Rica: Williams (1981)
	skin, purify blood, postpartum tonic	W Indies: Ayensu (1981)

Schizaceae

The stems of the climbing fern genus *Lygodium* are used for basket work in Southeast Asia (*L. circinatum* Sw., *L. scandens* Sw.), and in the Philippines *L. scandens* stems are used to make hats and to store betel-nuts (Uphof, 1959).

Smilacaceae

Smilax supplies the sarsaparilla of commerce, as well as many local medicinal uses. Sarsaparilla is a condiment from the dried rhizomes of some neotropical species (*S. aristolochiaefolia* Mill., *S. ornata* Lam., *S. regelii* Killip and Morton, *S. spruceana* A. DC.). Many other *Smilax* species provide edible rhizomes (e.g. in Java *S. megacarpa* DC., in South Asia *S. macrophylla* Roxb., in the Caribbean *S. havanensis*), edible shoots (e.g. in Southeast Asia *S. leucophylla*, in the Caribbean *S. havanensis*), and edible berries (e.g. in the Caribbean *S. beyrichii* and *S. havanensis*, in the Philippines *S. macrophylla*) (von Reis & Lipp, 1982; Kunkel, 1984). *Smilax macrophylla* is used for cordage and fencing by the Bontoc in the Philippines (Bodner & Gereau, 1988). A common medicinal application is in treating venereal diseases, whether in Mexico (*S. aristolochiaefolia*), Brazil (*Smilax* sp. indet.), Malaysia (*S. calophylla* Wall. and *S. leucophylla*), or the Philippines (*S. bracteata* Presl.) (Uphof, 1959; Perry, 1980; von Reis & Lipp, 1982).

Solanaceae

A few climbing *Solanum* species are used: *S. diffusum* R. and P. is used by the Siona in eastern Ecuador to treat stomach-ache and diarrhea (Vickers & Plowman, 1984), whilst the Waunana in the Colombian Chocó eat the fruits of *Solanum* aff. *anceps* R. and P. (Forero Pinto, 1980). *Solanum jasminioides* Paxton is cultivated as an ornamental in Brazil (Mabberley, 1987). *Solandra* is a source of sacred Mexican hallucinogens and *S. maxima* (Sesse and Mocino) P. Green is cultivated as an ornamental for its long yellow flowers (Mabberley, 1987).

Tropaeolaceae

South American *Tropaeolum* species are variously used as ornamentals (*T. majus* L., 'Common Nasturtium', and *T. peregrinum* L. are cultivated throughout the tropics), and as foods (in the Andes *T. majus* flower buds are pickled like capers, and the seeds and leaves are also eaten, *T. tuberosum* R. and P. is cultivated in the Andes for its edible boiled tubers) (Uphof, 1959; Kunkel, 1984; Mabberley, 1987).

Vitaceae

Given the vast number of species in this family that produce edible fruit, it is rather surprising that no serious effort seems to have been made to produce tropical wines from them! Certainly *Vitis vinifera* L., a temperate species, is unsuccessful in tropical climates (Williams, 1981). From Uphof (1959), Kunkel (1984), von Reis Altschul (1975), and von Reis & Lipp (1982), I compiled total tropical numbers of species with edible fruits for the following genera: *Ampelocissus* (14, mostly in tropical Africa and tropical Asia), *Ampelopsis* (1 in Mexico), *Cayratia* (3 in Southeast Asia), *Cissus* (19, in Africa, Asia, and America), *Rhoicissus* (6 in tropical Africa), *Tetrastigma* (6, in Southeast Asia and Mexico), and *Vitis* (11, in Africa, Asia, and America). Most reports are for fruits eaten raw, whilst in some cases they were more often preserved as jellies. *Cissus gongylodes* was only recently reported as an important domesticate for the Káyapo Indians in Brazil; the vine lives for up to 40 years, providing a steady supply of edible leaves and fruit from old swidden plots, an example which contradicts the canard that Indians abandon slash-and-burn fields after a few years (Kerr, Posey & Wolter Filho, 1978; Posey, 1983, 1984). The same plant is also a good example of the 'nomadic agriculture' of the Káyapo, who plant wild varieties alongside their forest path (Posey, 1983).

Medicinally, Vitaceae are also important. A strong theme is their use as treatment for swellings, boils, rheumatism, lumbago and headaches. For example, in the Philippines a poultice of *Ampelocissus ochracea* leaves is used for boils and swellings (von Reis Altschul, 1975), whilst in Tanzania *A. asarifolia* poultices are applied to sprains and boils (von Reis & Lipp, 1982); in Mexico *Cissus assamica* treats headaches, as does *C. triloba* in Indochina and *C. repens* in Malaysia (von Reis Altschul, 1975; Perry, 1980). Various vines are used also for strong cordage, e.g. *Tetrastigma papillo* in the Philippines (Bodner & Gereau, 1988), *Vitis discolor* in South America (Uphof, 1959), and *Cissus sicyoides* in Central America (Williams, 1981).

Ampelocissus	edible fruits	Philippines: Uphof (1959)
	edible fruits	Neotropics: Kunkel (1984), von Reis & Lipp (1982)
	swellings, boils	Philippines: von Reis Altschul (1975)
	sprains, cuts, boils	Tanzania: von Reis & Lipp (1982)
	cholera, rheumatism, lumbago	Malaysia: Perry (1980)
Ampelopsis	edible fruits	Mexico: von Reis Altschul (1982)
	cordage, millipede stings	SE Asia: von Reis Altschul (1975)

Cayratia	edible fruits	Paleotropics: Kunkel (1984)
	cordage	Philippines: Bodner & Gereau (1988)
	violent headaches	Philippines: Perry (1980)
Cissus	edible fruits	Tropics: Kunkel (1984), Posey (1984)
	sore eyes, headaches, snake-bites	Melanesia: von Reis Altschul (1975)
	cordage	Central America: Williams (1981)
	headache, boils, stomach-ache	SE Asia: Perry (1980)
Rhoicissus	edible fruits	Tropical Africa: Kunkel (1984)
	leprosy, swellings	Tanzania: von Reis & Lipp (1982)
Tetrastigma	edible fruits	SE Asia: Uphof (1959), Kunkel (1984)
	edible fruits	Mexico: von Reis Altschul (1975)
	cordage, fencing, foot-bridges	Philippines: Bodner & Gereau (1988)
	headache, fever, boils, dropsy	SE Asia: Perry (1980)
Vitis	edible fruits	Tropics: Uphof (1959), Kunkel (1984)
	leather preparation before tanning	Tropical Africa: Uphof (1959)

Discussion and conclusions

Tropical climbing plants thus seem to have a truly extraordinary range of uses. Although some of these are limited to indigenous societies, in many cases the use has reached the wider rural population. Several liana taxa are widely used throughout the world. All parts of tropical climbers – flowers, fruits, seeds, stems, leaves, roots and tubers – have important applications. I will briefly review the range of local uses, before considering those tropical vines that have been commercialized on a global scale.

In the broad category of arts, crafts, construction, and general utility, the most reported traditional uses are for cordage. Among other important uses, the leaves of an *Arrabidaea* species (Bignoniaceae) are an important source of dye through much of the Neotropics; the same genus, plus *Manettia* (Rubiaceae), is used by the Jívaro to protect against tooth decay, as is *Agelaea* (Connaraceae) in West Africa. Furthermore, several vines are used in Southeast Asia for soap (e.g. *Momordica* roots (Cucurbitaceae), and *Gouania* roots and shoots (Rhamnaceae)); seeds from the cucurbit liana genus *Fevillea* have the highest recorded oil content of any dicotyledon.

Among non-remedial biodynamic uses, the uses of tropical vines as poisons, hallucinogens and stimulants stand out. Parts used for hallucinogens in Central and South America range from the seeds of *Ipomoea* and *Rivea* (Convolvulaceae), through the woody stems of *Banisteriopsis*, and to a lesser extent *Mascagnia* and *Tetrapteris* (all in the Malpighiaceae), to the leaves of the bignoniaceous liana *Tanaecium*. Major stimulants include *Paullinia* in the Sapindaceae (both the seeds and the bark are used in Amazonia), the leaves and seed oil of *Piper* (Piperaceae) in tropical Asia, and the bark of *Mussatia* and *Clytostoma* (Bignoniaceae) in southwest Amazonia. Vines are also used locally as insecticides; for example, the seeds of *Ipomoea* in West Africa, and the roots of *Lonchocarpus* (Leguminosae) in Brazil. Important liana arrow poisons and fish poisons come from at least six families. In the Old World, *Strophanthus* (Apocynaceae) seeds and/or crushed stem and leaves yield an arrow poison; in Amazonia at least four menisperm genera are used in different 'curare' arrow poisons and the whole plant of *Lonchocarpus* is used as a fish poison. Another legume, *Derris*, is used pantropically as a fish poison. In Amazonia, *Martinella* (Bignoniaceae) and especially *Strychnos* (Loganiaceae) are arrow poison ingredients. Finally, both *Serjania* and *Paullinia* (Sapindaceae) are widely used in the Amazon as fish poisons.

Indigenous medicinal uses are astonishingly numerous and varied. One especially interesting theme is the use of climbers in snake-bite remedies. As I suggested for Aristolochiaceae, this may, at least in part, be a case of the doctrine of signatures based on vines' superficial resemblance to snakes. However, the roots of so many *Aristolochia* species are used in this way (especially in South America, but also in the Old World) that they should be investigated pharmacologically. Other vine parts used thus are the leaves of *Mikania* (Asteraceae) in Ecuador, and decoctions of the stems of several different *Passiflora* species in Panama.

Of foods provided by tropical vines among the most important are the tubers of *Dioscorea* (Dioscoreaceae) and *Ipomoea*; both genera are cultivated on a subsistence basis throughout the tropics. Less well known edible starchy tubers come from *Humiranthera* (Icacinaceae), *Anredera* (Basellaceae), *Tropaeolum* (Tropaeolaceae), and *Bomarea* (Amaryllidaceae), all in South America. Vine fruits tend to be wind-dispersed (many neotropical lianas), or small bird-dispersed berries or drupes, so most are relatively unimportant as human food. There are obvious exceptions, though, notably the universally important cultivated cucurbits and legumes, and also wild and cultivated *Passiflora* and Vitaceae. A wide variety of climbers are edible as vegetables, especially the young leaves which generally have a much lower concentration of defence compounds.

An impressive variety of tropical climbers have been widely commercialized.

Among foods, *Ipomoea batatas* stands out as one of the most successful crops, both within and outside the tropics (especially Japan). The trade based on *Piper nigrum* is worth tens of million of dollars annually; most is grown in Madagascar, Brazil, and its native South Asia. *Vanilla* (Orchidaceae), and to a lesser extent sarsaparilla (from *Smilax*, Smilacaceae), are also cultivated in several tropical countries for export as flavorings. *Passiflora* species are grown on a commercial basis pantropically, though mostly not for export.

The rattan trade, worth some US $4 billion annually in end products (Myers, 1984), is still largely based on wild-harvesting in Malaysia, Indonesia and the Philippines. Present levels of exploitation on larger diameter canes do not appear to be sustainable in the long term, although there is clear potential for developing techniques of sustained use.

Another predominantly Southeast Asian export, natural and biodegradable insecticides known as rotenones, is based on the cultivated vines of various species of *Derris*. In Mexico, *Dioscorea* is planted for its tubers which yield diosgenin, chemical precursors for oral contraceptives and a number of other steroidal compounds used in clinical medicine. The retail sales of the final products total well over $1 billion. Other, mostly wild-harvested, tropical vines from which drugs used in clinical medicine are extracted include *Chondrodendron* (Menispermaceae, yields tubocurarine, used especially in surgery), *Mucuna* (Leguminosae, yields L-dopa, the most prescribed drug to alleviate Parkinson's Disease), *Strophanthus* (Apocynaceae, yields ouabain, used mostly as a cardiac stimulant), and *Quisqualis* (Combretaceae, yields quisqualic acid, an anthelmintic).

Widely cultivated tropical vine ornamentals in the horticultural trade include the numerous varieties of Morning Glory (*Ipomoea* spp., Convolvulaceae), as well as *Clematis* (Ranunculaceae), and the Flame Lily (*Gloriosa superba*, Liliaceae).

It is largely thanks to the native people of the tropics, and the efforts of ethnobotanists past and present, that tropical climbers already contribute in major ways to everybody's welfare. Multi-billion dollar industries are based on *Calamus* and *Dioscorea*; innumerable lives have been saved or made more enjoyable through *Chondrodendron*, *Strophanthus*, *Dioscorea*, and *Mucuna*, to name but a few; and millions more are fed every day by *Citrullus*, *Cucumis*, *Cucurbita*, *Lagenaria*, *Passiflora*, *Ipomoea*, *Dioscorea*, *Phaseolus*, and *Vigna*. The use of other lianas such as *Banisteriopsis*, *Lonchocarpus*, and *Paullinia* has helped to shape whole cultures. And yet we have barely scratched the surface: tropical climbers could provide us with so much more in the way of foods, medicines, contraceptives, canes, insecticides and so on, that the current level of interdisciplinary ethnobotanical research is surely woefully inadequate. With the tragic destruction of both the tropical forests and, even more poignantly, of their indigenous peoples, the time to act is fast running out. The reasons for botanists to get more involved in saving both are growing daily more compelling, for if we choose not to respond to the challenges we may soon be reduced to writing mere historical footnotes.

Summary

For each vascular plant family with useful climbing tropical species I have reviewed the relevant literature in ethnobotany and economic botany. General patterns in vine use are discussed in an ecological and economic context. Selected important genera are discussed in more detail, documenting both the ways in which they already contribute to humanity's well-being, and the ways in which they might do in the future. Throughout, I have drawn attention to our continuing dependence in the fields of medicine, agriculture and industry, on tropical vines, and in particular on indigenous peoples' knowledge of these plants.

Acknowledgements

I thank both Alwyn Gentry for his original encouragement and advice concerning this manuscript, and Walter Lewis for his much appreciated comments. I also thank the editor, Francis Putz, for giving me many helpful suggestions for improving this review. I am indebted to many people for inspiring my interest in ethnobotany, but most especially Paul Wilkin and José Armas, as well as Didier Lacaze, Michel Alexiades, Benito Arevalo and Cesar Yoyaje of AMETRA 2001.

References

Albert-Puleo, M. (1979). The obstetrical use in ancient and early modern times of *Convolulus scammonia* or Scammony: another non-fungal source of ergot alkaloids? *Journal of Ethnopharmacology* 1: 193–5.

Ayala Flores, F. (1984). Notes on some medicinal and poisonous plants of Amazonian Peru. *Advances in Economic Botany* 1: 1–8.

Ayala Flores, F. & Lewis, W. H. (1978). Drinking the South American hallucinogenic ayahuasca. *Economic Botany* 32: 154–6.

Ayensu, A. S. (1978). *Medicinal Plants of West Africa*. Reference

Publications, Algonac, Michigan.
Ayensu, A. S. (1981). *Medicinal Plants of the West Indies*. Reference Publications, Algonac, Michigan.
Balick, M. J. (1984). Useful plants of Amazonia: a resource of global importance. In *Amazonia*, ed. G. T. Prance and T. E. Lovejoy, pp. 339–68. Pergamon Press, Oxford.
Basu, S. K. (1985). The present status of rattan palms in India – an overview. In *Proceedings of the Rattan Seminar*, ed. K. M. Wong and N. Manokaran, pp. 77–94. Rattan Information Centre, Kepong, Malaysia.
Bodley, J. H. (1978). *Preliminary ethnobotany of the Peruvian Amazon*. Reports of Investigations, No. 55, Laboratory of Anthropology, Washington State University, Pullman.
Bodner, C. C. & Gereau, R. E. (1988). A contribution to Bontoc ethnobotany. *Economic Botany* 42: 307–69.
Clarke, W. C. (1971). *Place and People: an ecology of a New Guinean community*. University of California, Berkeley.
Davis, W. E. (1983). The ethnobotany of chamairo: *Mussatia hyacinthina*. *Journal of Ethnopharmacology* 9: 225–36.
Davis, W. E. & Yost, J. A. (1983). The ethnobotany of the Waorani of eastern Ecuador. *Harvard University Botanical Museum Leaflets* 3: 159–211.
Dobkin de Rios, M. (1972). *Visionary Vine. Psychedelic healing in the Peruvian Amazon*. Chandler, San Francisco, Calif.
Dransfield, J. (1985). Prospects for lesser known canes. In *Proceedings of the Rattan Seminar*, ed. K. M. Wong and N. Manokaran, pp. 107–14. Rattan Information Centre, Kepong, Malaysia.
Eldredge, J. (1975). Bush medicine in the Exumas and Long Island, Bahamas: a field study. *Economic Botany* 29: 307–32.
Elliott, S. & Brimacombe, J. (1985). *The Medicinal Plants of Gunung Leuser National Park, Indonesia*. Expedition report, Edinburgh University.
Emboden, W. A. Jr (1972). *Narcotic Plants*. Macmillan, New York.
Food and Agriculture Organization (1983). *Food and Fruit-bearing Forest Species 1: Examples from Eastern Africa*. Forestry Paper 44/1, FAO, Rome.
Food and Agriculture Organization (1984). *Food and Fruit-bearing Forest Species 2: Examples from Southeastern Asia*. Forestry Paper 44/2, FAO, Rome.
Food and Agriculture Organization. (1986). *Food and Fruit-bearing Forest Species 3: examples from Latin America*. Forestry Paper 44/3, FAO, Rome.

Farnsworth, N. R. (1988). Screening plants for new medicines. In *Biodiversity*, ed. E. O. Wilson, pp. 83–97. National Academy Press, Washington, DC.

Farnsworth, N. R., Akerele, O., Bingel, A. S. Soejarto, D. D. & Guo, Z.-G. (1985). Medicinal plants in therapy. *Bulletin W.H.O.* **63**: 965–81.

Feeny, P. (1976). Plant apparency and chemical defence. *Recent Advances in Phytochemistry* **10**: 1–40.

Forero Pinto, L. E. (1980). Etnobotánica de las Comunidades Indígenas Cuna y Waunana, Chocó (Colombia). *Cespedesia* **9**: 115–306.

Garcia-Barriga, H. (1974). *Flora Medicinal de Colombia, Vol. 1.* Instituto de Ciencias Naturales, Universidad Nacional, Bogota.

Gentry, A. H. (1984). An ecotaxonomic survey of Panamanian lianas. In *Historia Natural de Panama*, ed. W. G. D. D'Arcy. Missouri Botanical Garden, St Louis, Missouri.

Gentry, A. H. (1986). Summario de patrones fitogeograficos neotropicales y sus implicaciones para el desarrollo en la Amazonia. *Revista Academia Colombiana Ciencias Exact.* **16**: 101–16.

Gentry, A. H. (1988). New species and a new combination for plants from trans-Andean South America. *Annals of the Missouri Botanical Garden* **75**: 1429–39.

Gentry, A. H. & Cook, K. (1984). *Martinella* (Bignoniaceae): a widely used eye medicine of South America. *Journal of Ethnopharmacology* **10**: 337–43.

Gentry, A. H. & Wettach, R. H. (1986). *Fevillea* – a new oil seed from Amazonian Peru. *Economic Botany* **40**: 177–85.

Gilbert, L. E. (1975). Ecological consequences of a coevolved mutualism between butterflies and plants. In *Coevolution of Animals and Plants*, ed. L. E. Gilbert and P. H. Raven, pp. 159–91. University of Texas Press, Austin.

Glenboski, L. L. (1975). *Ethnobotany of the Tukuna Indians, Amazonas, Colombia*. PhD thesis, University of Alabama.

Gottlieb, O. R. (1984). The chemical uses and chemical geography of Amazon plants. In *Amazonia*, ed. G. T. Prance and T. E. Lovejoy, pp. 218–38. Pergamon Press, Oxford.

Grenand, P. (1980). *Introduction à l'Etude de l'Univers Wayapi: Ethnoécologie des Indiens de Haut-Oyapock (Guyane Française)*. ORSTOM, Paris.

Grieve, M. (1967). *A Modern Herbal, Vols I and II*. Hafner, New York.

Jacobs, M. (1982). The study of minor forest products. *Flora Malesiana Bulletin* **35**: 3768–82.

Joly, L. G., Guerra, S., Septimo, R., Solis, P. N., Correa, M., Gupta, M.,

Levy, S. & Sandberg, F. (1987). Ethnobotanical inventory of medicinal plants used by the Guaymi Indians in western Panama, Part I. *Journal of Ethnopharmacology* **20**: 145–71.

Kerr, W., Posey, D. A. & Wolter Filho, W. (1978). Cupa, ou cipo babao, alimento do algunas indios amazonicos. *Acta Amazonica* **8**: 702–5.

Kreig, M. (1965). *Green Medicine*. Harrap, London.

Krukoff, B. A. (1965). Supplementary notes on the American species of *Strychnos*. VII. *Memoirs of the New York Botanical Garden* **12**: 1–94.

Krukoff, B. A. & Barneby, B. C. (1970). Supplementary notes on American Menispermaceae. *Memoirs of the New York Botanical Garden* **20**: 1–70.

Krukoff, B. A. & Moldenke, H. N. (1938). Studies of American Menispermaceae with special reference to species used in preparation of arrow poison. *Brittonia* **3**: 1–74.

Krukoff, B. A. & Smith, A. C. (1937). Rotenone-yielding plants of South America. *American Naturalist* **24**: 573–87.

Kunkel, G. (1984). *Plants for Human Consumption*. Koenigstein, West Germany.

Levin, D. A. (1976). Alkaloid-bearing plants: an ecogeographic perspective. *American Naturalist* **110**: 261–84.

Lewis, W. H. & Elvin-Lewis, M. P. F. (1977). *Medical Botany*. Wiley, New York.

Lewis, W. H. & Elvin-Lewis, M. P. F. (1984). Plants and dental care among the Jívaro of the upper Amazon basin. *Advances in Economic Botany* **1**: 53–61.

Lewis, W. H., Elvin-Lewis, M. P. F. & Gnerre, M. C. (1987). Introduction to the ethnobotanical pharmacopeia of the Amazonian Jívaro of Peru. In *Medicinal and Poisonous Plants of the Tropics*, compiler A. J. M. Leeuwenberg, pp. 96–103. PUDOC, Wageningen, The Netherlands.

Lopez Guillen, J. E. & Kiyan de Cornelio, I. (1974). Plantas medicinales del Peru, IV. *Biota* **10**: 28–56.

Luna, L. E. (1984). The concept of plants as teachers among four mestizo shamans of Iquitos, northeastern Peru. *Journal of Ethnopharmacology* **11**: 135–56.

Mabberley, D. J. (1987). *The Plant-Book: a portable dictionary of the higher plants*. Cambridge University Press, Cambridge.

McIntyre, A. R. (1947). *Curare: its History, Nature, and Clinical Use*. University of Chicago Press, Chicago, Illinois.

Manokaran, N. (1985). Biological and ecological considerations pertinent to the silviculture of rattans. In *Proceedings of the Rattan Seminar*, ed. K. M. Wong and N. Manokaran, pp. 95–105. Rattan Information

Centre, Kepong, Malaysia.
Martínez, M. H. (1984). Medicinal plants used in a Totomac community of the Sierra norte de Pueblo: Tozapan de Galeana, Puebla, Mexico. *Journal of Ethnopharmacology* 11: 203–22.
Morton, J. L. (1977). *Major Medicinal Plants*. Springfield, Illinois.
Myers, N. (1983). *A Wealth of Wild Species, Storehouse for Human Welfare*. Westview, Boulder, Colo.
Myers, N. (1984). *The Primary Source*. Norton, New York and London.
Naranjo, P. (1979). Hallucinogenic plant use and related indigenous belief systems in the Ecuadorian Amazon. *Journal of Ethnopharmacology* 1: 121–45.
Nuñez, M. E. (1975). *Plantas Medicinales de Costa Rica y su Folclore*. Universidad de Costa Rica, San José, Costa Rica.
Oblitas Poblete, E. (1969). *Plantas Medicinales de Bolivia: Farmacopea Callanaya*. Editorial Los Amigos del Libro, Cochabamba, Bolivia.
Ocana, G., Rubinoff, I., Smythe, N. & Werner, D. (1988). Alternatives to destruction: research in Panama. In *Biodiversity*, ed. E. O. Wilson, pp. 370–6. National Academy Press, Washington, DC.
Ohiri, F. C., Verpoorte, R. & Baerheim Svendsen, A. (1983). The African *Strychnos* species and their alkaloids: a review. *Journal of Ethnopharmacology* 9: 167–223.
Oliver-Bever, B. (1986). *Medicinal Plants in Tropical West Africa*. Cambridge University Press, Cambridge.
Perry, L. M. (1980). *Medicinal Plants of East and Southeast Asia: attributed properties and uses*. MIT Press, Cambridge, Mass.
Pio Correa, M. (1931). *Dicionario das plantas uteis do Brasil* 3: 33–5. Servicio de Informacão Agricola, Ministerio da Agricultura, Rio de Janeiro.
Plotkin, M. J. & Balick, M. J. (1984). Medicinal uses of South American palms. *Journal of Ethnopharmacology* 10: 157–79.
Plowman, T. (1980). Chamairo: *Mussatia hyacinthina* – an admixture to coca from Amazonian Peru and Bolivia. *Botanical Museum Leaflets, Harvard University* 28: 253–61.
Posey, D. A. (1983). Indigenous ecological knowledge and development of the Amazon. In *The Dilemma of Amazonian Development*, ed. E. F. Moran, pp. 225–57. Westview, Boulder, Colo.
Posey, D. A. (1984). A preliminary report on diversified management of tropical forest by Kayapo Indians of the Brazilian Amazon. *Advances in Economic Botany* 1: 112–26.
Prance, G. T. (1972). An ethnobotanical comparison of four tribes of Amazonian Indians. *Acta Amazonica* 2: 7–27.
Prance, G. T., Campbell, D. G. & Nelson, B. W. (1977). The ethnobotany

of the Paumari Indians. *Economic Botany* **31**: 129–39.
Prescott-Allen, R. & Prescott-Allen, C. (1988). *Genes from the Wild*. Earthscan Publications, London.
Purseglove, J. W., Brown, E. G., Green, C. L. & Robbins, S. R. J. (1981). *Spices, Vols 1 and 2*. Longman, New York.
Pushpangadan, P. & Atal, C. L. (1984). Ethno-medico-botanical investigations in Kerala I. Some primitive tribes of western Ghats and their herbal medicine. *Journal of Ethnopharmacology* **11**: 59–77.
Putz, F. E. (1983). Liana biomass and leaf area of a 'tierra firme' forest in the Rio Negro basin, Venezuela. *Biotropica* **15**: 185–9.
Rickard, P. P. & Cox, P. A. (1986). Use of *Derris* as a fish poison in Guadalcanal, Solomon Islands. *Economic Botany* **40**: 479–84.
Rivier, L. & Lindgren, J.-E. (1972). Ayahuasca, the South American hallucinogenic drink: an ethnobotanical and chemical investigation. *Economic Botany* **26**: 101–29.
Schultes, R. E. (1942). Plantae Colombianae II. Yoco: a stimulant of southern Colombia. *Botanical Museum Leaflets, Harvard University* **10**: 301–24.
Schultes, R. E. (1962). Plantae Colombianae XVI. Plants as oral contraceptives in the northwest Amazon. *Lloydia*: **26**: 67.
Schultes, R. E. (1970a). De plantis toxicariis e mundo tropicale commentationes VI. Notas etnotoxicologicas acerca de la flora Amazonica de Columbia. In *II Simposio y Foro de la Biologia Tropical Amazonica*, ed. J. M. Idrobo, pp. 178–96. Editorial PAX, Bogota, Colombia.
Schultes, R. E. (1970b). De plantis toxicariis e mundo tropicale commentationes VII. Several ethnotoxicological notes from the Colombian Amazon. *Botanical Museum Leaflets, Harvard University* **22**: 345–52.
Schultes, R. E. (1975). De plantis toxicariis e mundo tropicale commentationes XIII: Notes on the poisonous or medicinal malpighiaceous species of the Amazon. *Botanical Museum Leaflets, Harvard University* **24**: 121–31.
Schultes, R. E. (1977). De plantis toxicariis e mundo tropicale commentationes XVI: Miscellaneous notes on biodynamic plants of South America. *Botanical Museum Leaflets, Harvard University* **25**: 109–30.
Schultes, R. E. (1978). Evolution of the identification of the major South American narcotic plants. *Botanical Museum Leaflets, Harvard University* **26**: 311–32.
Schultes, R. E. (1979). De plantis toxicariis e mundo novo tropicale commentationes. XIX. Biodynamic apocynaceous plants of the

northwest Amazon. *Journal of Ethnopharmacology* **1**: 165–92.
Schultes, R. E. (1984). Fifteen years of study of psychoactive snuffs of south America: 1967–1982 – A Review. *Journal of Ethnopharmacology* **11**: 17–32.
Schultes, R. E. (1986). Recognition of variability in wild plants by Indians of the northwest Amazon: an enigma. *Journal of Ethnobiology* **6**: 229–38.
Schultes, R. E. & Hoffmann, A. (1980). *Plants of the Gods*. McGraw-Hill, New York.
Uphof, J.C.Th. (1959). *Dictionary of Economic Plants*. Engelmann, New York.
van den Berg, M. E. (1984). Ver-o-Peso: the ethnobotany of an Amazonian market. *Advances in Economic Botany* **1**: 140–9.
van Steenis-Kruseman, M. J. (1953). Select Indonesian medicinal plants. *Organization Scientific Research Indonesia, Bulletin* **18**: 1–90.
Vasquez, R. & Gentry, A. H. (1989). Use and misuse of forest-harvested fruits in the Iquitos area. *Conservation Biology* **3**: 350–61.
Vickers, W. T. & Plowman, T. (1984). Useful plants of the Siona and Secoya Indians of eastern Ecuador. *Fieldiana (Botany)* no. 15.
von Reis Altschul, S., (1975). *Drugs and Foods from Little-Known Plants*. Harvard University Press, Cambridge, Mass.
von Reis Altschul, S., & Lipp, F. J. Jr (1982). *New Plant Sources for Drugs and Foods from the New York Botanical Garden*. Harvard University Press, Cambridge, Mass.
Williams, L. O. (1981). The useful plants of Central America. *Ceiba* **24**: 3–381.

17

Biology, utilization and silvicultural management of rattan palms

STEPHEN F. SIEBERT

Introduction

Rattans are spiny climbing plants in the subfamily Calamoideae, the scaly-fruited palms (formerly known as the Lepidocaryoideae) (Uhl & Dransfield, 1987). They are a large, diverse group comprising approximately 600 species in 13 genera (Dransfield, 1988) that range from the West African coast to Taiwan and Fiji (Corner, 1966). The rattans, *Calamus* and *Daemonorops*, with 370 and 115 species respectively, are among the largest genera of palms (Tomlinson, 1979).

Rattans reach their greatest abundance and diversity in Southeast Asia, particularly the Malay Archipelago (Dransfield, 1981) where they are a prominent component of both the forest floor and the canopy. Rattans occur under a wide variety of climatic and edaphic conditions and are found from lowland swamp to upper montane and secondary forest formations (Dransfield, 1981).

To many inhabitants of Southeast Asia, rattans are an indispensible resource in daily life. Rattans are also well known in Europe and North America as the basic material in elegant furniture and handicrafts. The strong demand for rattan products has made this unique group of climbing palms the most important non-timber forest product in Asia and an important source of cash income for rural people (International Development Research Centre, 1980). The very popularity of rattan products, in conjunction with indiscriminate timber harvesting and forest conversion to agricultural uses, now threaten the existence of both rattans and the industry that is dependent upon them.

This chapter examines rattan biology, utilization, and silvicultural practices within the context of conservation and development. First rattan growth, climbing mechanisms, and their ecological role in the tropical forest biome are reviewed, followed by a discussion of the uses and socioeconomic importance of rattans in Southeast Asia. The current status of rattan resources and silvicultural efforts to cultivate and manage rattans on a sustainable basis are then examined.

Biology and ecology

For purposes of field identification, palms may be delineated into four basic growth forms: (i) tree palms (those that reach the forest canopy), (ii) shrub palms (those found in the forest understory), (iii) acaulescent palms (those with subterranean or very short aerial stems), and (iv) climbing palms (Dransfield, 1978). As is evident by these varied growth habits, palms occupy a wide variety of niches within forest communities.

Species of Calamoideae are readily distinguished from other palms by the closely overlapping scales which cover their ovaries and fruits (Uhl & Dransfield, 1987). Rattans are generally considered climbing palms and with the exception of *Calamus arborescens*, the tree rattan (Dransfield, 1969), all are climbers when reproductively mature. Nevertheless, rattan seedlings are a prominent component of the forest floor where they often persist as a basal rosette of leaves without climbing organs. Thus, in the forests of the Far East, rattans can be encountered underfoot (as juveniles) and at chest height and overhead (as mature individuals).

The large size and wide distribution of several rattan genera, notably *Calamus* and *Daemonorops*, and the high degree of morphological variability in the group suggest recent adaptive radiation (Tomlinson, 1979). For example, rattan stems may be creeping, erect, or climbing, and solitary, clustering, or branched aerially; leaves include both pinnate and semi-palmate forms; and inflorescences may be pleonanthic or hapaxanthic and large and profusely branched or small and with few branches (Uhl & Dransfield, 1987). *Calamus manan*, for example, is a massive rattan that produces one cane per plant. In contrast, *C. caesius* and *C. trachycoleus* are small, clustering rattans that produce numerous aerial stems (i.e. canes) from a single plant (Dransfield, 1979).

Rattans also exhibit a variety of morphological adaptations for climbing. Most commonly, the leaf rachis elongates into a 1–5 m long barbed whip, or 'cirrus', that is heavily armed with reflexed spines (Dransfield, 1978). An alternative climbing organ is developed in some *Calamus* species in which, instead of a cirrus, a 'flagellum' (actually a sterile infloresence) develops at the base of the leaf sheath (Dransfield, 1978).

Both cirrus and flagellum are light in weight and stand erect from the plant. These characteristics, in conjunction with reflexed spines, allow rattans to grasp virtually anything they encounter. With the cirrus or flagellum grasping supporting vegetation, upward growth occurs through internodal and leaf-tip elongation (Corner, 1966). Given this effective grappling ability, rattans are not limited to climbing on successively taller small diameter supporting vegetation as is the case with twining or tendril-climbing lianas (Putz & Chai, 1987).

While rattans do not occur in the Neotropics, there are New World counterparts to the Asiatic climbing Calamoideae. *Desmoncus*, a large genus of Arecoideae palms that range from Mexico to Brazil, closely resemble many rattans in that they produce an elongated cirrus with reflexed spines and reflexed hook-like leaflets (Uhl & Dransfield, 1987). Corner (1966) reports that *Desmoncus* are less effective climbers than Asiatic rattans due to their relatively poorly developed spines, but are nevertheless well suited to what he regards as the more broken and open canopy of South American forests. Another neotropical counterpart to rattan is *Chamaedorea elatior* which possesses perhaps the most simple adaptation to climbing: terminal leaflets that are reflexed and function as weak grapnels (Dransfield, 1978).

The length of time rattans remain as juvenile basal rosettes and the specific mechanism(s) that trigger shoot elongation and development of climbing organs are unknown for most rattan species. Increased light intensities are probably required to induce stem elongation in most rattans (Dransfield, 1979). However, light intensity responses appear to be species-specific and rattans exhibit the entire range of near complete shade tolerance to shade intolerance that is characteristic of woody perennials. For example, *C. manan* can survive for many years in the rosette stage in complete shade and then exploit small canopy gaps (e.g. branch falls). In contrast, *Daemonorops angustifolia* survives only under high light intensity and is thus restricted to large canopy gaps (e.g. treefalls), river banks, and disturbed forests (Dransfield, 1979).

The growth potential of many rattans is astonishing. For example, some *Calamus* species form internodes a meter or more in length and can produce 10 m canes from seed in 5–6 years (Corner, 1966). Canes of *C. trachycoleus* are reported by central Kalimantan villagers to grow at rates of up to 5 m per year (Dransfield, 1979). The ultimate length of rattan canes can also be impressive. For example, *C. manan* canes have been recorded up to 171 m long (Burkill, 1935), or almost twice the height of California redwoods (*Sequoia sempervirens*), a fascinating physiological feat for a plant lacking secondary vascular cambium.

The great diversity of ecological adaptations exhibited by rattans precludes a thorough discussion of them here. However, several aspects of rattan biology are particularly noteworthy, including flowering modes, pollinating vectors, and their relation to forest fauna.

Flowering in palms is most commonly pleonanthic (i.e. shoots flower continuously and do not die after flowering) (Uhl & Dransfield, 1987). However, some rattans, including all species of the wide ranging *Korthalsia* and *Plectocomia*, are hapaxanthic (i.e. shoots flower once and then the ramet dies: Dransfield, 1978). The evolutionary development of hapaxanthy is unclear. Dransfield (1976a) suggests that for at least one rattan, *Daemonorops calicarpa*, hapaxanthic flowering is derived and that hapaxanthy in general may be an adaptation allowing for increased colonizing potential in open habitats through the simultaneous production of large quantities of fruit (Dransfield, 1978).

Pollination in palms was originally thought to be exclusively by wind, although empirical studies now indicate that a variety of vectors is involved (Henderson, 1986). While the specific pollinators of most rattan species are not known, wasps and flies are likely pollinators of many *Calamus* species while bees and beetles have been observed on the flowers of *Plectocomia* (Dransfield, 1979).

In general, rattans produce great quantities of fruit (Dransfield, 1979). For example, *C. caesius* and *C. manan* plants can produce up to 2000 and 5000 seeds per stem, respectively, each year (Manokaran, 1979).

Rattans are a wide ranging group that occur from sea level to 3000 m elevation (Dransfield, 1981). Some species range widely, while others are restricted to specific elevations, forest types, or soil conditions. Ecologically, rattans are often considered weedy plants characteristic of wet and disturbed forest sites (Tomlinson, 1979). While this is certainly true for many rattans, some species are decidedly not weedy in habit.

Given the diversity of rattan species, it is not surprising that distinct rattan floras can be discerned. In Peninsular Malaysia, for example, Dransfield (1979) notes floristic changes in rattan flora associated with a variety of climatic, edaphic, and disturbance factors. The most prominent floristic change occurs at about 1000 m elevation where dipterocarp forests begin to be replaced by oak/laurel forests. Other distinct rattan floras are associated with peat swamps, steep ridgelines in hill dipterocarp forests, stream-sides in lowland dipterocarp forests, disturbed secondary forests, and in unique soil environments such as limestone outcroppings (Dransfield, 1979).

In general, palms (rattans included) are not common in upper montane forests (Dransfield & Whitmore, 1969) and are largely absent from truly xeric environments (Tomlinson, 1979). While the systematics of the Calamoideae are incomplete, it is clear that there is a high degree of endemism in the group. For example, about half of the rattan species in Peninsular Malaysia are endemics (Dransfield, 1979). A similarly high degree of endemism can be expected in other, as yet poorly described regions (e.g. the island of Sulawesi, which lies astride the important floristic transition zone of Wallacea).

Given the prominence of rattan in many Southeast Asian forests and their abundance on the forest floor, one is tempted to conclude that mature rattans are abundant as well. For example, as many as 320 individuals of *Calamus exilis*, a small clustering rattan, have been observed per hectare in Sumatra, although the majority of these were seedlings and vegetative clones (Siebert, 1989). Putz & Chai (1987) reported mean rattan densities of 40 and 60 large (i.e. climbing) individuals per hectare in valley and ridge sites, respectively, in Sarawak, Malaysia. Rattan population densities and seedling mortality rates are not known for most species (Dransfield, 1979). However, the population densities of large, solitary rattans, such as *C. manan*, may be similar to canopy tree species (i.e. densities greater than a few individuals per hectare would be unusual).

The widespread distribution of rattans and their copious production of fruit suggest that they could be an important food source for arboreal wildlife species. Monkeys (e.g. macaques) and hornbills have been observed feeding on rattan fruits (S. F. Siebert, personal observation) and the apex or 'cabbage' is eaten by elephants, wild cattle, pigs, and squirrels (Dransfield, 1979). Furthermore, both the fruits and cabbages of several species are consumed by rural people (Dransfield, 1979; Siebert & Belsky, 1985).

One form of rattan–fauna interaction that has been investigated in some detail is the association between certain rattan species and ants. Rattans exhibit a variety of adaptations that are used by ants, including (i) reflexed leaves that produce chambers, (ii) infloresences partially or wholly enclosed by bracts, (iii) auricles (i.e. extensions of the leaf sheaths) that enclose narrow tunnels, (iv) ocreas (i.e. swollen or elongated hollow chambers next to leaf sheaths), and (v) whorls of interlocking basal stem spines that form tunnels (Dransfield, 1979). While ants are thus provided with ideal nesting environments, it is unclear what advantages the relationship provides to rattans. One possible explanation is that ants deter herbivores (Dransfield, 1979). Another possibility is that ants facilitate nutrient uptake of throughfall water by producing a protective domicile around rattan stems (Rickson, 1984). Whatever the explanation, some rattans are so frequently associated with ants that they are known locally as 'rotan semut' (ant rattans) (e.g. *Korthalsia* spp.).

The overall importance of rattans to the fauna of Southeast Asia is virtually unknown, as are many other aspects of their natural history. These subjects represent important research needs in tropical ecology and wildlife biology.

Utilization and socioeconomic importance

The socioeconomic significance of rattans in Southeast Asia is difficult to exaggerate. Rattans have been used for centuries as cordage, binding, thatch, basketry, medicine, food, and toys (e.g. Dunn, 1975; Dransfield, 1981) and are now a major income source as well (IDRC, 1980). Indigenous forest inhabitants are generally most knowledgeable about and dependent upon rattan resources. For example, the Semai people of West Malaysia utilize rattan for a vast array of domestic purposes and collect at least four species for cash income (Avé, 1988). Similar reliance upon rattan by forest-dwelling people has been documented in Indonesia (Peluso, 1983; Weinstock, 1983), the Philippines (Schlegel, 1979; Conelly, 1985) and elsewhere in Malaysia (Dunn, 1975).

Many other inhabitants of Southeast Asia, in addition to forest-dwelling people, rely upon rattan products and income. In Leyte, Philippines, for example, rattan was found to be an indispensable source of cash income to lowland farmers, particularly among the poorest households (Siebert & Belsky, 1985) and a similar situation appears to exist in Kerinci, Sumatra (Siebert, 1989).

Rattan is also big business. The total value of world trade in rattan exceeds US $1.2 billion annually (Shane, 1977) and employs an estimated 500 000 people as collectors, traders, processors, and furniture manufacturers (IDRC, 1980). Indonesia controls approximately 90% of the world's supply of rattan cane, while Taiwan and Singapore are the world's leading makers and exporters of rattan furniture (*Far Eastern Economic Review*, 1988).

Indigenous forest people have uses for almost all locally available rattan species (Dransfield, 1981). However, in general, terminal flowering species such as *Plectocomia* and *Plectomiopsis* are not very useful because the cane splits too easily (Dransfield, 1976b). The most important rattan genera for both domestic and trade purposes are *Calamus* and *Daemonorops*.

Approximately 25 species form the basis of the international rattan furniture trade (Dransfield, 1988). The actual identity of all commercially important species is difficult to determine because (i) once rattans are cut and enter the market, species identification is difficult; (ii) many commercially important varieties are known only by their local or vernacular names; (iii) many major rattan producing regions are taxonomically poorly known (e.g. Sulawesi).

Commercial rattans can be divided into two broad categories based upon cane size and end use: (i) small diameter canes (i.e. less than 18 mm), which are used primarily in split form for woven cane seats, handicrafts, and baskets; (ii) large diameter canes (i.e. 25–40 mm), which are used in furniture construction (Dransfield, 1988).

The most important commercial rattan species include *C. manan*, *C. caesius*, and *C. trachycoleus* (Manokaran & Wong, 1983). *Calamus manan* is the premier furniture species, although a number of other large rattans, such as *C. tumidus*, *C. ornatus*, and *Daemonorops* spp. are increasingly used as substitutes as *C. manan* supplies are exhausted (Dransfield, 1979). *Calamus caesius* and *C. trachycoleus* are small diameter rattans.

At present, virtually all rattan collecting is from uncultivated wild plants. The only significant exception to this is the cultivation of *C. caesius* and *C. trachycoleus* in central Kalimantan (Dransfield, 1979). Rattan collectors, either indigenous forest inhabitants or villagers living adjacent to forests, typically gather rattan as a supplementary income source during periods of low agricultural labor demand. Rattan collecting involves lengthy trips into the forest to locate canes and then the transport of harvested cane to road or riverside collection points where they are purchased by traders.

Large rattans contain significant amounts of gums, resins, and water which must be extracted within about 15 days to insure good cane quality and color. Treatment for most large canes involves boiling in coconut oil, diesel fuel, or crank case oil (Dransfield, 1979; F. Putz, personal communication). In contrast, small diameter canes are usually treated through air-drying and are then split, either by hand or machine, and marketed in bundles.

Until 1988, most large rattan and bundles of split cane collected in Indonesia were shipped to furniture manufacturers in Taiwan and Singapore. However, in July 1988, Indonesia enacted a ban against the export of raw or polished canes. This legislation was adopted to maximize local value added and to generate employment opportunities for Indonesians (*Asiaweek*, 1987). At present, the Indonesian rattan industry is not capable of producing large quantities of export quality furniture. The current economic uncertainties in the furniture trade, in conjunction with the rapid rate of habitat destruction (i.e. forest conversion) and extensive overharvesting of many rattan species, casts serious doubt about the future of the Southeast Asia rattan industry.

Rattan collecting is a highly decentralized trade, but is nonetheless a prominent activity in forests throughout Southeast Asia. The once vast forests of Borneo, Sumatra, Sulawesi, Peninsular Malaysia, and the Philippines originally held tremendous quantities of rattan. Since the end of the Second World War, and particularly since the 1970s, rapid economic development, based largely upon the extraction of raw materials and the cultivation of cash crops for export, has dramatically transformed both the landscape and forests of Southeast Asia.

Tropical forests in the Philippines have been almost completely logged over and/or converted to agricultural uses. Forests in Thailand have experienced a similar fate. Huge areas of Peninsular Malaysia and Sumatra have been cleared of their original forests for rubber and oil palm monocultures and large-scale logging continues in both Malaysian and Indonesian Borneo. When forest habitat is destroyed, rattans that are unable to survive in secondary or disturbed forests are lost as well.

The overall status of rattan populations in Southeast Asia appears bleak.

Based on taxonomic accounts, Dransfield (1987) concludes that approximately 35% of the rattan species of Peninsular Malaysia, 25% of those in Sabah, and 30% of those in Sarawak are threatened with extinction. The status of rattan populations in other important areas, including Sumatra, Indonesian Borneo, and Sulawesi, is not known, but many rattans in these areas are certainly threatened as well.

In addition to habitat destruction, rattans are threatened by uncontrolled exploitation. This is particularly true for *C. manan* whose population is now estimated to number only a few thousand mature individuals (Dransfield, 1987). *Calamus manan* originally ranged throughout much of the Malay Archipelago and was reported to be 'extraordinarily abundant' in some locations such as the hill dipterocarp forests of west-central Sumatra (Dransfield, 1974). By the late 1980s, however, *C. manan* had been eliminated from west-central Sumatra (Siebert, 1989) and from many other regions.

The long-term availability of rattan resources has been a concern for decades. In the early 1900s, for example, Burkill (1935) reported that the more accessible rattans were rapidly being exhausted in Sulawesi as a result of overexploitation and habitat destruction. What makes the present situation unique is the fact that we are rapidly approaching the exhaustion of Southeast Asian forest resources: few large, unexploited forests remain. If logging, forest conversion to farms, and cane harvesting persist at current rates, rattan resources and the lucrative furniture industry are likely to vanish by the turn of the century.

Silviculture and management

The conservation of rattan resources will require simultaneous efforts in several areas, including: (i) controlling the exploitation of wild populations; (ii) identifying and developing alternatives to rattan for furniture and handicraft industries; (iii) cultivating rattans in forest preserves (Dransfield, 1981, 1987; Siebert, 1989). The Indonesian ban against the export of unfinished cane may temporarily ease pressure on rattan resources in that country, but will probably increase collecting pressures elsewhere (Dransfield, 1987). For example, Chinese manufacturers are now reportedly searching for new rattan suppliers in Papua New Guinea and Malaysia to replace Indonesian supplies (Xu Huanggan, unpublished report) and prices of *C. manan* canes have increased dramatically (*Asiaweek*, 1989). Furthermore, rattan exploitation and habitat destruction are likely to continue in Indonesia, threatening rattan resources there as well.

One largely unexplored avenue for conserving rattan resources is sustained yield harvesting in forest preserves. This approach is particularly appropriate with clustering rattans (i.e. those producing multiple canes), such as

C. caesius, *C. trachycoleous*, *C. pilosellus*, and *C. exilis*. Basic ecological data on the abundance, distribution, age-class structure, and cane resprouting rates would be required and then used to develop sustained yield harvesting rates in designated areas (Siebert, 1989). This system would be advantageous in that it would provide local populations with an economic incentive to preserve forest cover while maintaining viable populations of rattans in the wild.

Rattan conservation would also be aided by encouraging the use of substitutes. For example, buri palms (petioles of *Corypha utan*) have long been used in conjunction with rattan in furniture manufacturing in the Philippines and could be utilized in other regions of Southeast Asia where buri is found (Dransfield, 1987). Wood could also be substituted for rattan in some situations. For example, cultivated surian trees (*Toona sureni*, Meliaceae) are now used instead of *C. manan* as basket frames in Kerinci, Sumatra (Siebert, 1989). Similarly, less desirable, but still abundant rattans could be substituted for endangered species, such as the use of *Korthalsia rigida* in place of *C. manan* (Siebert, 1989). Ultimately, however, the viability of the Southeast Asian rattan industry will require the cultivation and management of commercially important species (Dransfield, 1987).

The cultivation of rattans is of growing interest throughout Southeast Asia. Rattan silvicultural research is now being conducted by the Malaysian Forest Research Institute, the Sabah Forestry Development Authority (Malaysia), the Research Institute of Tropical Forestry in China, the Philippine National Development Corporation and the Indonesian Forest Research and Development Centre. Interestingly, the most successful and long-lived rattan cultivation efforts have been the work of indigenous inhabitants in central Kalimantan, Indonesia where *C. trachycoleus* and *C. caesius* have been grown on a wide scale for over 100 years (Dransfield, 1979).

The success of the central Kalimantan rattan plantings results, at least in part, from the favorable growth characteristics exhibited by *C. trachycoleus*, the principal species. *Calamus trachycoleus* is an invasive rattan that is capable of producing canes at a rapid rate, in part, because aerial stems and root stolons are not crowded at the base of the plant; thus cane growth is uninhibited and stolons do not become dormant (Dransfield, 1988).

The cultivation of *C. trachycoleus* in central Kalimantan involves fruit collection and seed germination, transplanting of seedlings at about one month of age to nurseries that are established in light shade along river banks, the maintenance of nursery stock for about 14 months (at which time the plants have stems 1 m long and 1–2 stolons), and finally transplanting at approximately 20 × 10 m spacing in permanent plantations. The first cane harvest can usually be made about 7–10 years after planting, with subsequent harvests possible every 1–2 years thereafter (Dransfield, 1979; Manokaran,

1984). Detailed production data are unavailable for the central Kalimantan plantations; however, entire villages depend upon cultivated rattan as a primary means of subsistence and reportedly maintain a relatively high standard of living (Dransfield, 1979).

Another indigenous system of rattan cultivation has been reported from east Kalimantan, where *C. caesius* seedlings are planted in the regrowth of abandoned swidden fields (Weinstock, 1983). In this system, rattan matures during the customary 10–15 year fallow period and is harvested before the secondary forest is cut and burned and the farms re-established.

Recent attempts by government agencies to cultivate rattan have focused primarily on species important to the furniture industry, notably *C. manan*, *C. caesius* and *C. trachycoleus*. This work has culminated in a significant number of published papers on such topics as seed germination and storage (Manokaran, 1978; Mori, Rahman & Tan, 1980), tissue culture (Umali-Garcia, 1985), nursery techniques (Darus & Aminah, 1985), seedling survival and growth rates (Manokaran, 1981a, b, 1982a, b; Aminuddin, 1985; Shim & Muhammad, 1985) and the effects of variable light regimes and soil moisture conditions on growth and yield (Mori, 1980; Nainggolan, 1985). The cumulative results of this work suggest that while commercial rattan cultivation appears promising, significant problems remain, particularly with large-caned species such as *C. manan*.

Collection and germination of rattan seeds present difficulties in that seed germination rates vary not only among species, but among samples of the same species. For example, seed germination in *C. manan* varied from 0.2 to 83% in one study, presumably because of inherent genetic differences and variability in seed ripeness (Manokaran, 1978). *Calamus manan* seed viability is also very sensitive to soil moisture and temperature conditions (Mori *et al.*, 1980). Nevertheless, enough is known about germination requirements of *C. manan* to produce commercial quantities of seedlings at germination rates of approximately 75% (J. Dransfield, personal communication). In general, rattan seeds cannot withstand drying (Dransfield, 1979).

Rattan propagation through tissue culture techniques could eliminate problems of seed storage and germination. Umali-Garcia (1985) produced potentially viable transplanting material (i.e. callus) from 11 species of *Calamus* and two species of *Daemonorops*. Attempts to encourage further plant development (i.e. of leaf petioles), however, were unsuccessful. Clearly, much remains to be done before tissue culture can be used to propagate rattan on a commercial scale. Furthermore, given careful nursery techniques, tissue culture is probably unnecessary.

Another potential means of cultivating rattan is through vegetative propagation of root suckers or rhizome segments (Aziah & Manokaran, 1985). This approach eliminates the need to germinate seeds and reduces the length of time from transplanting to cane production. Vegetative propagation appears well suited to clustering rattans (e.g. *C. caesius*, *C. trachycoleus*, *C. exilis*, and *C. pilosellus*: Aziah & Manokaran, 1985; Siebert, 1989). However, large solitary canes, including most premier furniture species (e.g. *C. manan*) cannot be propagated vegetatively and only small quantities of seedlings of clustering species can be produced in this manner.

The growth performance of rattans in cultivation trials reveals a great deal of variability both among and within species and high sensitivity to light regimes and soil moisture conditions. For example, *C. manan* seedlings exhibit optimal growth at about 50% shade (Mori, 1980), but rarely grow beyond the rosette stage in deep shade (Dransfield, 1979). *Calamus caesius* appears to grow best on poorly drained soils under high light conditions (Manokaran, 1982a), but suffers high seedling mortality when flooded (Dransfield, 1988). *Calamus trachycoleus* can survive seasonal flooding, but requires very high light intensities (Dransfield, 1988).

The most impressive attempts to cultivate rattans on a large scale are presently being conducted by the Sabah Forestry Development Authority at the Batu Putih Estate Rattan Plantation in Sabah. The Estate has over 4000 ha of rattan under cultivation (2834 ha of *C. trachycoleus* and 1234 ha of *C. caesius*) (Dransfield, 1988). Detailed growth data from these trials have not been published, but commericial cane harvests were scheduled to begin in 1989. Dransfield (1988) reports, however, that growth and cane production rates appear encouraging. For example, at 2.5 years some *C. tracycoleus* individuals had produced 10 aerial stems and at 4 years some plants had over 40 aerial stems and were beginning to flower. Estate managers estimate that if harvesting begins at 11 years, *C. trachycoleus* can be expected to produce approximately 2.5 tonnes of canes per hectare per year (Dransfield, 1988), which represents an annual return of US $1456 per hectare. If the canes are treated (i.e. cleaned and smoked in sulphur: Dransfield, 1988), annual returns increase to US $2729 per hectare.

While commercial cultivation of small and medium-sized rattans appears very promising, the situation with large diameter canes (whose production is essential for the furniture industry) is less sanguine. Cultivation of *C. manan* may prove economically viable in situations where manipulation of the canopy can provide optimal light regimes, but its cultivation will remain problematic because: (i) plants must be replanted after every harvest (i.e. they do not resprout when cut), and (ii) well developed secondary or primary forests will be required to support the massive canes. Nevertheless, economic projections by Salim (1982) (based on growth, yield and labor input data) indicate favorable returns on investments with *C. manan* harvested after 15 years, assuming cane growth rates of 1 m per year.

Given the limitations inherent with *C. manan* cultivation (and with other solitary rattans), Dransfield (1985, 1988) has advocated increased taxonomic inventorying of rattans to identify large, clustering canes of silvicultural potential. He reports that *C. merrilli* (the premier furniture rattan in the Philippines) and *C. zollingeri* (a relative of *C. merrilli* from Sulawesi) fit this criterion very well. Assuming that large rattans can be successfully cultivated, the next step would be to encourage enrichment plantings in protected forests (Dransfield, 1981) and, where appropriate, large-scale plantations.

Conclusion

Rattans are a prominent and integral component of the forests of Southeast Asia. They are also an important and in some cases irreplaceable income source to rural and forest-dwelling inhabitants. Growing human populations and resource extraction pressures now threaten not only the existence of many rattan species, but the forest habitats in which they occur. Overexploitation and habitat conversion also imperil the lucrative rattan trade.

The destruction of rattan resources is regrettable, not only because of their ecological and socioeconomic significance, but because rattan cultivation and management represents one of the most promising ways of utilizing tropical forests on a sustainable basis. Managed rattan collection and enrichment plantings could provide local people with an economic incentive to preserve forests (rather than clear them for the cultivation of cash crops), while simultaneously conserving wild rattan populations and other forest flora and fauna.

The long and lucrative history of *C. trachycoleus* and *C. caesius* cultivation by indigenous people in Kalimantan should serve as a model for contemporary development activities. Ample silvicultural information already exists to ensure successful cultivation of two species of small and medium-sized rattans. Increased research and taxonomic work could lead to the successful cultivation of large diameter canes as well.

Unfortunately, time is running short for the forests and rattans of Southeast Asia. The preservation of productive, functioning forests with their full complement of flora and fauna will require an immediate and major commitment by tropical nations and international development agencies. Failure to develop and manage rattan resources will result in the collapse of the furniture industry, severe economic hardship, and the loss of an unparalleled incentive for forest conservation.

Acknowledgements

The author gratefully acknowledges the assistance of F. Putz, University of Florida; J. Dransfield, Kew; and J. Belsky, J. Lassoie, and N. Uhl, Cornell University for their valuable comments on an earlier draft of this chapter.

References

Aminuddin, M. (1985). Performances of some rattan species in growth trials in Peninsular Malaysia. In *Proceedings of the Rattan Seminar*, ed. K. M. Wong and N. Manokaran, pp. 49–56. Rattan Information Centre, Kepong, Malaysia.

Asiaweek (1987). Riches of the Rainforest. July 5, pp. 54–5.

Asiaweek (1989). No Easy Chair for Furniture Men. May 5, p. 55.

Avé, W. (1988). Small-scale utilization of rattan by a Semai community in West Malaysia. *Economic Botany* 42: 105–19.

Aziah, B. H. & Manokaran, N. (1985). Seed and vegetative propagation of rattans. In *Proceedings of the Rattan Seminar*, ed. K. M. Wong and N. Manokaran, pp. 13–21. Rattan Information Centre, Kepong, Malaysia.

Burkill, I. H. (1935). *A Dictionary of the Economic Products of the Malay Peninsula, Vols 1 and 2*. Crown Agents for the Colonies, London.

Conelly, W. T. (1985). Copal and rattan collecting in the Philippines. *Economic Botany* 39: 39–46.

Corner, E. J. H. (1966). *The Natural History of Palms*. Weidenfeld and Nicholson, London.

Darus, H. A. & Aminah, B. H. (1985). Nursery techniques for *Calamus manan* and *C. caesius* at the Forest Research Institute nursery, Kepong, Malaysia. In *Proceedings of the Rattan Seminar*, ed. K. M. Wong and N. Manokaran, pp. 33–40. Rattan Information Centre, Kepong, Malaysia.

Dransfield, J. (1969). Palms in the Malayan forest. *Malayan Nature Journal* 22: 144–51.

Dransfield, J. (1974). Notes on the palm flora of central Sumatra. *Reinwardtia* 8(4): 519–531.

Dransfield, J. (1976a). Terminal flowering in Daemonorops. *Principes* 20: 29–32.

Dransfield, J. (1976b). Palms in the everyday life of west Indonesia. *Principes* 20: 39–47.

Dransfield, J. (1978). Growth forms of rain forest palms. In *Tropical Trees as Living Systems*, ed. P. B. Tomlinson and M. H. Zimmermann, pp. 247–68. Cambridge University Press, Cambridge.

Dransfield, J. (1979). *A Manual of the Rattans of the Malay Peninsula*. Malayan Forest Records No. 29. Forest Department, West Malaysia.

Dransfield, J. (1981). The biology of Asiatic rattans in relation to the rattan trade and conservation. In *The Biological Aspects of Rare Plant Conservation*, ed. H. Synge, pp. 179–186. Wiley and Sons, New York.

Dransfield, J. (1985). Prospects for lesser known canes. In *Proceedings of the Rattan Seminar*, ed. K. M. Wong and N. Manokaran, pp. 107–14. International Rattan Centre, Kepong, Malaysia.

Dransfield, J. (1987). The conservation status of rattans in 1987: A cause for great concern. Paper presented at the International Rattan Seminar, 12–14 November 1987, Chiangmai, Thailand.

Dransfield, J. (1988). Prospects for rattan cultivation. *Advances in Economic Botany* **6**: 190–200.

Dransfield, J. & Whitmore, T. C. (1969). Palm hunting in Malaya's national park. *Principes* **13**: 83–98.

Dunn, F. L. (1975). Rain-forest collectors and traders: a study of resource utilization in modern and ancient Malaya. *Monograph of the Malaysian Branch of the Royal Asiatic Society* **5**: 78–103.

Far Eastern Economic Review (1988). July 14, p. 8.

Henderson, A. (1986). A review of pollination studies in the Palmae. *Botanical Review* **52**: 221–59.

International Development Research Centre (1980). *Rattan: a report of a workshop held in Singapore, 4–6 June 1979*. IDRC, Ottawa, Canada.

Manokaran, N. (1978). Germination of fresh seeds of Malaysian rattans. *Malaysian Forester* **41**: 319–24.

Manokaran, N. (1979). A note on the number of fruit produced by four species of rattan. *Malaysian Forester* **42**: 46–9.

Manokaran, N. (1981a). Survival and growth of rotan sega (*Calamus caesius*) seedlings at 2 years after planting. I. line-planted in poorly-drained soil. *Malaysian Forester* **44**: 12–22.

Manokaran, N. (1981b). Survival and growth of rotan sega (*Calamus caesius*) seedlings at 2 years after planting. II. line-planted in well-drained soil. *Malaysian Forester* **44**: 464–72.

Manokaran, N. (1982a). Survival and growth of rotan sega (*Calamus caesius*) seedlings at 2 years after planting III. group-planted in poorly drained soil. *Malaysian Forester* **45**: 36–48.

Manokaran, N. (1982b). Survival and growth of rotan sega (*Calamus caesius*) seedlings at 5⅓ years after planting. *Malaysian Forester* **45**: 193–202.

Manokaran, N. (1984). *Indonesian rattans: Cultivation, production and trade*. RIC Occasional Paper No. 2. Rattan Information Centre, Forest Research Institute, Kepong, Malaysia.

Manokaran, N. & Wong, K. M. (1983). The silviculture of rattans – an overview with emphasis on experiences from Malaysia. *Malaysian Forester* **46**: 298–315.

Mori, T. (1980). Growth of rotan manau (*Calamus manan*) seedlings under

various light conditions. *Malaysian Forester* **43**: 187–92.
Mori, T., Rahman, Z. & Tan, C. H. (1980). Germination and storage of rotan manau (*Calamus manan*) seeds. *Malaysian Forester* **43**: 44–55.
Nainggolan, P. H. J. (1985). Preliminary observations of the effect of different canopy and soil moisture conditions on the growth of *Calamus manan* (manau). In *Proceedings of the Rattan Seminar*, ed. K. M. Wong and N. Manokaran, pp. 73–6. International Rattan Centre, Kepong, Malaysia.
Peluso, N. L. (1983). Markets and Merchants: The Forest Product Trade of East Kalimantan in Historical Perspective. MS Thesis, Cornell University, Ithaca, NY.
Putz, F. E. & Chai, P. (1987). Ecological studies of lianas in Lambir National Park, Sarawak, Malaysia. *Journal of Ecology* **75**: 523–31.
Rickson, F. R. (1984). Ant facilitated uptake of nutrient rich throughfall by *Daemonorops verticillaris* and *Daemonorops macrophylla* (Angiospermae, Palmaceae) in Malaya. *American Journal of Botany* **71**: (5, Part 2): 88.
Salim, A. R. (1982). A financial appraisal of rotan manau (*Calamus manan*) cultivation at the Bukit Belata forest reserve. *Malaysian Forester* **45**: 476–82.
Schlegel, S. A. (1979). *Tiruray Subsistence: From Shifting Cultivation to Plow Agriculture*. Ateneo de Manila University Press, Quezon City, Philippines.
Shane, M. (1977). The economics of a Sabah rattan industry. In *A Sabah Rattan Industry*, ed. W. Y. Wong and M. Shane, pp. 43–53. Transcript. SAFODA, Kota Kinabalu, Malaysia.
Shim, P. S. & Muhammad, A. M. (1985). A preliminary report on the growth forms of *Calamus caesius* and *C. trachycoleus* in SAFODA's Kinabatagan rattan plantation. In *Proceedings of the Rattan Seminar*, ed. K. M. Wong and N. Manokaran, pp. 63–71. International Rattan Centre, Kepong, Malaysia.
Siebert, S. F. (1989). The dilemma of a dwindling resource: rattan in Kerinci, Sumatra. *Principes* **33**: 79–87.
Siebert, S. F. & Belsky, J. M. (1985). Forest-product trade in a lowland Filipino village. *Economic Botany* **39**: 522–33.
Tomlinson, P. B. (1979). Systematics and ecology of the Palmae. *Annual Review of Ecology and Systematics* **10**: 85–107.
Uhl, N. W. & Dransfield, J. (1987). *Genera Palmarum. A Classification of Palms Based on the Work of Harold E. Moore, Jr*. Allen Press, Lawrence, Kansas.
Umali-Garcia, M. (1985). Tissue culture of some rattan species. In

Proceedings of the Rattan Seminar, ed. K. M. Wong and N. Manokaran, pp. 23–31. International Rattan Centre, Kepong, Malaysia.

Weinstock, J. A. (1983). Rattan: ecological balance in a Borneo rainforest swidden. *Economic Botany* 37: 56–68.

Xu, Huanggan (n.d.) Rattan research in China (unpublished report). Research Institute of Tropical Forestry, Chinese Academy of Forestry, Guangzhou, China. 11 pp.

18
Silvicultural effects of lianas

FRANCIS E. PUTZ

Introduction

In forests managed for timber production, lianas can be a serious nuisance (e.g. Dawkins, 1958). Lianas rely on trees for mechanical support, often display their foliage above the leaves of their host trees, and root in more or less the same soil volumes as trees. It is not surprising, therefore, that lianas have repeatedly been reported to compete with trees for water, light, and nutrients. The deleterious effects of lianas on trees can include stem breakage and deformation, increased accessibility to herbivorous animals, and decreased growth rates. In this chapter I review evidence for the negative effects of lianas on trees, discuss the characteristics of lianas that may enhance their competitive ability, and outline silvicultural practices used to minimize the ill effects of lianas on trees. Beneficial effects of lianas on ecosystems, for example their roles in reducing soil erosion and contributing to the maintenance of animal diversity, are not considered because the focus of this chapter is on maximizing the production and quality of timber.

Tree stem deformations due to lianas

In the process of climbing up to and through forest canopies, lianas often damage the trees on which they rely for physical support. Trunk deformations are of particular concern to silviculturalists because trees with poor stem form are physically difficult to handle during logging and generally are not saleable other than as pulpwood. Furthermore, deformed trees are more susceptible to pathogens and breakage (e.g. Shigo, 1986).

Although lianas are often observed in the process of girdling tree stems, the hypothesis that the incidence of stem deformations is higher among liana-laden than liana-free trees is supported by few quantitative data. Some evidence was provided by Lowe & Walker (1977) for natural tropical rainforest in Nigeria where a high proportion of crooked and defective trees was infested with lianas. Japanese honeysuckle, introduced into the USA, causes considerable damage to trees both in the understory and in the canopy (Little, 1961; Thomas, 1980). Several years after selective logging of lowland dipterocarp forest in East Malaysia, stem deformation was much more common among liana-laden than among liana-free trees (Putz, Lee & Goh, 1984). Observations of liana-caused damage are made even more consequential by the additional observation that liana infestations and stem deformation are both less common on weed trees than on more slow growing, shade-tolerant, and long-lived commercial timber species (Putz *et al.*, 1984).

Interpretation of liana infestation and tree deformation contingency tables is fraught with difficulties. First of all, deformed trees grow more slowly than straight-boled trees (Lowe & Walker, 1977) and thus may be particularly susceptible to liana infestation (Putz, 1984). Secondly, light-demanding and weak-wooded trees may not survive liana infestations long enough to display stem deformations. Liana species also differ in their likelihood of girdling or otherwise damaging tree stems. If one disregards the types of climber involved in liana–tree interactions, effects may be obscured that otherwise would be clear.

The diameter of stems and branches that are susceptible to liana damage obviously depends on the size of supports to which different types of climbers can attach. For example, stem twiners successfully climb larger diameter trees than tendril-climbers; tendrils of few species can cling to branches > 5 cm in diameter whereas twining lianas can ascend stems up to 20 or even 30 cm in diameter (Putz, 1984; Putz & Chai, 1987). Examples of the differential distribution of damage by tendril-climbers and twiners are provided for *Clematis* and *Lonicera* in France (Falcone *et al.*, 1986), *Vitis* and *Celastrus* in the eastern USA (Lutz, 1943), and various liana species in *Chamaecyparis* plantations in Japan (Suzuki, 1984).

Lianas damage tree branches and stems by interfering with translocation processes. Generally the angle at which lianas spiral precludes loosening to accommodate growth in tree stem diameter. Although there is little active 'strangulation' by lianas, this is effectively what happens because trees must grow to replace dysfunctional xylem and phloem tissues. In response to blockage of downward translocation of photosynthates, liana-entwined stems generally swell above each coil. The swollen areas are characterized by phloem and xylem cells oriented parallel to the spiral of the liana (see photographs in Lutz, 1943 and Suzuki, 1984); although weakened, entwined trees often persist for many years. Lianas with tendrils that become woody (e.g. *Vitis*), in contrast, sometimes completely girdle the twigs to which they attach, leading to breakage at the constriction point.

In addition to girdling, liana-laden trees are particularly susceptible to other forms of damage. Trees heavily liana-infested in the northeastern USA, for example, suffer greater damage during ice storms than liana-free trees (Siccama, Weir & Wallace, 1976). During controlled understory burns in northern Florida, pines with lianas, particularly *Smilax* spp., are more likely to suffer crown scorch than liana-free trees; flames can be watched as they climb up liana tangles into the canopy (personal observation).

Competition between lianas and trees

Other than requiring mechanical support, lianas resemble trees in all other basic resource requirements. The two life forms may differ in their optimal resource ratios (Tilman, 1982) but some competition is still expected for light, water, and mineral nutrients. There is considerable evidence, both experimental and otherwise, in support of the obvious prediction that lianas suppress tree growth. It is not clear, however, whether liana–tree competitive interactions are more intense above ground or below ground, nor whether liana species vary in their effects on trees. In this section evidence for the deleterious effects of lianas on trees will be reviewed, followed by a discussion of the nature of the competitive interactions.

Most liana species are light-demanding and, when given the opportunity, display their foliage above that of their host trees. Blanketing of forest canopies by liana leaves is quite apparent when looking down from a tall tower or low-flying plane. Generally the pedestrian observer can only see liana stems disappearing into the canopy and recognizes the abundance of liana leaves only on forest edges (e.g. Gysel, 1951), in artificial clearings, (e.g. Boring, Monk & Swank, 1981) or treefall gaps (e.g. Putz, 1984). Canopy research, litterfall studies (e.g. Hladik, 1974), leaf area calculations based on allometrical equations (Putz, 1983), and total biomass inventories in felled plots (Klinge & Herrera, 1983) all indicate that lianas can be major contributors to total forest leaf production. It follows that competition for light between lianas and their supporting trees can be intense.

By relying on trees for support, lianas economize on biomass allocations for stem girth increments. There is no reason to believe that lianas and trees differ in root uptake resistances, mycorrhizal efficiency, or root depletion zone volumes. The economics of below-ground functions in lianas and trees may differ, however, even though both lianas and trees increase their access to water and minerals by root extension growth through the soil. In particular, freedom from the need to be self-supporting may influence below-ground root allometry just as it does above ground.

Because they do not require roots suitable for mechanical support, for a given resource allocation below ground, lianas may increase root system length (and hence surface area) more than trees. Emerging from the base of trees are large diameter and rapidly tapering roots that support the tree against the forces of gravity and wind-induced sway. A substantial proportion of the below-ground biomass of trees is accounted for by large diameter roots. Liana root systems are quite different; lack of a requirement for large diameter structural roots probably results in root systems more efficient in absorption of water and nutrients from soil. Tree vs. liana root competition studies designed to test this hypothesis are currently under way (F. Putz, unpublished data). Preliminary results indicate that rates of extension growth are generally higher in liana than tree roots. Consequently, liana roots colonize resource-rich volumes of soil more rapidly than tree roots. Rapid soil colonization rate, although only one characteristic of a successful below-ground competitor, is likely to be important because most soils are extremely heterogeneous both spatially and temporally.

Evidence for the deleterious effects of lianas on tree growth has been provided by both correlative and experimental studies. In the former, growth rates of trees with and without lianas, or with different degrees of liana coverage, are compared. For example, Featherly (1941) reported that grape-vine-infested hardwood trees in Oklahoma had narrower annual rings than grape-free trees. Diameter growth rates of *Sterculia rhinopetala* trees in Nigeria were negatively correlated with estimates of crown coverage by lianas (Lowe & Walker, 1977). Total basal area of lianas in the crowns of *Luehea seemannii* trees in Panama was also negatively correlated with tree diameter growth rate (Putz, 1984). In addition to the problem of inferring a causal relationship from these sorts of correlative data (Stevens, 1987), slow growing trees are particularly susceptible to lianas (Putz *et al.*, 1984).

Liana removal experiments confirm that lianas can reduce tree growth rates (Patterson, 1974; Putz *et al.*, 1984; Whigham, 1984) as well as tree fecundity (Stevens, 1987). Although the adverse effects of heavy crown coverage by lianas are obvious, Whigham (1984) provides some experimental evidence for the importance of liana–tree root competition.

Effects of lianas on felling damage

When felled, liana-laden trees do more damage to the residual forest than liana-free trees of the same size (Fox, 1968; Appanah & Putz, 1984). The logging damage problem is particularly acute among larger diameter trees; in the understory the probability of damage increases with tree size. In many cases trees tens of meters from the felling site are damaged when inter-crown liana connections with the felled tree fail to break. Although the deleterious effects of lianas during logging are quite evident (e.g. Dawkins, 1958), studies of felling damage have been fraught with difficulties and need to be replicated.

Plot-based studies of felling damage in selectively logged forests are plagued by high among-plot variance in logging intensity. The best predictor of damage to advanced growth is often the basal area of timber removed from an area (Nicholson, 1958), and thus either logging needs to be controlled or plots need to be large enough to be broadly representative. To test the hypothesis that felling damage is decreased by cutting and poisoning lianas prior to felling, in a lowland dipterocarp forest in Sabah, Fox (1968) cut and applied sodium arsenite solution to the cut ends of lianas in five randomly selected 80 × 200 m plots and left the lianas intact in adjacent plots of the same size. Trees unlikely to be cut were discretely marked (but not mapped) and the entire area was commercially logged. After logging the marked trees were sought out and classified as undamaged, lost, fallen or broken off, or having suffered bark or crown damage. Trees in plots in which lianas were cut several months before logging were less likely to be broken or completely lost than trees in the control plots. Pre-felling climber cutting had little effect on the incidence of bark or crown damage. Interpretation of the results, however, is difficult because the control plots were more heavily logged than the treated plots.

In a lowland dipterocarp forest in Peninsular Malaysia, Appanah & Putz (1984) conducted a felling damage study similar to that of Fox (1968). Initially they attempted to overcome the problem of high variance in basal area felled by using paired plots (five pairs of treated and control 1 ha plots). Although the plot pairs were nearby one another, intra-pair variability in felling intensity was overwhelming (unpublished data). In place of a plot-based approach, Appanah & Putz (1984) selected 25 liana-laden trees in control plots and 25 trees of similar size in plots in which all lianas were cut 9 months prior to logging. As in Fox's (1968) study, neighboring trees were classified as having received bark damage, crown damage, or having been pulled down altogether. Neither bark nor crown damage were more frequent in the control plots but liana-laden trees pulled down nearly twice as many neighbors as liana-free trees (means of 3.9 and 7.2, respectively). Furthermore, compared with liana-free trees, liana-laden trees inflicted damage on larger diameter neighbors.

Silvicultural control of lianas

Lianas are among the most serious obstacles to timber management, from the Solomon Islands (Neil, 1984) to West Virginia (Trimble & Tryon, 1974). By inhibiting tree growth, exacerbating logging damage, and causing tree stem deformations, lianas are a silvicultural nuisance in mature as well as regenerating stands. Prescriptions for liana control have consequently been suggested for implementation before felling, after felling, and throughout the regeneration process.

Basic ecological features of lianas suggest that pre-felling liana cutting is the silviculturally most sensible control method. To start with, by cutting lianas at least several months or years prior to logging, damage to advanced growth can be substantially reduced. Furthermore, although most cut liana stumps sprout, the sprouts generally grow slowly and fail to grow back up into the mature forest canopy. If left uncut, many lianas survive falling from the canopy. In selectively logged lowland dipterocarp forest in Malaysia, for example, about half of the canopy lianas survived falling to the ground during logging (Appanah & Putz, 1984). Six years after logging in the same forest reserve, slightly more than half of the lianas in the canopy of the regenerating forest were sprouts from the procumbent stems of former canopy individuals. Anatomical features that allow liana stems to survive mechanical trauma and repair damaged tissues have only recently been elucidated (see Fisher & Ewers, Chapter 4; Putz & Holbrook, Chapter 3).

Pre-felling liana cutting is not an expensive silvicultural treatment, especially if applied selectively around crop trees. Cutting and poisoning the stumps of about 250 lianas >2 cm dbh per hectare in Malaysia requires approximately one day of labor. Foresters in West Virginia cut about 32 grapevines per hour; one person can treat approximately 5 ha per day (Smith & Smithson, 1975). Although pre-felling liana cutting is generally an effective method for reducing post-felling liana infestations after selective logging, if the canopy is drastically opened, prolific sprouting of liana stumps reduces the value of the treatment (Trimble & Tryon, 1974).

Given the propensity for lianas to sprout, treatment of cut stems with herbicide may be necessary where logging is intensive. This is unfortunate because herbicides are both environmentally hazardous and expensive. Sprouted lianas, however, have been observed to dominate logged-over forests in North Carolina, USA (Boring *et al.*, 1981), Amazonian Brazil (Gentry, 1978), and lowland forest in Malaysia (Appanah & Putz, 1984). Rates of extension growth for stump sprouts of cut lianas ranging from 5 m per year for *Clematis vitalba* in New Zealand (Ryan, 1985), to 5–6 m per year for *Vitis* spp. in West Virginia (Trimble & Tryon, 1974), to more than 6 m per year for several species in Malaysia (Barnard, 1953).

Silvicultural treatment of lianas after logging is difficult because of their proliferation and the difficulty of access amongst the fallen branches and downed trees. Broadcast spraying of herbicides is generally not recommended owing to the interference of logging debris with movement of equipment and personnel as well as the great potential for damage to non-target species. Seed production by light-demanding lianas, like the convolvulaceous climbers that plague foresters in the Solomon Islands, however, can be reduced by spraying along roads and other openings (Chaplin, 1985). Foresters in the Solomon Islands have also experimented with controlled burns and cattle but encouraging rapid tree growth seems to be the best way to reduce infestations (for a review see Chaplin, 1985). Manual cutting of liana stems is fairly labor intensive and often inspection of purportedly 'treated' areas reveals the presence of numerous uncut lianas. The cost of treating large areas can be substantially reduced, however, by cutting only vines interfering with the growth of potential crop trees, leaving unstocked areas untreated. Biological control of lianas using insects, or pathogens holds considerable promise but to date little progress has been made with this approach (but see Baloch, Arif & Irshad, 1985).

Lianas can be controlled in regenerating forests through a combination of cutting and canopy manipulation. Because most lianas are light-demanding, encouraging the development of a closed canopy is a good first step towards liana control. Even the sparse shade cast by pioneer trees can suppress the vigor of liana growth (Putz *et al.*, 1984). Weed trees, like *Cecropia* spp. or *Macaranga* spp. might be looked upon as allies of the forester because during their short life spans they may benefit shade-tolerant commercial tree species by discouraging lianas.

Summary

Through the combined effects of shade, mechanical damage, and root competition, vines can have serious deleterious effects on trees. Silvicultural problems with vines are common in the tropics but also occur in temperate zone forests. Tree growth rates and stem quality can be enhanced by pre-felling vine removal in selectively logged forests and proper cultivation techniques in plantations. The substantial, if poorly studied, role of lianas in natural forest ecosystems, however, should be considered when prescribing vine removal treatments.

References

Appanah, S. & Putz, F. E. (1984). Climber abundance in virgin dipterocarp forest and the effect of pre-felling climber cutting on logging damage. *Malaysian Forester* 47: 335–42.

Baloch, G. M., Arif, M. I. & Irshad, M. (1985). Natural enemies associated with high altitude forest weeds in Pakistan. *Pakistan Journal of Forestry* 35: 105–11.

Barnard, R. C. (1953). Woody climbers, cutting and recovery. *Headquarters Bulletin of the Malayan Forest Department, July 1953*: 73–4.

Boring, L. R., Monk, C. D. & Swank, W. T. (1981). Early regeneration of a clear-cut southern Appalachian forest. *Ecology* 62: 1244–53.

Chaplin, G. E. (1985). An integrated silvicultural solution to weedy climber problems in the Solomon Islands. *Commonwealth Forestry Review* **64**: 133–9.

Dawkins, H. C. (1958). The management of natural tropical high forest. *Imperial Forestry Institute Paper 34*, Oxford.

Falcone, P., Keller, R., Le Tacon, F. & Oswald, H. (1986). Facteurs influençant la forme des feuilles en plantations. *Revue Forestière Française* **38**: 315–23.

Featherly, H. I. (1941). The effect of grapevines on trees. *Oklahoma Academy of Science Proceedings* **21**: 61–2.

Fox, J. E. D. (1968). Logging damage and the influence of climber cutting prior to logging in the lowland dipterocarp forest of Sabah. *Malaysian Forester* **31**: 326–47.

Gentry, A. H. (1978). Diversidade e regeneracao da capoeira do INPA, con referencia especial as Bignoniaceae. *Acta Amazonica* **8**: 67–70.

Gysel, L. W. (1951). Borders and openings of beech–maple woodlands in southern Michigan. *Journal of Forestry* **49**: 13–19.

Hladik, A. (1974). Importance des lianes dans la production foliare de la forêt equatoriale du nord-est du Gabon. *Comptes Rendus, Académie des Sciences (Paris), sér. D* **278**: 2527–30.

Klinge, H. & Herrera, R. (1983). Phytomass structure of natural plant communities on spodosols in southern Venezuela: the tall Amazon Caatinga forest. *Vegetatio* **53**: 65–84.

Little, S. (1961). Recent results of tests in controlling Japanese honeysuckle. *Hormolog* **3**: 8–10.

Lowe, R. G. & Walker, P. (1977). Classification of canopy, stem crown status and climber infestation in natural tropical forests in Nigeria. *Journal of Applied Ecology* **14**: 897–903.

Lutz, H. J. (1943). Injuries to trees caused by *Celastrus* and *Vitis*. *Bulletin of the Torrey Botanical Club* **70**: 436–39.

Neil, P. E. (1984). Climber problems in Solomon Islands forestry. *Commonwealth Forestry Review* **63**: 27–34.

Nicholson, D. I. (1958). An analysis of logging damage in tropical rain forest, North Borneo. *Malaysian Forester* **21**: 235–45.

Patterson, D. T. (1974). The ecology of oriental bittersweet, *Celastrus orbiculatus*, a weedy introduced ornamental vine. PhD dissertation, Duke University, Durham, NC.

Putz, F. E. (1983). Liana biomass and leaf area of a 'tierra firme' forest in the Rio Negro Basin, Venezuela. *Biotropica* **15**: 185–9.

Putz, F. E. (1984). The natural history of lianas on Barro Colorado Island, Panama. *Ecology* **65**: 1713–24.

Putz, F. E. & Chai, P. (1987). Ecological studies of lianas in Lambir

National Park, Sarawak, Malaysia. *Journal of Ecology* 75: 523–31.

Putz, F. E., Lee, H. S. & Goh, R. (1984). Effects of post-felling silvicultural treatment on woody vines in Sarawak. *Malaysian Forester* 47: 214–26.

Ryan, C. (1985). Pests and problems – old man's beard. *Soil and Water* 3: 13–17.

Shigo, A. L. (1986). *A New Tree Biology: facts, photos, and philosophies on trees and their problems and proper care.* Shigo and Trees, Durham, NC.

Siccama, T. G., Weir, G. & Wallace, K. (1976). Ice damage in a mixed hardwood forest in Connecticut in relation to *Vitis* infestation. *Bulletin of the Torrey Botanical Club* **103**: 180–3.

Smith, H. C. & Smithson, P. M. (1975). Cost of cutting grapevines before logging. *USDA Forest Service Research Note NE-207.*

Stevens, G. C. (1987). Lianas as structural parasites: the *Bursera simaruba* example. *Ecology* **68**: 77–81.

Suzuki, W. (1984). Tree damage caused by vines in hinok (*Chamaecyparis obtusa*) plantations. *Bulletin, Forestry and Forestry Products Research Institute Japan* No. 238, pp. 145–55.

Thomas, L. K. Jr (1980). The impact of three exotic plant species on a Potomac island. *National Park Service Scientific Monograph Service No. 13*, US Department of the Interior, Washington, DC.

Tilman, D. (1982). *Resource Competition and Community Structure.* Princeton University Press, Princeton, NJ.

Trimble, G. R. Jr & Tryon, E. H. (1974). Grapevines, a serious obstacle to timber production on good hardwood sites in Appalachia. *North. Logger and Timber Proc.* **23**: 22–3, 44.

Whigham, D. (1984). The influence of vines on the growth of *Liquidambar styraciflua* L. (sweetgum). *Canadian Journal of Forest Research* **14**: 37–9.

Taxonomic Index

General Index

Note: for full listing of species see Taxonomic Index

General index